AF749061

THE U

LANDMARK COLLECTOR'S LIBRARY

JAMES WATT

Volume 1:
His time in Scotland, 1736-1774

Rev. Dr. Richard L. Hills

LANDMARK COLLECTOR'S LIBRARY

JAMES WATT

Volume 1:
His time in Scotland, 1736-1774

Rev. Dr. Richard L. Hills

Landmark Publishing

Published by

LANDMARK
Publishing Ltd

Ashbourne Hall, Cokayne Ave
Ashbourne, Derbyshire DE6 1EJ England
Tel: (01335) 347349 Fax: (01335) 347303
e-mail: landmark@clara.net
web site: www.landmarkpublishing.co.uk

UNIVERSITY OF BIRMINGHAM LIBRARY

24181897

1st edition

ISBN 1 84306 045 0

© Rev. Dr. Richard L. Hills 2002

The rights of Rev. Dr. R. L. Hills as author of this work
has been asserted by him in accordance with the Copyright,
Design and Patents Act, 1993.

All rights reserved. No part of this publication may be reproduced,
stored in a retrieval system or transmitted in any form or by any means,
electronic, mechanical, photocopying, recording or otherwise without
the prior permission of Landmark Publishing Ltd.

British Library Cataloguing in Publication Data: a catalogue
record for this book is available from the British Library.

Printed by CPI, Bookcraft, Midsomer Norton, Somerset

Design & reproduction by James Allsopp

Cover captions:

Front cover & page 3: James Watt, F.R.S., 1736 – 1819. (Smiles, *Boulton & Watt*)

Back cover: Detail of the Newcomen engine model, showing the cylinder and boiler fittings.

Contents

WATT'S LIFE

		Personal	*Civil Engineering*	**Steam Engines**
1736	19 January	Birth in Greenock.		
1739	1 September	Birth of brother John, 'Jockey'.		
1753	23 January	Death of mother, Agnes.		
1753	April	Living in Glasgow, acting as agent for his father.		
1755	7 June	Watt and John Marr set out for London.		
	18 June	They arrive in London.		
	3 July	Watt working informally with John Morgan, scientific instrument maker, agreement for Watt to be trained by him reached in August.		
1756	19 August	Watt said he would leave London on Monday.		
1756	26 October	Asked by Glasgow University to restore the MacFarlane astronomical instruments.		
1756	December	Returns to Greenock.		
1757	3 January	Starts 'Waste Book' recording business transactions.		
1757	July	Moves to the College, Glasgow, and sets up workshop there.		
1757	October	Makes furnace doors for Jospeh Black.		
1757	November		*Assists on Clyde survey with Wilson and Barrie.*	
1758	March	Starts to prepare Uncle John's Map of the Clyde for publication.		
1758				**Makes model steam engine with Robison.**
1759	June	Goes to London to buy scientific instruments to sell in his shop in the College.		
1759	14 September	Printed copies of Uncle John's map sent from Edinburgh to be sold in the College shop.		
1759	7 October	Partnership formed with John Craig.		

		Personal	*Civil Engineering*	**Steam Engines**
1760		Increase in numbers of people employed in Watt's workshop. Costs of musical instruments start appearing in the ledgers.		
1761 - 2				**Experiments with Papin digester and small high pressure engine.**
1762	30 October	Death of 'Jockey'.		
1763	1 December	Advertisement for Watt's shop being moved from the Sautmercat to the Trongate - range of goods sold increased enormously.		
1763 - 4	Winter			**Repairs Newcomen engine model for the University.**
1764				**Experiments on model engine with wooden cylinder, boiler, etc. Experiments to determine volumes of steam, heat capacities of materials, latent heat, etc.**
1764	16 July	Marries Margaret Millar, 'Peggy'.		
1764	Autumn	Starts advertising his Perspective Drawing Machines.		
1765	Spring			**Before 29 April, conceives idea of separate condenser.**
1765	18 July	Birth of John, who died in late November or early December.		
1765	Summer			**Conducts experiments on separate condenser.**
1765	August			**Explains steam engine experiments to Roebuck.**
1765	August			**Starts designing Newcomen engine for mines at Leadhills which was not accepted.**

		Personal	*Civil Engineering*	**Steam Engines**
1765	September	By this time, dealing in salt with John Roebuck. Experiments to make alkali from salt start at same time.		
1765	Autumn			**Experiments on properties of steam.**
1765	November			**Plans for large experimental engine at Kinneil which was never finished.**
1765	December	Death of John Craig.		
1765	December			**Possible Newcomen engine for Carron Wharf.**
1766	19 February			**Tells Roebuck about his steam wheel and rotary engines.**
1766	Spring			**Trials stop on experimental engines.**
1766		Numbers of men employed by Watt begin to decrease.		
1766	May			**Erects Newcomen engine for Lord Kennet.**
1766	Summer		*Possibly establishes office as surveyor in King Street, Glasgow.*	
1767	January	Surveying partnership with Robert Mackell.		
1767	Winter		*Surveys 'Ditch' canal with Mackell.*	
1767	February			**Erects Newcomen engine at 'Newton of Air'.**
1767	Spring			**Erects Newcomen engine for Carron Company.**
1767	April		*Goes to London to procure Act for 'Ditch' canal.*	
1767	May	Meets William Small in Birmingham.		
1767	Summer		*Mackell and Watt survey River Devon.*	
1767	Summer	Birth of daughter Margaret, died June 1799.		
1767	July - August		*Watt surveys the Lomond Passage.*	
1767	September		*Survey of the tidal Forth.*	

		Personal	Civil Engineering	Steam Engines
1767	Autumn	Partnership with Roebuck probably formed.		
1768	January			**Trials start again on separate condenser, possibly at Delftfield Pottery, which continue into late summer.**
1768	15 Feb.	Has small share in the Delftfield Pottery.		
1768	August			**Goes to London to obtain steam engine patent.**
1768	August	Meets Matthew Boulton for first time.		
1768	Autumn			**Erects small Newcomen engines for Peter Colville at Torryburn.** **Experiments start again on separate condenser which continue well into 1769.**
1768	Autumn			**Starts designing large engine with separate condenser for Kinneil.**
1769	5 January			**Patent granted for separate condenser.**
1769	5 February	Birth of son James jun., died 2 June 1848.		
1769	Summer			**Erects large Kinneil engine, first unsuccessful trial in early September. Trials continue until April 1770.**
1769		Shop probably closed during this year but certainly packed up in May 1770.		
1769	August		*River Clyde survey.*	
1769	Autumn		*Report on Port Glasgow.*	
1769	26 October		*Start of Monkland Canal survey.*	
1769	November			**William Small agrees to construct a steam wheel at Birmingham.**

		Personal	*Civil Engineering*	**Steam Engines**
1770	March		*Caldwell Canal survey for Baron Mure.*	
1770	May		*Surveys of Strathmore and Crieff Canals.*	
1770	May		*Designs waterwheel at Clunie coal mine.*	
1770	7 June		*Surveys site for Hamilton Bridge.*	
1770	13 June			**Last trials on Kinneil engine.**
1770	August		*Designs crane for the Broomielaw, Glasgow.*	
1770	12 October		*Dumbarton Road survey.*	
1770	Autumn			**Erects Newcomen engine for Robert Montgomerie at Ayr.**
1770	24 Dec.	Birth of daughter Agnes, died 15 Feb. 1772.		
1771	March		*Proposals for waterwheel at Dalnotter Iron Works.*	
1771	May		*Inspection at Port Glasgow.*	
1771	May		*Survey of Ayr Harbour.*	
1771	July		*Surveys for Crinan and Tarbert Canals.*	
1771	August		*Surveys for site of tide mill at Culross.*	
1771	September		*Gartsheary Road survey.*	
1772	March		*Renfrew Canal survey.*	
1772	November		*Surveys for Greenock harbour and town water supply.*	
1773	7 May		*Measures completed section of Monkland Canal.*	
1773	May			**Watt dismantles Kinneil engine and sends parts to Birmingham.**
1773	July		*Estimates for Hurlet to Paisley waggon way.*	
1773	July		*Survey of Clyde and Argyll streets, Glasgow.*	
1773	1 September		*Leaves Schaw Park to commence Caledonian Canal survey.*	
1773	24 September	Death of Peggy.		

		Personal	*Civil Engineering*	**Steam Engines**
1773	October		*Surveys for canals around Stirling.*	
1773-4	Winter			**More experiments on steam.**
1774	Spring			**Designs and advises on erection of Newcomen engine for Lord Cochrane at Culross.**
1774	9 March		*Cathcart Road survey.*	
1774	19 April		*Drawing up plans for Rutherglen bridge.*	
1774	31 May	Watt arrives in Birmingham.		

Abbreviations

Argyll and Bute Record Office,	ARO
Birmingham Central Library,	BCL
Boulton and Watt Collection,	B. & W. Col.
James Watt Papers,	JWP
Matthew Boulton Papers,	Boulton
Muirhead Papers,	M
National Library of Scotland,	NLS
Scottish Record Office,	SRO
Strathclyde Regional Archives,	SRA

Illustrations, Photograpic Section

Illustrations, Early Life.

ILLUSTRATIONS, CIVIL ENGINEER.

Illustrations, Steam Engine

Watt's drawing for the proposed Leadhills engine with the domed top of the boiler on the left and the framing and cylinder above it to the right. (JWP, 4/52.5 By permission of Birmingham Central Library) p354

Plan of the Carron engine with the pump cylinder on the left and the steam cylinder on the right. The steam pipe enters from the bottom of the picture. (JWP, 4/52.3 By permission of Birmingham Central Library) p357

Watt's drawing of his first separate condenser which was probably prepared for one of the patent trials in the 1790s. (Based on B. & W., Portfolio 1381) p367

Watt's earliest sketch of the piston in his engine, showing it had an inverted cylinder. (JWP, 4/32.13, 12 Oct. 1765) p371

This is the most complete surviving sketch of any of Watt's early experimental models. It shows an inverted cylinder 'a', a fulcrum 'c' for a horizontal beam 'b' along which a weight 'd' could be moved to vary the load like a set of scales. (JWP, 4/32.12, 16 Oct. 1765) p372

The scale drawing which Watt sent to Roebuck in November 1765 for a cylinder presumbably to be cast at Carron. The inside length of the cylinder was to be '7 foot, diam. 24 in. to be made as truely round & equally wide as possible'. The hole for the piston rod was to be 2 $^{3}/_{4}$ ins. diameter. (redrawn from JWP, 4/32.34, 10 Nov. 1765) p375

This engine must be the first one planned for Kinneil in 1766 but was never erected because the cylinder was so badly bored. (Muirhead, (1854), Vol III, plate I) p376

No mention can be found of any engine with a separate condenser and wheel instead of a beam having been erected in Scotland. Perhaps Watt drew this as an advertisement to show the layout of his new engine. (Muirhead, (1854), Vol III, plate I) p377

A piston with pasteboard packing. (JWP, 4/32.10, 29 April 1768) p383

The first sketch of a condenser with twin air pumps having solid pistons. Steam entered the condenser through the pipe 'c' and the condensate was ejected through the vertical pipe 'e'. (JWP, 4/32.9, 10 May 1768) p385

The drawing proposed for the 1769 patent of the reciprocating engine has been annotated with identification letters. (based on JWP, 6/42) p394

This is the earliest drawing with flap valves in the bucket of the air pump 'a'. The condenser pipe was surrounded by an outer pipe 'b' so cold water ran between them. At 'c' a valve opened when the air pump bucket was at its lowest position so cold water ran in and helped to flush out the hot. (Boulton 348.5, 22 Feb. 1769) p395

Twin air pumps worked by a rocking beam. The valves in the bottom of the condenser were opened by triggers so that cold water rushed in when the appropriate piston was at the bottom of its stroke, giving rapid condensation. (Boulton 348.9, 28 May 1769) p396

Watt's workshop behind Kinneil House. (Smiles, (1878), p. 99) p399

Each condenser had eight vertical pipes, 18 inches long, $^{3}/_{4}$ in. diameter, $^{1}/_{2}$ in. apart. The appropriate pump drew the water and condensate down them 16 inches. (Boulton 348.13, 3 Nov. 1769) p403

An improved condenser with air pumps 5 inches diameter. 'a' and 'b' are cooling boxes with flaps at the side, joined by a copper box 'c'. One pump was worked ahead of the other so at 'd' were wooden partitions to act as separators. (Boulton 348.19, 16 March 1770) p404

A drum condenser with horizontal pipes. Air and condensate were driven out through valves 'c' by the respective pumps 'd'. Valves 'e' prevented the condensate flowing back into the drum. (Boulton 348.19, 16 March 1770) p405T

The rocking beam operating the air pumps is here driven through a crank, necessitating some form of rotary motion. (Boulton 348.13, 3 Nov. 1769) p405B

Matthew Boulton, F.R.S., 1728 - 1809. (Smiles, (1878), p. 110) p414

Soho Manufactory, (Smiles, (1878), p. 118) p416

The drawing proposed for the 1769 patent of Watt's steam wheel has been annotated with identification letters. (based on JWP, 6/42) p425

Small's sketch of his method of steam admission and exhaust through pipes and a flat rotary valve uncovering ports. (Boulton 340.4, 12 or 19 Feb. 1769) p427

(A) Watt's suggestion for a mercury trap with steam entering through 'B' and out to the condenser at 'A'.
(B) Watt thought spiral pipes might prevent mercury escaping. (Boulton 348.17, 25 Feb. 1770) p428

Flap valve and frame for the main valves in the steam wheel. (Boulton 348.17, 25 Feb. 1770) p429

Watt's sketch of his 'spiral oar'. (Muirhead, (1854), Vol II, p. 8, No. 85, 30 Sept. 1770) p430

Small's explanation of his proposal for cutting off the spigots on the ends of the rim sections of the steam wheel to give easier access to the valves. (Boulton 340.16, 3 Feb. 1771) p431

Boulton's suggestion for 'spiral arms'. (Boulton 348.57, 24 Feb. 1776) p435

Watt's semi-rotative steam engine with a vane 'A' that was pushed by the steam round to 'E,D' and then back again. (Muirhead, (1854), Vol II, p. 155, 24 Sept. 1782) p436

A presumably conjectural drawing of the semi-rotative engine connected to a condenser. (Muirhead, (1854), Vol III, p. 84, plate XVIII) p437

A rotary engine with three movable vanes 'A,B,C' on the rotor 'G'. The steam enters at 'E'. (Muirhead, (1854), Vol II, p. 157, 24 Sept. 1782) p438T

A rotary engine with one fixed vane 'D' on the rotor and a hinged valve 'C'. The steam enters at 'E'. (Muirhead, (1854), Vol II, p. 158, 24 Sept 1782 and Vol III, p. 85, plate XIX) p438B

I
Introduction

Soon after the death of James Watt in 1819, John Rennie suggested to James Watt junior that 'possibly a book should be published' and that Watt junior should collect the documents together. (1) Watt junior made investigations but could find very little about his father's early years so he reported to Rennie,

> I expect to be able to make out everything very completely from the year 1770 and tolerably well from the year 1760... I think there should be a volume of his life, and another volume to contain his Reports as a Civil Engineer, his Patents and other such Inventions and Documents as it may be deemed useful to preserve. (2)

The format of this book might have followed that of the earlier three volumes of the *Reports of the Late John Smeaton, F.R.S.* which had been published in 1812. Watt junior only found time to write an obituary for the *Encyclopaedia Britannica* which was published in 1823/4. Business commitments prevented him writing a fuller account so he later delegated the task of an 'official' biography to James Patrick Muirhead, a distant cousin, who had translated M. Arago's *Historical Eloge of James Watt.* (3)

The earliest biography written in Watt's own lifetime was published in *Public Characters 1802 - 3*. The compiler, Richard Phillips, wrote to Watt in May 1803 after the publication of the first edition, asking if Watt wanted to make any changes or alterations for a new edition. (4) There is no reply preserved in Watt's out-going Letter Books so we have no way of knowing if he approved of the article. This was the first of many similar articles to appear over the next few years in biographical collections and the like. Watt himself passed on few comments about his early childhood which appear to have been made mostly only to members of his own family. There are few archival papers before 1754 to throw light on the early period of his life and it is evident that Watt junior went through the later papers. Just what he removed is now impossible to determine but few papers covering financial details have survived.

When M. Arago, perpetual secretary to the Academy of Sciences in Paris, was searching for material for his *Eloge,* Watt junior gave him every assistance both by sending copies of manuscripts and approaching some of his father's friends. A Dr. Gibson sent some anecdotes which he had heard from Miss Jane Campbell. But then Watt junior discovered that these stories, which he himself had heard, had been written down by Jane Campbell at her mother's dictation in 1798. 'That lady, Mrs. [Marion] Campbell, was first cousin to my father, being a daughter of his mother's brother Mr Muirheid'. (5) Mrs. Campbell was about his father's age and they had been brought up a good deal together in the houses of their respective parents. Watt junior described her as 'a person of strong sense and clear judgement and possessing an affectionate esteem for my father' and that, even in 1798, she was probably the only person capable of giving an account of his early years. (6) Watt junior ordered Arago 'to obliterate what you have already written of this period of my father's life, and to substitute, if you please, this entire narrative in its place'. (7) Unfortunately the original document of Mrs. Campbell does not seem to have survived but, having been written fifty years after the events, may not have been entirely reliable.

The biographical section in Arago's *Historical Eloge*, published in 1839, is quite short, leaving scope for something more detailed. Muirhead also had access to Watt's papers and in 1854 produced his three volume work, *The Origin and Progress of the Mechanical Inventions of James Watt illustrated by his correspondence with his friends and the specifications of his patents,* (1854). The first volume is composed mostly of a biographical section containing long extracts from documents and, as the title implies, concentrates mostly on the steam engine. The rest of this volume and the whole of the second are filled with correspondence. While these letters have been a major source for later historians, they have been edited to show Watt in the best light and many have been omitted completely. The third volume reprints the text of Watt's patents. So this publication fulfilled to a certain extent Watt junior's suggestion. Four years later, Muirhead published an enlarged version of his biographical section in *The Life of James Watt, with selections from his correspondence*, (1858). This suffers from having the same approach as the earlier work so that large sections of Watt's life are omitted.

George Williamson, in his discursive *Memorials of the Lineage, Early Life, Education, and Development of the Genius of James Watt*, (1856), concentrated on Watt's early years in Greenock but probably never had access to the family papers. At least Samuel Smiles is known to have carried out extensive research on these papers for his *Lives of the Engineers. The Steam Engine, Boulton and Watt*, (1878). While this contains much original information, it suffers from Smiles' preconceived ideas on biographical writing stemming from his Victorian virtues of 'Self-help'. The most important book published since then is that by H.W. Dickinson and Rhys Jenkins, *James Watt and the Steam Engine*, (1927). While the family papers were consulted and there is a biographical section, the greater part of this work is concerned with Watt's later development of the steam engine. In addition, there have been many shorter biographies published as well as entries on the internet which have drawn their sources mainly from those listed above. Collections of Watt's papers have been printed which expand on those printed by Muirhead but in no way cover the vast amount of material available. In addition, while articles have been written on various aspects of the firm of Boulton & Watt, there has been nothing new recently published on the life of Watt himself.

The present study arose through the realisation that even Watt's most important invention, his separate condenser for the steam engine, had not been investigated in any detail in these earlier biographies. While researching this, permission was given by Lord Gibson Watt to consult the microfilms, then in the National Library of Wales, of Watt's outgoing Letter Books from 1782 until his death in 1819. These confirmed that a more comprehensive biography remained to be written. The task has been made easier by the purchase for the Nation of the Doldowlod Papers and their housing at Birmingham Central Library, thus bringing together the major collections on Matthew Boulton, James Watt and the firm of Boulton and Watt as the Archives of Soho. Yet it was difficult to convince people that Watt deserved a new biography. The Leverhulme Foundation rejected an application for funding and the research for the present work was made possible only through grants from the Royal Society. The manuscript letters have been checked against the printed versions in such works as Muirhead mentioned above and in E. Robinson and D. McKie, *Partners in Science, James Watt and Joseph Black*, (1970). These sources have been quoted in the references because they are more readily available than the originals. However, where, for example, Muirhead has changed the original or omitted part, the original reference is given.

Additional assistance has been received from very many people and institutions. I wish to thank them all and apologise to those I have omitted in the list below. The archive and other staff at Birmingham Central Library have been particularly helpful in obtaining my many requests from their collections. Other places have been:- in Edinburgh, the National Archives of Scotland, Register House, The National Library of Scotland and Edinburgh University Library; in Glasgow, the Mitchell Library and the Hunterian Museum; in Greenock, the Library and the Museum; in London, the Science Museum, the Royal Society and the Map Library at the British Library; at Aberystwyth, the National Library of Wales as well as libraries in Inverness, Manchester, History of Science and Technology Department, UMIST, and Record Offices at Lochgilphead, Stirling and Truro. In addition there has been the help of the staff at museums, particularly those at the Science Museum, Michael Wright, Michael Harding and Jane Insley as well as those at the National Museum of Scotland, Alison Morrison-Low and David Bryden. I also wish to thank the many other people who have given me advice over the last ten years or so, including the late Professor Donald Cardwell, and Professors Jennifer Tann, Robert Fox, and John Black. Mr. Joe Marsh at UMIST has been most helpful as have Dr. Fred Page and Mr.

Bernard Champness who have made comments on large sections of the manuscript. There are also the editors of the various journals which have published my articles about Watt. Their help has been invaluable but the faults and shortcomings are mine.

Richard L. Hills.

References

1 JWP, C6/10, J. Rennie to J. Watt jun., 7 Oct. 1819.

2 Ibid, C6/10, J. Watt jun. to J. Rennie, 14 Oct. 1819.

3 M. Arago, *Historical Eloge of James Watt*, J. Murray, London, 1839, translated by J.P. Muirhead.

4 JWP, 6/18.22, R. Phillips to J. Watt, 3 May 1803.

5 Ibid, W/10.38, J. Watt jun. to M. Arago, 22 Jan. 1835.

6 Ibid.

7 Ibid, see also JWP, W/10.36, J. Watt jun. to M. Arago, 8 Nov. 1835, 'I wish you to adopt implicitly the narration of Mrs Campbell, as I can *prove* that all the other traditional stories wherever they differ from hers are not founded in fact'.

1
The Formative Years

The Family Background

James Watt, 1736 - 1819

> I believe there has not, in our times, existed any individual so thoroughly conversant with the practical operations, as well as the scientific theories, of every Art & Trade. The motto he had chosen for his own seal... was "observare", and whatever he observed he instinctively improved upon. (1)

This was written in 1834 by James Watt junior to M. Arago. It outlines the essential component of his father's character which was his keen memory and ability to apply it to science and engineering. James Watt was born in Greenock on 19 January 1736 where he spent the early part of his life. Following training as a maker of mathematical instruments in London he returned to Greenock where he set up his first business in this profession. After nearly a year there, he moved to Glasgow in 1757, establishing himself in the University precincts. This venture prospered so that he moved into premises in the town and then branched out into a shop selling a wide variety of goods. While he was occupied with these enterprises, he not only conducted experiments in chemistry such as those for the Delftfield pottery but also on the model Newcomen atmospheric steam engine belonging to the University which led to his famous invention of the separate condenser. Then around 1770, he gave up the scientific instrument business and the shop to pursue the career of engineer, covering both what today we would divide into civil and mechanical branches. His major civil work was the construction of the Monkland Canal but the failure of the Ayr Bank in June 1772 led to an economic depression with consequently few openings in this profession. Experiments on steam engines with his separate condenser had to be terminated through the bankruptcy of his partner, Dr. John Roebuck, but Watt had managed to erect several Newcomen engines. Watt must have felt that prospects were dim in 1773, especially after the death of his first wife. He could not have anticipated how all this would change after his arrival in Birmingham on 31 May 1774 where he developed his steam engine partnership with Matthew Boulton. This marked the second part of his career and Birmingham remained his home for the rest of his life until his death on 19 August 1819. With hindsight, we can see that the different activities during his time in Scotland were laying the foundations for the more famous part of his life in England with his own steam engine. His motto, 'observare', was particularly appropriate because he had little formal education and based his inventions on what he was able to teach himself. Whether through reading literature or through watching others, he was able to develop knowledge of quite remarkable breadth in many different subjects, some of which he gained from his own family background.

Grandfather Thomas Watt, 1639/42 - 1734

The Watt family connection with Greenock began with Thomas, Watt's grandfather. (2) Watt junior wrote about his great grandfather who was born at Aberdeen in either 1639 or more likely 1642,

> Thomas Watt... left an infant [on his father's death] was brought up by relations, and having a turn for the Mathematical Sciences, made such proficiency by his own exertions, under very untoward circumstances occasioned by the persecutions of the times, as to enable him at a later date and quieter period to establish himself at Greenock as an instructor of them and of the dependent arts of surveying and navigation. (3)

Why Thomas should have chosen to settle at Crawfordsdyke close to Greenock remains unknown, neither is it known when he moved there. (4) What is clear is that there must have been sufficient people at both places needing instruction in the art of navigation for him to have remained, prospered and to have become an influential person in the locality. He had been appointed Bailie to the Barony of Greenock by 1696. (5) Then he became an Elder of its church and in 1708 Treasurer. (6) In 1711, he was appointed Session-clerk. He was involved with repairing the church, widening the bridge and was 'highly esteemed among his worthy brother-citizens of the burgh. (7) He died on 27 February 1734, aged, according to the Register, 'about 95 years'. (8) In the Register, he was called 'Teacher of Navigation', while on his tombstone he was 'Profesor [sic] of the Mathematicks'. (9) The importance of Thomas for the later history of civilization and industry may lie in the legacy he left posthumously to his grandson. Not only the memory of his profession but, more tangibly, his books and manuscripts on engineering and mathematics were passed down to his descendants. (10) Thomas also taught these subjects to his sons. There were three surviving children, John, James the father of our Watt, and Jean. (11) Jean married John Cochrane, the minister of nearby Inverkip. She died in 1771, having brought up seven children. Watt continued to keep in touch with some of these cousins even after he had moved to Birmingham.

Uncle John Watt, 1694 - 1737

Uncle John Watt was the elder surviving son who 'was brought up a mathematician and settled first at Ayr as a surveyor of land, and afterwards at Glasgow, where he was in much repute'. (12) At Glasgow, he also taught 'arithmetic, book-keeping, navigation and the other

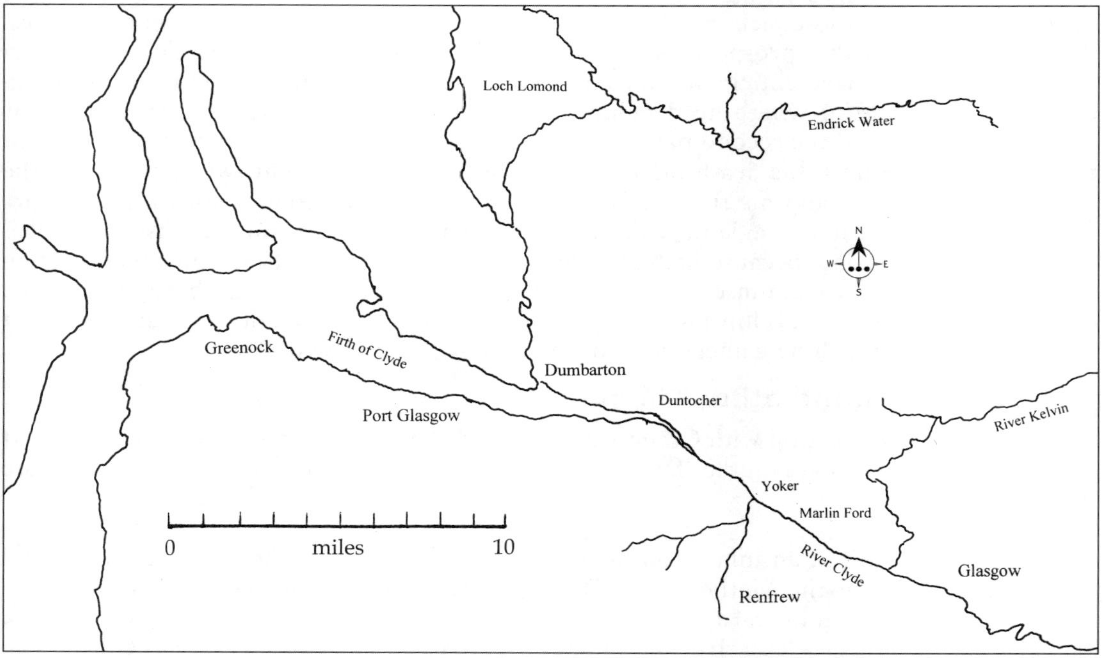

The Clyde Estuary.

parts of the mathematick' and was well liked because the merchants and shipowners asked that he might be given an annual salary of £5 as he deserved encouragement. (13) He received commissions from Glasgow Council to carry out a survey of the lands of Provan, make plans of Port Glasgow in 1732 for which he received £15, measure the Gorbals lands as well as many more. (14) Those plans which bear his name are carefully drawn, the lettering being particularly neat, 'an attainment not very common place in those days'. (15) Other maps which have been attributed to him include a survey of the River Clyde around Hamilton and a *Section of the Clyde* showing the river bed and depths of water below Glasgow. (16) If the date of 1734 and later attribution are correct, then he made one of the earliest proposals for deepening that river up to Glasgow. There are also two maps of the Clyde estuary and the west coast of Scotland which must have been drawn as the result of a survey in the same year. (17) James Watt senior acquired these maps on his brother's death and Watt himself knew about them because he was instrumental in publishing the last two as his uncle's 'Map of the Clyde' in 1759. (See p94) It is possible that Watt senior also collected his brother's surveying instruments which Watt could have seen when a youth, rousing his interest in this subject. Uncle Watt died in 1737, 'having obtained a good fortune for that time of day'. (18) Watt was only a year old so could never have remembered him, but it is highly likely that the knowledge of his uncle's surveying career played an important role in influencing the nephew's decision to take up the profession of 'engineer'.

The Clyde Region

Grandfather Thomas and Uncle Watt were able to develop successful careers in their respective professions because the Clyde region was advancing economically very quickly at this period. Watt senior and Watt himself would also be intimately involved with activities linked to the River Clyde so it is necessary to understand why that river was so important in their lives. Rising in the Lowther Hills in the southern part of Scotland, the Clyde flows north, then roughly west past Glasgow. The river is still tidal above Glasgow bridge. Twenty-two miles down stream from Glasgow lies Greenock, where the channel begins to swing southwards and the estuary becomes the Firth of Clyde until it reaches the Atlantic Ocean past the Isle of Arran. The land between the estuaries of the rivers Forth and Clyde forms the narrowest part of mainland Britain and has long been a route linking the east and west coasts. Conquerors and traders followed the Clyde as a natural gateway into the Lowlands of Scotland. Although there were fords and ferries across the Firth and the estuary, Glasgow was the first point where the Clyde could be crossed by a bridge and so Glasgow soon became an important centre for this region.

The waters of the Firth of Clyde are very deep because its bed has been gouged out by glaciers. At its mouth between Ayr and Arran, in places a depth of 200 to 300 ft. is quite common. This deep water extends into the adjoining sea lochs, but, with an abruptness unique among British estuaries, it ends suddenly as the Firth bends east past Gourock. Off Greenock, the sea bed starts to rise with such steepness that, within a distance of less than half a mile, the deep water turns into shoals and banks that may be quite dry at low tide. (19) While at high tide, the Clyde Estuary gives an impression of a continuation of the Firth's deep water, the river between Dumbarton and Glasgow could not be navigated by ocean-going vessels until well after 1760. The Clyde has been slowly filling up its own estuary by bringing down up to 200,000 tons of sand and silt in floods each year. (20) The inhabitants of Glasgow did not help themselves either because they discharged their rubbish and sewage directly into the river which caused it to silt up further and increased the effects of periodic floods. Like his uncle, Watt would be involved with schemes to improve the navigation of the Clyde up to Glasgow.

The Trade of Glasgow

Until at least the middle of the eighteenth century, Scotland was economically very much the poor relation compared with England. The trade of Glasgow started from small beginnings. A report on the revenue of the Excise and Customs of Scotland, dated 1656, stated about Glasgow that,

> The Inhabitants, all but the Students of the College which is here, are Traders and Dealers; Some for Ireland, with small smiddy Coales in open boats from Foure to Ten Tonnes, from whence they bring Hoops, Ronges [rings], barrel staves, Meal Oats, and Butter; Some for France with Pladding [plaiding], Coales and Herring,... for which they return saltpetre, Rozin, and prunes; Some to Norway for timber. (21)

Coal, herring and salt long formed basic commodities of the region in which Watt senior traded later. In 1651, some traders had ventured as far as Barbados but had not met with much success. At that time, the merchants of Glasgow owned twelve ships with an aggregate burden of 957 tons. But even these were too large to reach the town itself and all their cargoes had to be transhipped lower down the Clyde in open roadsteads, at Dumbarton or other ports in Ayrshire. Then the goods had to be carried to Glasgow either on pack-horses or in very small boats of three or four tons up the river.

Port Glasgow

Yet the mercantile genius of the Glasgow people remained strong. (22) To develop their trade further, the Glasgow merchants decided to construct their own harbour with access to deep water. Dumbarton lay on a lee shore and, in any case, the Glasgow merchants could not have control over it. In 1667, the Dumbarton residents refused to sell land to Glasgow for a harbour. A similar approach to purchase land at Greenock was also refused. However, in 1668, land was acquired at Newark which, in the following year, parliament approved should be incorporated into the port and harbour of Glasgow. (See map on p30) Seven years later, work commenced on the construction of a quay at what later became known as Port Glasgow. (23) In 1710, just three years after it had absorbed the neighbouring village of Newark within its bounds, Port Glasgow received official recognition of its growing importance by being appointed the principal customs house port on the Clyde. (24) Watt would carry out improvements to this harbour.

The Effect of the Act of Union, 1707

The Act of Union in 1707, from which at first the Glasgow merchants expected no good, in fact opened up new opportunities for them because they were able to trade with the British colonies in North America and the West Indies. However, to begin with, they were unable to take full advantage of this new situation because their ships could not venture on the transatlantic crossings and they had to charter others from ports in northern England, particularly Whitehaven. (25) It was not until 1716 that a ship of only 60 tons was constructed at Greenock, which was small in comparison with English vessels, but her safe arrival in Virginia in 1718 made her the first Glasgow-owned vessel to reach North America. The Clyde estuary was well suited for trading to America and, together with increased sailings to Scandinavia and the Continent, there resulted a steady rise in the number and size of ships sailing to and from the river.

There was also the more local trade within the Firth itself, to the Western Isles and to Ireland. In about the year 1730, there were upwards of 2,000 boats, manned by 14,000 men, engaged in the herring fishery in the Firth and along the west coast. (26) In 1735, there were

67 vessels of 5,600 tons aggregate belonging to and trading from the Clyde, of which one third were involved in the colonial trade. By 1783, this number had risen to 386 vessels of 22,896 tons. (27) Watt senior was involved in similar trading and here must have been a good market for navigation instruments which Watt exploited.

Greenock

Since few areas of Scotland are far from the sea, coastal shipping provided the usual means of communication for most local trade. This was where Watt's place of birth, Greenock, was well situated. There was some sort of harbour there at least by the middle of the seventeenth century and the people of the burgh owned some small ships by 1700. (28) During the latter part of the seventeenth century, the land on which Greenock now stands belonged to the Shaw (or Schaw) family. Sir John Shaw, the third of that name, began to improve the harbour at Greenock in 1707. (29) He had applied for the provision of a customs house at his new harbour but was turned down in favour of Port Glasgow as Greenock was only a burgh of a barony and not a royal burgh as was Glasgow. Nevertheless, the harbour was opened in 1710. It consisted of two arms jutting into the estuary, the East and West Quays, and was divided into two parts by a small projecting pier, the Mid Quay. This was a formidable work for those days, costing more than 100,000 Scottish marks. (30) Shipbuilding soon commenced and one of the most famous Clyde shipbuilders, the Scotts of Greenock, was already building fishing and coastal vessels there by 1711. (31) In 1740, the inhabitants numbered around 3,000. (32)

Increasing Trade

With trade developing rapidly, it was not long before the quays at both Greenock and Port Glasgow were filled to capacity. Virtually all of the tobacco imported into the west of Scotland was landed from the ocean-going vessels at these two ports. Sheds and warehouses adjacent to the quays were quickly filled. It seems that the Watt family was only incidentally involved in this trade for its control was in the hands of a small influential group of merchants. But its importance lay in the general economic stimulation which helped to boost the fortunes of the whole region. It was a trade which developed quickly after the Act of Union but presented difficulties to finance in what was then a poor country. It focused attention on problems of finance, the management of money and banking. At first, great profits were ploughed back into the growing business and others associated with it. (33)

In 1724, 4,192,576 lbs. of tobacco were brought into the Clyde, of which 3,053,570 lbs. were re-exported. By 1771, the total import of tobacco into Scotland had reached 47,268,873 lbs., of which 45,588,720 lbs. were re-exported, mainly to France. An allied trade was that in sugar. Between 1700 and 1770, sugar consumption in England rose fourfold. It was a trade concentrated in the hands of a few merchants to whom it brought great wealth. (34) In 1771, 6 $^1/_2$ million pounds of sugar were imported to Scotland as well as 826,741 gallons of rum, a considerable amount of which was re-exported. (35) One effect of this great increase was that the size of ships doubled in the quarter century after 1750. (36)

In order to carry on the tobacco trade, it was not only ships that were needed but also European consumer goods to exchange in Virginia and Maryland. So some tobacco merchants started their own ventures such as tanneries, cloth print works, iron works, bottle works and other industries in Glasgow and its environs, not only to diversify their own operations but also to have their sources of supply under their own control. (37) One such industry was the Delftfield pottery where Watt became a partner through acting as technical adviser. These entrepreneurs were important in laying the foundation of industrial development which otherwise might not have evolved either so early or so rapidly.

'In that sense, the later industrial pre-eminence of the region within the Scottish economy was at least in part due to their initiatives'. (38) Others lent money to fellow merchants, or, more commonly, to local landowners in need of cash to develop their properties. Sir John Shaw, Lord Cathcart (see p186) and the Earl of Dundonald, were among such landed gentry who feature in this biography. (39)

The Effects on Greenock

No wonder the roads between Greenock and Glasgow became choked with pack-horse trains and waggons. (40) In spite of this, Greenock flourished and, by the end of the 1750s, threatened to eclipse Port Glasgow. Nearer to the ocean, and with a more vigorous management, Greenock made handsome profits for Shaw and his fellow shareholders. A large part of these profits was employed in deepening and extending the harbour which further strengthened Greenock's competitive position over Port Glasgow where siltation was threatening its very existence. (41) In 1741 and again in 1751, Shaw obtained charters for Greenock which gave the feuars and burgesses large powers of legal and civil administration under a free electoral system. When a new municipal constitution was introduced in 1751, Watt senior was appointed one of the two Baillies to administer the burgh's affairs, a position reflecting the importance of his business activities in this rapidly expanding town. (42) Such was the place in which Watt was born and brought up. His family shared in the prosperity and activity which this growth brought.

Notes: The Family Background

1 JWP, W/12, J. Watt jun. to M. Arago, 2 Sept. 1834.
2 Dickinson & Jenkins, (1927), facing p. 10 for the Watt family tree.
3 JWP, 3/26, J. Watt jun., 'Memoir of James Watt', p. 1.
4 Williamson, (1856), p. 18.
5 Ibid, p. 29.
6 Ibid, p. 48.
7 Muirhead, (1858), pp. 7 - 8.
8 Register House, Edinburgh, 'Greenock Registers, Old or West, Births & Deaths, 1698 - 1756', 564, 3/1, Feb. 1734.
9 Williamson, (1856), p. 60.
10 JWP, C6/10, J. Watt jun. to J. Rennie, 14 Oct. 1819.
11 See JWP, C6/9.11, 'Pedigree of Mrs. Birnie - 1833', written by J. Watt jun. Dickinson & Jenkins (1927), facing p. 10, show this Jean in 'Pedigree of Watt Family' as brother of James Watt which must be wrong.
12 JWP, C6/10, J. Watt jun. to J. Rennie, 14 Oct. 1819. See also Williamson, (1856), p. 14.
13 Dickinson & Jenkins, (1927), p. 9 and R. Renwick, *Extracts from the Records of the Burgh of Glasgow, 1718 - 38*, p. 76.
14 Moore, (1996), p. 9 and JWP, C7/11 - 18 for more of John Watt's surveys.
15 JWP, 3/26, J. Watt jun., 'Memoir of James Watt', p. 2.
16 B. & W. Col., MIII, 5/8, 'Section of River Clyde'.

17 B. & W. Col., MIII 5/9, Map of the West Coast of Scotland and MIII 5/12, Map of the Firth of Clyde. See Williamson, (1856), p. 52 f. and Muirhead, (1858), p. 10.

18 JWP, C6/10, J. Watt jun. to J. Rennie, 21 Oct. 1819.

19 Riddell, (1979), p. 4.

20 Ibid, p. 6.

21 See Williamson, (1856), p. 122 and Bremner, (1869), p. 59.

22 Williamson, (1856), p. 123, note 1, 'I once asked the late provost Cochrane of Glasgow... to what causes he imputed the sudden rise of Glasgow. He said it was all owing to four young men of talents and spirit who had started at one time in business, and whose success gave example to the rest'.

23 Riddell, (1979), p. 10, but see Hamilton, (1932), p. 241, for Port Glasgow being laid out in 1662.

24 Riddell, (1979), p. 12.

25 Ibid, p. 12.

26 Bremner, (1869), p. 516.

27 Hamilton, (1932), p. 4.

28 Williamson, (1856), pp. 75 & 77.

29 Snoddy, (1950), p. 6.

30 Williamson, (1856), p. 78 and Riddell, (1979), p. 13.

31 Hamilton, (1932), p. 222.

32 Williamson, (1856), p. 79.

33 Devine, (1975), p. 46 and Devine & Jackson, (1995), p. 155.

34 Nugent & Walker, (1996), p. 3.

35 Hamilton, (1932), pp. 4 - 5.

36 Devine & Jackson, (1995), p. 151.

37 Devine, (1975), p. 46.

38 Ibid, p. 48.

39 Ibid, p. 13.

40 Riddell, (1979), p. 17.

41 Ibid, p. 18.

42 See Williamson, (1856), pp. 86, 90 & 91.

The Immediate Family

Father James Watt senior, 1698 - 1782

James Watt senior was the younger son, born at Cartsdyke on 18 January 1698. His father gave him some instruction in mathematics, including mensuration and geometry, even though it would appear that it was never intended that he should become either a teacher or a surveyor for he was apprenticed to a carpenter and shipwright. (1) He settled at Greenock where his father owned some property and where he remained for the rest of his life. He pursued various businesses, such as merchant and builder, but was better known as ship builder, ship chandler and ship owner. In 1734, he purchased land bordering onto the sea in Greenock harbour by the Mid Quay where he built a house and workshop with access to deep water. (2)

Watt senior joined up with various trading partners in his merchant's business. An important one was David Cation while another was his brother-in-law, John Muirheid. His pocket account books show that he traded to the Western Isles of Scotland, the Isle of Man, Dublin, Liverpool and Bristol as well as more rarely the West Indies. (3) The trade in salt and herrings with people on the west coast of Scotland started from at least 1751; there was a cargo from the West Indies in 1752; and he was obtaining his chandler's stock from Liverpool in 1755. (4) Two of his ships were the *Ronak* and the *Crawford*. (5) The Irish connection may have been very important because it supported the large number of middling merchants who emerged in the eighteenth century, among whom Watt senior must rank, and who provided a further stimulus to Glasgow's infant industrialisation. (6) Watt himself would take advantage of his father's contacts in these places to advance his own scientific instrument business.

Watt senior was also something of an inventor. In his carpenter's shop, he designed a special line in pulley blocks for ships (7) and commented in 1754 that 'I have more blocks than I can sell to my customers in a year'. (8) But he was still manufacturing them ten years later when there was some dispute over a patent and he sent to Watt in Glasgow some big double tackle blocks. (9) The first crane built at Greenock for unloading tobacco was designed and built by Watt senior. (10) The ship chandlering business covered supplying and repairing navigation instruments. This must have been how Watt became acquainted with quadrants which would play an important role in determining the direction of his early training and career. While Watt was allowed the run of his father's workshop, there is no tradition of his learning how to handle a boat.

Watt senior would have inherited part of the estates of both his father and brother but, sometime probably in the 1750s, he suffered a severe financial loss although the precise nature is not known. (11) This seems to have been after the death of his wife Agnes. Today it can only be speculation about any role which she may have played in the family business and possibly her death removed her firm guiding hand. However Watt senior retained the respect of his fellow citizens for, from 1751, he 'for upwards of twenty years was a member of the Town council, a magistrate and a zealous promoter of the improvements of the town'. (12) He enlarged the western front of Shaw's mansion. In 1753, he obtained land for the town waterworks from Cathcart who had inherited the Shaw estates. Then in 1758, Cathcart gave land for a new church. Its design was based on St. Martin-in-the-Fields, London, and Watt senior superintended its construction as well as contributing financially. Watt may well have engraved the copper medals which were placed in its foundation. (13) Watt senior was the architect and contractor of the Town hall in 1765. (14) Clearly he held an influential position in Greenock and so was able to support his son during the early years of his training and career. Their relationship remained close even after Watt had settled permanently in Glasgow. Watt senior offered advice which seems to have been accepted and there was no breakdown in this rapport as there was later between Watt and Watt junior.

Mother Agnes Watt, c.1701 - 1753

The Greenock Register of Marriages show that the banns for the marriage of 'James Watt Wright in Gr[eenock] and Angess Muirhead' were entered on 9 April 1728. (15) Agnes was the daughter of Robert Muirheid, a merchant in Hamilton, and was about 25 years old when she married. The Muirheids were an ancient Scottish family but nothing has been written about her father. Her brother, John, was a 'builder and timber merchant in Glasgow'. (16) Accounts survive of his trading with his brother-in-law as partner in salt and herrings (17) and his Glasgow connections were important to both Watts, father and son. On his death in 1769, Uncle Muirheid was succeeded by his son, Robert, with whom Watt would maintain a lengthy correspondence until his death.

Agnes was described as 'a fine-looking woman, with pleasing, graceful manners, a cultivated mind, an excellent understanding, and an equal cheerful temper'. (18) She also had great sense and judgement. The Greenock baptismal registers show a son, Robert, was born on 13 February 1730 but he died only two and a half months later. Then, in November 1731, there followed a daughter, Margaret, who lived for just over a year. Thomas, born in June 1733, died on 30 August 1734. (19) We can well imagine the anxiety of a mother, who, having lost her first three children in infancy, found her next child, James, born in January 1736, was also sickly. The final son, John, born in 1739, appears to have been a normal healthy child. Agnes herself died on 23 January 1753. (20) There is nothing in Watt senior's account books to indicate that she had been ill immediately before her death although, after the funeral, he noted the cost of crêpe and handles for the coffin. (21) Her death would have been severely felt by a sensitive seventeen year old, especially in a family where relationships were warm and cordial. This, and possibly his father's changed circumstances, may have been the reason why Watt went to Glasgow for a prolonged stay.

Brother John Watt, 1739 - 1762

John had a very different personality from his elder brother. He was born on 1 September 1739 (22) but then our records are silent until after the death of his mother. He was usually called "Jockey" and appears to have had a lively, enterprising character, quite different to the apparently quiet and rather retiring Watt. His father gave him 5s. to go to Glasgow in May 1753, when Watt was there already. (23) He was used by Watt to carry a letter home in February 1755 (24) and the brothers appear to have had great affection for each other. Later, when Watt was learning his trade in London, he sent Jockey some 'red chalk... to draw with' (25) and intended to make a rule or ruler as a Christmas present in 1755 but probably failed to complete it through pressure of other work. (26) Watt regretted that both he and Jockey were away from home that Christmas at a time when his father could 'so little spare us'. (27) We know that Jockey was trained as a bookkeeper (28) and that he was expected to follow his father's business. (29)

After Watt's return from London, both brothers were in Glasgow in the autumn of 1756. (30) When Watt decided to change his place of business from Greenock to Glasgow in the summer of 1757, Jockey was involved with the move. (31) Jockey also seems to have helped with accounts, whether those of his brother or father is not clear, perhaps both. However, that December, he went to Liverpool in all probability to obtain a job. His father's letter to a John Rigbey, possibly a trading acquaintance, states that John might stay some time so Rigbey could show his friendship and any expenses would be met by his father. (32) Watt felt that, through his brother's training, there should be no need to pay a fee for Jockey to be accepted in some post. It is most likely that Jockey failed to obtain suitable employment (33) and it is possible that he may have tried to start trading on his own account. A little later, he considered seeking work abroad in some trainee capacity which his elder brother thought was a foolish notion. (34) At about this time, he helped to re-survey part of his Uncle Watt's 'Map of the Clyde'. (35. See p95)

The next we know is that Jockey was in Bristol by March 1760 where he may have remained until early 1762. (36) He was trading on his own account as well as taking commissions from others. Watt senior had a contact there, John Lean, whom Jockey met and with whom Watt had dealings over many years. Lean was a merchant dealing in linen among other articles. Jockey acted as a sort of agent for his brother, selling the quadrants and flutes produced in Glasgow. (37) He also bought supplies and materials for his brother such as sheets of brass, ivory teeth and barometer tubes. The brothers met at Bristol when Watt made a sales trip to England in October 1761 and had to borrow money from Jockey in order to help his return home after his horse became unfit. (38. See p112)

At Bristol, Jockey ran into problems with a cargo of wheat sent back to Glasgow in the autumn of 1761. The recipient, a Mr. Morison, at first disputed having to pay for the bags and then tried to avoid making any payment at all. Jockey may have borrowed money on dated bills to finance the deal which he found were becoming due for payment but he had no means of meeting them. Watt tried to sort out these problems by raising nearly £100 in Glasgow on behalf of his brother. (39) Obviously their father was very worried. This fiasco may have changed Jockey's mind about becoming a merchant, because he returned home in January 1762, saying that he was willing to accept orders for any goods wanted in Bristol by his acquaintances. (40)

We have only a few comments about some of Jockey's financial dealings during the earlier part of 1762 until the announcement that he was to be chief mate on the *George* and was sailing to Havana to purchase some vessels. Should they purchase two, he was to have command of one. (41) How or where Jockey obtained the skills of a seaman is totally obscure. At any rate, he asked that all his debts should be settled, presumably those incurred by his trading ventures. He must have set sail in the late summer or early autumn of 1762 because his boat was shipwrecked on the Bahamas Bank on 30 October. It broke up and only three people got ashore. (42) The news did not begin to filter through until the following March (43) and the poor father waited in agony for nearly a month before Jockey's death was confirmed. Watt senior wrote to his only surviving son, 'I pray God my comfort you under the loss of so Deir a brother who if he hade been spaired would have been a credit & suport to both you & me but since Providence has deprived us of that comfort we must indevor to suport under it & confort on another'. (44) Watt junior described his uncle as 'a youth of promising abilities, who was lost at sea soon after he came of age'. (45) When Watt's first son was born in July 1765, he and his wife Margaret decided to call him John 'for poor Jockey thinking it would be more agreeable to you to call him by his name than your own as I still remain your namesake'. (46) Other than this, we have no reaction from Watt about the death of his brother with whom he had a close relationship.

NOTES: THE IMMEDIATE FAMILY

1 Williamson, (1856), pp. 56 - 7.
2 Ibid, p. 83.
3 JWP, C7/50 - 52, 'Pocket Account Books of J. Watt sen., 1752 - 5'.
4 JWP, C4/A11, Letter Book of J. Watt sen., 1751 - 58, 7 Feb. 1755.
5 JWP, C4/A3 'Orders to Build the *Ronak*' and C4/A10, 'Account for the ship *Crawford*'.
6 Devine & Jackson, (1995), p. 215.

7 JWP, C4/A20, 'Account of Lads Work, 1767 - 8'.
8 JWP, C4/A11, Letter Book of J. Watt sen.,9 Feb. 1754.
9 JWP, 4/7.55, J. Watt to J. Watt sen., 21 March 1765; 4/11.119, 2 March 1765; 4/11.124, May 1765.
10 Crowther, (1962), p. 108.
11 Muirhead, (1858), p. 16 and JWP, C6/10, J. Watt jun. to J. Rennie, 21 Oct. 1819.
12 JWP, 3/26, J. Watt jun., 'Memoir of James Watt', p. 2.
13 JWP, 4/11.49, J. Watt sen. to John Watt, 17 Mar. 1759.
14 Williamson, (1856), pp. 97 & 114.
15 Register House, Edinburgh, 'Greenock Registers, Marriages', 564, 3/4.
16 JWP, W/10.38, J. Watt jun. to M. Arago, 22 Jan. 1835.
17 JWP, JW/19, see for example No. 43, 'Copy of James Watt Salt Account, 15 Oct. 1765 - April 1766'.
18 Muirhead, (1858), p. 13 and Williamson, (1856), p. 128.
19 Register House, Edinburgh, 'Greenock Old or West, Births & Deaths, 1698 - 1756', 564, 3/1, Aug. 1734.
20 B. & W. Col., MIV, Box 11, Folder 4.
21 JWP, C7/50, Pocket Account Book of J. Watt sen., 1750 - 3, 27 Feb. 1753.
22 Register House, Edinburgh, 'Greenock Old or West, Births & Deaths', 564, 3/1, Sept. 1739.
23 JWP, C7/50, Pocket Account Book of J. Watt sen., 1750 - 5, 15 May 1753.
24 JWP, 4/11.157, J. Watt to J. Watt sen., 4 Feb. 1755.
25 JWP, 6/46.26, J. Watt to J. Watt sen., 9 Aug. 1755.
26 JWP, 6/46.17, J. Watt to J. Watt sen., 29 Nov. 1755.
27 JWP, 6/46.16, J. Watt to J. Watt sen., 27 Dec. 1755.
28 JWP, 4/11.14, J.Watt to J. Watt sen., 9 Jan. 1758.
29 Muirhead, (1858), p. 17, see JWP, C6/10, J. Watt jun. to J. Rennie, 21 Oct. 1819.
30 JWP, 4/11.3, J. Watt to J. Watt sen., 18 Oct. 1756.
31 JWP, 4/11.4, J. Watt to J. Watt sen., 29 July 1757.
32 JWP, C4/A11, J. Watt sen. to J. Rigbey, 5 Dec. 1757.
33 JWP, 4/11.14, J. Watt to J. Watt sen., 9 Jan. 1758.
34 JWP, 4/11.32, J. Watt to J. Watt sen., 31 May 1758.
35 NLS, Rennie Papers, 19824, No. 16, J. Watt to J. Rennie, 2 Dec. 1794.
36 JWP, 4/11.57, J. Watt to John Watt, 3 March 1760.
37 JWP, 4/11.59, J. Watt to John Watt, 15 May 1761 and C1/1, 'Accounts as Mathematical Instrument Maker', p. 19, 24 Oct. 1761.
38 JWP, 4/11.62, J. Watt to John Watt, 1 Oct. 1761.
39 JWP, 4/11, see letters from J. Watt to John Watt, Nos. 65, 66, 67, 68, 72, 76, Dec. 1761 - Jan. 1762.
40 JWP, 4/7.70, J. Watt sen. to J. Watt, 25 Dec. 1761 and 4/11.81, J. Watt to J. Watt sen., 21 Jan. 1762.
41 JWP, 4/7.69, J. Watt sen. to J. Watt, 6 Dec. 1762.
42 JWP, 4/7.65, J. Watt sen. to J. Watt, 28 March 1763.
43 JWP, 4/7.68, J. Watt sen. to J. Watt, 5 March 1763.
44 JWP, 4/7.65, J. Watt sen. to J. Watt, 28 March 1763.
45 JWP, 3/26, J. Watt jun., 'Memoir of James Watt', p. 2.
46 JWP, 4/11.127, J. Watt to J. Watt sen., 18 July 1765.

Youth in Greenock

Watt's Early Life

'I look upon him, considering both the magnitude and the universality of his genius, as perhaps the most extraordinary man that this country has ever produced'. William Wordsworth

'Mr. Watt's abilities and acquirements placed him next, if not superior, to Newton'. William Withering. (1)

These views of two of his contemporaries show the high esteem in which Watt was held later in his life but at first he showed few signs of this promise. The Parish Register of Baptisms in Greenock has the entry, 'Watt, James son lawful to James Watt wright Gr[eenock] and Agnes Muireheid his spouse was born the 19th and baptised the 25th' of January 1736. (2) He was baptised into the Presbyterian Church of Scotland, in which religious faith he remained for the rest of his life, even after moving to Birmingham. The earliest biography about him stated that he was 'descended not from affluent but truly respectable parents who took special care of his education'. (3) We can be certain that these respectable parents would have brought up their son with the rudiments of the Christian faith since both grandfather and father were involved with the Kirk in Greenock. For many years after the death of his father in 1782 until at least 1801, Watt continued to pay for renting a pew, presumably in the Mid Kirk which his father had helped to build. (4) Another religious influence may have been Aunt Jean through her marriage with the minister of Inverkip. (5) While there seems little doubt that young Watt's home had at the very least a conventional Christian background, his relations might not have approved of the colourful way he used Biblical expressions later in his correspondence.

The surviving evidence shows that young Watt was a sickly child who must have caused his mother great concern. Watt junior wrote to John Rennie that 'The state of his [father's] health was such as rendered it very improbable that he should have been put out to any trade after leaving school'. (6) This may be the reason why Jockey was trained as a bookkeeper, yet we shall see later that Watt acted on behalf of his father in his father's business while in Glasgow and London. (See p48 & 53) Subsequently, Watt suffered from severe headaches, almost certainly what today would be called migraine. Mrs. Campbell suggested that these started in his youth but she may have been mistaken. (7) His ill-health may have aroused an interest in anatomy and medicine, particularly from the age of fourteen, and he once told Watt junior that he would have become a surgeon if he could have withstood the sight of the sufferings of patients. (8)

For his early general education, Watt was taught by his mother. Then he was sent to the ordinary local school and after that to the grammar school. Watt junior mentioned his father's early ill-health at this period. 'He received the rudiments of his education in the public schools of his native town, but from the extreme delicacy of his constitution was with difficulty enabled to attend the classes, and owed much of his acquirements to his studious habits at home'. (9) Watt employed the time he had to spend away from school through ill-health in reading a great deal and, later in life, he commented to a friend, 'I have never yet read a book, or conversed with a companion, without gaining information, instruction, or amusement'. (10) We will find, time and time again throughout his life, that, when he was tackling any new subject, he read all the appropriate books he could find and then turned to practical experiments.

Watt junior had it impressed on his mind that his father 'acquired his first knowledge of mathematics and engineering from his Grandfather's and Uncle's books & Manuscripts. His father too had a turn that way and no doubt encouraged his studies'. (11) Watt's interest in mathematics can be traced back to an early age, possibly influenced by portraits of John Napier and Sir Isaac Newton inherited by his father from his grandfather. (12) There is a crude set of Napier's 'bones' (a calculating device) in Watt's Garret Workshop now in the Science Museum, London. One of Mrs. Campbell's stories was that, at the age of six, Watt was observed drawing with chalk on the hearth. A visitor thought that he was wasting his time and should be sent off to school, but his father pointed out that he was working out some mathematical problem. (13) Dr. Gibson elaborated this, saying that Watt senior found how his son had covered not only the hearth but also the walls of his bedroom with calculations because he had left his slate behind at school. (14) On the other hand, it has been stated that Watt did not shine at school until he was thirteen or fourteen, when he entered the mathematical class where his interests were at last aroused. (15) According to Mrs. Campbell, he had twice read s'Gravesande's *Elements of Natural Philosophy* before he was fifteen.

There are two Note Books in Watt's hand of passages which he may have copied from his grandfather's and uncle's books. One, 'Notes on Surveying of Land', (16) outlines how to prepare to carry out a survey, measure different shaped pieces of land, level between two places as well as how to use the 'Azimuthe' compass and 'plain' table. The other, 'Notes on Dailing and Making Instruments', (17) contains instructions for making scales, dialing, and, what would prove to be of great significance for his later life, 'How to make a sea quadrant. This instrument is ye most easiest divided instrument of any for it consists all of equall parts'. Here surely we can see two influences at work on the boy from his family background, that of the mathematician and land surveyor from his grandfather and uncle and that of the interest in navigation instruments through his father's shipping connections. While later Watt would stress the importance of mathematics as part of Watt junior's education, he himself admitted in 1814, 'I was always averse to labour, and never was a mathematician'. (18)

Early Interests in Chemistry and Mechanics

Another of Mrs. Campbell's stories was about the young boy watching the kettle boil and condensing the steam against a cup or silver spoon. She placed that event at the home of Watt's aunt, Mrs. Muirheid, her mother, (19) while Gibson says it was when he had been sent into the country to recover his health. (20) This interest in experiments covered ones in chemistry. When Watt was declining the offer of employment in Russia in 1771, he wrote to their Ambassador, 'I am a person of no great learning But I have had from my infancy a propensity to mechanics & Chemistry, and have tryed many experiments in both these sciences. What little knowledge I have is the fruit of these experiments'. (21) Unfortunately, we have no accounts of any chemical experiments which he may have carried out as a boy or indeed of any until after his investigations into the steam engine which led to his invention of the separate condenser in 1765.

Watt's interest in mechanics has attracted most attention. His father's shipyard must have been irresistible. (22) There were, of course, the navigation instruments such as quadrants, with which he quickly became familiar. In addition, Watt junior said that his father 'learnt the use of tools in his father's workshops who carried on the business of a ship chandler & builder in addition to that of merchant'. (23) Mrs. Campbell said that Watt senior gave his son a set of tools (24) and set up a small forge for his use. (25) With these, Watt definitely made some models. Many years later, he asked a Mr. R. Barclay to search the attics of his father's house, which Barclay did but replied, 'The only models I could find are the model of a house and a crane or something of that nature'. (26) In 1834, Watt junior wished to introduce Arago to Peter Ewart in order to confirm his father's practical abilities. Arago was preparing his *Eloge* of Watt and Ewart was

> ...the only surviving pupil of my father, [who] will be able to afford you the most authentic information respecting him and particularly upon the point of his skill in all handicraft works and operations. I believe there has not, in our times, existed any individual so thoroughly conversant with the practical operations as well as the scientific theories of every Art & Trade. (27)

This was how Watt would secure the title of 'Craftsman'.

Watt's Character

Watt has often been described as a solitary lad, preferring his own company to that of his contemporaries. This has been deduced from his love of fishing because apparently he used to sit for hours on one of the quays at the back of his father's house and workshop in Greenock harbour. (28) On the other hand, this might have been an excellent place from with to observe the activity of a busy little port. That he tended to be a 'loner' is confirmed by the first biography written in 1803.

> In his early days, young Watt was considered a reserved lad often separating himself from the companions of his youth, and devoting many of those hours to the improvement of his mind, which by others were dissipated in childish sports. He was not, however, wanting in sociability, but frequently entered into all the amusements of his school-fellows, ever assuming a will of his own when any matter in dispute led him to exert that spirit of high independence which has so steadily marked his character and conduct. (29)

In contrast to Watt's solitariness and independence, this passage equally highlights his capacity for sociability. Another of Mrs. Campbell's tales shows Watt himself in the position of story-teller. When he was fourteen, his mother had taken him to Glasgow to stay with a friend. But this person soon asked for him to be taken home again because he had kept them in 'breathless attention' with his stories until the small hours of the morning and they were worn out. (30) Later in his life, Watt formed deep friendships with people like Joseph Black, John Robison, Erasmus Darwin, William Small and, above all, Matthew Boulton. They supported him in his steam engine ventures when these were little more than schemes. To have convinced such people that he was worth backing, Watt must have had an attractive personality as well as a convincing manner and this childhood story may give us the first glimpse into part of his character which would play a vital role later in his life.

We have looked at a few of the events in Watt's youth which may have laid foundations for his later career. It seems as if he began to develop around the age of fourteen and to assume those characteristics which he would retain for the rest of his life. Therefore it is appropriate to note a few extracts from the obituary written by Lord Jeffrey soon after Watt's death to compare how he had developed over the intervening years.

> Independently of his great attainments in mechanics, Mr. Watt was an extraordinary, and in many respects a wonderful man. Perhaps no individual in his age possessed so much and such varied and exact information, - had read so much, or remembered what he had read so accurately and well. He had infinite quickness of apprehension, a prodi-

gious memory, and a certain rectifying and methodising power of understanding which extracted something precious out of all that was presented to it. His stores of miscellaneous knowledge were immense...
It is needless to say, that, with those vast resources, his conversation was at all times rich and instructive in no ordinary degree... No man could be more social in his spirit, less assuming or fastidious in his manners, or more kind and indulgent towards all who approached him. He rather liked to talk...
His talk, too, though overflowing with information, had no resemblance to lecturing or solemn discoursing, but on the contrary, was full of colloquial spirit and pleasantry. He had a certain quiet and grave humour, which ran through most of his conversation, and a vein of temperate jocularity, which gave infinite zest and effect to the condensed and inexhaustible information, which formed its main staple and characteristic. (31)

Notes: Youth in Greenock

1 Crowther, (1962), p. 95.
2 Register House, Edinburgh, 'Greenock Old or West, Births & Deaths', 564, 3/1, Sept. 1736.
3 Phillips, (1803), p. 527.
4 JWP, 6/35.31, J. Rankin to J. Watt, 12 April 1799 and 6/33.11.2, J. Walkinshaw to J. Watt, 2 May 1801.
5 JWP, C6/9.11, 'Pedigree of Mrs. Birnie - 1833' by J. Watt jun.
6 JWP, C6/10, J. Watt jun. to J. Rennie, 14 Oct. 1819.
7 Muirhead, (1858), p. 23.
8 Ibid.
9 JWP, 3/26, J. Watt jun.,'Memoir of James Watt', p. 3.
10 Muirhead, (1858), p. 24.
11 JWP, C6/10, J. Watt jun. to J. Rennie, 14 Oct. 1819.
12 Muirhead, (1858), p. 29.
13 Ibid, p. 18.
14 JWP, 4/83, Dr. Gibson to J. Watt jun., 8 Oct. 1834.
15 Smiles, (1878), p. 1.
16 JWP, C4/B27, Notes on 'Surveying of Land' by J. Watt.
17 JWP, C4/B33, Notes on 'Dailing and making instruments', by J. Watt.
18 Robinson & Musson, (1969), pp. 2 - 3 and Muirhead, (1854), Vol II, p. 357, Letter 395, J. Watt to Dr. Brewster, 7 June 1814.
19 Muirhead, (1858), p. 21.
20 JWP, 4/83, Dr. Gibson to J. Watt jun., 8 Oct. 1834.
21 Royal Society, London, James Watt Papers, PH 13, No. 71, J. Watt to Russian Ambassador, 7 Aug. 1771.
22 Muirhead, (1858), p. 19.
23 JWP, W/10.67, J. Watt jun. to M. Arago, 6 Oct. 1834.
24 Muirhead, (1858), p. 19.
25 Ibid, p. 27.
26 JWP, 4/18.1, R. Barclay to J. Watt, 6 April 1801.
27 JWP, W/12.84, J. Watt jun. to M. Arago, 2 Sept. 1834.
28 Muirhead, (1858), p. 25.
29 Phillips, (1803), p. 527.
30 Muirhead, (1858), p. 21.
31 Arago, (1839), 'Lord Jeffrey's Character of Mr. Watt', pp. 177 - 180.

The First Period in Glasgow, 1753 – 1755

Town and Gown

Although Watt had visited his Uncle Muirheid in Glasgow during his childhood, it was not until after the death of his mother that he made a prolonged stay. That city would have seemed a much more interesting and stimulating place than Greenock because it was the centre for the region and was growing rapidly. In 1708, its population was 2,766; in 1755, 23,546; and in 1785, 45,889. (1) This expansion necessitated the construction of suburbs around the city which reflected a change in the style of the wealthier people. When Adam Smith was a student at Glasgow University before 1740, the merchants lived in flats in the city but when he returned just over ten years later, the tobacco magnates had begun to build villas outside. (2) By that time, numerous coffee houses and clubs had come into existence because the rapidly growing trade, particularly with the New World, stimulated the outlook and imagination of the inhabitants. People became emancipated from the ancient Scottish poverty and provincialism which was reflected in the development and growth not only of the town itself but also of its University.

Glasgow University is one of the ancient British universities, having been founded in 1451. Although medical degrees had been awarded since the fifteenth century, it functioned largely as a regional university, where the teaching concentrated principally on the liberal arts until the early part of the eighteenth century. (3) Stemming from a commission into its running and an Act passed in September 1727 regulating its affairs, the University was transformed over the next forty years into one of the most intellectually active institutions ever known in Scotland. (4) Of particular relevance to the career of Watt was the appointment of Robert Simson in 1711 to the Chair of Mathematics. He was one of the professors who befriended Watt.

Although there had been some teaching of Natural Philosophy before 1710, this subject was given a firmer foundation through the appointment of the elder Robert Dick as professor around 1725. He was succeeded by his son, also Robert, in 1751, who also befriended Watt. Lists of demonstration apparatus survive from 1725 which include an air pump, apparatus for demonstrating pneumatics and mechanics, a hydrostatic balance and other balances, thermometers and barometers, three telescopes, five microscopes and in all about one hundred distinct groups or sets of apparatus. (5) In 1731, Henry Drew was employed as the first laboratory assistant or steward to take care of this apparatus, as would Watt himself later. Then there was Professor William Cullen, who developed the teaching of chemistry, apparently informally at first. Building a proper laboratory was started in 1747 followed by Cullen commencing formal classes around 1749. (6) In 1755, he was appointed professor of chemistry in Edinburgh and was succeeded by Black, who became one of Watt's closest friends.

Adam Smith also knew and helped Watt and could have influenced the way Watt organised his workshop. Smith had learned in his boyhood about the effectiveness of the division of labour to increase production. He had watched people making pins and realised the importance of organisation in production. He recognised that sub-division of labour could be applied to science as well as in manufacture. We may note what he said about the introduction of different forms of power.

> He who contrived that the outer wheel [of a corn mill] should go by water was a philosopher, whose business it is to *observe* [author's underlining] everything. They must have extensive views of things, who, as in this case, bring in the assistance of new powers not formerly applied... Fire-machines, wind and water-milns were the inventions of philosophers, whose dexterity too is increased by the division of labour. They all divide

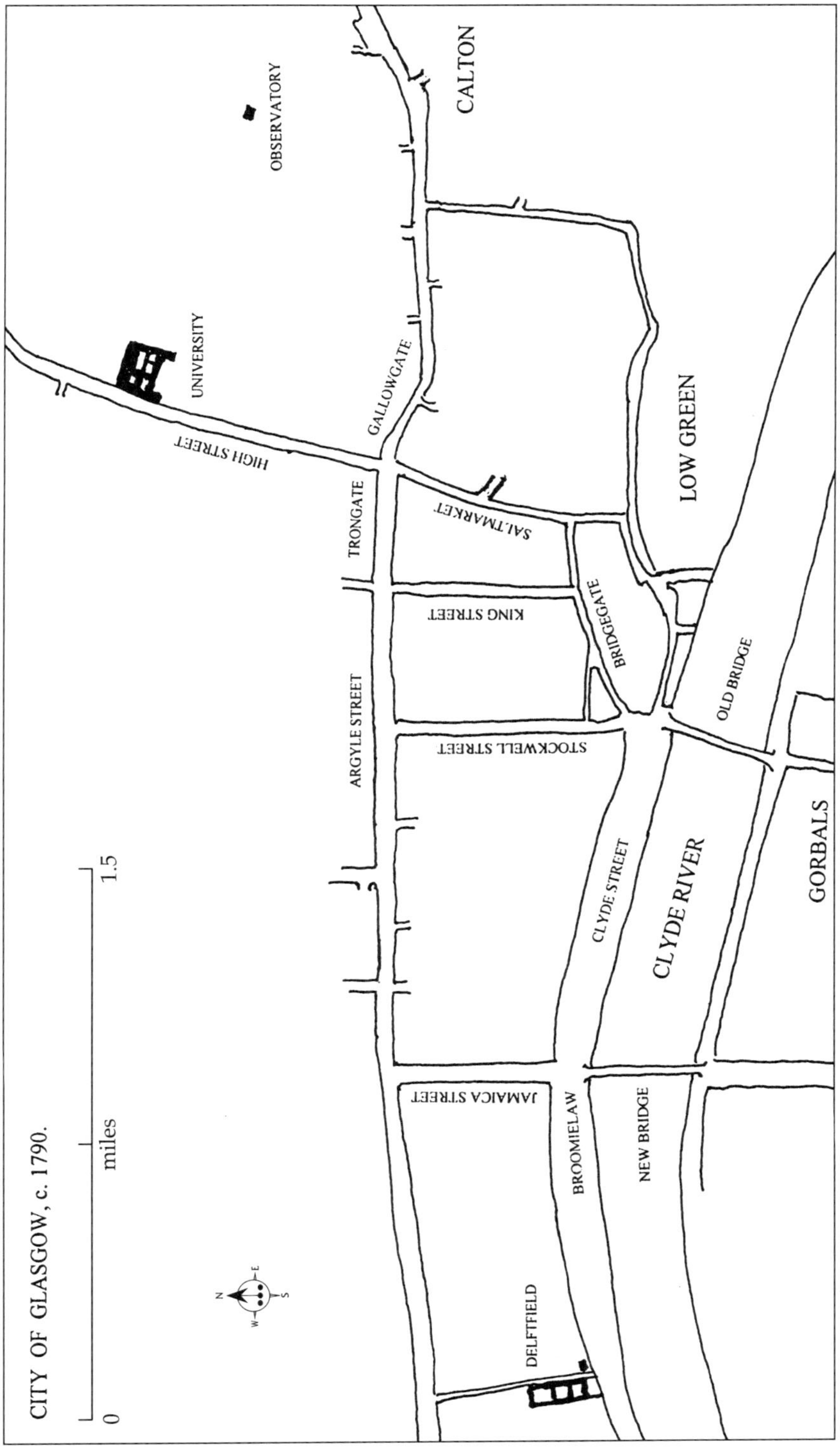

Sketch Map of Glasgow showing some of the places associated with Watt.

> themselves, according to the different branches, into the mechanical, moral, political, chemical philosophers. (7)

Watt aspired to be a philosopher and perhaps Smith was the source of his motto, 'Observare'.

Watt's First Residence in Glasgow

An entry in his father's Pocket Account Book for April 1753 states 'Sent by post a 20s not[e] to son James to buy sundrie small things' followed by son John being given 5s. that May for going to Glasgow. (8) This was just after the death of their mother and suggests that Watt was sent to his Uncle Muirheid to obtain some relief from his bereavement. The entries continue in August with 2s. 6d. for a black 'gravat' for Watt who must have been still in mourning. His father continued to send other small sums over the next couple of years for clothing and sundries including 5s. on 31 December 1754 for New Year's gifts for the two sons. (9) Watt's surviving letters home do not start until June 1754 and among them is a list of clothes and other items which he took to Glasgow. (10) The date when his father started to send money shows that Watt spent over two years away from home at this time and not the one given in earlier biographies.

The tradition is that Watt was sent to Glasgow to stay with his uncle to be apprenticed to learn the trade of mathematical instrument maker. (11) Indeed, our earliest biography of Watt states, 'Having finished his mathematical studies, and laid a solid foundation in the several branches of useful and important knowledge, he was, at the age of sixteen, articled as an apprentice to learn the art of an instrument maker'. (12) But Watt junior wrote in 1819, 'I never heard of my father's being articled as an apprentice, nor have any of his surviving friends... nor do I find any record of it among his surviving papers'. (13) Samuel Smiles gave the reason for there being no formal apprenticeship.

> On making inquiry for a proper person, under whom to learn the business of mathematical instrument making, it was found that there was no such person in Glasgow. There was, however, a mechanic in the town, who dignified himself with the name of "optician", under whom Watt was placed for a time. He was a sort of Jack-of-all-trades, who sold and mended spectacles, repaired fiddles, tuned spinets, made and repaired the simpler instruments used in mechanical drawing, and eked out a slender living by making and selling fishing-rods and fishing tackle. (14)

Surely Uncle Muirheid, living in Glasgow, would have known that there was no suitable master for his nephew.

However Watt may have received some informal tuition from various sources. One could have been John Carlisle who possibly was the 'optician' mentioned by Smiles. He ran a shop at the Bell's Wynd where he sold many of the items in which Watt dealt later. Selected from a long list in an advertisement of 1755 are just a few such as

> Fine Violins, German and common Flutes of all Kinds... Hadley's Quadrants, and common Quadrants, Gunter's Scales and Compasses, complete sets of Mathematical and Surgeons Instruments; Everard's, Brenaun's, and Leadbetter's Rules, Ivory Foot Rules, Bradford's new-invented Sliding-rules... best Spectacles, Reading-Glasses of all Kinds, Telescopes, Microscopes, and Prospects, Thermometers;... (15)

Carlisle did not say that he could repair any of these instruments and we have no further details. This was the type of merchandise which Watt sold first soon after he started as

Mathematical Instrument Maker to the University and then later in a much larger shop in the Trongate. (See p115)

Recently it has been suggested that there were two men under whom Watt might have been given some instruction, Henry Drew and George Jarden. (16) Both acted as laboratory stewards to the University and both were members of the Incorporation of Hammermen with special competence in horological work. No mention has been found of Drew in Watt's papers so Jarden was the more likely of the two. One of his tasks was to clean and keep in repair the old Tolbuth clock set up on the Trone steeple which included painting the dial plates. (17) Watt's 'Waste Book' gives details of financial transactions with Jarden. For example, in May 1757, Watt paid Jarden 10s for Mr. Gray's quadrant. Watt made some prototype instruments for John Gray and this may be the cost of having them divided by Jarden. (18) Watt purchased a rest for his lathe from Jarden who also fabricated other iron work for him. (19) The May 1757 transaction confirms that Watt knew Jarden before he settled permanently in Glasgow during the summer of that year, pointing to the possibility that Jarden could have helped him during his first period of residence there.

Watt the Craftsman

We can be certain that Watt did not abandon his practical interest in carpentry while first in Glasgow. This can be seen when we study the list of clothes he took with him. (20) This mentions '1 Lether Apron' and *Tools* among which were '1 Jack & 1 Hand plain 1 saw 1 large former 1 small ditto 4 Chissals 4 Gamlets & 1 Quadrant & Vanes 2 Fyles 1 Handsaw & 1 Tenon'. These could have been the carpentry tools given to him by his father. Robert Muirheid described to Watt junior how he remembered Watt as a youth.

> I know that when I was very young he used to work in my Fathers shop in the Gorballs for his own amusement & at making or contriving some mathematical or mechanical articles having seen him often there in his shirt sleeves & his appearance at that work & undress had the effect of making me copy him as far as in my power having procured a hand plane, chissels, saw, etc. and always in my shirt sleeves which I then considered absolutely necessary. (21)

Robert Muirheid thought this was in about 1758 before he went to College but it was probably earlier because, by that date, Watt had his own workshop in Glasgow University.

Watt's dexterity as a craftsman in wood was recognised by his friends. Then there was his interest in navigation instruments shown by the quadrant listed as being taken to Glasgow. From the mention of vanes, the quadrant was most likely the simpler backstaff or Davis variety rather than the later Hadley's. In addition, there was his interest in mathematics. All this pointed to the recommendation that he ought to receive proper training to manufacture mathematical instruments, particularly those used in navigation. This would help the ship chandlering side of his father's business. But it is doubtful whether this was the initial reason why he was sent to Glasgow.

Watt the Trainee Merchant

We must investigate further the list of clothes taken to Glasgow and the letters to understand why Watt's tuition to become a mathematical instrument maker may not have been the prime cause of his being sent to that city. At this period, it would have been normal for the eldest son to have followed in his father's footsteps. We can realise that Watt's weak constitution would have prevented him taking up the career of either a building contractor or shipwright. But Watt senior was also a merchant. Was Watt so sickly that he could not have assisted his father in that capacity? By being sent to Glasgow, young Watt could have received instruction as a merchant from his Uncle Muirheid. In Glasgow, Watt could act as his father's agent, so extending the family business there. He would have needed suitable clothes which we find in the list that he took with him. (22) It included one pair of almost new shoes, various pairs of 'Stockens', listed as grey blue, plain, black, blue and white thread and two pairs of silk ones. He had 'Rufled' and plain shirts, 8 Holland night caps, coloured and silk napkins, a black silk cravat, one black and one blue coat, two black waistcoats as well as cut velvet and tartan ones. The list continued with black stocking and blue cloth breeches and a hat with mourning 'crape'. All these black items suggest that he was still in mourning for his mother. Watt certainly liked fine garments and these are not the clothes of a poor apprentice but of the son of a reasonably prosperous merchant.

Watt's letters also point to this being the reason for his being sent to Glasgow. The early ones contain a few pieces of family news but consist mostly of transactions which he was carrying out for his father. For example, in the first one dated 11 June 1754, Watt says, 'have bot. all the things I could get I went with the 7£ to Mr Pagan & he took it very Discreetly & said if you could get him the rest this week you would oblige him but I can get no money from either Mr Cation or Mrs. Berries as yet'. (23) Watt added an invoice of things which he had purchased such as skins, chest 'lox' [locks], padlocks, hooks, marble paper and pencils. Scientific instruments are never mentioned. This contrasts with his later letters from London which contain many descriptions of the pieces of apparatus on which he was working as well as his progress as an instrument maker.

Watt senior's Pocket Account Books show that he had accepted bills of exchange with Cation for £58 in August 1753 and for £80 in February 1754. (24) That June there were transactions involving Uncle Muirheid. Further letters show that Watt was still trying to obtain money from Cation and other people in addition to buying a variety of other items. In fact, these details of attempts to settle financial transactions and the purchase of bits and pieces form the main content of these letters to his father. (25) Watt continued to act as an agent for his father during his year in London and also when he had moved permanently to Glasgow after 1757. If learning to become his father's agent were the main reason for Watt being sent to Glasgow in 1753, we can now understand why no name of the person who was meant to have been his master for an apprenticeship in making mathematical instruments has ever been discovered. This evidence suggests there was none.

The rapidly developing young Watt would have found life in Glasgow more stimulating than in the small port of Greenock. Mrs. Campbell said that

> Under Mr. Muirheid's roof he met with good society, and formed friendships with several intelligent and well-educated young men; they had frequent evening meetings to give or receive information. These gentlemen acknowledged and appreciated Mr. Watt's superior abilities. (26)

Watt himself later denied that his abilities were superior to those of his friends, in fact rather the reverse. He described them as 'much my superiors, I never having attended a College & being then a mechanick'. (27) We have seen that there were many clubs being formed in Glasgow at this time and Watt may have frequented some of them. (See p44) But we must doubt whether Watt and his friends formed any club as early as this and it is more likely that one mentioned by Mrs. Campbell may have been started when Watt returned from London and would have been better known in the University.

However, Watt was introduced to several members of the University through George Muirheid who became Professor of Oriental Languages and Latin at about this time. (28)

Among these was the younger Dick. We can realise the attraction a Professor of Natural Philosophy would have for Watt. Dick recognised the abilities of young Watt and recommended that he be sent to London to receive better training in the manufacture of mathematical instruments than could be obtained anywhere in Scotland at that time. Dick also provided a letter of introduction to James Short, the famous maker of telescopes. We have no details of the discussions which led to Watt setting out for London in June 1755, this time definitely to search for tuition in the sphere of mathematical instrument making.

NOTES: THE FIRST PERIOD IN GLASGOW

1 Hamilton, (1932), p. 5.
2 Crowther, (1962), p. 62.
3 Simpson, (1982), Anderson, p. 7, rt.
4 Simpson, (1982), Swinbank, p. 23, lft.
5 Ibid, p. 30, rt.
6 Ibid, p. 32, rt.
7 Crowther, (1962), p. 64.
8 JWP, C7/50, Pocket Account Book of J. Watt sen., 1752 - 3, 24 April and 15 May 1753.
9. JWP, C7/52, Pocket Account Book of J. Watt sen., 1754 - 5, 31 Dec. 1754.
10 JWP, 4/11, Letters 155 - 160.
11 Muirhead, (1858), p. 31.
12 Phillips, (1803), p. 527.
13 JWP, C6/10, J. Watt jun. to W. Playfair, 20 Oct. 1819.
14 Smiles, (1878), p. 23.
15 Bryden, (1972), p. 19, from *Glasgow Courant*, 22 Sept. 1755.
16 Smith, (1994), Bryden, p. 10, rt.
17 SRA, C1.1.32, Glasgow Council Act Book, Sept. 1755 - Jan. 1763, p. 249, 11 April 1758, 'Treasurer to pay George Jarden Smith in Glasgow Two pounds six shillings sterling for work at cleaning and repairing the clocks and furnishings thereto' and p. 274, 29 Sept. 1758, 'George Jardin four pounds sterling in part of his account... for repairs on the old Tolbuth clock set up in the Trone steeple and painting the dyall plates'.
18 Bryden, (1972), p. 11.
19 B. & W. Col., MIII 3/1, 'Waste Book 1757 - 63', 17 May, 8 Oct. 1757, 27 Jan. 1758 and 15 June 1759.
20 JWP, 4/11.161, 'A List of James Watt Clothes taken to Glasgow'.
21 JWP, C6/10, R. Muirhead to J. Watt jun., 13 Oct. 1819.
22 JWP, 4/11.161, 'A List...'.
23 JWP, 4/11.160, J. Watt to J. Watt sen., 11 June 1754.
24 JWP, C7/51, Pocket Account Book of J. Watt sen.,1753 - 1754.
25 See JWP, 6/46.36, J. Watt to J. Watt sen., 7 May 1755 for the last of these early Glasgow letters.
26 Muirhead, (1858), p. 23.
27 JWP, Letter Book 4, p. 107, J. Watt to J. Craig, 30 April. 1805, see also Muirhead, (1854), Vol II, p. 299, No. 354 .
28 Muirhead, (1858), p. 33.

The Year in London, 1755 – 1756

The Journey to London

Watt's letters from Glasgow to his father do not contain any suggestions either for training as a craftsman or that such training might best be secured in London. There is a hint that sending him there in June 1755 was arranged in a hurry (1) and that his father agreed possibly because John Marr, a distant relative through marriage and also probably the son of Watt's mathematics teacher in Greenock, (2) was going to London to sit an examination to become an instructor in the navy. Marr acted as a sort of guardian but had passed this exam by 21 July (3) and soon went to sea, although he and Watt were to remain in touch for many years afterwards.

Because the pair would make the journey on horseback, Watt's chest was despatched to Leith on 6 June to be sent by sea to London where it was directed to the care of Captain William Watson at the Hermitage. (4) It arrived eight days before the youths did. (5) They set out on 7 June and went via Coldstream where at that time there was no bridge so the Tweed had to be forded on horseback and then on to Newcastle. At Durham, they visited the cathedral where Marr was impressed with the architecture if not the form of worship.

> The Cathedral of Durham far surpassed I had seen before both for the magnificance of the structure and the ridiculous manner of the worship for the whole of the Prebends and Canons could not refrain from laughing the time they were addressing the most high in prayer and confession and tho I am not very squeamish I was quite shocked at their behaviour'. (6)

This is one of the few occasions when we know Watt may have attended a divine service but his reactions have not been recorded. Probably with his Scottish Presbyterian background, he would have had a similar reaction to that of Marr. The cathedrals, at first Durham and then York, must have been impressive to a person who had not strayed beyond Glasgow before, but all we have about his first impressions of England are a brief scrawl at the end of Marr's letter that he liked 'the country very much but think the people are very sharp'. (7)

They continued their journey from York along the Great North Road. They ran into heavy rain between Ferrybridge and Doncaster and then passed through Newark, which Watt liked, and Biggleswade. Arriving in London on 18 June, they found lodgings at 'John Stewarts Blue Anchor, Little Britain'. (8) Watt sold his horse at Smithfield, where it fetched less than £4 because it was too small, but he may well have not been adept at bargaining. (9) He started exploring London immediately and found Wapping 'an ugly and confused place', but, while the streets of London were mostly 'very crooked and dirty', there were 'the Finest shops here I ever saw'. (10)

A Brief Chronology of Watt's Stay in London

Watt remained at the Blue Anchor or some other lodging house for the first few weeks. He was much older than the customary age for apprentices and it proved more difficult to find a suitable master than had been expected. Before 24 June, Marr was able to arrange for Watt to have some temporary work with Mr. Neale, a watchmaker, (11) but by 3 July he was working informally for John Morgan who finally would be his master. It was not until after the first week of August that Watt wrote home outlining the terms which seem to have been arranged. Morgan would be paid 20 guineas for nine months instruction which did not include either board or wages. (12) These terms were sent to Greenock and about a week later Watt seems to have left his lodgings to move into Morgan's house. Discussions continued and Short was drawn in to help settle the details. There was probably never any formal agreement but by 21 September Morgan had been paid an initial £15 and Short agreed to act as surety. (13) It should be noted that Watt's training would last for only nine months and not the year as has often been stated.

Watt continued to work for Morgan until the middle of June 1756 but there is no mention of any formal termination of his employment. In November 1755, there were fears of an invasion and preparations were made for war with France. This may have brought more trade for Morgan because Watt was kept very busy for the next four months. In the middle of March he became ill with what he described as rheumatism and was unable to shake it off until the middle of June. By this time, his thoughts were turning towards home and his father asked him to buy 'sundries', but whether for Watt himself or his father is not clear. Watt drew up a list of tools he needed to purchase in order to establish himself in his own business in Greenock. (14) He spent most of the next two months buying this equipment and does not seem to have been in a great hurry to leave London. One excuse which he made was that he was reluctant to buy a horse at Smithfield through his inexperience. He dallied a further three weeks until his letter of 19 August stated, 'I design to set out on Monday having now got all my things'. (15)

Watt's Health in London

Too often the more dramatic passages in Watt's letters have been taken in isolation, presenting a very distorted impression of his time in London. This has been especially true of his health. His first letter home after arriving there stated that he was in good health. (16) Perhaps he had benefitted from the fresh air and exercise through horse-riding. We know that he had been ill during the previous winter because he left Glasgow in a hurry without paying the doctor's bill for bleeding him twice. (17) After he had been with Morgan for a few days, he apologised to his father, 'This is very bad write my hand shaking after working'. (18) This might well be expected after starting a new job which demanded considerable manual dexterity. His hand writing matured during his time in London from a youthful scrawl into a neat style which he retained to the time of his death. Life in London must have suited him for he reported early that October, 'I keep my Health better than ever I did'. (19) At last he was growing out of his childhood ailments.

There is no record that Watt was unwell in London until the middle of March 1756 when he wrote that he was recovering from a fit of rheumatism which had been very troublesome though not so bad as to confine him to the house. He expected that it would go away with the approaching warm weather but that 'In other respects I never had my health better since I remember'. (20) By the end of March, he found himself daily growing better and expected to be quite well again soon. (21) However this illness continued which prevented him working for about three weeks and hindered him from improving his skills. (22) He gave his father a description of it.

> It came first like a weariness all over me and a continual gnawing pain in ye small of my back but since it has gone about into my armes legs & almost every bone in my body, but by Bleeding & drinking the antimonial wine which was prescribed by one Dr Bevis a Physician of my Master's acquaintance I am now so well as to be able to work 10 hours in the day'. (23)

This could well have been some form of post-viral arthritis. (24) It persisted throughout May and into June, 'but I have my health very well in other respects which makes it the more tolerable'. (25) The last reference to it occurred on 19 June when Watt reported, 'The Rheumatism continues much as it was when I wrote to you last But I am in hopes that my Native air & the riding will produce a change to ye better'. (26)

William Playfair wrote in his obituary of Watt that this illness had been caused by sitting in winter near the door of the workshop where he caught a severe cold. It caused the headaches which continued to afflict Watt until he was sixty. (27) Watt junior questioned this but Playfair replied, 'As to the contracting the Head-Achs from sitting by a door, I had it from himself & therefore there could be no mistake in that'. (28) Watt's own account of this illness quoted above does not mention headaches. Nothing has been found in Watt's papers to confirm Playfair's story and, as Watt junior had not heard it, we must assume that Playfair was mistaken. In Watt's final letters from London, there are no comments about any ill-health so he must have recovered which again puts into doubt the version concocted by Smiles, 'Distressed by a gnawing pain in his back, and greatly depressed in spirits, he at length with his father's sanction, determined to return to Greenock, to seek renewal of health in his native air'. (29) Romantic but quite untrue. Watt's good health continued for some years, first in Greenock and then in Glasgow except for a few minor ailments.

Social Life in London

The letters that Watt sent home mention a large number of people whom he had seen or with whom he had been in contact. For example, he told his father that Marr had been in Lisbon at the time of the disastrous earthquake. (30) Cathcart was at his country house at Petersham, twelve miles from London, but it is doubtful if Watt actually went there. (31) Some people about whom Watt wrote were known to his father through his mercantile and shipping interests and these Watt visited either socially or for business reasons. Two such were the Henderson brothers but, as they lived more than two miles away, Watt thought he would not have the opportunity of seeing them frequently. (32) Captains of ships trading with Greenock such as James Fleming would have been well known to him. (33) In particular Watt mentioned the wife of Capt. Watson, presumably the person to whom his chest had been sent, as being 'one of the kindest persons I have met with in London'. (34) Therefore it is doubtful if Watt were isolated in London especially as we will see shortly that he was acting as agent for his father.

We have few details about his domestic arrangements. He gave the name of the Blue Anchor as the place to which his father should direct mail (35) where he may have stayed until terms had been agreed with Morgan. A note book meant to contain his expenses was soon abandoned. (36) Being in lodgings of this type may explain the comment he sent his father in the middle of August,

> I find now I am pretty well acquainted [hole] this place that I cannot live under 8/- a week without pinching my Belly & live as frugal a manner as I can. I can win some little money for myself before I fall to my masters work in the morning but at night am so wearied that I get to bed as soon as I can. (37)

But before the end of that month, he must have moved into Morgan's house although he still had to provide his own food. 'I wrote you that Mr Morgan does not board me or do anything but teach me for the twenty guineas... I have the conveniency of his maids buying my meat & Dressing it for me, which saves me near one half at Dinner'. (38) He still complained at the beginning of September that London was very expensive and that he would soon be short of money but since this is the last time he mentioned the high cost of living he must have been able to reduce his outgoings to a reasonable level as he became better acquainted with the town. (39)

There were a few lighter moments. Watt sent home graphic accounts of the King's return to London in September 1755 (40) and the heralds announcing the formal declaration of war with France in May 1756. (41) The threat of war caused him anxiety because press-gangs roamed the streets looking for able-bodied men to seize for the navy. To avoid this risk and partly owing to his illness, Watt seldom went out during March. (42) If he had been taken in the City of London, he would have been brought before the Lord Mayor.

> Unless one be either an a'prentice or a Creditable Tradesman there is scarce any getting off again & if I was carried before my Lord Mayor I durst not avow that I wrought in the city it being against their laws for any unfreeman to work even as a Journeyman within the liberties. (43)

The subject was not mentioned again so the threat must have soon passed. But the comment about apprentices being able to escape the press-gangs confirms that Watt had no formal apprenticeship.

His Father's Agent

While Watt's visit to Wapping on the day after his arrival in London may have been to search for his chest, he was quickly involved with his father's affairs. Watt senior almost certainly never visited London. The opportunity of his son's stay in the capital proved convenient because Watt could act as a personal representative, carrying out some business transactions. We do not know whether Watt's becoming his father's agent was part of the scheme for broadening his education and experience. Nor do we know whether he had been instructed to deal with the primary producers rather than the wholesalers and middlemen with whom Watt senior had dealt so far. Also Watt may have hoped that, by helping his father in this way, he could make some restitution for the expenses he was incurring. He was very conscious of this especially as trade in Greenock was dull that September and October so that Watt senior may have been financially a little straitened. However, Watt wrote in September 1755, 'I will very soon want money. My living here is very hard on you especially as trade is so dull at present but am improving all I can that I may be the sooner able to do for myself'. (44) Because Watt was buying goods for his father and the financial records are incomplete, it has not been possible to determine the cost of his time in London.

Watt senior dealt with Greg and Campbell who must have been wholesalers or middlemen. We see the businessman in Watt because he knew the prices his father was paying and compared them with those at other places in London. In July 1755, he told his father, 'I find we have been overcharged by Greg and Campbell for most of things in ye way'. (45) If his father would send a list of things needed, he would 'see to buy as cheap as possible'. (46) Watt soon discovered the person who supplied Greg and Campbell with compasses, what sort we are not told, and purchased some from this source. In this he may have been helped by Morgan 'who has been of great use in telling me where to get the things cheap'. (47) Otherwise this supplier would have charged the same price as Greg and Campbell and Watt commented that it seemed to be 'a maxim with most Tradesmen to charge country shopkeepers the same or rather more than the wholesale people will'. (48) In January 1756, Watt had to buy some more compasses from Greg and Campbell because he did not have the cash to buy them from the manufacturer but was able to arrange 'to have your last two orders at 3 months credit & at the same price as he sells them to Mr Campbell'. (49) Watt also discovered that William McDowall, a Greenock bookseller, was buying 'Books draughts & other things' cheaper from Mount and Page than Campbell was supplying his father. He was afraid that his father would be undersold and suggested that his father should write to Mount and Page to see if they would offer similar credit terms as Campbell's. (50) McDowall was later a purchaser of Watt's quadrants and other items and presumably at this period he had been buying some from London. It is surprising to discover that Watt had been haggling with these merchants because, afterwards in his civil engineering undertakings and in his steam engine business, he appeared reluctant to face up to people.

The country merchant was at a disadvantage when compared with his counterpart in London. Watt reported that, 'Tho this is not a good place for hardware yet I find that I can buy some things Cheaper in Retail than you use to pay for them in Wholesale & the shopkeepers here are said to take as great profits in retail as anybody whatever'. (51) It looks as if London traders took advantage of those in the provinces who might not know the prices and this may be why Watt himself went to London in later years to buy stock for his shops. One problem for the country trader was how to make payment. Finance bills and credit were frequent topics in the letters Watt sent home. Ready cash usually bought better terms. Watt purchased some instruments for his father.

> The Mathematical part I have bought myself & paid the most part of them out of the £10 but cannot pay the rest of them without straining myself. Therefore you'll please advise next post whether I will draw upon you or if youll send up a bill to pay them because I having bought them very cheap the people I got them from wont give Long Credit'. (52)

Giving credit was difficult for the small trader. 'You can have no credit here for any of the things anywhere else [than Greg and Campbell] the workmen being all so poor that they can't want the money a minute & a great many of the Wholesale dealers deal for so little profit that they cannot give credit'. (53) His father wanted some stoves but Watt told him that they 'can be got a good deal cheaper from the Maker but not without ready money which everybody here wants Greatly as well as with you'. (54) Shortly before he was due to leave London, he told his father that Morgan was willing to buy anything his father wanted on commission and that he did 'not believe that you can meet with anybody that will deal with you more honestly'. (55) It must have been very important to have reliable trading connections across the country who would be willing to look after your interests. Watt had to find such people as he tried to expand his own later enterprises.

Watt's letters home mention some of the things he bought for his father. He found scales at two prices and would send samples of both. (56) The first shipment at the beginning of August must have contained Hadley's quadrants because Watt said that the next he sent would be finished by himself. (57) Stoves and compasses have been mentioned earlier. He was unsuccessful in finding ivory rules and the 6d. pencils went missing on board ship. (58) Later Watt spent some time searching out drinking and beer glasses. The wooden punch bowls and hat hooks which his father wanted could not be found because they had gone out of fashion. (59) His final letter announced that he had bought two 3 ft. long telescopes at 14s. each, but one of these was probably a commission for somebody at Greenock. (60)

From these letters, we can see that Watt was busy on his father's business until the New Year of 1756 when he gave an account of money spent. (61) Whether a trade recession, pressure of work or his father's misfortunes caused little activity until after the spring we cannot tell but Watt was busy seeking purchases for his father in the couple of months before he left London. The first batch of orders was sent direct to Greenock by ship under Captain Hyndman. Because some items were not ready in time, Watt had to dash down river to Deptford and load them there. (62) In November, one ship went to Borrowstowness or Bo'ness (63) while at the end of June, Captain Erving was sailing to Greenock. (64) Watt must have taken advantage of any ship that was bound for Scotland.

The year in London gave Watt valuable training in the world of business. He gained financial expertise and was quite shrewd in ascertaining the best prices. He must have known how to handle bills of exchange, credit facilities, ship's invoices and much more. In addition, he found his way around London so that he discovered various manufacturers and sellers of a wide range of goods because those mentioned in his letters must be like the tip of the proverbial iceberg. His experience in London was far wider than just learning how to make mathematical instruments and stood him in good stead when he came to set up his own business in Scotland.

Training as Mathematical Instrument Maker

Watt had arrived in London with his letter of introduction to Short but we do not know when they first met. Short had been born in Edinburgh in 1710 where he started his career as scientific instrument maker but had moved to London by 1738. (65) He specialised only in telescopes for which he became world famous. In London, he worked alone and organised a system of out-workers who provided the constituent parts while he concentrated on his speciality, the mirrors. (66) In Short, we find an extreme case of specialisation in manufacture which was becoming more general in the scientific instrument trade. It is doubtful whether he ever engaged any apprentices and so would not have accepted Watt. For his part, Watt must have realised that working under Short would not give him the broader training he was seeking and that sales of telescopes would be very limited back in Greenock.

Both Short and Watt were attracted to London because it was the world centre for the manufacture of scientific instruments. This trade had been stimulated in the second half of the seventeenth century through the beginning of the scientific movement in England with, for example, the founding of the Royal Society, as well as English maritime supremacy. English instrument makers benefitted from abundant basic materials such as steel, brass, copper and tin of consistently high quality so that English instruments were for a time unsurpassed. From the end of the seventeenth until the beginning of the nineteenth centuries, the English were the acknowledged leaders and pioneers. (67) By about 1745, the workshop of George Adams the elder, followed later by that of John and Peter Dollond, drew customers from all over Europe. For a considerable period, neither workshop had any serious competition from other countries and their products circulated throughout Europe.

Yet the eighteenth century was marked by a complete transformation in the methods of instrument construction through the introduction of new techniques. Earlier craftsmen concentrated on fashioning a single instrument at a time but the manufacture of scientific instruments began to take on the character of genuine industrial enterprise where the greater productivity through division of labour enabled parts of instruments to be manufactured in batches and sold in large quantities. Numbers of artisans were employed who were simply routine workers, trained to perform specific repetitive operations at which they became highly proficient. An instrument would be assembled from parts produced by these workers. Towards the end of the century, there was a rapid evolution in the diversity of instruments produced.

Watt's arrival in London coincided with the beginning of this transformation from craftsman to industrial production. With his background as craftsman, he appears to have been seeking to be trained both in the manufacture of such navigation instruments as he could expect to sell in Greenock as well as receiving more general instruction in metal- and woodworking techniques. The navigation instruments would include both those for determining longitude as well as others for plotting the position afterwards. Watt certainly did not want to be limited to making a single part and this was probably one reason for his difficulty in finding a master. He aimed to return to Greenock with sufficient skill to make a variety of instruments but he could not have anticipated how quickly specialised production techniques would be introduced so that the old methods of manufacture became uneconomic. This could be one reason why he sold his instrument-making business later.

But all this lay in the future and he was able to find some temporary work with a watch-maker, Mr. Neale, with whom Marr had dealt previously for some small patent globes. Here Watt began to divert himself in cutting letters and Figures etc. (68) But then Marr and Watt began to discover the problems which would face him in trying to find somebody who would be willing to give him instruction. Marr told Watt senior that Neale was willing to introduce Watt to instrument makers in Westminster 'who will be more easily dealt with than those in Town every one of which insist on his binding an Apprentice tho at the same time the indenture may be bought up some years before the seven expire but this he cannot do according to your plan'. (69) We are not told why those in the City of London were preferred. A comment scribbled by Watt senior at the bottom of this letter gives us a hint at what this plan was; 'Wrote to him July 5th not to agree as aprentice but reather gow to some common workman or stay 6 months to improve in generall etc.' (70) It looks as if his father wanted Watt to remain in London for a much shorter period than a year and certainly not for a full seven year apprenticeship.

Marr hoped that Neale would accept Watt especially after having seen his abilities, but he warned Watt senior that 'The Tradesman who employs him will expect a pecuniary Gratification for their Trouble'. (71) Marr remained confident that he would soon send proposals but this proved more difficult than he expected. We can imagine Watt's discouragement when he wrote after a couple of weeks, 'I have not yet got a master, we have tryed several but they all make some objection or other & I find that if any of them agree with me at all it will not be for less than a year & even at that time will be expecting some money'. (72) His father noted that he had replied;

> I wrote him July 8th to agree wt a good man for any time under 12 months & to give a ?word illegible? of any sume under 10 guineas but to advise me first of ye mans character ye kinds of worke he commonly makes & what he proposed to lern him besides paying his bording. (73)

Watt later told his father that 'I now find that there are not above 5 or 6 [people] that could have taught me all I wanted'. (74) From this, it sounds as if he wanted to be trained in making specific instruments.

At last on 3 July, Watt had started work with 'Mr John Morgan Mathematical Instrument Maker in Finch Lane Corn hill who I have been close with ever since... he is a very good natured man & is very ready to show me anything I want to know'. (75) Watt later described him as 'a man of as good a Character, both for accuracy in his Business, & good morals as any in his Way in London'. (76) This was for a trial period and no bargain had been made. At this point, Short became involved concerning the terms and thought that Morgan should be paid 20 guineas. (77) Marr tried to haggle for 15 guineas but Morgan remained firm. Both Morgan and Short said that no other workman in London would teach Watt 'because it is giving one an Opportunity of running away with a Deal of their Customers'. (78) The fear of training a potential rival must have been very real.

Marr did not think this demand extravagant because, besides the insights and skills that Watt would learn, '... he will in a very few commissions either on your account or his own save more than Triple of the sum demanded and within the Bill of mortality he could not have found a man better recommended for good nature and ingenuity than Morgan'. (79) Watt himself confirmed this when he told his father, 'I have conversed with several of my

Friends here upon who considering what he is to learn me think it not extravagant However I can find no other in London of any character that will undertake it on any terms so that there is no choice'. (80) He also commented that

> If Mr Short had not spoke to Mr Morgan I could not have got one in town to take me on these terms they scrupling much at the shortness of the time & very few have known any more than how to make a Rule others a pair of dividers & such like which they serve a seven years apprenticeship to & give perhaps 20 guineas prentice fee. (81)

Here we may note that these other potential masters were quibbling at the shortness of the time Watt would serve and not at his age. Normally apprenticeships were started at the age of fourteen so that the person would be qualified by the time he reached twenty-one. John Smeaton and Jesse Ramsden, both near contemporaries of Watt, were past the normal age for apprentices when they started so perhaps training older people was not that uncommon.

Morgan's terms were 20 guineas for 9 months (82) for which he was 'to teach me how to do & let me work at any work that comes about the shop' (83) and that 'he promises to teach me all he knows of Mathematical Instruments making'. (84) Not hearing from his father, Watt felt obliged to put off signing the agreement as long as he could. After Short and Morgan had become suspicious, he signed at the end of August and both agreements were lodged with Short. He again told his father, 'If it had not been for Mr Short I would not have got a man in London that would have undertaken to teach me'. (85) His father did not reply until 9 September when he agreed to the terms proposed. (86) An initial £15 was paid and, as a formal indenture would have cost a guinea, one was never drawn up. Therefore Watt was never formally apprenticed but Short undertook to act as surety for Morgan performing his part of the bargain. (87)

Watt's letters home tell us a little about the sort of training he received. We have seen that he started with Neale by cutting letters and figures. Watt told his father that Morgan 'Tho he works chiefly in the Brass way yet he can teach me most Branches of the Business such as Rule's scales quadrants etc & is almost always in the shop'. (88) Already by 15 July Watt reported that he was 'working at the Brass part of Hadleys quadrants'. (89) It is not clear whether he was working at the same time on some fine sea compasses which Morgan was making from some newly invented metal as 'Brass is found to attract the Needle' but which Watt thought his father would find too expensive. (90) By the first week in August, Watt had made a brass parallel ruler 18 ins. long and a brass scale of the same length but had not done the dividing which Morgan was going to teach him. He also hoped that the next quadrants which he sent to his father would be his own making. (91) Soon he was finishing quadrants at 35s. each (92) and indeed on 23 August he boasted to his father 'I have done Hadleys Quadrant better than his 'prentice who has been two years with him'. (93) Later he offered to plane out the engraving on some bad quadrants which his father had bought and divide them again. (94)

At the end of October, Watt was working until nine o'clock every night except Saturdays but thought he was improving. (95) This pressure of work continued into the spring. It prevented Watt searching for goods for his father and finishing some rules or rulers which he intended as Christmas presents for his uncle and Jockey. (96) Part of the rush was caused by making 'azimuth compasses for Eastindiamen' with which Watt is likely to have been involved. (97) However his illness in March hindered him from improving as much as he wished. (98) He hoped to finish the rules in April. In that month, he had in hand 'a Brass sector & a Theodolite'. By this time he thought 'I shall be able to get my bread anywhere as I am now able to work as well as most Journeymen tho I am not so quick as many'. (99) In

May, he was to begin a theodolite and 'some few other instruments of the better sort'. (100) His final comment in June was that he could 'make a brass sector with a French joint which is reckoned as nice a piece of framing work as is the trade'. (101) Taken all together, these letters show that Watt had gained experience in a fair range of different instruments and had attained reasonable competence. In addition, he had become aware of prices and of the quality which those prices ought to purchase. He could tell a good quality instrument from a bad and, of course, had learnt how to repair them, all of which would be invaluable when he returned home.

The Return to Scotland

Watt's formal employment with Morgan must have ceased sometime in June 1756 when his thoughts were already beginning to turn towards his journey home. He asked his father,

> Whether or not do you think it proper that I should write to professor Dick at Glasgow (who recommended me to Mr Short without whose help I believe I would not have got a master) thanking him for his letter & to know whether there is anything that I could do for him here before I come down. (102)

Perhaps he was already looking towards Glasgow as his place of work. He had obviously decided that he would not remain in London, like Short and Ramsden, even though London was the international centre for the manufacutre of scientific instruments. He clearly had sufficient confidence in his own abilities to establish a business in either Greenock or Glasgow where there was not so much competition and where wages and production costs were cheaper.

His father told him to buy sundries, to come home at his convenience and to draw £20 on George Anderson (103) who seems to have been a banker because Watt senior frequently drew bills on him. Watt made out a list of tools and materials costing £23.12s.6d. which he wanted to take back to Greenock. (104) His list included a standing vice and hand vices, one large and one small turn bench, anvils, hammers of various sizes, nearly one hundred files, pliers, pincers, a brace and bits, broaches, an alphabet of letters (stamps for marking metal) and three screw plates, to name a few of the items. We can see what he intended to manufacture from the list of cast brass for rule joints and Hadley's quadrants, bubbles for spirit levels and tubes for barometers. Then there were materials, such as brass plate of various thicknesses, cast steel, wire, box and other hard woods, emery and other polishing materials. He aimed to give a professional finish to his work by using tortoise-shell and mother-of-pearl. He overspent, but whether this was for his own tools or for purchases on behalf of his father is not clear, so he had to borrow a further £5 from George Anderson. (105)

Near the end of the list, Watt included 'the Wooden part of 3 Hadleys Quadrants, £1 11 6'. (106) He thought that Hadley's quadrants would be one of his main lines in his career as mathematical instrument maker. The back-staff had been transformed into the octant by John Hadley who published his invention in the *Philosophical Transactions of the Royal Society* in 1731. (107) The name 'quadrant' causes confusion because this instrument can measure 90°, a quadrant of a circle, but the arc through which the arm moves occupies only one eighth of a circle, an octant. (108) Hadley's novelty was the use of a mirror mounted over the pivot of a radial arm that moved over a graduated arc. Viewing was through a pinhole sight at a half-silvered mirror. The horizon could be seen through the clear half, while the sun was reflected in the mirror. The arm was moved, thus 'bringing the sun down' to the horizon. The quadrant did not present any great difficulties in construction other than the need for accurate division of the arc and the production of truly parallel mirrors. There followed a gradual simplification of its lines and greater accuracy in the manufacture of its

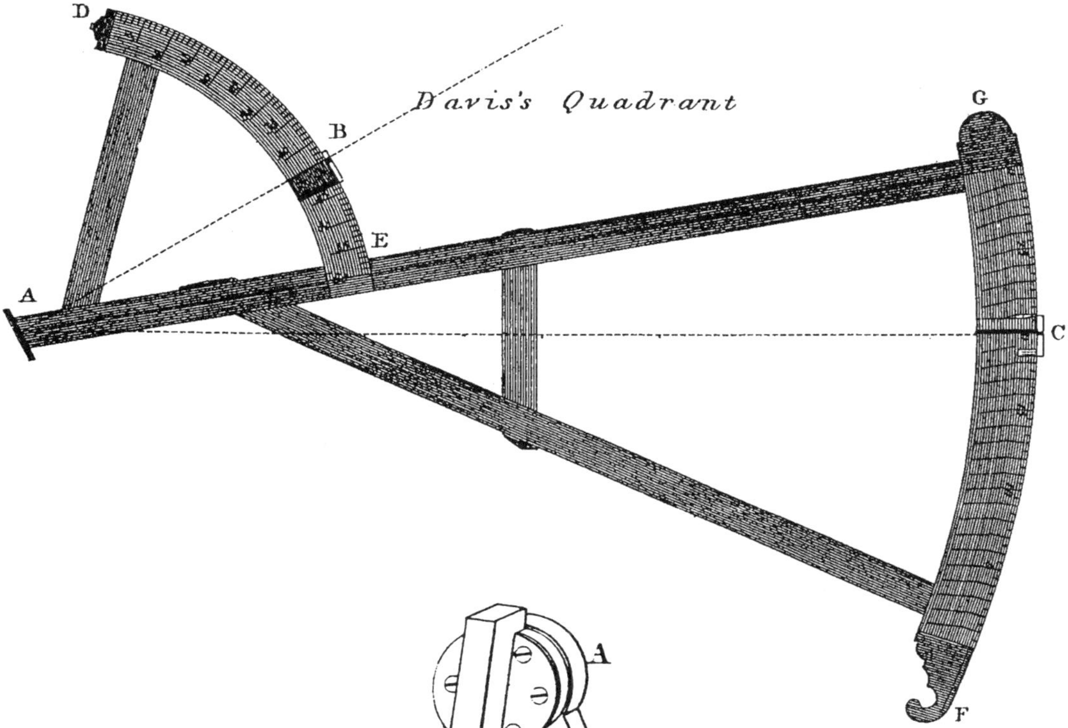

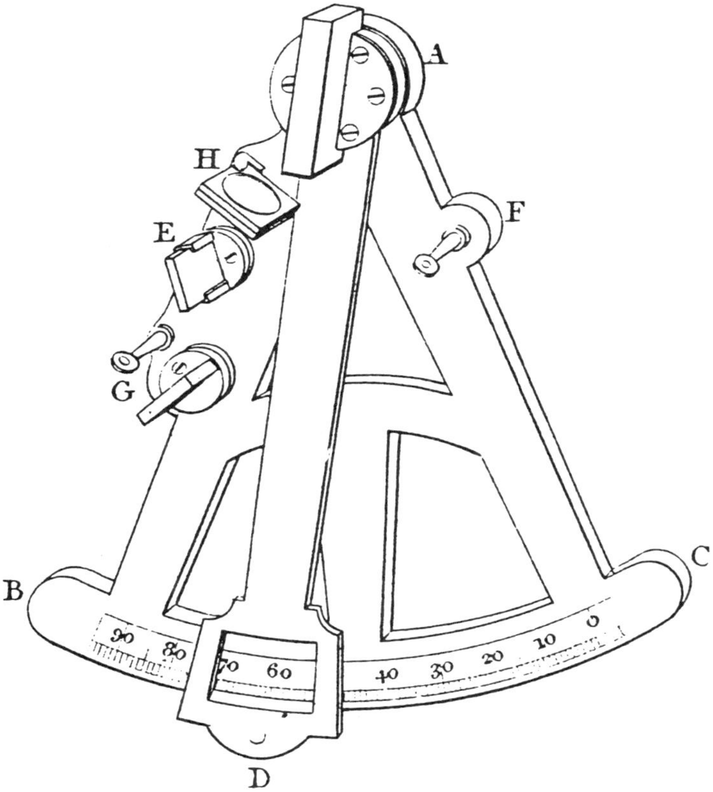

Examples of the Davis's Backstaff or Quadrant and Hadley's Octant or Quadrant.

parts. The use of ebony resulted in a fairly bulky instrument so that the wood was replaced by brass later in the century, resulting in lighter, more accurate instruments. Hadley's quadrant came into general use after 1750 and Watt clearly identified the commercial prospects for selling it to the increasing number of ships sailing in and out of the Clyde.

This is confirmed by Watt telling his father,

> There is people that make it their Business to make the wooden parts of Hadleys Quadrants & as I never wrought at them I design to bring down 3 of 3 different sorts as patterns & get my Master to divide them as perhaps I may have an opportunity of selling 2 or 3 before I can get a dividing plate made as there is a good many tools that I must make myself which will take me up a good deal of time as I do not think it will be proper to buy any Goods to sell again. (109)

It is strange that Watt said he never made wooden parts for Hadley's quadrants, perhaps Morgan made only high quality brass instruments. But we must notice Watt's ambitions to sell instruments he had made himself. He obviously wished to be a maker and not just a retailer. To do this, he realised that he would need specialised tools which, with his newly acquired skills, he could make himself. For example, in a list of his stock dated October 1759, there is a dividing plate for a quadrant. (110) He was willing to copy instruments made by others which we will see that he did after establishing himself in Glasgow. (See p84) He purchased 'A Book called Bions construction of Mathematical Instruments at 25s-' which he took home to Scotland. (111) Here we see Watt reading the available literature on the subject, but, although Bion covered most of the instruments in current use, he seems to have avoided giving exact details. The book was intended for the amateur public rather than as a manual for an able workman to construct accurate instruments. (112) Therefore when Watt left London, he returned home not only as a skilled craftsman able to make a variety of mathematical instruments, but he took with him both the tools and materials with which to make these instruments. He also had a plan of action, which was to make those he thought would be most profitable. But by the time he left London, he had drawn £30 on George Anderson. (113)

NOTES: THE YEAR IN LONDON

1 JWP, 6/46.32, J. Watt to J. Watt sen., 1 July 1755.

2 Muirhead, (1858), p. 34.

3 JWP, 6/46.30, J. Watt to J. Watt sen., 21 July 1755.

4 JWP, C7/52, Pocket Account Book of J. Watt sen., 1754 - 5, 2 June 1755.

5 JWP, 6/46.31, J. Watt to J. Watt sen., 15 July 1755.

6 JWP, 6/46.35, J. Marr to J. Watt sen., 12 June 1755. Don Wilcock wrote a letter in 1999; 'The time of his visit apparently coincided with the departure of Canon Leslie, an Irishman and Canon of the 8th stall (1743 - 1755), who had been appointed Bishop of Limerick, and who was giving his final address to his fellow canons. In the course of which he told the story of having the weather vane on his house fixed so as to indicate an East wind; a wind which would have speeded up his journey to Ireland. However such a wind in the North East indicates cold, damp and cheerless weather, so a fellow canon recovering from his illness, kept looking out of his window and on seeing Leslie's weather vane,

refused to venture forth until the wind changed'.
7 See end of JWP, 6/46.35, J. Marr to J. Watt sen., 12 June 1755.
8 JWP, 6/46.34, J. Watt to J. Watt sen., 19 June 1755.
9 JWP, 6/46.31, J. Watt to J. Watt sen., 15 July 1755.
10 JWP, 6/46.34, J. Watt to J. Watt sen., 19 June 1755.
11 JWP, 6/46.33, J. Marr to J. Watt sen., 24 June 1755.
12 JWP, 6/46.26, J. Watt to J. Watt sen., 9 Aug. 1755.
13 JWP, 6/46.23, J. Watt to J. Watt sen., 2 Sept 1755 and 6/46.21, J. Watt to J. Watt sen., 21 Sept. 1755.
14 JWP, 6/46.6, J. Watt to J. Watt sen., 19 June 1756.
15 JWP, 6/46.3, J. Watt to J. Watt sen., 29 July 1756 and 6/46.1, J. Watt to J. Watt sen., 19 Aug. 1756.
16 JWP, 6/46.34, J. Watt to J. Watt sen., 19 June 1755.
17 JWP, 6/46.32, J. Watt to J. Watt sen., 1 July 1755.
18 JWP, 6/46.30, J. Watt to J. Watt sen., 21 July 1755.
19 JWP, 6/46.20, J. Watt to J. Watt sen., 9 Oct. 1755.
20 JWP, 6/46/11, J. Watt to J. Watt sen., 18 March 1756.
21 JWP, 6/46.10, J. Watt to J. Watt sen., 31 March 1756.
22 JWP, 6/46.9, J. Watt to J. Watt sen., 20 April 1756.
23 JWP, 6/46.9, J. Watt to J. Watt sen., 20 April 1756.
24 I wish to thank Dr. Jur Kingma for suggesting this diagnosis.
25 JWP, 6/46.8, J. Watt to J. Watt sen., 18 May 1756 and 6/46.7, J. Watt to J. Watt sen., 8 June 1756.
26 JWP, 6/46.6, J. Watt to J. Watt sen., 19 June 1756.
27 JWP, C6/10, *The Monthly Magazine*, 1 Oct. 1819.
28 JWP, C6/10, W. Playfair to J. Watt jun., 22 Oct. 1819.
29 Smiles, (1878), p. 27.
30 JWP, 6/46.9, J. Watt to J. Watt sen., 20 April 1756, see 6/48.18, J. Watt to J. Watt sen., 25 Nov. 1755 for the first news of this earthquake.
31 JWP, 6/46.30, J. Watt to J. Watt sen., 21 July 1755.
32 JWP, 6/46.31, J. Watt to J. Watt sen., 15 July 1755.
33 JWP, 6/46.22, J. Watt to J. Watt sen., 4 Sept. 1755.
34 JWP, 6/46.23, J. Watt to J. Watt sen., 2 Sept. 1755.
35 JWP, 6/46.34, J. Watt to J. Watt sen., 19 June 1755.
36 JWP, 3/21, 'Household Expenses 1757 - 64'.
37 JWP, 6/46.25, J. Watt to J. Watt sen., 14 Aug. 1755.
38 JWP, 6/46.24, J. Watt to J. Watt sen., 23 Aug. 1755.
39 JWP, 6/46.23, J. Watt to J. Watt sen., 2 Sept. 1755.
40 JWP, 6/46.21, J. Watt to J. Watt sen., 15 Sept. 1755.
41 JWP, 6/46.8, J. Watt to J. Watt sen., 18 May 1756.
42 JWP, 6/46.11, J. Watt to J. Watt sen., 18 March 1756.
43 JWP, 6/46.10, J. Watt to J. Watt sen., 31 March 1756.
44 JWP, 6/46.19, J. Watt to J. Watt sen., 25 Oct. 1755, see also 6/46.23, J. Watt to J. Watt sen., 4 Sept. 1755.
45 JWP 6/46.30, J. Watt to J. Watt sen., 21 July 1755, see also 6/46.26, J. Watt to

J. Watt sen., 9 Aug. 1755, 'Whatever you take from Greg and Campbell I find is much dearer than I can buy them elsewhere except the ship chandlery things which I don't know where to get so cheap'.

46 JWP, 6/46.30, J. Watt to J. Watt sen., 21 July 1755.
47 JWP, 4/46.29, J. Watt to J. Watt sen., 5 Aug. 1755.
48 JWP, 6/46.26, J. Watt to J. Watt sen., 9 Aug. 1755.
49 JWP, 6/46.14, J. Watt to J. Watt sen., 5 Jan. 1756.
50 JWP, 6/46.26, J. Watt to J. Watt sen., 27 Dec. 1755.
51 JWP, 6/46.19, J. Watt to J. Watt sen., 25 Oct. 1755.
52 JWP, 6/46.29, J. Watt to J. Watt sen., 5 Aug. 1755.
53 JWP, 6/46.26, J. Watt to J. Watt sen., 9 Aug. 1755.
54 JWP, 6/46.17, J. Watt to J. Watt sen., 29 Nov. 1755.
55 JWP, 6/46.2, J. Watt to J. Watt sen., 27 July 1756.
56 JWP, 6/46.30, J. Watt to J. Watt sen., 21 July 1755.
57 JWP, 6/46.29, J. Watt to J. Watt sen., 5 Aug. 1755.
58 JWP, 6/46.27, J. Watt to J. Watt sen., 9 Aug. 1755.
59 JWP, 6/46.17, J. Watt to J. Watt sen., 29 Nov. 1755.
60 JWP, 6/46.1, J. Watt to J. Watt sen., 19 Aug. 1756.
61 JWP, 6/46.14, J. Watt to J. Watt sen., 5 Jan. 1756.
62 JWP, 6/46.27, J. Watt to J. Watt sen., 9 Aug. 1755.
63 JWP, 6/46.18, J. Watt to J. Watt sen., 25 Nov. 1755.
64 JWP, 6/46.5, J. Watt to J. Watt sen., 29 June 1756.
65 Bryden, (1968), pp. 3 - 4.
66 Clarke, (1989), p. 2.
67 Daumas, (1989), p. 119.
68 JWP, 6/46.33, J. Marr to J. Watt sen., 24 June 1755.
69 Ibid.
70 Ibid.
71 Ibid.
72 JWP, 6/46.32, J. Watt to J. Watt sen., 1 July 1755.
73 Ibid.
74 JWP, 6/46.23, J. Watt to J. Watt sen., 2 Sept. 1755.
75 JWP, 6/46.31, J. Watt to J. Watt sen., 15 July 1755 and 6/46.30, J. Watt to J. Watt sen., 21 July 1775.
76 JWP, 6/46.30, J. Watt to J. Watt sen., 21 July 1755.
77 JWP, 6/46.29, J. Watt to J. Watt sen., 5 Aug. 1755.
78 JWP, 6/46.28, J. Marr to J. Watt sen., 6 Aug. 1755.
79 Ibid.
80 JWP, 6/46.26, J. Watt to J. Watt sen., 9 Aug. 1755.
81 JWP, 6.46.28, J. Marr to J. Watt sen., 6 Aug. 1755, J. Watt's note at end.
82 JWP, 6/46.26, J. Watt to J. Watt sen., 9 Aug. 1755.
83 JWP, 6/46.24, J. Watt to J. Watt sen., 23 Aug. 1755.
84 JWP, 6/46.23, J. Watt to J. Watt sen., 2 Sept. 1755.
85 Ibid.
86 JWP, C4/A11, J. Watt sen. Letter Book 1751 - 58, 9 Sept. 1755.

87 JWP, 6/46.21, J. Watt to J. Watt sen., 21 Sept. 1755.
88 JWP, 6/46.30, J. Watt to J. Watt sen., 21 July 1755.
89 JWP, 6/46.31, J. Watt to J. Watt sen., 15 July 1755.
90 Ibid.
91 JWP, 6/46.29, J. Watt to J. Watt sen., 5 Aug. 1755.
92 JWP, 6/46.26, J. Watt to J. Watt sen., 9 Aug. 1755.
93 JWP, 6/46.24, J. Watt to J. Watt sen., 23 Aug. 1755.
94 JWP, 6/46.14, J. Watt to J. Watt sen., 5 Jan. 1756.
95 JWP, 6/46.19, J. Watt to J. Watt sen., 25 Oct. 1755.
96 JWP, 6/46.17, J. Watt to J. Watt sen., 29 Nov. 1755.
97 Ibid.
98 JWP, 6/46.9, J. Watt to J. Watt sen., 20 April 1756.
99 Ibid.
100 JWP, 6/46.8, J. Watt to J. Watt sen., 18 May 1756.
101 JWP, 6/46.6, J. Watt to J. Watt sen., 19 June 1756.
102 JWP, 6/46.7, J. Watt to J. Watt sen., 8 June 1756.
103 Ibid, see note at bottom by J. Watt sen.
104 JWP, 6/46.6, J. Watt to J. Watt sen., 19 June 1756. See Daumas, (1989) p. 116 for note on development of hand tools.
105 JWP, 6/46.5, J. Watt to J. Watt sen., 29 July 1756.
106 JWP, 6/46.6, J. Watt to J. Watt sen., 19 June 1756.
107 Daumas, (1989), p. 181.
108 Turner, (1980), p. 34.
109 JWP, 6/46.6, J. Watt to J. Watt sen., 19 June 1756.
110 JWP, C1/1, 'Accounts', f. 1 Inventory 7 Oct. 1759.
111 JWP, 6/46.6, J. Watt to J. Watt sen., 19 June 1756.
112 Daumas, (1989), p. 80. The title in French is N. Bion, *Traite de la Construction et des Principaux Usages des Instruments de Mathematiques*, 1709, which was translated into two English editions.
113 JWP, 6/46.1, J. Watt to J. Watt sen., 19 Aug. 1756.

Back Home in Greenock, 1756 – 1757

Glasgow and the MacFarlane Instruments

If Watt's journey back to Greenock took as long as the one up to London, then he could not have arrived home until the end of August or early September 1756. This could not have allowed him much time to recover from the strains and stresses of his stay in the capital, if indeed this were ever necessary, because we have a letter written in Glasgow on 27 September. (1) From it, and another written on 18 October, (2) we discover that he was once again acting on behalf of his father in Glasgow trying to sort out errors in accounts and obtain money from a Mr. Mcaul as well as others. Watt was hoping to return home to Greenock before the end of September either by horse or boat but must have stayed on, presumably with his Uncle Muirheid, because he asked in the second letter for a tenon saw to be sent by the first boat.

Watt's return to Greenock was further delayed because he wrote on 25 October, 'There are some instruments come from Jamaica that Dr. Dick desired that I would help unpack which are expected to day'. (3) These instruments were the bequest to Glasgow University of Alexander MacFarlane, a merchant who had resided in Jamaica for a long time where he had set up an observatory. He purchased his instruments from Colin Campbell who had 'furnished himself with an apparatus of instruments not unworthy of the observatory of a Prince'. (4) Among them were an astronomical clock by George Graham, a 4 ft. brass arch by Jeremiah Sisson, a 12 ins. Gregorian reflecting telescope by Short, three refracting telescopes, three transit instruments of varying sizes and a 5 ft. portable zenith sector. (5)

Watt must have been very excited to have been asked to help unpack this fine collection, only to discover that the instruments had suffered during the voyage. The minute of a University meeting on the following day reads,

> Several of the instruments from Jamaica having suffered by the sea-air, especially those made of iron, Mr. Watt, who is well skilled in what relates to the cleaning and preserving of them, being accidentally in town, Mr. Moor and Dr. Dick are appointed to desire him to stay some time in town to clean them, and put them in the best order for preserving them from being spoiled. (6)

The request was quickly accepted and the task finished by the beginning of December when Watt returned to Greenock. The University agreed 'to pay James Watt five pounds sterling for cleaning and refitting the instruments lately come from Jamaica'. (7) This may have been the first money which Watt actually earned with his new skills and was a large sum for only a month's work.

John Robison, 1739 - 1805

The MacFarlane instruments were a very important collection which members of the University wished to see. Among them was most likely John Robison, son of a prosperous merchant, who was born in 1739 and educated at Glasgow grammar school before going to the University in 1750, taking the degree of Master of Arts in 1756. He was a bright pupil with strong interests in mathematics and mechanics but, in the following year, was considered to be too young to be assistant to Dick. (8) In 1757, he published a design for a Newcomen engine with an inverted cylinder in the *Universal Magazine*. (9) At this period of his life, he was probably far better educated than Watt and it would be Robison who first introduced Watt to the steam engine. (See p299)

Plate 1: Beam compass with interchangeable heads, one of which can scribe eleven lines at once, in the garret workshop. (see page 85)

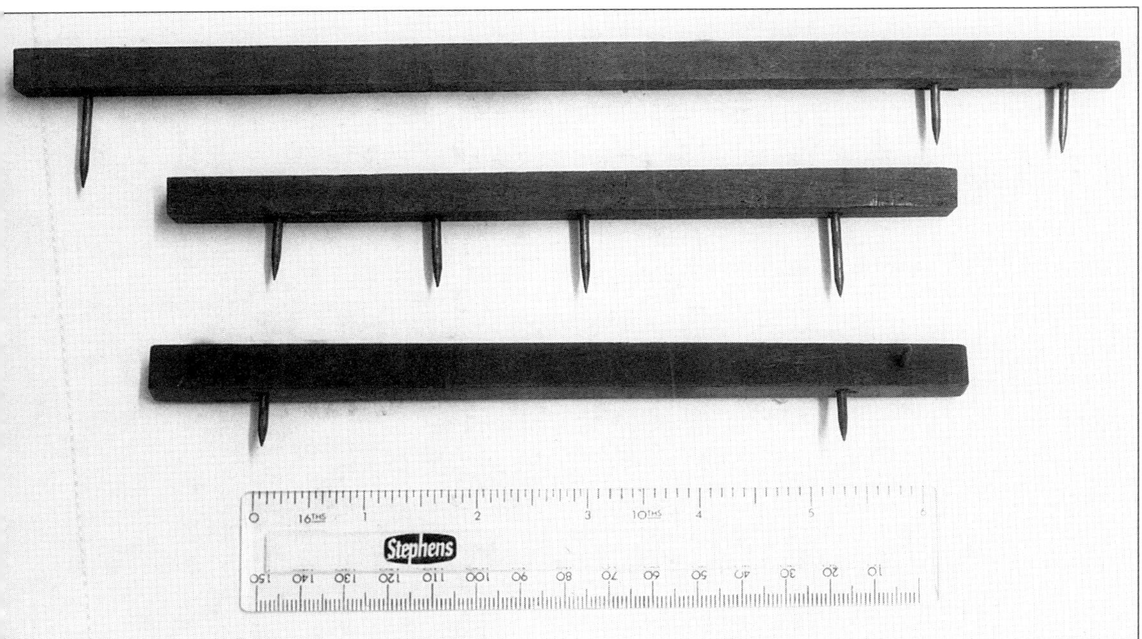

Plate 2: Pin gauges in Watt's garret workshop. (see page 85)

Plate 3: Printing plates for Watt's paper barometer scales, in the garret workshop. (see page 88)

Opposite page, left; Plate 4A: Stick barometer sold by Watt, in the National Museums of Scotland, Edinburgh. (see page 88)

Opposite page, right; Plate 4B: Head of the Watt barometer. (see page 88)

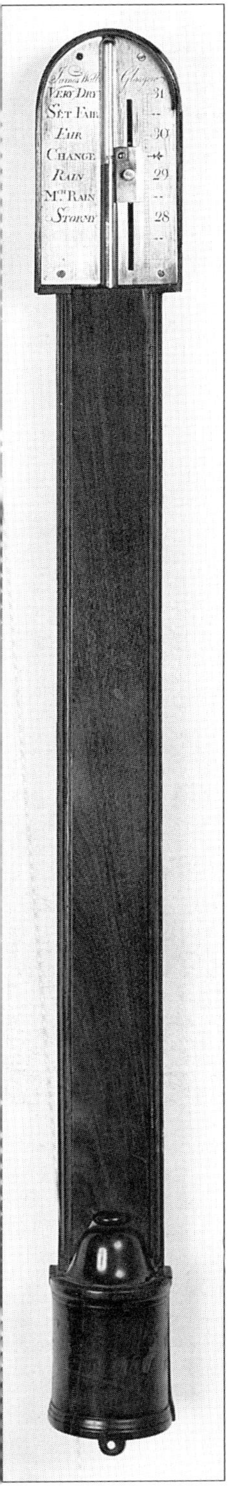

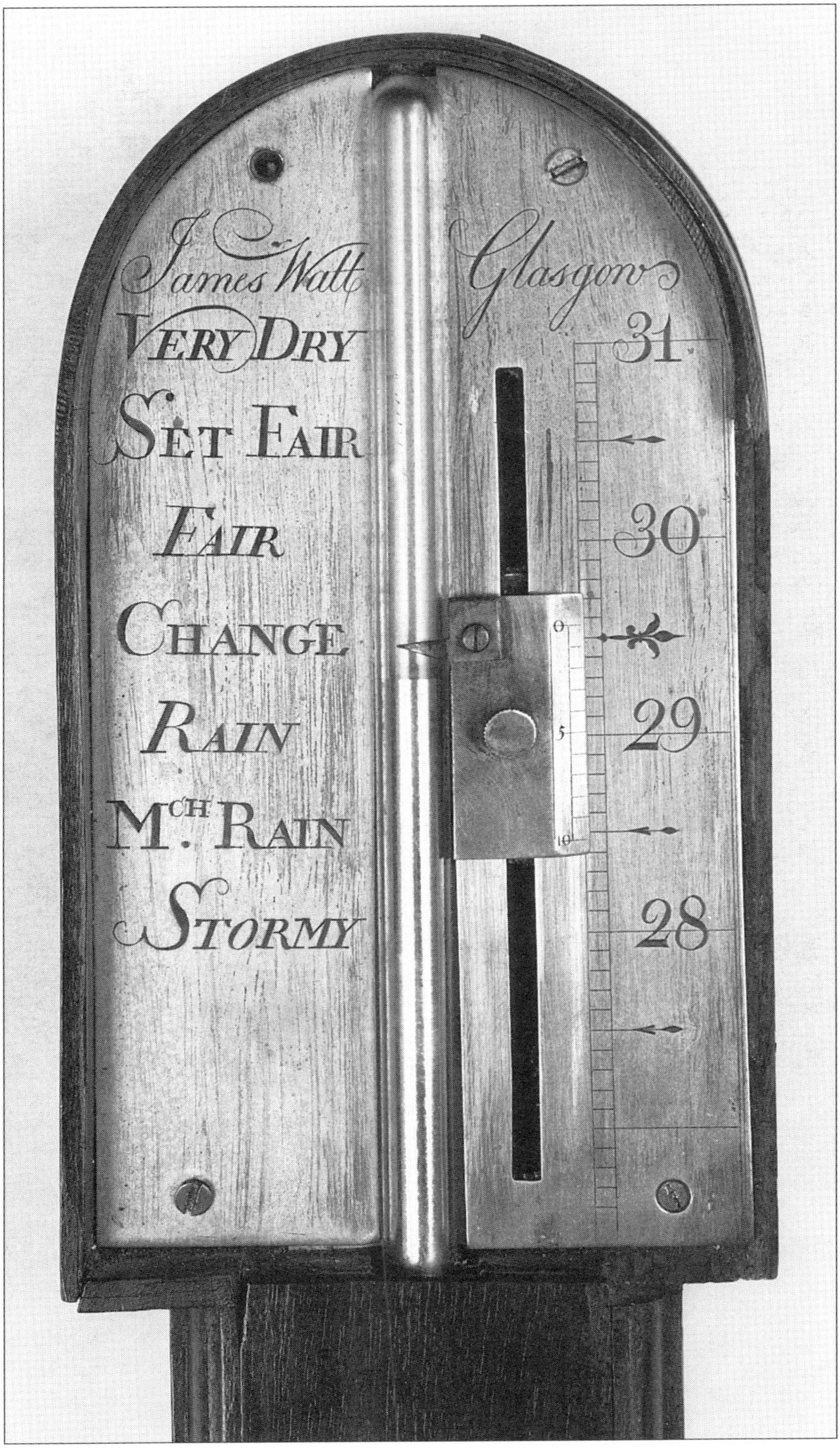

James Watt
Glasgow
VERY DRY
SET FAIR
FAIR
CHANGE
RAIN
M.CH RAIN
STORMY
31
30
29
28
0
5
10

Plate 5: The mercury cistern of an incomplete barometer in the garret workshop. (see page 89)

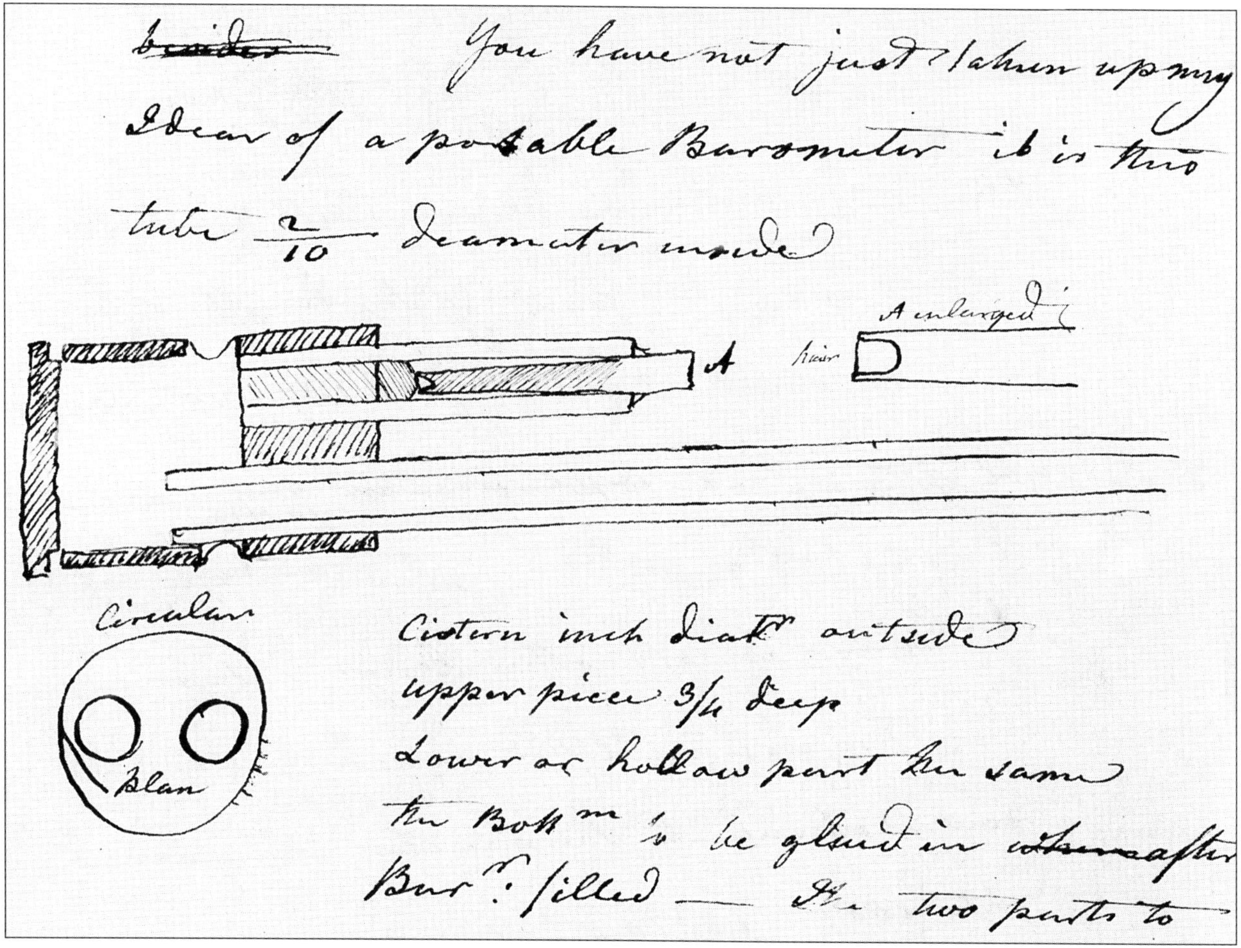

Plate 6: Watt's drawing of his cistern where the level of the mercury could be adjusted accurately.
(see page 89)

Plate 7: Parts of a flute and a rotary file for making three swells at once, in the garret workshop. (see page 85 & 112)

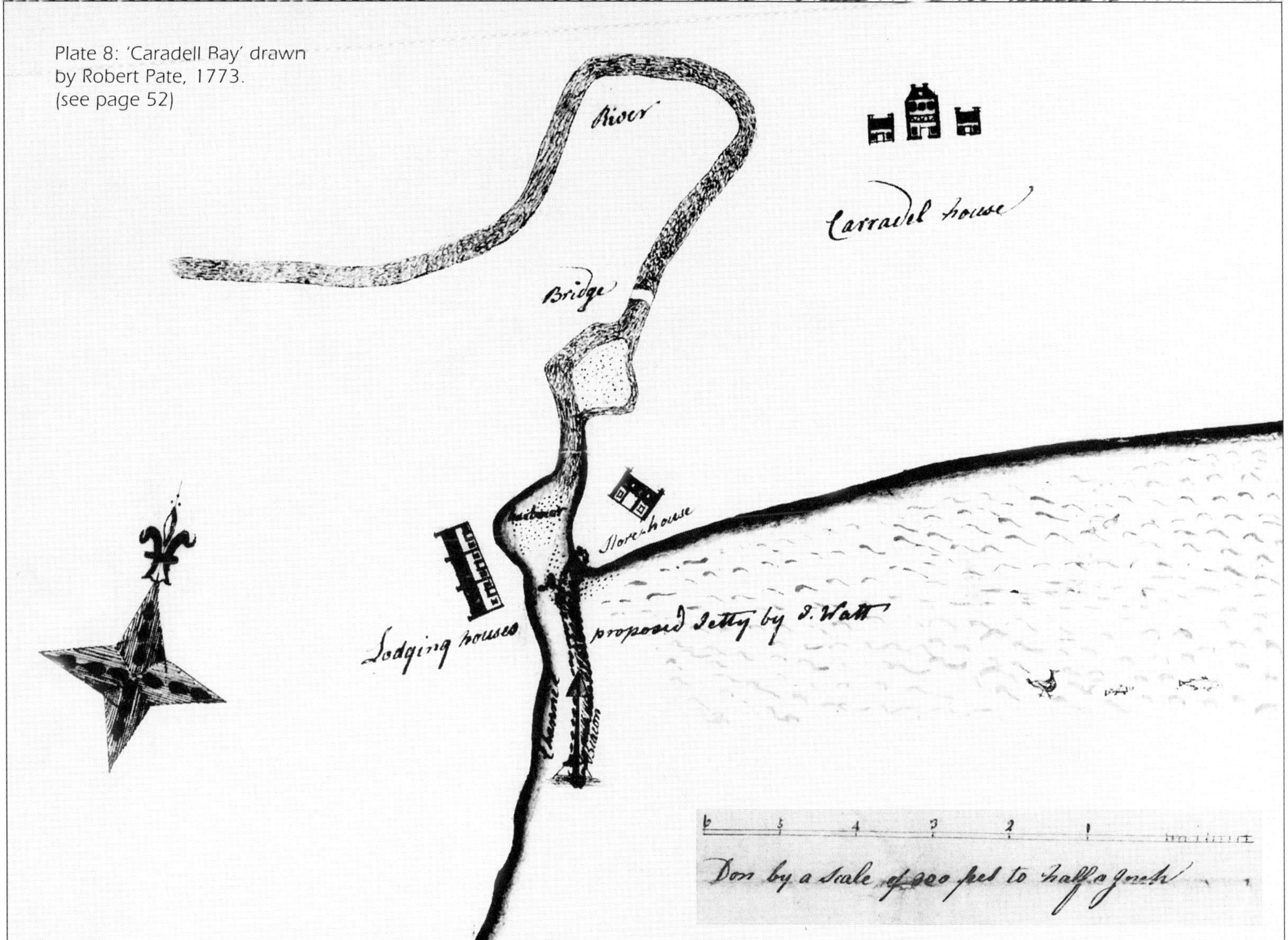

Plate 8: 'Caradell Bay' drawn by Robert Pate, 1773.
(see page 52)

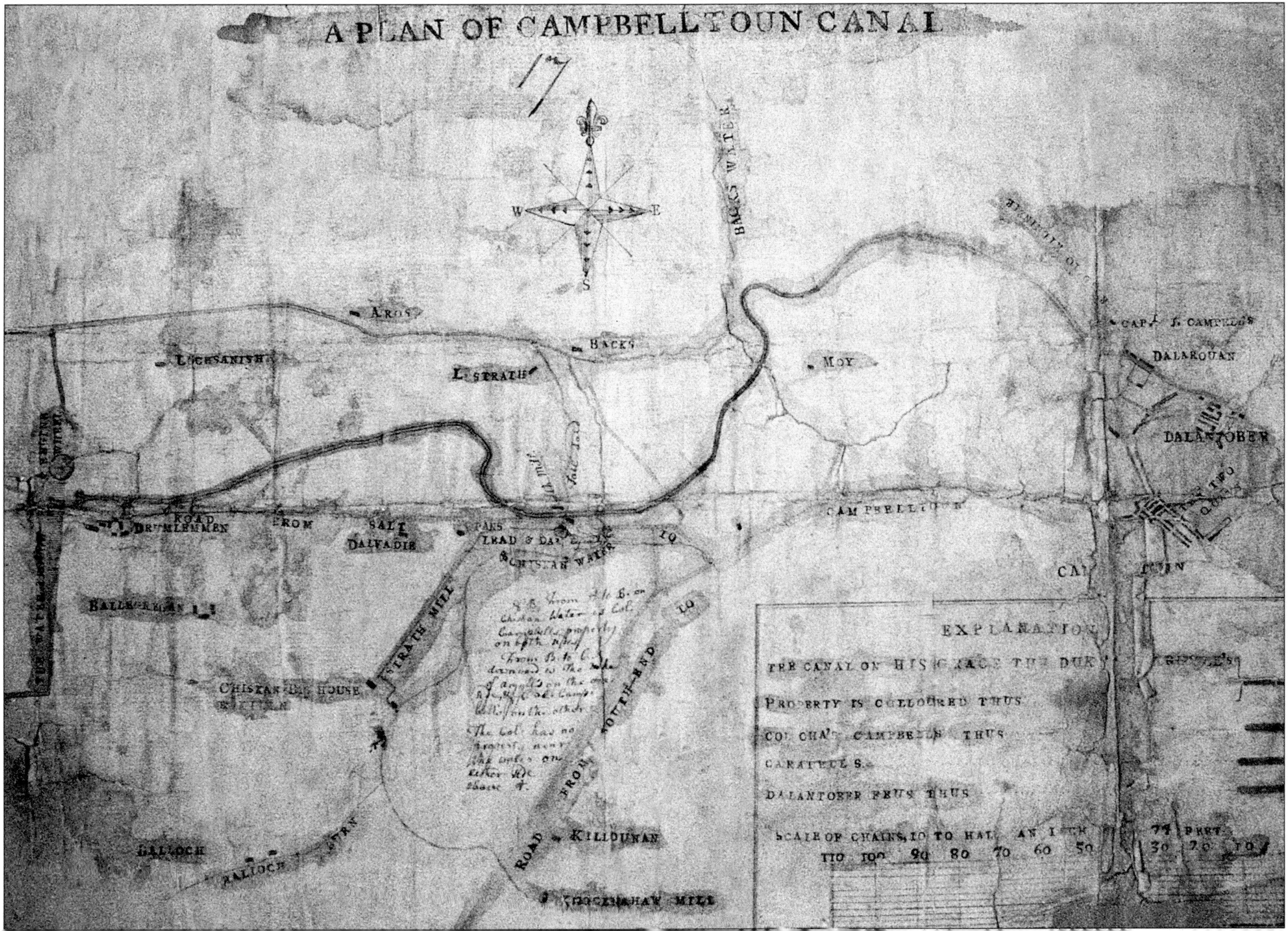
A PLAN OF CAMPBELLTOUN CANAL
W
E
S
BACKS WATER
AROS
LOCHSANISH
BACKS
L STRATH
MOY
CAPT. L CAMPBELL'S
DALAROUAN
DALANTOBER
CAMPBELLTOUN
ROAD
DRUMLEMMEN
FROM
SALT
DALVADIE
PANS
LEAD & DAL
CHISKAN WATER
TO
STRATH MILL
CHISKAN HOUSE
BALLEGREGAN
SOUTH-END
TO
FROM
ROAD
KILLDUNAN
DALLOCH
BALLOCH
KNOCKNAHAW MILL
EXPLANATION
PROPERTY IS COLLOURED THUS
DALANTOBER FEUS THUS
110 100 90 80 70 60 50
30 20 10

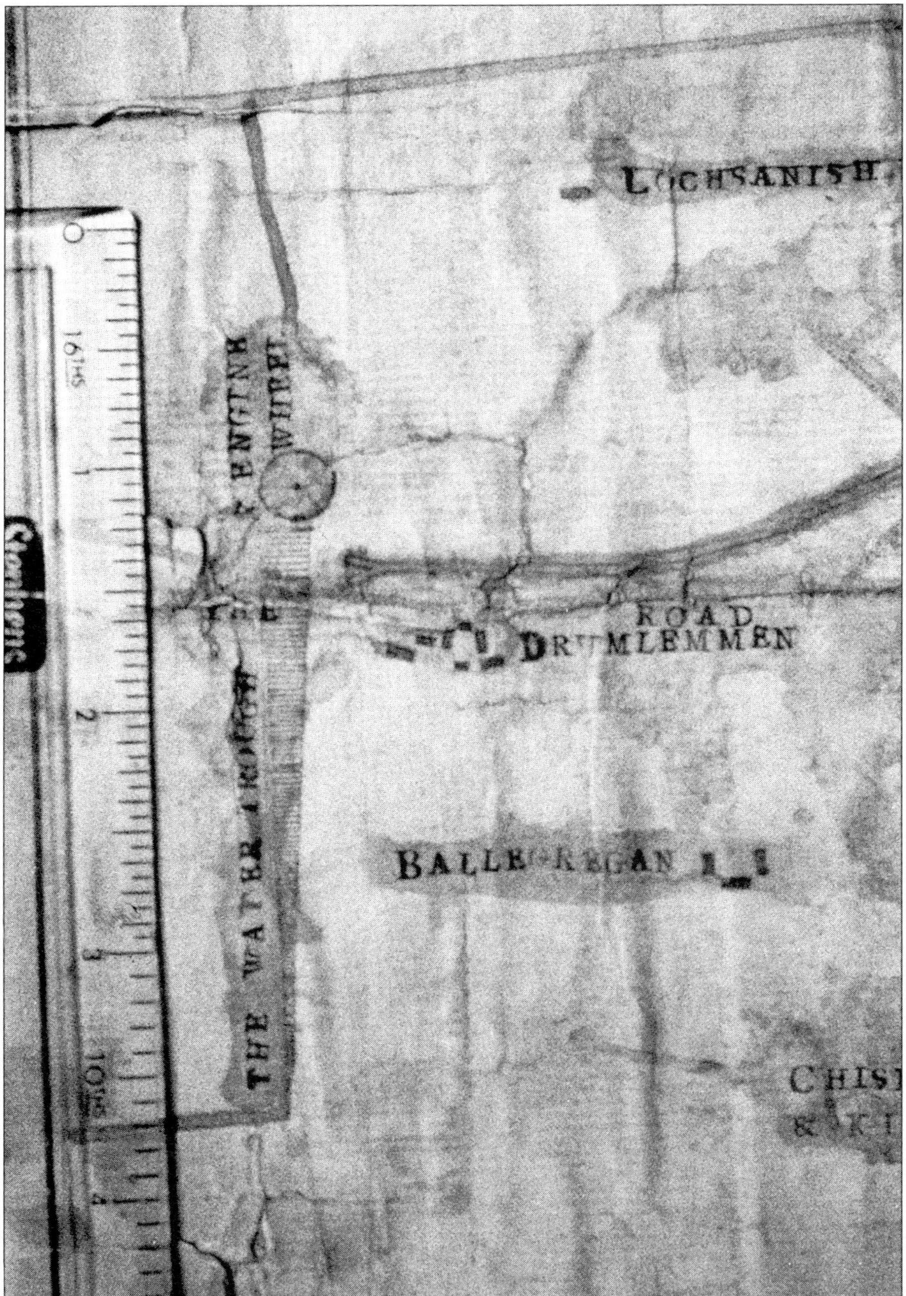

Above; Plate 10: Detail of the plan of the Campbeltown canal showing the waterwheel at the Drumlemble coal mine. (see page 251)

Opposite page; Plate 9: A hand-drawn plan of the Campbeltown canal, no date. (see page 248)

Plate 11: The Newcomen engine model tested by Watt now displayed in the Hunterian Museum, Glasgow.
(see page 312)

Plate 12: Detail of the Newcomen engine model, showing the cylinder and boiler fittings. (see page 312)

Opposite page; Plate 13: Detail of the Newcomen engine model, showing the cylinder and part of the valve gear. (see page 312)

Plate 14: The heavy lead weight on the piston rod of the water pump in the Newcomen engine model (see page 316)

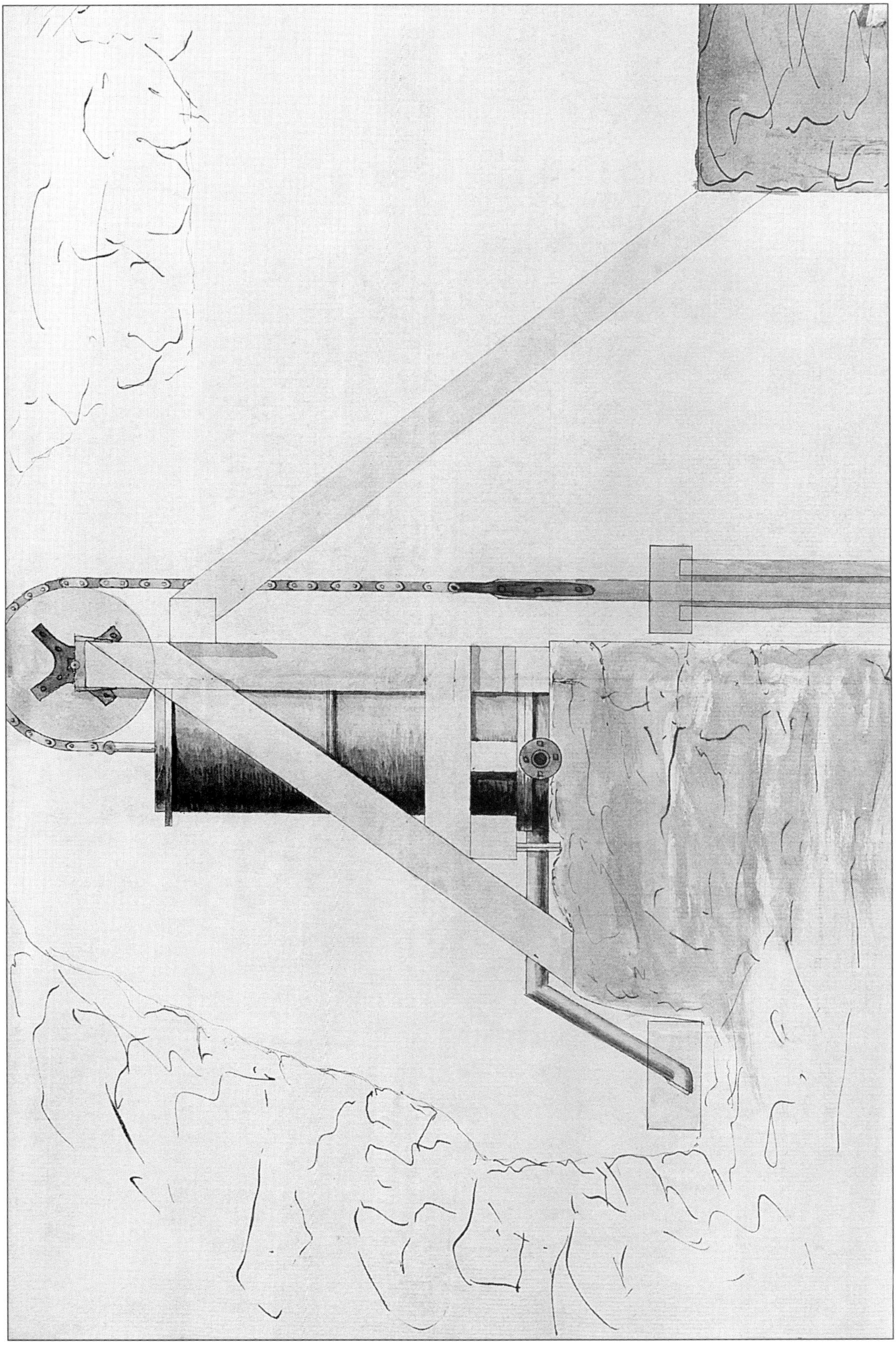

Plate 16: Watt's drawing of the Newcomen engine for the Carron Iron Works. (see page 357)

Opposite page; Plate 15: Watt's drawing for the Newcomen engine he proposed to install down the Leadhills mine with a wheel instead of a beam. (see page 354)

Plate 17: Watt's double conical apparatus for ascertaining the sensible and latent heats of steam in the garret workshop. (see page 373)

Plate 18: Watt's workshop and a cylinder of a Newcomen engine behind Kinneil house still standing. (see page 399)

John Robison, F.R.S.E., 1739 - 1805.

On Robison's death in 1805, Watt wrote, 'Our acquaintance began in 1756 or 7, when I was employed by the University of Glasgow to repair and put in order some astronomical instruments bequeathed to the University by Mr. Macfarlane of Jamaica'. (10) Watt's recollections of this first meeting were more accurate than Robison's and it is necessary to treat Robison's statements made in the 1790s about Watt with great care because they are often inaccurate. In this instance, Robison thought their first meeting was in 1758 when Watt had a workshop within the College. Robison wished to extend his interest in the natural sciences and mechanical philosophy into astronomy and was encouraged by 'the celebrated Dr. Simson, Professor of Geometry, Dr. Dick, Professor of Natural Philosophy, and Dr. Moor, Professor of Greek', who took him along to see the MacFarlane instruments on which Watt was working. 'After feasting my eyes with the view of fine instruments, and prying into everything, I conversed with Mr. Watt'. (11) As Dick took Robison to see these instruments when Watt was cleaning them, it must have been at the end of 1756 because Dick died the following May and was replaced as Professor of Natural Philosophy by John Anderson. But the seeds of the lifelong friendship between Watt and Robison were sown at this first meeting and would blossom when Watt returned to the University in the second half of 1757.

Robison says that he went to sea at the end of 1758 and in February 1759 sailed in HMS *Neptune* for Canada where he was present at the capture of Quebec. (12) He was chosen to be the representative of the Board of Longitude during the trials of Harrison's fourth chronometer when it was placed on HMS *Deptford* and was sent to Jamaica in the autumn of 1761. (13) But illness forced him to abandon his naval career so he returned to Glasgow in late 1763 where he was able to resume his friendship with Watt. About this he commented,

> [Through these experiences] I reckoned myself more on a par with Mr. Watt, and hoped for a closer acquaintance. Nor was I disappointed. I found him as good and kind as ever, as keen after the acquisition of knowledge, and well disposed to listen to the information I could give him concerning things which had not fallen in his own way. But I found him continually striking into untrodden paths, where I was always obliged to be a follower. (14)

From these remarks by Robison, it is evident that Watt had developed and broadened his knowledge during the years leading up to 1764. But when we read Robison's accounts of Watt's erudition, we must bear in mind that they were written many years afterwards and Robison may have confused three periods in their lives. The first was that single month in the autumn of 1756 when Watt was cleaning the MacFarlane instruments. The second was after Watt set up his workshop in the College. We know they met then because Watt's own records show that he sold Robison a microscope frame in January 1758 and the eyepiece of a microscope that March. (15) Also Robison told Watt how Black had caught him imitating his whistling and banned all students from Watt's workshop, but of course they soon returned. (16) Then there would have been the third period after 1763 which led up to Watt's invention of the separate condenser.

But Robison's comments, however confused, show that Watt achieved a remarkable grasp of a wide range of subjects and that he studied them in depth. Robison said about their first encounter,

> I conversed with Mr. Watt. I saw a workman, and expected no more; but was surprised to find a philosopher, as young as myself, and always ready to instruct me. I had the vanity to think myself a pretty good proficient in my favourite study, and was rather mortified at finding Mr. Watt so much my superior. But his own high relish for those

> things made him pleased with the chat of any person who had the same tastes with himself, or his innate complaisance made him indulge my curiosity, and even encourage my endeavours to form a more intimate acquaintance with him. I lounged much about him, and, I doubt not, was frequently teazing him. (17)

There were other students in addition to Robison who were attracted to Watt's workshop and who may have gone there partly to see the MacFarlane and other instruments for the observatory, because it is unlikely that Watt could have afforded to stock many instruments himself at this period. Robison said,

> I was not singular in this attachment. All the young lads of our little place that were any way remarkable for scientific predilection were acquaintances of Mr. Watt; and his parlour was a *rendezvous* for all of this description. Whenever any puzzle came in the way of any of us, we went to Mr. Watt. He needed only to be prompted; everything became to him the beginning of a new and serious study; and we knew that he would not quit it till he had either discovered its significance, or had made something of it. No matter in what line, - languages, antiquity, natural history, - nay, poetry, criticism, and works of taste; as to anything in the line of engineering, whether civil or military, he was at home, and a ready instructor. (18)

Robison's recognition of Watt's sociability and range of knowledge reminds us of what Jeffrey said in his obituary of Watt. (See p42/3) Watt sought perfection in whatever he studied. Another person who was attracted to Watt like Robison was A. Simple who was a student at Glasgow and later had some medical post in the *Rupert* at Plymouth during the 1760s. Letters reveal that he was also well known to Black and had been selling some of Watt's quadrants. (19)

Watt Stays in Greenock

After cleaning the MacFarlane instruments, Watt returned to Greenock. Living at home meant that he had no need to write to his father and so his letters cease until the end of July 1757 when he had returned to Glasgow to establish himself there as Mathematical Instrument Maker to the University. For this time in Greenock, the most important surviving record is Watt's 'Waste Book, 3 January 1757 - May 1763'. (20) The January date may show when he commenced his own business for in it he kept a record of at least some of his transactions and started it with 'An Inventory of the Goods Money Debts etc. Belonging to me James Watt junr also what I owe to others'. At least Watt intended to keep proper accounts but, if these are all the dealings in which he was involved, then they seem remarkably few and he must have struggled to make any sort of a living. This is where the assistance he gave to his father may have been much more important in helping him develop his own business than has been realised before but once again the records are so incomplete that it is impossible to draw any firm conclusions.

That January, Watt totalled the assets of his tools and materials as £21.4s.2d. which included £5 ready money. There was one complete Hadley's Quadrant at £1 17s. and wooden and brass parts for several more. His hopes for quadrant sales are revealed in the second part of this inventory where we discover that he had 300 copies of a book on Hadley's quadrant printed the previous October by the Glasgow printer, Robert Urie, at a cost of £2.5s. They consisted of three half sheets of demy, so would be considered as a leaflet or pamphlet today but had some illustration. (21) Additional sales lines were telescopes, rules and protractors. Watt seems to have launched into a form of batch production for he listed 'six sliding rules framed & part divided'. A 'Brass pattern for rules' and joints may show an early special production tool. There were also parts for telescopes, prospects and glasses. There were still eight glass tubes for barometers but those for spirit levels are not mentioned. His six magnets in one case and a couple of larger ones in another were probably for making magnetic compasses for use in navigation.

Watt continued to add tools since that January he purchased some iron work 'for a Great Lathe, 2/6' (22) and more pieces a month later. This is the first mention of such a machine tool which he considered an important addition to have been building it so early. (23) He lent his father £5 on January 19 but the first recorded sale was not until January 26 when he sold an object glass for a telescope for 1s. The first sale of work carried out by himself was for cleaning and making a new eye piece probably for the same telescope at 1s. 6d. A couple of days later, he sold a Hadley's quadrant at £2.3s. (24) followed by more later. Making sights, silvering glasses, forging magnets, mending a telescope, making a punch for sockets of beam compasses were among the tasks he carried out over the next few months. At the end of April, he sold 'William M'Dowal a Hadleys Quadrant with one observation to be paid in a month, £1.10s'. (25) This was the bookseller in Greenock who took more instruments later.

We do not know why Watt went to Glasgow in May. Probably it was in connection with the graphometer he was making for Gray and nothing to do with Dick's death. Business must have been increasing because, at the end of May, we find him employing Alex McKenzie for a day and three quarters on grinding lenses for Hadley's quadrants. Intermittent payments to McKenzie continue for the rest of Watt's time in Greenock. (26) June saw him at Largs Fair where he bought ribbons for his 'hair & neck', which, with other entries for clothes, suggests he took great care over his personal appearance. In July, Watt purchased more hand tools such as pliers, files, broaches, draw plates, shears and brace and bits. During these months, he laid the foundations of a small enterprise in mathematical and navigational instruments with outlets through his father, McDowall and others in Greenock. If he could move to Glasgow and its University, not only could these initial contacts be retained but he would have realised that he could expand into new lines with new contacts in that city at the heart of its region.

NOTES: BACK HOME IN GREENOCK

1 JWP, 4/11.1, J. Watt to J. Watt sen., 27 Sept. 1756.
2 JWP, 4/11.3, J. Watt to J. Watt sen., 18 Oct. 1756.
3 JWP, 4/11.2, J. Watt to J. Watt sen., 25 Oct. 1756.
4 *Notes and Records of the Royal Society*, Vol 24 (1970), D.J. Bryden, 'The Jamaican Observatories of Colin Campbell, F.R.S. and Alexander MacFarlane, F.R.S.', p. 261.
5 Bryden, (1970), p. 267.
6 Muirhead, (1858), p. 42.
7 Ibid, p. 42; the minute was passed on 2 Dec. 1756.
8 Chambers, (1834), Vol IV, p. 161.
9 *Universal Magazine of Knowledge & Pleasure*, Letter by J. Robison, Nov. 1757, p. 229.
10 Muirhead, (1854), Vol II, pp. 293 - 4, 'J. Watt's Recollections of his Friend Dr. J. Robison', April 1805.
11 Ibid, Vol I, pp. xlii - xliii.
12 Robinson & McKie, (1970), p. 321, Letter 213, J. Robison to J. Watt, 18 Dec. 1799 and Dickinson & Jenkins, (1927), p. 18.
13 Sobel, (1996), pp. 119 - 20.
14 Muirhead, (1858), p. 62. See B. & W. Col., MIII, 3/6, 'Ledger of Personal Accounts', f. 7, Mr. John Robison, 1763, Dec. 'To a lib. hair powder, 9d.; a stick wax 1/-'.
15 B. & W. Col., MIII, 3/1, 'Waste Book', 27 Jan. and 20 March 1758.
16 Robinson & McKie, (1970), p. 321, Letter 213, J. Robison to J. Watt, 18 Dec. 1799.
17 Muirhead, (1854), Vol I, p. xliii.
18 Ibid, Vol I, pp. xliv - v.
19 JWP, 3/41.6, A. Simple to J. Watt, 25 April 1762, see also 19 Sept. 1761 and 9 Aug. 1766.
20 B. & W. Col., MIII, 3/1, 'Waste Book', 2 January 1757 - May 1763.
21 B. & W. Col., MII, 8/7.1, Invoices from R. Urie, 26 Oct. 1756, see also C1/1, 'Accounts', Oct. 1759; 'A copper plate for Hadleys Qut Books 10/-'.
22 B. & W. Col., MIII, 3/1, 'Waste Book', 4 Jan. 1757.
23 See Daumas, (1989), p. 115 for importance of the lathe.
24 B. & W. Col., MIII, 3/1, 'Waste Book', 28 Jan. 1757.
25 Ibid, 30 April 1757.
26 Ibid, 28 May, 3, 9, 17 June & 26 July 1757.

2
From Instrument Maker to Engineer

Inner quadrangle, Glasgow College.

Mathematical Instrument Maker to the University, 1757 – 1759

The Move to Glasgow

On 29 July 1757, Watt sent a letter to his father, addressed from the 'College, Glasgow'.

> We got well up about 10 o'clock Wednesday night & got all my things up here yesterday. Jockey is just now out with the accts. George Jarden... has got payment of the bill which shall give Mr. Brown... I shall want $^1/_2$ a Doz afternoon China tea cups a stone tea pot not too small a sugar box & slop bowl as soon as possible Letters directed for me at the College will come to me again ten o'clock in the morning. (1)

This letter corrects some earlier misconceptions about Watt's move to Glasgow. He had transferred his workshop equipment and tools from Greenock to Glasgow by boat with the help of his brother and entered the expenses in his 'Waste Book'.

2 Aug. 1757. Expenses removing to Glasgow	
Boat freight 5/-, porterage 1/6, Cartage 1/2	0.7.8
Town dues 6d., Brushes 1/- & pr Buckles 6/-	0.7.6
Pocket expenses 1/-, Physick & drawing a tooth 3/6	0.4.6
20 Aug. Postage 8d., Expenses cleaning my shop 1/-	0.1.8 (2)

In his accounts, Watt made no distinction between what we should consider items for his business, such as the cost of hiring the boat, and personal expenditure, such as drawing a tooth. The fact that this letter was written from the College, and that others should be sent there, confirms that Watt moved his workshop directly into rooms in the University and did not have a temporary place elsewhere as suggested by D.J. Bryden. (3) The confusion seems to have arisen from Watt's 'Waste Book' where, with good Scottish economy, he continued making entries to the bottom of the page and it was only on 6 December that he turned over to the next where he put the heading 'Mathematical Instrument Maker to the University'.

Much has been made about the opposition from the Guild of Hammermen to Watt's setting up his business in the city of Glasgow itself. This originated with Black's evidence at the trial of Watt's patent in 1796.

> Mr. Watt came to settle in Glasgow... but being molested by some of the corporations, who considered him as an intruder on their privileges, the University protected him by giving him a shop within their precincts, and by conferring on him the title of Mathematical-instrument-maker to the University'. (4)

Lumsden demonstrated in 1911 that there was no evidence whatsoever from the records of the Incorporation of Hammermen, or the Dean of the Guild Court, that any attempts had been made to exclude Watt from the city itself. (5) The letter quoted at the beginning of this chapter shows that, as soon as he had arrived in Glasgow, Watt was dealing with Jarden, a member of the Incorporation of Hammermen, which suggests that Black was mistaken. There were precedents for tradesmen being given space within the College buildings; for example, the University printers, Robert and Andrew Foulis, had rooms in the Academy of Fine Arts. (6)

This still begs the question of how Watt was able to find accommodation within the University precincts especially as one of his main supporters, Dick, had died in the previous May. Recently it has been suggested that,

> Having a craftsman with Watt's skill based in the College had advantages for the professoriate. There was the potential requirement to assist with the new astronomical instruments being acquired for the proposed observatory, further work on the MacFarlane bequest, plus the routine maintenance of the apparatus used in teaching natural philosophy and chemistry. (7)

Moor and Simson had visited Watt while he was cleaning the MacFarlane instruments initially. Just over a year later when a ship loaded with various delicacies from India arrived at Greenock, Watt asked his father to obtain some special requests for these two professors and Black because 'I need not tell you I am very much obliged to each of these gentlemen & would be glad you would answer their Comissions'. (8) Perhaps Watt's obligation to the first two was due to their having supported his move into the University. Another supporter at this time was Adam Smith but Black had still to take up residence in Glasgow.

John Anderson, 1726 - 1796

Anderson was appointed to fill Dick's position as Professor of Natural Philosophy in 1757. He stressed the usefulness of science in helping to bring about and perfect

> ... all those improvements in our habitations, manufacturers, and machines, which render polished society so comfortable, by supplying our wants, and by diminishing our labours... How beneficial is the contrivance, for instance of Spectacles,... of Telescopes, Microscopes, Gunpowder, Steam Engines, the Mariner's Compass, an Electrical Apparatus, large Ships and, to name no more, Graving and Letter Printing, which have given new Powers, and important knowledge, to all the people who possess them. (9)

Anderson was a brother of one of Watt's school fellows and may have been important first through helping to introduce Watt to the University and second through expanding Watt's knowledge of natural philosophy and mechanics. He engaged Watt to repair demonstration apparatus and also actually to demonstrate that apparatus.

Watt would become involved over the years with many of the items mentioned by Anderson but it was not until 15 May 1758 that we find the first entry in his 'Waste Book' relating to Anderson when he altered a 'Francinis pump' and a model of a comet. (10) That November, there was a long list of work done for Anderson's class which included cleaning various pieces of demonstration apparatus such as a jack, a machine to show composition of motion, another to show the force of wheels and pinions, a model of a crown wheel and pinion as well as two balances. Similar work followed later, such as making a pendulum 39 ins. long, cleaning and putting hooks in balls, cleaning a balance, a piece of brass for discharging electricity, a brass rake [for an electrical machine], cleaning magnets, mending a pair of bellows, an artificial fountain, an air gun, an elliptical compass and a pipe belonging to the air pump as well as much more. (11) There was a third list in September 1762 which included mending a barometer and apparatus for a wheel barometer, cleaning and varnishing a quadrant by John Bird, work done on the air pump and 19 days spent on an electrical machine as well as hydrostatic bellows, a water pump and a centrifugal machine. (12)

The variety of work repairing and maintaining Anderson's apparatus covered most of the demonstration equipment used to explain the basic principles of physics, mechanics, pneumatics and electricity. This taught Watt a great deal, augmenting his rudimentary education. In addition, there are charges in the accounts for Watt sometimes acting as Anderson's assistant in lectures. (13) No doubt Watt would have listened attentively but, if he were also demonstrating the apparatus, he needed to know how it actually worked. The manipulation of say the small pulleys and their weights which had to be hooked onto arms of levers, the adjustment of pendulums with small threaded brass nuts, and similar operations, demanded considerable adroitness and ingenuity on the part of the demonstrator. (14) All this practical experience would have been of considerable value when he came to devise experiments on models of Newcomen engines. No wonder it was said that Watt 'was a person of truly philosophical mind, eminently conversant in all branches of natural knowledge'. (15)

While Watt's association with Black is famous, that with Anderson has generally been very inadequately appreciated. Though Watt frequently spent evenings at Anderson's house where scientific books were freely available to him, (16) there never seems to have developed such a close relationship compared with that between Watt and Black, possibly because their political views differed. However, this did not prevent Anderson turning to Watt in 1763 when he wanted a model of a Newcomen atmospheric steam engine put into proper working order. This request not only changed Watt's career, but also the history of world civilization through his improvement of the steam engine with the separate condenser which changed the steam engine into an economical and reliable source of power for driving all types of machines.

Alexander Wilson, 1714 - 1786

Another person whose influence on Watt has not been recognised was Alexander Wilson, Professor of Practical Astronomy and Observer at Glasgow University from 1760, but who had an informal association with the University much earlier. He had set up a type foundry at St. Andrews in 1742 but moved near to Glasgow a couple of years later. He won acclaim for the Greek type which appeared in the books of the Foulis brothers, University printers, and became called 'egregius typorum artifex' through the type he produced for an edition of Homer. (17) The Duke of Argyll had not supported Anderson for the Chair of Natural Philosophy but favoured Wilson who shared his interests in chemistry and who had long supplied him with scientific instruments. The Duke was able to have a Chair of Astronomy created for Wilson and George II issued a Royal Warrant in January 1760 establishing it with an annual allowance of £ 50. (18) During his leisure moments, Wilson took great delight in constructing reflecting telescopes, which must have been one of the accomplishments which led to his appointment as Professor.

Wilson and Watt were closely connected through a friendship which seems to have escaped notice previously. At the beginning of October 1786, Patrick Wilson wrote to Watt that his father, 'Watt's old friend', was very ill. (19) Wilson died on the sixteenth of that month. (20) Some years afterwards, Patrick confirmed their early connection when he told Watt,

> My brother and I are truly sensible of your invariable and active friendship - a friendship highly valued by us for its own sake; but doubly grateful to us, knowing so well it originated in your early and strong affection for our excellent parents. No one knows better than you how much our beloved father has done for us. (21)

Wilson was among those who had been quick to recognise the young Watt's talents and tried to advance him. This could have been because he saw how Watt cleaned the MacFarlane instruments.

In 1758, Wilson received £9.16s.0d. expenses and £20 gratuity from the Glasgow magistrates who had commissioned him to make a survey of the Clyde in order to make it navigable. (22) He involved Watt who was well aware of the deficiencies of the river for navigation through using it himself, such as the time when he transferred his workshop equipment from Greenock to Glasgow. Although there had been earlier proposals for its improvement, serious action began to be taken after the warning was given to a meeting of Glasgow Council in 1751 that their harbour at Port Glasgow might silt up. Therefore James Stirling, manager of the mines at Leadhills, was invited to suggest how the river might be made properly navigable up to Glasgow itself. He made a quick survey in the summer of 1752 and suggested that a series of weirs and locks should be built in the river to, in effect, turn it into a series of deeper reaches, which would contain sufficient depth to float coastal vessels at all times. (23)

Stirling gave only a general outline scheme without any costs and discussion continued for a further three years until Smeaton, who would play an important role with Watt in both his civil engineering and steam engine ventures, entered the scene. It seems that it was by chance that Smeaton met Andrew Cochrane, a former Lord Provost of Glasgow, who was in London to try and obtain the approval of Parliament for a number of bills on behalf of Glasgow, including one to improve the Clyde. (24) Smeaton had already commenced his career as a civil engineer since he had designed a masonry bridge for Blackfriars in 1754 and, that September, he had visited Scotland to draw up a scheme for draining the Lochar Moss near Dumfries. (25) He had met Cochrane before he left on 15 June 1755 for a tour of the Low Countries, visiting parts of Belgium and the Netherlands. On his return, a gale prevented his boat entering harbour, so, eventually, he was forced to land on 19 July from a pilot vessel at Blyth. Because he found himself nearer to Edinburgh than his home at Leeds, he set off straight away for Scotland. (26)

Smeaton carried out an accurate survey of the Clyde and found that the depth at low water was, in places, less than 18 ins. He recommended that a dam be put across the river at the Marlin Ford and that a channel be cut past it on the south side in which a lock should be built to enable ships to pass. The lock was to be 18 ft. clear breadth to take a vessel 70 ft. long of 100 tons burden. The locks on the later Forth and Clyde Canal, which were designed to enable coastal ships to pass across the isthmus, would take vessels 19 ft. 8 ins. wide and 66 ft. long. (27) No doubt Smeaton thought his scheme more than adequate for the usual vessels sailing on the Firth of Clyde and to the Western Isles for it gave a minimum of 4 ft. 6 ins. of water over the fords and shoals.

The request for Wilson's survey was the result of a meeting in the Merchant's House on 14 June 1757, when it was unanimously agreed 'that Glasgow apply forthwith for a bill to allow the building of a lock and dam'. (28) On 24 November 1757, Watt told his father, 'I am going out of town to the River' (29) which could be the time when he helped Wilson. On 15 May 1758, Watt received from 'Mr. Wilson for attendance of the Leveling the river, £3 15s'. (30) Among the plans and maps collected by Smeaton, there is a manuscript 'Plan of the River Clyde in the County of Clydesdale from Dalmuir Burn to the Green of Glasgow', which has been signed by 'James Barry and Alex. Wilson'. (31) Barrie was a surveyor who often carried out work for Glasgow Council and assisted Watt in his civil engineering career. (See p. 182) Some of the lettering on their map has been impressed with printer's type. It showed the depth at Marlin Ford as 1 ft. 9 ins., 'hard gravel' and they commented, 'By raising a dam at Marlin ford seven and one half feet high, no grounds are overflowed beyond what they are at Common Spring tydes but all contain'd within the ordinary Banks of the River: the banks all being pretty high'. (32) Certain stylistic features on this and other plans for Smeaton suggest that Watt may have helped to prepare them. It was at this period that he was organising the publication of his Uncle Watt's 'Map of the Clyde'. Wilson assisted in Watt's own survey of the Clyde in 1769 (33) and in these ways helped him lay the foundations for his career as a civil engineer.

Wilson presented papers about his various experiments to a group of professors who had formed a Philosophical and Literary Society which held weekly meetings. Much earlier, Colin Campbell, whose instruments formed the basis of the MacFarlane collection which contained a clock by Graham, had made a series of observations on the isochronous nature of a clock's pendulum. (34) Perhaps Wilson found some of this equipment among the instruments at Glasgow and, after experiments, made this the subject of a paper. Patrick Wilson said,

> In the year 1758, he read another discourse to the same Society upon the motion of pendulums. On this occasion he exhibited a spring clock of small compass, which beat seconds by means of a new pendulum he had contrived, upon the principle of the balance, whose centres of oscillation and motion were very near to one another. At one of the trials it performed so well as not to vary more than a second in about forty hours, when compared with a very exact astronomical clock near to which it was placed. It was some view of rendering much more simple and cheap the machinery of ordinary movements, by the slow vibrations of such a pendulum, which induced Mr. WILSON to prosecute these experiments. (35)

This coincides with a partnership formed between Wilson, Watt and Black involving clocks and it is most likely that to Watt fell the task of producing the apparatus described above. Watt entered into an initial partnership with Wilson in October 1758 'to make some experiments on clocks, To Wilson and Watt a Time Piece with Wooden Frame Plates, £1 5s'. (36) Black joined in November when Watt wrote, 'We stand each one third concerned in profit or loss' and charged the partnership £7 for an astronomical clock. (37) Later Watt charged £8. 3s. 4d. for 70 days work at 2s. 4d. per day, which suggests that one of his 'lads' had been carrying out the actual construction. Watt also recorded woodwork for a 'cutting engine', a machine for cutting teeth on gearwheels. These were common tools among clock-makers for forming these particular parts. The clock enterprise ended successfully in May 1759 when Watt received £18 in settlement (38) and it may have encouraged Black to form other partnerships with him later.

In the 1740s, Black had corresponded with Wilson about thermometers. Wilson had been making instruments thermometers and barometers since the 1730s when he had assisted George Martine in his experiments on heat. (39) Patrick wrote that his father had, not long after 1758,

> ... put in execution a remarkable improvement of the thermometer, which consists in having the capillary bore drawn very much of an elliptical form, instead of being round. By this means the thread of quicksilver upon the scale presents itself broad, and much more visible than it does in a cylindrical bore of the same capacity. (40)

Watt recorded that, in May 1759, he 'paid Alex. Wilson a thermometer 10/6' (41) and also in March 1760 that he sold Wilson four and a half dozen thermometer cases. (42) But it was not until ten years later that we find that he bought 'an Elliptical Boiling Water Therm[ometer]' for £2. 2s. from Wilson. (43) Watt charged Wilson £10. 1s. 6d. in April 1763 for various instruments including a barometer (44) so could easily have sought Wilson's advice on thermometers for his tests on steam engines which he began in that year. Moreover Wilson had experimented with a thermometer to record changes in temperature of the boiling point of water under different pressures. This raises the question whether Watt based his experimental equipment on that used previously by Wilson. (45)

Joseph Black, 1728 - 1799

The most important of the University professors who helped Watt was Joseph Black. The pair developed a life-long friendship. Black was born in Bordeaux in 1728 where his father, an Ulsterman, was a general merchant with a large trade in wine. Black remained there until he was sent to school in Belfast at the age of twelve. He had a severe infection in childhood, possibly bronchiectasis, which affected him for the rest of his life. (46) He commenced studying the arts course at Glasgow University in 1744 where the elder Robert Dick was one of his tutors before changing to anatomy and medicine under Cullen who was the major influence on his subsequent career. (47)

Joseph Black, 1728 - 1799.

Cullen was the first teacher of chemistry in Scotland to extend its scope beyond application to pharmaceutical preparation. He was interested in tackling problems in such areas as agriculture, bleaching, brewing and purification of common salt and was closely involved with the improvement of chemical processes in emerging Scottish industries. (48) According to Robison, Cullen 'quickly succeeded in taking chemistry out of the hands of the artists, the metallurgists, and pharmacists, and exhibited it as a liberal science; the study of a gentleman'. (49) This was reflected in his teaching course and no doubt encouraged his pupil to take an interest in the industrial application of chemistry. He employed Black as his laboratory assistant for his lectures so, at an early stage in his medical career, Black was plunged into practical work with his teacher who shunned the narrow approach of pharmaceutical chemistry. This influence was continued through Black to Watt.

As Black intended to follow a medical career, he moved to Edinburgh in the autumn of 1752 where he presented his M.D. dissertation in 1754 on *De humore acido a cibis orto, et magnesia alba* (On the acid humour arising from food and on magnesia alba). He carried out further experiments on this subject, which he published in 1756, and his work at this period has been claimed as laying the foundation of modern chemistry. (50) He demonstrated the relationship between quicklime, fixed air (carbon dioxide) and chalk, showing that fixed air was chemically distinct from atmospheric air. This recognition that there was more than one species, as distinct from variety, of gas in nature, was the starting-point of the spectacular development in the later part of the eighteenth century of pneumatic chemistry, the chemistry of gases. In 1757, he said of the discovery of fixed air that 'a new and perhaps boundless field seemed to open before me. We know not how many different airs may be thus contained in our atmosphere, nor what may be their separate properties'. (51) Although Black himself never followed up his discoveries, later Watt was involved with Joseph Priestley and others in pneumatic chemistry. Black exhibited an extraordinary combination of talents as a chemist. Throughout this and later work, he took pride in using simple apparatus, the balance playing a crucial role in his discoveries. Inventing the use of weighing as a logical instrument in making chemical analyses was, in itself, a major innovation, (52) one which Watt would use in his development of the separate condenser.

At the same time, Black was also studying the properties of heat, which he probably saw as a more attractive subject for a philosophical chemist. He may have looked to heat, and ultimately to the force of repulsion, as fundamental concepts in chemical theory. In his search for general laws, which would explain changes in the states of matter, he realised that specific substances had particular properties. This knowledge enabled one to apply the general laws of heat, flow, change of state and elective attraction to particular cases. (53) He provided the science of heat with a foundation of exact experiment and theory which gave Watt the basis and example for his own discoveries. Black's contributions to the understanding of theories of heat will be examined later.

Black remained in Edinburgh till 1756, after which he spent a few months in London. He was appointed to the Chair of Medicine at Glasgow on 4 April 1757 but it is not known precisely when he moved back there. (54) When Robison returned to Glasgow in 1763, he found Black in high repute as a professor and a favourite physician in the city. 'His sweetness of manner, which the dullest eye must have perceived to be free from all studious endeavour to please, and the evident concern which he took in the cases under his care, could not but make him a most welcome visitor in every family'. (55) He was an accomplished musician, was well versed in all the elegant accomplishments of life and was a highly respected man of science. In fact, Black's strenuous life strained his health soon after his return to Glasgow. It was his attractive personality which drew students to attend his classes as well as his methods of presentation. There is no doubt that Watt formed a very deep, life-long relationship with this remarkable person and Black's move in 1766 to become professor of medicine and chemistry in Edinburgh must have been a severe loss.

Black stated that Glasgow University conferred on Watt the title, 'Mathematical-instrument-maker to the University', (56) but it was in the role of making and repairing lecturing apparatus that Watt served Black. The first equipment made there by Watt was 'To the College, turning 9 doors for furnaces for Dr. Black's class, 2s' on 8 October 1757. (57. See p79) Watt had therefore met Black soon after setting up his workshop and not when first cleaning the MacFarlane collection. (See p81) Black stated,

> I became acquainted with Mr. James Watt in the year 1757 or 1758... I soon had occasion to employ him to make some things which I needed for my experiments, and found him to be a young man possessing most uncommon talents for mechanical knowledge and practice, with an originality, readiness, and copiousness of invention, which often surprised and delighted me in our frequent conversations together. I also had many opportunities to know that he was as remarkable for the goodness of his heart, and the candour and simplicity of his mind, as for the acuteness of his genius and understanding. I therefore contracted with him an intimate friendship, which has continued and increased ever since that time. (58)

These feelings were reciprocated, for Watt had an equal admiration for Black. On Black's death in 1799, he wrote to Robison, 'Like you I may say I owe in great measure my being what I am, he taught me to reason & experiment in natural philosophy & was always a true friend & adviser whose loss will always be lamented while I live'. (59) A few years later, Watt again stressed the importance of his learning from Black how 'to reason & experiment in natural philosophy' when he wrote to Dr. D. Brewster about his invention of his separate condenser.

> Although Dr. Black's theory of latent heat did not suggest my improvements on the steam-engine, yet the knowledge upon various subjects which he was pleased to communicate to me, and the correct mode of reasoning, and of making experiments of which he set me the example, certainly conduced very much to facilitate the progress of my inventions: and I still remember with respect and gratitude the notice he took of me when I very little merited it, and which continued throughout my life. (60)

Black kept systematic notes on his researches (61) and it was his 'sober plan of experimental inquiry', (62) which he constantly promulgated, that would be so important in systematizing Watt's own researches and experiments. Watt and Black would become 'Partners in Science' with profound implications for the development of the steam engine.

It is now impossible to know when Watt started to learn about science and, in particular, chemistry, from Black. Robison later wrote,

> When I returned to Glasgow, after four years absence [1763], I became Dr. Black's scholar. He was then engaged in his speculations about what he termed *latent heat*. I renewed my acquaintance with him, and with Mr. James Watt, the celebrated engineer. He had already heard two courses of the Professor's lectures, and was completely master of the subject. Dr. Black was pleased to admit me to the greatest intimacy of acquaintance, and few days passed in which I was not in company with these eminent philosophers, both of them in the prime of their scientific energy, and the period of their keenest research'. (63)

But Robison is an unreliable source and Watt had to correct him, denying that he ever attended any of Black's lectures; (64) perhaps Robison confused them with Anderson's. Furthermore, Watt was certainly not master of latent heat when he stumbled upon this phenomenon during his experiments on Newcomen engine models. He was also very busy with his own enterprises so that he may not have been able to spend much time with Black and Robison in the period before he was asked to repair the University model Newcomen engine.

Watt claimed that he was interested in chemistry from his youth (See p41) but Black also taught him a great deal. Robison pointed out to Watt soon after Black's death, 'You was much with the Dr. during his most important researches, and know dates and anecdotes which will support his Claims to originality and priority; and you can recollect some of the vehicles, by whom some of his thoughts became known to others'. (65) Robison thought that Watt could understand Black's thoughts because he was 'profoundly versed in the chemical science', (66) but Watt later wrote that the knowledge he learnt from Black came through informal contacts such as during conversations and doing odd jobs for him. (67) Little has survived in the archives concerning this period of Watt's life so we cannot trace what he learnt.

Our first known contact between these two was Watt's task of turning the doors for Black's furnaces. (See p78) Black had been interested in the design of furnaces for both laboratory and industrial operations since about 1753 when he first designed a portable coal-burning laboratory furnace. (68) It would be a portent of the future that this first connection concerned heat and furnaces. Later Watt consulted Black about improving the design of kilns for the Delftfield Pottery in Glasgow and went on to develop for his steam engine boilers which consumed their own smoke. In the meantime, Watt continued to be associated with Black's furnaces and must have made them. He wrote to Roebuck in October 1765, 'I fancy Black's furnace is finished by this time & brought to tryal'. (69) Watt should have completed a furnace for Boulton in the early part of 1769 but was late in supplying it. (70) He sent Small sketches and advice on how to properly lute or seal it. (71) It was typical of Watt that he chose to improve the design of one for Dr. William Irvine in 1769. (72) As late as 1798, Dr. Thomas Beddoes, from Clifton near Bristol, asked Watt, 'I shall be much obliged by the drawings of Black's furnace as soon as you can send them - I can get it made here - I do not know what is the ordinary size - but I have seen some too small, I think for evaporation'. (73)

Black's Portable Furnace.

Meanwhile, in Glasgow, Watt continued making small parts for Black, such as a condensing syringe, a screwed head, 3 copal moulds and 4 pistons, but, of greater importance, on 17 August 1758, 'An alarm clock, £2. 5s'. (74) We do not know if Watt actually made or merely sold this clock to Black, but, with his bent for mechanical ingenuity, clock mechanisms held a fascination for him. (75) This surfaced in the partnership with Wilson and Black for the astronomical clock. A more detailed account will be given later of another partnership between Black, Watt and John Craig, Watt's partner in the instrument business, to launch a hardware shop in the Trongate, Glasgow, in 1763. As well as helping to finance this venture, (76) Black further supported it by purchasing goods which included one of Watt's first perspective drawing machines, (see p109) a memorandum book, a set of pokers and tongs, candlesticks and snuffers, knives and forks, a pack of cards, wafers and a stick of wax as well as what we might expect from a mathematical instrument maker, spring dividers, a ruler, a hand vice and a 'Diamant pencil'. This list ends in December 1766 with Black being charged ten pence for a packing box; this coincides with with his move to Glasgow. (77) A third partnership between Black, Watt and Roebuck for producing alkali from salt began either late in 1765 or early 1766 and will be examined later. (See p156)

These partnerships involved lending money. Black lent Watt four guineas in the summer of June 1759. (78) Four years later in June 1763, there was an initial £150 for the shop followed by £4.15s. in August. (79) In February 1766, there was a further £50 (80) and £22.10s. in June the same year. (81) These last two sums coincide with Watt developing his steam engine and starting his surveying business. We can trace two more loans in 1775, the earlier for £20 when Watt had returned to Scotland and may have been short of cash (82) and a sum of £142 which could have been rescheduling bills of exchange. (83) We therefore have a total of about £230 lent before Black left for Edinburgh in 1766, but there is a hint in one of his letters to Watt in 1774 that there could have been another £100 borrowed in March 1772 and possibly more. (84)

Interest was paid on these sums but we do not know the rate. Watt paid some interest in June 1766 followed by £50 in March 1772. (85) It appears that interest had amounted to £66.12s. on loans of £222.10s, so that Watt was left owing £239.2s. in 1772. (86) Black wrote to Watt in February 1774 asking that bills about to become outdated should be renewed and interest paid on some as he was moving house and needed money. (87) Watt may have repaid £118.2s.6d. in October 1775. (88) Existing records show only another repayment of £40 in February 1781. (89) The account was not finally cleared until the outstanding balance of £18.8s.6d. was paid on 7 June 1784. (90) Watt sent a covering letter and said, 'It has grieved me much to have been obliged to have been so long owing you; but it is very lately that the sun began to shine on me, and I could not think of paying one friend by borrowing from another'. (91) It was Black's support in these and other ways that would assist the 'sun to shine' on Watt through the development of his steam engine.

The Workshop in the University

In August 1757, the observatory to house the MacFarlane instruments was no longer merely proposed but a start had been made on its construction. Watt wrote home, 'I have been so busy these 2 posts making medals to lay in the Foundation of the Observatory that I could not possibly get writing you sooner'. (92) After Watt's initially cleaning these instruments, somebody was needed to keep them in good condition until the new building was ready. It is possible that, in return for free accommodation, Watt took care of them during this time. There is no mention in his accounts, which admittedly are very fragmentary, of any rent being paid for his workshop in the University. It can be inferred from a statement by Robison that Watt continued to look after them.

> My first acquaintance with Dr. Black began in your rooms, when you were rubbing up M'Farlane's instruments. Dr. Black used to come in, and, standing with his back to us, amuse himself with Bird's quadrant, whistling to himself, in a manner that thrilled me to the heart. (93)

Bird's quadrant is not mentioned in the list of instruments owned by MacFarlane in Jamaica so must have been acquired by the University from elsewhere. It would have been an impressive piece of apparatus because Bird perfected the technique of dividing instruments by geometric methods. He was also the most expert maker of astronomical quadrants so we should not be surprised that Black was admiring it. (94) We find Watt looking after items for the observatory in May 1758 when he cleaned an astronomical clock and a gridiron pendulum. (95) It is quite possible that he continued to care for the MacFarlane bequest until Wilson was appointed Professor in 1760.

We have no knowledge of where Watt resided after moving to Glasgow. There is no tradition of his lodging with his Uncle again for, while his letters mention members of the family, they do not suggest that he was actually staying with them. Robison said that Watt 'had apartments and a workshop within the College'. (96) According to Smiles, the workshop, which was about twenty feet square, was situated on the first floor of the range of buildings forming part of the inner quadrangle of the University buildings. (97) Whatever space he was given may have been partly furnished for all that Watt requested from his father were a few small extras such as tea-brewing equipment. His father noted the value of some of the things he sent, such as 'a marled sugar box 8d, 1 doz. knives & forks 3s'. (98) Each New Year, Watt recorded 'gifts to servants' which may mean that he was responsible for employing them. (99)

Watt may have lived in his workshop. When he had a cold in the autumn of 1758, his father appears to have suggested sending a proper bed because Watt replied, 'I am pretty well recovered from the sore throat & hope to be out if Good Weather the Latter end of this week there will be no occasion for sending up the bed as I sleep quite well now & could get one here if there was occasion'. (100) It was not until two years later that he told his father, 'I am now furnishing my room myself & shall want sheets Blankets towels etc'. (101) Once again, we do not know where this was but an inventory has survived, listing among his possessions, '6 Elm chairs and 2 Elbow ditto... a walnutree desk... tea kettle... an earthenware and a stone tea pot... a hair matress' as well as blankets, sheets, etc. but no bed or cooking utensils. (102)

Workshop Equipment and Employees

An undated list of tools and books which may have been those taken to Glasgow shows a valuation of £26.5s.1¼d. His father helped to equip the workshop further by giving 'a wheel 26 diameter 4 ½ thick, a plank for a turn bench... a grindston & trackboards". (103) The wheel could have been a flywheel for Watt's great lathe for which he bought a rest from Jarden. Watt took a couple of weeks to sort everything out and erect his tools so that, when he wrote home on 21 August, he had to apologise that he had been unable to finish a quadrant. (104) Equipping the workshop continued well into September when Uncle Muirheid sold him a 'bench with double screws' and gave him a vice. (105) Watt also constructed a glass furnace because he had ambitions to make the glass parts of his quadrants and possibly lenses but had to admit that he had been unable to bring this to perfection. (106) The furnace was sold very quickly to Robert Foulis for £1.5s. (107) and Watt contracted out or bought in glass items at this period although later he did make some thermometers for Black. (108)

Illustrations from Diderot's Encyclopaedia and Rees Cyclopaedia showing a lathe and a few typical tools of the mid eighteenth century.

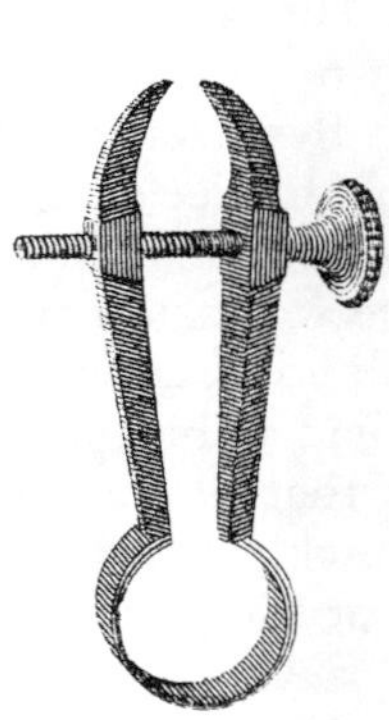

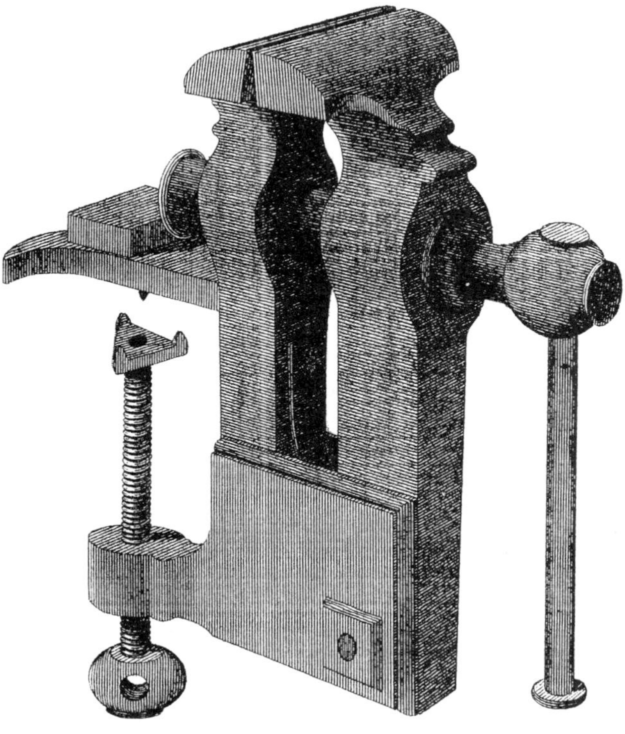

Even while setting up his workshop, Watt was once again acting as his father's agent, continuing to collect debts as well as purchasing anything needed. This is apparent from that first letter written after his arrival in Glasgow and further instances can be found in most of the others he sent home. How much of his own time this took is not clear, neither do we know whether he received any remuneration. He drew up a statement of the accounts with his father which had reached a total of £120 up to the end of 1759. It included seven Hadley's quadrants at their full selling price of around £1.15s. (109) These letters show the shortage of cash and capital which both Watts faced in their businesses. In September 1757, Watt wrote to Jockey to ask his father to send 20s 'as I cannot spare so much till I get in someway or other'. (110) That November, Watt reported to his father that his Uncle was 'in as great straits for money as you are'. (111) The following June, Watt could not afford to repay Marr £3 (112) and wondered if his father could lend this. He expected that he would soon receive some glasses from London which would enable him to finish and sell some quadrants 'as I lying out of a good deal of money just now having 4 on hand ready for the dark Glasses besides 6 other that are near finished'. (113) Watt clearly had great difficulty in properly financing his production until he could sell it as well as diversifying and increasing his stock of items on offer. We can see a classic case of an undercapitalised small business.

Yet Watt was able to expand and engage employees. It has always been assumed that, when he first moved to Glasgow, he was working on his own. The following February, he found that he had a great deal of work on hand but could not get any help. Therefore he asked his father to approach two people, one of them in Edinburgh, to see if they knew of 'any lad that can file tolerably well', particularly in brass, who would be willing to work in Glasgow. (114) Watt used the term 'lad' for a trained journeyman and 'boy' for a trainee. At the beginning of March 1758, he told Jockey that his business went pretty well but he still had no assistance. (115) He remained very busy during May but by then had got a lad who was skilled at making quadrant dark glasses and was about to get a boy whom he intended teaching how to make dividers. (116) The boy was Dougal McMillan who was engaged initially for three months at 14s. per week but who stayed for a year before returning to Campbeltown.

(117) Probably a journeyman and an apprentice were employed regularly from December 1758 with others taken on for short-time contracts. (118)

Products and Production Techniques

Although Watt's workshop was within the University precincts, he was able to continue with his own instrument business. He continued to supply quadrants to his father and wrote to him in October 1757,

> There is no occasion to send to London either for Davis or Hadley's Quadrants as hope to have some of both soon. If you get a good well shaped Davis quadrant that has been approved by some Experienced seaman send it up & shall return it in a week. At any rate send the best you can get. (119)

In spite of copying the best he could find, Watt must have produced some incorrectly because he wrote in August 1758, 'I have found out the Fault of some of the Quadrants & shall send you down some of them first Boats'. (120) Living and working in Glasgow, he was far away from the centre of the trade in London and no doubt had difficulty learning about the latest improvements in scientific instruments. Also in a provincial centre, there would not have been the same demand as in the capital. His later visits there would have kept him abreast of developments but, even so, his position as an isolated manufacturer was difficult. The same problems were faced by the Delftfield Pottery with which Watt became associated.

Watt must have thought that the quadrant still held out his best business prospect for he wrote in a much quoted letter,

> As I have now had a year's trial here I am able to form a judgement of what may be made of this Business & find that unless it be the Hadleys Insts. there is little to be got by it as at most other Jobs I am obliged to do the most of them myself & as it is impossible for one person to be Expert at everything they very often cost me more time than they should do However if there could be a ready sale procured for H. Quadrants I could do very well as I & one Lad can finish 3 in a week easily selling them at 28/.6d which is vastly below what they were ever sold at before I have 40/ clear on the three so it will be absolutely necessary that I take a trip to Leverpool to look for Customers & hope that the profits of what I shall be able to sell there I can go to London in the spring where I make no Doubt of selling more than I can get made all which I want your advice on, and if that does not succeed I must fall into some other way of Business as this will not do in its present situation'. (121)

This letter is worth studying in greater detail for it reveals Watt's early ambitions and business acumen. He appreciated that 'one off' jobs did not pay; they took valuable time to organise and execute and, if he had to do them himself, he could not be attending to other tasks. He had determined that quadrants were his most profitable line. It is possible that labour costs were cheaper in Glasgow than elsewhere but it would seem that Watt achieved his reduction in costs through introducing new manufacturing methods. Here we see the first appearance of his mechanical ingenuity and inventiveness. This letter also shows that he recognised the need for specialisation and concentration on specific lines in manufacturing. By producing three quadrants in a week, it appears that he had introduced some form of specialised or batch production. If his business were to survive, he would have to meet the competition from the large producers in London who were increasingly introducing early batch production and specialised production techniques at this period.

Watt made for himself special tools to help speed production. In his first inventory of January 1757, there is an item, 'Sundrie Tools for making mathematical instruments', (122) and that of October 1759 included 'A dividing plate for quadrants'. (123) In his Garret Workshop, now preserved in the Science Museum in London, a number of pieces of equipment show his ingenuity in what today might be termed 'production engineering'. There is a beam compass for drawing circles with interchangeable heads, one of which can be used to scribe eleven lines at once for making scales on instruments. There are strips of wood with nails driven through them which would have acted as pin gauges for making identical parts. For making flutes, there is a rotary 'power operated' file or gouge with three separate cutting heads which can form three swells beneath the finger holes at once. (See p112/Plate 7) All these little devices would have helped to lessen production costs and given Watt the edge over his competitors. Of course, he may have copied these tools from others and we have no idea of their dates, but they do show his willingness to experiment and adapt. It is evident from incomplete examples of his perspective drawing apparatus that he must have manufactured these, and so probably his other instruments, in batches.

Having attained a price advantage with his production techniques, Watt was faced with the problem of the disposal of his output. Throughout the whole of his career, he manufactured what might be termed capital goods for which there was a relatively small demand and which might be expected to last a long time. Therefore the immediate market in any particular region would soon be sated. He quickly recognised that he must look further afield than the Glasgow area in order to maintain his sales of quadrants. One answer was to develop new products which may explain his diversification into musical instruments. But he soon found himself in the same situation with those, as indeed he would later with his steam engines. At the same time, we should note Watt's confidence and optimism of 'no doubt selling more than I can get made'. (124) Although the sales trip to Liverpool did not materialise at this period, Watt continued with outlets for his quadrants through his father, Jockey (125) and McDowall in Greenock. His accounts show that many more were sold and repaired but no details are given of where or to whom. However, when he ordered 300 copies of his pamphlet on the quadrant to be printed in October 1756, he was being exceedingly optimistic because, by the time he formed his partnership with Craig three years later, there were still 242 left. (126)

Expanding the variety of instruments offered for sale by copying other people's products continued to be a feature of Watt's strategy. In August 1758, he told his father, 'If you'll please send up a Binacle Compass & the exact measure of all the different sizes, shall get you some made'. (127) His Inventory of October 1759 does not make clear what type of compasses were listed at ten different prices, with a dozen of each. From their position among rules, scales and cases for instruments, it is more likely that they were drawing rather than mariners' compasses. But there were 'Sundrie Brass plates partly engraved for Compass cards' showing that Watt must have at least intended to produce magnetic compasses. Watt also asked his father for 'Brass Dividers. I wish you would send up a pair of each sort & an acct. of the London prices as I am getting a Boy who I intend to set making them to bring in his hand I believe the wholesale prices are 1d pr inch'. (128) These were the dividers which McMillan had to reproduce but Watt was careful to note their cost because it was no use manufacturing something that would not be competitive.

A glance at a few other items during this first period of Watt's career as a mathematical instrument maker shows the great range of articles he undertook to make and the need to be an expert in all of them. There was brass for a syringe, a lid for a digester, a parallel ruler, dividing a theodolite, mending a telescope, cleaning an astronomical clock and silvering its dial, cleaning, polishing and varnishing the bars of a gridiron pendulum, 'cleaning a watch and helping it', glass lamp for an electrical machine and 'To His Grace Duke of Montrose 2 Barometer Tubes. (129) The Duke was Chancellor of Glasgow University. To satisfy those who had helped him gain entry into the University for his workshop, Watt had to be a 'jack of all trades', yet he recognised this did not pay and it would seem that, at this period, his ambition was still to be a true manufacturer of mathematical instruments.

Barometers

The only instrument where we can trace Watt's improvements through surviving artefacts is the barometer which he believed he could sell successfully on his return from London. He also used barometers and mercurial manometers (in the form of pressure gauges) as crucial scientific instruments to determine the relationship between the temperatures and pressures of steam in connection with various model Newcomen atmospheric steam engines and related experiments which led to his invention of the separate condenser. Of course, it was through his invention of the separate condenser that he achieved the title natural philosopher or scientist, but, nevertheless, his improvements to the barometer demand close attention. They were adopted later by John Gardner, Watt's chief technical assistant, who took over Watt's instrument business around 1770.

During the eighteenth century, the mercurial barometer was gradually improved to give greater precision in readings taken from it and also to make it portable. Through increasing wealth, more people could afford to purchase one for domestic use as a means of predicting changes in the weather, so that, by the middle of that century, there was something of an explosion in the demand for barometers and all sorts of other mathematical and scientific instruments. (130) The simplest type of barometer was the so-called Torricellian with a straight vertical glass tube, closed at the top with the lower end immersed in a cistern of mercury. For support, it was mounted on a wooden backboard and termed a 'stick' barometer. While considered the most accurate for normal purposes, should one have to be moved, the mercury might spill out of the cistern or surge up the tube and break the glass so it was not portable. Yet barometers of this type continued to be constructed into the third quarter of the eighteenth century mainly because they could be read easily. The atmospheric pressure was determined by measuring the height of the column of mercury in the glass tube above the top of the level of the mercury in the cistern. However, as the mercury rose and fell in the tube according to variations in atmospheric pressure, so more or less mercury would be taken out of the cistern with the result that its level continually fluctuated. The smaller the cistern containing the mercury, the more pronounced would be this fluctuation, which affected the reading.

One way of reducing this inaccuracy was to have a large reservoir of mercury in the cistern. John Smith, writing in 1688, said that he had discovered by trial and error that a tube of 0.2 ins. bore needed a cistern of at least 3 ins. diameter and that of 0.25 ins. bore needed one of 3.5 ins. to minimize the effect of the changing cistern level and preserve any degree of accuracy. (131) The article in Rees's *Cyclopaedia* stated that the diameter of the cistern should be ten times larger than the bore of the tube so that 'the addition or subtraction of the mercury, contained between the greatest and least altitudes, may not sensibly affect its depth'. (132)

Two barometers made by Watt's contemporaries are preserved in the Science Museum in London. Both have large bowls for their cisterns. One by Sisson, which is probably the earlier, has a cistern 4.75 ins. diameter and a tube with bore about 0.33 ins. The other by Ramsden has a cistern diameter of 3.3 ins. and the bore about 0.35 ins. Therefore both these barometers should have been reasonably accurate yet both have devices to show the correct level. Ramsden's is the simpler being just a pointer whereas Sisson's has an ivory slider with a vernier for really accurate measurement of the mercury level in the cistern as well as a microscope for ascertaining the level in the tube. (133) However in neither case can the level of the mercury be altered, so on the Ramsden instrument, the pointer will show only that the level is incorrect while on the Sisson, the reading of the lower vernier must be added or subtracted to that of the upper one. In neither case is the method satisfactory. Furthermore, the mercury is in open cisterns so that it could be spilt easily if either instrument had to be moved. In addition, such a large volume of mercury would add to the cost of these instruments. Jeremiah Sisson died in 1770 so Watt may have seen his instruments. He knew Ramsden well enough to have his letters directed to 'Ramsden' when staying in London in 1767 and must have been aware of his improvements. (134)

Most common barometers at the time when Watt was an instrument maker between 1757 and 1770 would have been a form of Torricellian barometer with the lower end of the tube glued into a small wooden cistern like an inverted cup. The mercury would be contained either in a leather bag within this cup or leather would be stretched over the bottom opening which allowed enough flexibility to cope with changing atmospheric pressures. Wood, such as box, was sufficiently porous to allow air through to maintain the internal pressure but dense enough to prevent the mercury escaping. This made such barometers more portable but they still suffered from two disadvantages. First, the mercury could surge up and break the tube if the instrument had to be carried. Second, and more serious for a scientific instrument but something which did not seem to concern the ordinary householder, the level of the mercury could not be determined accurately and, because these cisterns were invariably smaller than Smith's proportions, the level must have fluctuated a great deal and hence the readings would have been less accurate.

Before 1700, Daniel Quare had some method of squeezing the leather bag so that the mercury filled the tube to prevent it surging up and possibly breaking the tube when the barometer was being carried. (135) Soon after this, there certainly appeared a screw device which could be turned to push up the leather base of the cistern and so fill the tube but it is not known who introduced it. By the second decade of the century, this method had achieved widespread acceptance. (136) Although this gave reasonable portability, it was impossible to see the lower level of the mercury and therefore such instruments would not give very accurate readings, certainly not accurate enough if barometers were to be carried up hills or mountains to ascertain heights. A true portable barometer where the mercury level could be properly adjusted to its correct level still needed to be invented.

Watt's Manufacture of Barometers

In June 1756, Watt sent to his father an estimate of the costs of tools and materials which he wanted to purchase in London for equipping himself at Greenock. (137. See p58) Out of the total £23.12s.6d, he allowed £1 for barometer tubes and £3 for 'box and other hard woods'. (138) Some tubes were purchased for, while barometers are not given a separate heading in his 'Ledger, January to April 1757', (139), his 'Waste Book' on the third of January the same year has the entry,

8 glass tubes for barometers	9.0
Several bits of Plate Brass for various uses	10.0
$^{1}/_{2}$ lb. Quicksilver	2.9 (140)

Eight glass tubes are also listed in an undated inventory which most likely was compiled when Watt moved his workshop to the College in Glasgow at the end of July 1757. Two tubes were sold for 3s. to his Grace the Duke of Montrose in June 1759 when mercury was bought at the same time at 5s.6d. per pound. (141) Presumably this sale was for repair to some instruments. A further inventory drawn up on 7 October 1759 appears to be very detailed but contains no mention of barometers. This suggests that Watt had disposed of the remaining tubes either by repairing other barometers or making some himself.

From the two surviving barometers which may be ascribed to Watt and the printing plates for paper scales for barometers preserved in the Garret Workshop, it is safe to assume that Watt concentrated on the stick or portable types. Records show that he sold a barometer in January 1761 for £1.1s. and another in May for £1.17s. (142) Later Dr. Wilson, presumably the Glasgow Professor of Astronomy, bought one in 1763 for £1.15s. (143) Sales must have been satisfactory because Jockey purchased eighteen 'Barrometer Tubes' for him at Bristol in October 1761. (144) Watt's prices compare with those of Henry Pyefinch who charged £1.11s.6d. and Adams who charged £2.2s. for common barometers in 1769 in London. (145) These prices suggest that Watt either aimed his barometers at the lower end of the market or could produce them at lower costs.

We have one piece of evidence to show that he was involved with another type. The invoice which Anderson paid in September 1762 for Watt's work on behalf of the University's Natural Philosophy class mentioned 'Mending a Barrometer, 6d.' and 'apparatus for a wheel Barrometer, 2/.6d'. (146) The wheel barometer was probably the type with a dial and a hand pointing to the pressure. When Watt went to London in the summer of 1763 to select goods for selling in his shop, barometers were among the items purchased. This may indicate a change in policy on his part through a recognition that it was cheaper to buy wholesale rather than try to make a wide range of instruments himself. So far Watt was involved with barometers only as a manufacturer and retailer of the common types.

Surviving Barometers

In the Garret Workshop, there are two plates for printing barometer scales on paper. We know that Watt was using paper scales, a cheaper form of construction than brass, from a letter to his father in October 1757. (Plate 3)

> I have no papers for Barometers but the inches are counted from the surface of the mercury... there is a box with Barometer tubes in the press at the West room Bed head which pack well & send up first opportunity but not by Willie Crawford as he's an extravagant rascal. (147)

This letter shows that Watt was fully aware of the need to measure the column of mercury from the level of the top of the mercury in the reservoir at the bottom. The wording on one of the scales must have been designed for Scottish weather as it reads on one side, 'Very dry; Set Frost; Frost; [change]able; Snow; Much Snow; Stormy'. The scales are set close to the tube itself and there is no room for a vernier scale and slider which would have soon worn away the paper. (Plate 3)

The only complete instrument surviving from Watt's Greenock and Glasgow days is a stick barometer now displayed in the National Museums of Scotland in Edinburgh. The straight glass tube is concealed behind a flat mahogany casing. It is probable that Watt bought in partly made wooden parts for barometers as he did for some of his musical instruments. There is no glass window to protect the scales at the top which are engraved in brass with a semicircular top. This style became common by the third quarter of the century. (148) It is surprising that this barometer, although it has the small wooden cistern with leather bottom, has no screw for adjusting the level. This type persisted for a long while in Scotland. (149) With a cistern diameter of 2 $^{1}/_{8}$ ins. on the outside and a tube bore of 0.35 to 0.40 ins., some form of compensation would have been necessary for accurate measurements. The outer cover for the mercury cistern has been machined on a lathe from a piece of mahogany to form an ornate cylinder, closed at the bottom and parallel outside but slightly tapered inside. Such a cylinder could be cut in half lengthways, giving two for the price of one. This is an engineer's solution rather than a cabinet maker's. The boss on the top has been turned in the same way.

The engraving on the scales follows the more usual wording of 'Very Dry; Set Fair; Fair; Change; Rain; Mch. Rain; Stormy', with 'James Watt Glasgow' in his typical sloping style. This may well have been added to plates bought in from another maker as the lettering is so different. There is a vernier slider which, as there is no lower compensation adjustment, would have been more use as a marker to show changes in level rather than giving accurate readings. Most barometers sold from about 1770 were fitted with verniers, (150) and it can be imagined that this would be a device that appealed to Watt with his mathematical interests. (Plate 4b)

There is also an incomplete barometer in Watt's Garret Workshop. The tube has been broken and the upper part is missing but it was exposed, being mounted on a round-topped backboard covered with mahogany veneer with patterned edging. Although the brass scales are missing, they were rectangular with provision for a vernier slider. A rectangular hinged box also veneered covers the cistern which is an unusual pattern. The bottom of the tube is secured in a wooden rectangular receptacle or tray for holding the mercury. This tray, of which the internal dimensions are 2.87 ins. by 0.65 ins. by 0.7 ins. deep, connects with the lower part of the cistern through a small hole behind the tube so the level of the mercury could be easily observed in the tray. The lower part has been turned into a cylinder whose outer diameter matches that of a wooden cup underneath, 1.35 ins. outer diameter. The two are connected by bellows made from gut secured by string. The inside diameter of the tube is 0.3 ins. so that adjustment of the mercury level would be necessary for accurate measurements. The adjustment of the level of the mercury can be effected by raising or lowering the cup by means of a hand screw. The screw presses against the wood of the bottom of the cup instead of the leather in other barometers so that this part was strengthened for frequent use. The bellows and screw adjustment must have been devised to allow the level of mercury in the rectangular container to be adjusted but there is no indicator to give the precise level needed for accurate measurements. Perhaps any pointer has been lost over the years. With an open tray of mercury, the barometer was not easily portable. (Plate 5)

While this barometer cannot be dated, it is likely that it was a first attempt by Watt to make one which could be adjusted to give accurate readings when he was carrying out experiments on steam pressures and temperatures in the second half of 1773 and early 1774. He needed to know if the atmospheric pressure varied during the course of his trials because this would affect his results. It was a period when other people were also seeking to improve the barometer. Adams had developed an ivory float as a level indicator around 1769 (151) and then, in about 1770, Ramsden had made a much improved barometer which had both an ivory pointer over the mercury cistern and also a leather bottom to the cistern with a screw so that the level of the mercury could be adjusted to the pointer. This was contained in a casing with a hole for viewing the mercury, the hole being normally sealed by an ivory plug. (152) Such barometers would not appear to have been easily portable. Jean A. de Luc, who later became a close friend of Watt, published his *Recherches sur les Modifications de l'Atmosphère* in 1772 which contained a study of the barometer. Watt was making a barometer to de Luc's instructions in August 1773 which he broke during boiling the mercury when filling it. (153).

Watt's Portable Barometer

We do not know the extent to which Watt was aware of these improvements but he kept himself informed through, for example, knowing Ramsden and reading published accounts. These may have stimulated him to improve the barometer and try to make it fully portable as well as more accurate to read. The first description we have is in his *Journal* for February 1774. 'Contrived a new portable baromt. the cistern in 2 pieces joined by 3 ply of gut - the lower surface of the mercury to appear in a bit of open tube fixed in the top of cistn. and to be brought in contact with a black hair stretched in a Ivory fork'. (154) It would be interesting to know if the bellows arrangement for the cistern was Watt's invention. He wrote to James Lind on 7 February about the 'method of calculating Barm. by Farenheit', (155) and started to make a new barometer the same day with a cistern of this type. He fixed the barometer tube in the cistern on the 16th. and began to fill it but some air got in and he spoilt the cistern when he tried to boil it out. (156) It was only on the 25th. that he divided the barometer scale '... with new dividing screw - 332 teeth of the leading screw correspond to $^{1}/_{20}$ inch exactly & 241 to one revolution of the screw this screw seems to be very exact it is 9 inches long and is made of steel'. (157) He fitted the gut to the cistern and presumably filled it satisfactorily for he commenced some experiments on water boiling in a vacuum on the following day.

It is from a letter to Lind written on 22 February that we can follow what Watt was doing. He first wrote about experiments de Luc carried out on a 27 ins. column of mercury to see how much it expanded when heated from freezing to boiling point of water. Although the expansion was 0.5 ins., Watt eventually decided that in a thermometer, this 'would not sensibly affect the result of the experiment'. (158) Also the temperature of any barometer recording barometric pressure would remain near enough constant during his experiments so he did not need to worry about the expansion of mercury in barometers either. He went on to sketch the cistern of his portable barometer in which the bellows resembled those of the one in the Garret Workshop. For his portable barometer, the whole of the upper part for the cistern was circular. Into the hole through which the mercury would be raised, Watt proposed to insert a short glass tube so that the mercury could be seen in it. The top of this tube was sealed with a wooden plug glued in. If this plug were made of fir, he thought it would be porous enough for the air to pass through according to changes in pressure. To make quite sure, a small hole could be drilled down it which normally would be blocked by a wire that could be drawn in or out. The plug was to be flattened into a wedge ending in two points like a fork. Across the ends of the fork would be stretched a black hair to show the correct level of the mercury. 'I have made the Barometer and it seems to answer' was Watt's final comment. (159)

While this was the end of the development of barometers as far as Watt was concerned, it was not the end of the development by others of his portable barometer. The bellows cistern soon became popular especially in Scotland. One by 'J. Gardner, Glasgow', which suggests a date before 1799 when Gardner was joined by his son, (160) has a bellows cistern similar to Watt's drawing but without any level indicator. Stott of Dumfries had a similar type. (161) The diameter of Watt's cistern was quite narrow, possibly to save the cost of some mercury but, following Smith's principle, the mercury in the tube would have fluctuated more as the pressure changed. Miller, an Edinburgh manufacturer around 1800, was producing barometers with bellows cisterns shallower but broader which would have been more accurate. (162)

Further developments lay in the methods of determining the level of mercury in the cistern more accurately. We have seen that this was being attempted by manufacturers in

London but the problem also interested Watt's scientific friends. Small wrote from Birmingham at the end of March 1774,

> In one of my barometers, the scale floats within the tube, and is of such specific gravity, and of so accurately expansile matter, that it constantly shows the weight of the atmosphere correctly. This is difficult to be made. The other is very easy. The basin is large; a float of metal, with a hole for the tube in its middle, lies upon the surface; a scale stands upon the float, and expands and contracts just as much as mercury. You can easily imagine how both may be rendered portable. (163)

Watt raised some objections to it which he may have related to Small when he moved to Birmingham at the end of May 1774. (164) Here is another example of a float which would have been known to Watt and so possibly through him to Gardner.

A little later, another barometer of Watt's type was made by Gilbert Hamilton, Watt's brother-in-law. Hamilton married the sister of Watt's second wife and acted as their Scottish agent after Watt moved to Birmingham. He became one of the leading Glasgow merchants and was associated with a wide range of businesses, among them becoming agent in Glasgow for the Bank of Scotland and the Carron Company. He was also a partner in the Delftfield Pottery, which he directed for many years. He was elected Provost of Glasgow in 1792 to 1793 and died on 30 November 1808. He thought that such instruments might be a commercial proposition and suggested they might be made by Boulton's craftsmen in Birmingham,

> I have one made upon your construction but greatly wider in the tube than yours but answers exceedingly well - I can always set it up to the same height. A Professor Nin Hill has been deiring [sic] me to write to you that he can get you orders for near a dozen of Barometers if you will execute them to some of our Dons [hole] Cuningham etc. who wont grudge a price [hole] have them exact & handsome & wishes you could get some neat contrivance in order to make them in the form of a Corinthian pillar, the Base to open for adjusting them & on the top of the Pillar which should be at 27 inches to have a neat brass or bronze figure holding the scale. As the workmen with you are more in the way of fancy things they could hit it off better ther [sic] than could be done here or perhaps give you a better contrivance. (165)

None is known to have been built to such a fancy design. Watt was just starting out on his career as designer of steam engines and barometers disappear from his correspondence but that is not quite the end of the story.

De Luc published rules for the construction and use of the mercury barometer. With his formula for relating pressure to altitude, the barometer enabled surveyors to obtain heights more rapidly and simply than by trigonometrical methods. Gardner was quick to exploit this since he placed an advertisement in the *Glasgow Mercury* in 1779, part of which stated that he sold

> Barometers on the most improved principles, so as to admit of being carried to any distance with the greatest safety. Likewise Barometers, for measuring the heights of hills, which have a peculiar adjustment to regulate the lower surface of the Mercury in one place, and have the mercury boiled in the tube. In this branch of business he has been favoured with particular methods and directions by some Gentlemen of this place of undoubted knowledge and skill in regard to this instrument. (166)

Watt and Hamilton are likely to have been the gentlemen who helped Gardner develop this instrument. Another later barometer signed 'Gardner Glasgow' in private possession has Watt's type of bellows cistern and, in addition, an ivory float to indicate the level of the mercury. (167) The float moves in an ivory guide and is protected by a small glass dome which can be unscrewed, allowing the level to be determined accurately. This is in the same position that Watt drew for his glass tube and hairline indicator and must have been developed from Watt's design. The lower end of the float is tulip-shaped so it will seal the opening as the level of the mercury is raised by operating the screw at the bottom. Therefore this instrument is truly portable. There are two ivory scales with verniers which give readings on one side for 'Night' and the other for 'Day' with no wording. The scales read down to 25 ins. This barometer must have been specially ordered but it does confirm that Watt's method of accurately determining the mercury level was improved and resulted in an instrument which was sold commercially. If so, this was thirty years before Fortin developed his more famous method.

The First Shop

During the summer of 1759 Watt made a trip to London. While there is no mention of him selling quadrants, he made many purchases with which he could stock his new venture, a shop in the College. His original workshop, being on the first floor, would not have been very convenient. J.P. Muirhead says that Watt originally had received permission in 1757 to occupy an apartment and to open a shop within the precincts of the College. (168. See p71 & 80) Yet it could well be that it was not until 1759 that he expanded his business into a shop selling scientific instruments purchased wholesale as well as some made in his own workshop. We know that his shop was open in the College precincts at least from October through an advertisement for his Uncle Watt's map of the Clyde. (See p96) The reason for suggesting that this was when Watt started his shop is because in June he borrowed £4.4s. from Black and £70 from George Anderson on a three months loan, presumably to finance new stock. (169) In addition, he received £12.12s. from Hugh Brown and £5.5s. from John Carlisle (See below) to buy things for them in London. George Anderson was repaid £53 on 7 October, coinciding with the beginning of the partnership with Craig, although this sum may have been borrowed from Uncle Muirheid who lent a further £17.3s. in January 1760. (170)

For his journey to London in the middle of June 1759, Watt spent £7 on a mare and took with him £19 in gold. (171) He returned by the middle of August; his outward trip costing £4.6s.3d. and the return £4.18s.$6^{1}/_{2}$d. with £5.6s.$10^{1}/_{2}$d. expenses there. No doubt the contacts he had made during his earlier year in the capital helped him again for he bought sundry mathematical instruments from Christopher Stedman at the Globe, Leadenhall Street for £19.15s; optical instruments from James Tomlinson, Golden Spectacles, Aldersgate Street, £17.18s.3d. and from Charles Lincoln, optician near St. Bartholomew's Hospital, £6.7s; sundry rules from Richard Rust in the Minories, £14.1s.6d; sea compasses from J. Jarman, £4.8s.6d; globes from Nathaniel Hill, £3.2s.6d; fiskin cases from Richard Bounds, case maker, Shoe Lane, £2.16s. and spectacles from J. Oakely, £1.7s. (172) He continued to deal with some of these people over the next few years who supplied him with complete instruments. This indicates the sort of merchandise he expected to sell and suggests that he recognised that he could not manufacture everything himself. His 'Waste Book' shows he sold 'bottle' microscopes, a globe, Davis's and Hadley's quadrants, Gunter's scales and sea compasses soon after his return.

Uncle John Watt's Map of the Clyde

At the same time as he was developing his shop, Watt was launching out into another venture, that of publishing the maps of the Clyde Estuary which his Uncle Watt had surveyed.

(173) He commented to his father on 22 February 1758 about the Clyde navigation , 'The lock I hear is absolutely laid aside for this year as the Bill came too late for the Parliament'. (174) So it was not until 31 May 1759 that the first Clyde Navigation Act received the Royal Assent. Construction under Smeaton's direction of a dam and a lock began at the Marlin Ford in the spring of 1760, witnessed by a large crowd of people. After a postponement during the winter months, work commenced again in the spring of 1761, but the sands and silt underlying the site of the lock proved to be incapable of bearing the great weight of the walls. After a partial collapse that year and further failures in 1762, the whole scheme was abandoned in the autumn, leaving the Clyde unnavigable. Therefore schemes for any canal to link the rivers Forth and Clyde had to remedy this deficiency by extending their line past Glasgow and on towards Dumbarton. Watt would be involved with these.

In the meantime, Watt took the opportunity of the excitement and enthusiasm for the project to have printed the survey of the Firth of Clyde carried out by his Uncle Watt. Watt wrote home on 3 March 1758, 'The sooner that you can get the Map sent the Better that if possible the parts that are wrong might be helped before I took it in'. (175) Three days later he wrote that he was going to Edinburgh with Johnie Craig and hoped for the map that Tuesday night. He and his father were also involved with a map of Glasgow, but comments about size of paper and printing in another letter must refer to the Clyde map. This letter also begins to show some of the problems Watt encountered.

> As to the other Map [his uncle's] it must not exceed 22 inches or the Paper will be double price that at 22 in being £4 pr Ream & that at 24 £8 besides there are no Roller presses in this country that will print a plate above 22 $^1/_2$ inches besides as it is a sea chart there must be a new drawing made & the Meridian placed paralel [sic] to the side of the paper as there is none mde [sic] otherwise. (176)

Watt went to Edinburgh presumably because there was no printer in Glasgow equipped with a roller press to take such large copper plate engravings. Even the one that Andrew Bell had was limited in size. The high cost of the paper should be noted because, in 1795, Withering paid only one pound a ream for printing Demy paper, which was probably 22 $^1/_2$ ins. by 17 $^1/_2$ ins. (177) It was during this Edinburgh visit that some other problems with the map were discussed and Watt told his father:

> I showed the Plan to Mr Bell who says that if it be above an Inch Bigger it cannot be done in one plate & to do it in two would considerably increase the Expence he thinks that it should have dregrees [sic] about the edges he says that it the Plate & Graving will cost about eleven pound & that they will cost about 5/- a hundred printing. (178)

Bell was quite right about naval charts needing degrees of latitude and longitude, but more important was the problem of the meridian pointed out earlier. The convention was and is to place north at the top of the map and to have the north-south meridian running vertically.

So far, we have talked about 'John Watt's Map'. It is in fact two maps printed on the same sheet. The lower right part of the sheet contains one small scale map covering, as its title indicates, 'Entry to the River and Firth of Clyde and Adjacent Islands'. The tip of Northern Ireland protrudes at the bottom; a little part of Mull appears in the top left corner; and the coast of Ayrshire forms the right boundary while the River Clyde disappears behind the shield carrying the title at the top right. This map is bordered at its top and to its left by the second, entitled, 'The River Clyde Surveyed by John Watt Senr. 1734'. This covers the Firth of Clyde from the south of Bute, and the River Clyde itself past Glasgow to Hamilton Farm.

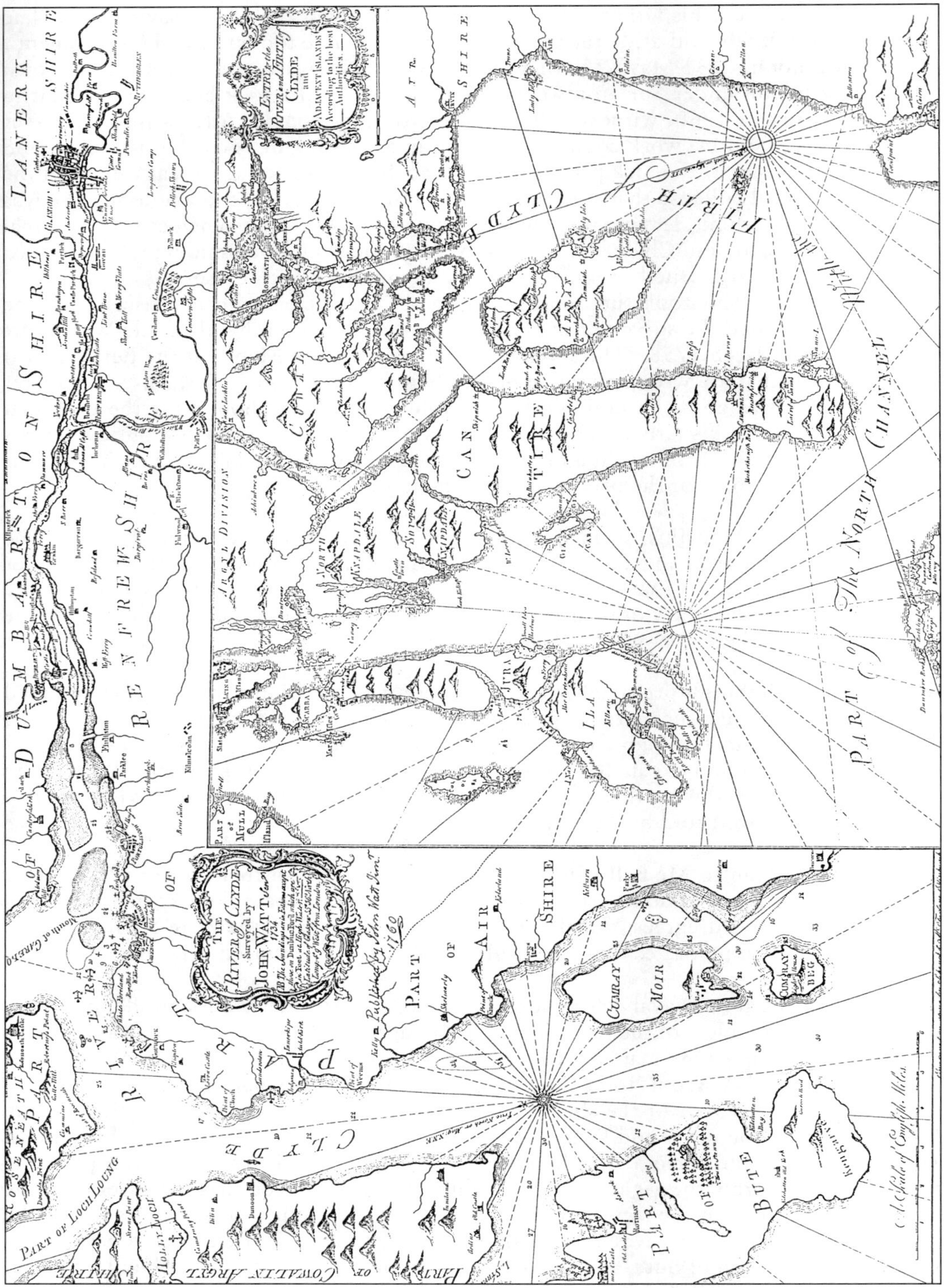

John Watt's Map of the Clyde.

It must have been to fit this map onto the paper with the other that the north was twisted round to the left of the top of the map. Otherwise Glasgow would have come down too far across the sheet into where the second map was to be drawn. The original of the map now at Birmingham has been drawn properly with north at the top, but has lines pencilled across it at an angle to show how it is to be transferred askew onto the copper plate. (179) Presumably the smaller scale map of the islands and 'Cantire' peninsula was also twisted round to correspond with the other. (180)

But there were greater faults than these. If the published map is compared with a modern one, not only will it be seen that the River Clyde does not turn far enough towards the south in its course up to Glasgow, but large sections of the coastal outlines are quite inaccurate as well. Watt himself quickly realised some of these faults for he wrote home:

> I have got the Map pretty well set to rights but cannot believe that the Island of Mull is right placed as its at least 8 mile farther south than I can find it in any other Draught there is none of Dorets Maps that I can find in Glasgow... I send away the Plan tonight but shall desire Mull not to be Graved till we are more certain about it. (181)

Many years afterwards, Watt said that his uncle had made part of the survey and his father and brother added the remainder 'from the best authorities they could collect'. (182) Part of this redrafting, which Watt himself later admitted was 'not over accurate', must have been completed by the middle of April when Watt wrote to his father:

> I recd yours. I wish your alterations are not already too late however shall write Mr. Bell to stop working at it tonight. I would not be too ready altering the Parts that you mention as I compared Jura etc with several Draughts & they agreed pretty well & people sailing by a Coast are very apt to have mistaken notions of the shape of it therefore would only trust them as to the Bearings and not too far to that as sailors are not generally over exact. (183)

Further investigations and possibly surveys may have taken place during the summer of 1758, for the map is not mentioned again in Watt's letters home until the following December, when we find changes were still being made.

> I... have altered the Map. I have done a little to Ila as it was pretty near the thing already except the Loch & the Long nose which I have cut a good piece off. Your drawing of Mull was rather too far in & have likewise altered Colonsa etc both as to their shape & situation. I cannot get to Edinburgh myself my lad being sick & several things in hands but intend to send the Map & proper instruction tonight & will go in myself as soon as possible. (184)

He must have gone to Edinburgh three days later, but in the meantime heard about a possible rival map being produced by the Glasgow magistrates. Eventually, he found that this map extended only as far as Dumbarton from Glasgow and did not cover the whole of the Firth so it would not compete with theirs and in any case was not to be offered for sale but given only to Members of Parliament. (185)

Real progress must have been made during this visit to Edinburgh for the alterations had been completed on the little map and the whole was to be finished by New Year's Day. (186) In the middle of December, Watt was able to send his father a proof and apologised for tearing it as he put it into the case before it was dry. He also said:

I altered little else of the Island of Ila excepting Loch in Dual but you will see a considerable alteration of Oransa & Collansa. As Mcdowal had likewise surveyed them I thought it was proper to lay them down in their right shapes tho I believe that his survey is not much to be depended on which was the cause I did not rub out Ila altogether. It still very different from his but have such manifest Errors in his Bearings that there is no knowing the situation of any place. (187)

So, with islands changing their shapes like amoebae floating around off the Scottish coast, the map was finally engraved and printed in Edinburgh.

The finished copies were sent off on 14 September 1759 with a covering letter and the account.

"You will receive by the Carrier 396 Impressions of your map, with the Remainder of the Rimm[ream] of paper, fifty six sheets being kept for the paper formerly furnished. What paper sent unprinted is part of the Cording Quires what would not do. The following is a copy of the charge of the whole...

1758 March 13 paid for 6 sheets Imperial paper for you	2. 6
28 paid carriage of white iron boxes	1. 6
1759 March 12 To engraving a Map of the River Clyde etc	10.10. 0
To a copper plate for Do. lb. 14 at 2/6 pr. lib	1.15. 0
Sept 14 To printing fifty four Impressions with 396 now sent in all 450 at 15 sh pr Hd	3. 7. 6
To a Rimm of Imperial paper	4. 0. 0
To the packing Box	2. 0
	£19.18. 6
By cash received	5. 0. 0
Ballance	£14.18. 6

P.S. We would have sent you the Copper-plate but thought proper to wait your orders in case you should nead [sic] any more printed. (188)

Watt senior paid £13.10s.6d. on 22 March 1760 to settle this account. (189)

The only known copy of this map is one bound with Smeaton's other maps of the Clyde, now in the library of the Royal Society in London,(190) which presumably he bought during his time in Scotland. The size of the paper is 28 $^{1}/_{2}$ ins. by 21 $^{4}/_{10}$ ins., confirming that the original sheet was Imperial, a size standardized at 30 x 22 inches, quite large for handmade paper. (191) The paper is laid, as one would expect for this period, with the laid and chain lines showing clearly. The watermark is a fleur de lys or Strasbourg lily, with the initials LVG underneath, which should indicate the Dutch papermaker, Lubertus Van Gerrevink. However the countermark is intriguing for it consists of the initials 'JW'. It is most unlikely that this paper was produced specially for Watt. These initials were used by the famous papermaker, James Whatman, in Kent, who must have made this paper. (192)

Watt placed an advertisement in the *Glasgow Courant* on 15 - 22 October 1759,

Just published and to be sold by James Watt, at his shop in the College of Glasgow, price 2s. 6s. [sic] A Large Sheet MAP of the River CLYDE, from Glasgow to Portincross, from an actual survey. To which is added, A DRAUGHT of PART of the NORTH CHANNEL, with the FRITH of Clyde, according to the best authorities. (193)

His father also sold some because Watt wrote in October 1760 telling him to keep the money since he had helped to finance the project. (194) John Willison at Port Glasgow bought at least eighteen, some at a wholesale price of 1s.11d. (195) Barrie was charged £1.4s. for a dozen in December 1761. (196) While sales of a few other maps can be traced, (197) Watt's surviving accounts are too fragmentary to discover what happened to all 450 printed. Yet most were sold because in April 1762 Watt reported to his father, 'You have also inclosed the acct of the Map of the Clyde which you see is near Ballance'. (198) Unfortunately this account has not been preserved but the venture must be considered a success even though Watt never attempted a similar publication again while he lived in Scotland.

Many years later a second edition was printed. This happened after Rennie visited Edinburgh in 1794 when he was engaged on his own surveying business. He enquired about plans of the River Clyde and was told about John Watt's map, which he thought had been surveyed by Watt himself. Somehow he discovered the original copper plate at Bell's, who had repeatedly denied to Watt that he had still got it. (199) A friend of Rennie asked Bell to run off a few copies, which Bell refused, so Rennie wrote to Watt. (200) Watt immediately wrote to Bell and ordered more copies to be printed, but unfortunately in the press copy of the letter retained by Watt, the number has not been included. (201) The copies were to be sent to Hamilton at Glasgow whom Watt instructed 'to send a couple of dozen to Mr Cochrane & to put some into Mr. Dunlop's shop for sale'. The price was to be raised to three shillings 'or perhaps more'. Another couple were to be sent to Rennie, and Watt's son Gregory, then at school in Glasgow, was to bring a dozen with him when he returned to Birmingham. (202) We are given no indication of the success of the sales of this second edition. Probably it did not do as well as the first. In October 1817, Alex Wilson returned seventeen copies because he was giving up bookselling (203) and in 1819 John Smith of Glasgow said that those in his stock had not been sold. (204) One copy has been preserved in the Map Library of the British Library. It is printed on laid paper with the watermark 'J. Whatman', which shows it must be the re-print of 1794 - 95. (205)

NOTES: MATHEMATICAL INSTRUMENT MAKER

1 JWP, 4/11.4, J. Watt to J. Watt sen., 29 July 1757.
2 B. & W. Col., MIII, 3/1, 'Waste Book' 2 Aug. 1757.
3 Smith, (1994), Bryden, p. 10 rt.
4 Muirhead, (1854), Vol I, p. xxxv.
5 Smith, (1994), Bryden, p. 11 lft., see H. Lumsden in *Glasgow Herald*, 26 Dec. 1911.
6 Ibid, p. 18 rt., note 17 and Crowther, (1962), p. 111.
7 Ibid, p. 11 lft.
8 JWP, 4/11.44, J. Watt to J. Watt sen., Dec. 1758.
9 Devine & Jackson, (1995), p. 342.
10 B. & W. Col., MIII, 3/1, 'Waste Book', 15 May 1758.
11 B. & W. Col., MIII, 3/1, 'Waste Book', 17 Nov 1758 and 18 May 1759.
12 JWP, C1/1, 'Accounts', f. 25, 12 Sept. 1762.
13 See B. & W. Col., MIII, 3/1, 'Waste Book', 18 May 1758 and JWP C1/1, 'Accounts', f. 25, 12 Sept. 1762.
14 Daumas, (1989), p. 139.

15 Musson & Robinson, (1969), p. 80, quoted from *Encyclopaedia Britannica*, 3 rd. edn., 1797, J. Robison, 'Steam Engine'.
16 Williamson, (1856), pp. 159 - 64.
17 Goring, 1992, entry for A. Wilson and *Transactions of the Royal Society of Edinburgh*, Vol 10, 1826, P. Wilson, 'Biographical Account of Alexander Wilson', pp. 283 - 4.
18 Simpson, (1962), Swinbank, p. 29 lft and A. Morrison Low, 'Scottish Barometers', a talk given to the Royal Meteorological Society Historical Group, London, 8 March 2000.
19 JWP, JW/22.25, P. Wilson to J. Watt, 3 Oct. 1786.
20 *Trans. Royal Society of Edinburgh*, Vol 10, 1826, p. 297, Biographical Account of A. Wilson.
21 JWP, 4/13.30, P. Wilson to J. Watt, 25 May 1791.
22 Riddell, (1979), p. 2, note 2 and *Extracts from the Burgh Records of Glasgow, 1739 - 59*, Glasgow, 1911, p. 521.
23 Riddell, (1979), pp. 18 - 23.
25 Ibid, p. 131 - 2.
24 Skempton, (1981), p. 12.
26 Smeaton, (1755), p. 60.
27 Riddell, (1979), p. 25, note 2 and Lindsay, (1968), p. 213.
28 Riddell, (1979), p. 25.
29 JWP, 4/11.11, J. Watt to J. Watt sen., 24 Nov. 1757.
30 B. & W. Col., MIII, 3/1, 'Waste Book', 15 May 1758.
31 Royal Society, Smeaton, *Designs*, Vol VI, Nos. 57 & 58.
32 Smeaton, (1950), p. 158, the original plan is in Vol VI, ff. 52v and 53.
33 B. & W. Col., MI, 3/16, Surveying Note Book.
34 Bryden, (1970), p. 263.
35 *Trans. Royal Society of Edinburgh*, Vol 10, 1826, p. 287 - 8, 'Biographical Ac count of A. Wilson'.
36 B. & W. Col., MIII, 3/1, 'Waste Book', 26 Oct. 1758.
37 Ibid, 17 Nov. 1758.
38 Ibid, 10 May 1759, see Swinbank, (1968), p. 7 and Swinbank, (1969), typescript.
39 Swinbank, (1969) p. 5.
40 *Trans. Royal Society of Edinburgh*, Vol 10, 1826, p. 288, 'Biographical Account of A. Wilson'.
41 B. & W. Col., MIII, 3/1, 'Waste Book', 18 May 1759.
42 B. & W. Col., MIII, 3/4, 'Personal Ledger of James Watt, Oct. 1759 - Dec. 1764', f. 2, 1 March 1760.
43 Robinson & McKie, (1970), p. 33, letter 26, J. Watt to Prof. Russell, 2 Sept. 1765 with added note on 28 Jan. 1769.
44 JWP, C1/1, 'Accounts', f. 32, 23 April 1763.
45 *Trans. Royal Society of Edinburgh*, Vol 10, 1826, pp. 288 - 9, 'Biographical Account of A. Wilson'.
46 Simpson, (1962), Doig, p. 37 rt.
47 Donovan, (1975), p. 221.
48 Simpson, (1962), Anderson, p. 8 lft.
49 Crowther, (1962), p. 26.

50 Ibid, p. 33.
51 Ibid, p. 40.
52 Ibid, (1962), p. 35.
53 Donovan, (1975), p. 221.
54 Ramsay, (1918), p. 32.
55 Crowther, (1962), p. 38.
56 Muirhead, (1854), Vol I, p. xxxv.
57 B. & W. Col., MIII, 3/1, "Waste Book", 8 Oct. 1757.
58 Muirhead, (1854), pp. xxxv - vi.
59 JWP, Letter Book 3, J. Watt to J. Robinson, 16 Dec. 1799.
60 Brewster, (1822), Vol II, p. ix, J. Watt to D. Brewster, May 1814.
61 Donovan, (1975), pp. 222 - 3 and Ramsay, (1918), p. 16).
62 Robison, (1803), Vol I, p. iii.
63 Ibid, p. viii.
64 Robinson & McKie, (1970), p. 416, Letter 259, 'Copy of Mr. Watt's Remarks on Mr. Robison's Edition of Dr. Black's Lectures, communicated to Mr. Playfair, 4 Jany. 1809'.
65 Ibid, p. 322, Letter 213, J. Robison to J. Watt, 18 Dec. 1799.
66 Robison, (1803), p. ix.
67 Robinson & McKie, (1970), p. 416, Letter 259, "Copy of Mr. Watt's Remarks...", 4 Jan. 1809.
68 Donovan, (1975), p. 176.
69 JWP, 4/32.35, J. Watt to J. Roebuck, 11 Oct. 176[5].
70 Boulton 348.2, J. Watt to M. Boulton, 12 Dec. 1768 and 348.9, J. Watt to W. Small, 28 May 1769.
71 Boulton 348.10, J. Watt to W. Small, 5 July 1769 and 348.15, J. Watt to W. Small, 3 Jan. 1770.
72 Boulton 348.14, J. Watt to W. Small, 12 Dec. 1769.
73 JWP, W/9.6, T. Beddoes to J. Watt, 9 Feb. 1798.
74 B. & W. Col., MIII 3/1, 'Waste Book', 27 Jan. 1758, 27 Dec. 1757 and 17 Aug. 1758.
75 See JWP, 4/11.38, J. Watt to J. Watt sen., 28 June 1758. 'This comes by Mr. Craig who likewise brings Thos Cochrans watch It was the spring broke & not the chain the common price is 5/- but to him shall charge only what it costs me together with the putting in to wit 2/6'.
76 B. & W. Col., 3/4, 'Personal Ledger', 8 June 1763.
77 B. & W. Col., MIII, 3/6, 'Ledger of Personal Accounts, Jan. 1764 - May 1769', f. 5, Oct to Dec. 1766.
78 B. & W. Col., MIII, 3/1, 'Waste Book', 14 June 1759.
79 JWP, 3.21, 'Household Expenses', 9 June and Aug. 1763.
80 B. & W. Col., MI, 2/3, 'Cash Memorandum of James Watt, April 1765 - I Jan. 1767', 19 Feb. 1766.
81 Robinson & McKie, (1970), p. 40, Letter 34, J. Black to J. Watt, 23 Feb. 1774.
82 B. & W. Col., MI, 2/6, 'Cash Memorandum of James Watt, March 1774 - 26 Dec. 1775', 21 Aug. 1775.
83 Ibid, 16 Oct. 1775.
84 Robinson & McKie, (1970), p. 28, Letter 23, J. Black to J. Watt, 22 Mar. 1772

and p. 40, Letter 34, J. Black to J. Watt, 23 Feb. 1774.
85 B. & W. Col., MI, 1/18, 'Journal', 21 Mar. 1772.
86 Robinson & McKie, (1970), p. 28, Letter 23, J. Black to J. Watt, 22 Mar. 1772.
87 Ibid, p. 40, Letter 34, J. Black to J. Watt, 23 Feb. 1774.
88 B. & W. Col., MI, 2/6, 'Cash Memorandum', 16 Oct. 1775.
89 B. & W. Col., MI, 2/11, 'Cash Book, Cornwall, 1778 - 82', 12 Feb. 1781.
90 JWP, C3/8, Diary 1784, 8 June 1784 and B. & W. Col., MIII, 3/4, 'Personal Ledger', f. 31, 'This amount was settled & Dr. Black paid in full since J.W. came to England about 1785'.
91 Robinson & McKie, (1970), p. 142, Letter 99, J. Watt to J. Black, 6 June 1784.
92 JWP, 4/11.6, J. Watt to J. Watt sen., 21 Aug. 1757.
93 Muirhead, (1858), p. 61, see Robinson & McKie, p. 321, Letter 213, J. Robison to J. Watt, 18 Dec. 1799.
94 Daumas, (1989), pp. 92 & 189 - 90.
95 B. & W. Col., MIII, 3/1, 'Waste Book', 31 May 1758.
96 Muirhead, (1854), Vol I, p. xlii.
97 Smiles, (1878), p. 29.
98 JWP, 4/11.5, J. Watt to J. Watt sen., 5 Aug. 1757, see note added by J. Watt sen.
99 JWP, 3/21, 'Household Expenses', 7 Jan. 1758, 'New Years gifts to servants etc. 6/8'; 17 Jan. 1759, 'New Years gifts to servants 5/-'; 11 Jan 1760, 'New Years gifts to Servants, 6/6'.
100 JWP, 4/11.43, J. Watt to J. Watt sen., 18 Oct. 1758.
101 JWP, 4/11.55, J. Watt to J. Watt sen., 24 Oct. 1760.
102 JWP, 3/21, 'Household Expenses', 3 Dec. 1760.
103 JWP, 4/11.5, J. Watt to J. Watt sen., 5 Aug. 1757, note added by J. Watt sen.
104 JWP, 4/11.6, J. Watt to J. Watt sen., 21 Aug. 1757.
105 B. & W. Col., MIII, 3/1, 'Waste Book', 24 Sept. 1757.
106 JWP, 4/11.8, J. Watt to John Watt, 14 Sept. 1757.
107 B. & W. Col., MIII, 3/1, 'Waste Book', 6 Dec. 1757.
108 Robinson & McKie, (1970), p. 23, Letter 18, J. Black to J. Watt, 20 Dec. 1769.
109 JWP, 4/11.56, 'Current Acct. with J. Watt sen.', Dec. 1760.
110 JWP, 4/11.8, J. Watt to John Watt, 14 Sept. 1757.
111 JWP, 4/11.10, J. Watt to J. Watt sen., 28 Nov. 1757.
112 JWP, 4/11.34, J. Watt to J. Watt sen., 6 June 1758.
113 Ibid.
114 JWP, 4/11.21, J. Watt to J. Watt sen., Feb. 1757 [presumably 1758].
115 JWP, 4/11.23, J. Watt to John Watt, 6 March, 1758.
116 JWP, 4/11.30, J. Watt to J. Watt sen., 15 May 1758.
117 B. & W. Col., MIII, 3/1, 'Waste Book', 13 May 1758 and 18 May 1759.
118 Ibid, 'Pd. James McBriar for 9 days work', 8 Oct. 1757.
119 JWP, 4/11.9, J. Watt to J. Watt sen., 10 Oct. 1757.
120 JWP, 4/11.40, J. Watt to J. Watt sen., 16 Aug. 1758.
121 JWP, 4/11.41, J. Watt to J. Watt sen., 15 Sept. 1758.
122 B. & W. Col., MIII, 3/1, 'Waste Book', p. 2.
123 JWP, C1/1, 'Accounts', Inventory, 7 Oct. 1759. I wish to thank Michael Wright and Michael Harding for showing me Watt's Garret Workshop.
124 JWP, 4/11.41, J. Watt to J. Watt sen., 15 Sept. 1758.

125 B. & W. Col., MIII, 3/1, 'Waste Book', 30 Sept. 1758. 'Send a Hadleys Quadrant for John Watt to sell'.
126 JWP, C1/1, 'Accounts', Inventory, 7 Oct. 1759.
127 JWP, 4/11.40, J. Watt to J. Watt sen., 16 Aug. 1758.
128 JWP, 4/11.30, J. Watt to J. Watt sen., 15 May 1758.
129 B. & W. Col., MIII, 3/1, 'Waste Book', various entries.
130 Goodison, (1992), pp. 64 & 66. A longer version of this section was published in *Scientific Instrument Society Bulletin No. 60,* March 1999, pp. 5 - 10.
131 Goodison, (1992), p. 38.
132 Rees, (1819), see article 'Barometer', c. 1804.
133 Goodison, (1992), pp. 224 & 240. I wish to thank Miss Jane Insley at the Science Museum, London, for showing me these barometers.
134 JWP, 4/4.20, J. Watt to Margaret Watt, 11 April 1767.
135 Middleton, (1964), p. 151.
136 Goodison, (1992), pp. 46 & 49.
137 JWP, 6/46.6, J. Watt to J. Watt sen., 19 June 1756.
138 Ibid.
139 B. & W. Col., MIII, 3/3 'Ledger, January to April 1757'.
140 B. & W. Col., MIII, 3/1, 'Waste Book', 3 Jan 1757. Dr. F.G. Page has pointed out that half a pound of quicksilver is a small amount, about 16 cc., while around 72 cc. would be required to fill a barometer.
141 Ibid, 15 June 1759.
142 JWP, C1/1, 'Accounts', f. 14, 30 Jan. 1761 and f. 16, 25 May 1761.
143 Ibid, f. 32, 23 April 1763.
144 Ibid, f. 19, 24 Oct. 1761.
145 Goodison, (1992), p. 82.
146 JWP, C1/1, 'Accounts', f. 25, 12 Sept. 1762.
147 JWP, 4/11.9, J. Watt to J. Watt sen., 10 Oct. 1757.
148 Goodison, (1992), p. 71.
149 Ibid, p. 77.
150 Ibid, pp. 25 & 71.
151 Ibid, pp. 79, 123 & 224 - 5.
152 Middleton, (1964), p. 207.
153 B. & W. Col., MI, 1/20, 'Journal', 11, 23 & 28 Aug. 1773.
154 Ibid, 3 Feb. 1774.
155 Ibid, 7 Feb. 1774.
156 Ibid, 16 Feb. 1774.
157 Ibid, 25 Feb. 1774.
158 JWP, C1/15, J. Watt to J. Lind, 22 Feb. 1774.
159 Ibid.
160 Clarke, (1989), p. 164.
161 Goodison, (1992), p. 76.
162 I wish to thank Miss A.D. Morrison-Low for her help with showing me the collection of Scottish barometers in the National Museum of Scotland.
163 Muirhead, (1854), Vol II, p. 77, Letter 133, W. Small to J. Watt, 29 Mar. 1774; Boulton 340.48.
164 Ibid, Vol II, p. 78, Letter 134, J. Watt to W. Small, 29 April 1774; Boulton 348.42.

165 JWP, 4/20.17, G. Hamilton to J. Watt, 14 June 1775.
166 *Glasgow Mercury*, 1 July 1779, see Clarke, (1989), p. 164.
167 I wish to thank Andrew Foott for borrowing this barometer and bringing it to my home so I could inspect it.
168 Muirhead, (1858), p. 43.
169 B. & W. Col., MIII, 3/1, 'Waste Book', 14 & 15 June 1759.
170 Smith, (1994), Bryden, p. 19, note 38, borrowed on 6 Oct. 1759 and 25 Jan. 1760.
171 B. & W. Col., MIII, 3/1, 'Waste Book', 16 June 1758.
172 Ibid, various entries.
173 A longer account concerning the publication of this map was printed in *Notes and Records of the Royal Society, London*, Vol 52 (1), 1998, pp. 63 - 71, as 'John Watt's Map of the Clyde'.
174 JWP, 4/11.20, J. Watt to J. Watt sen., 22 Feb. 1758.
175 JWP, 4/11.22, J. Watt to J. Watt sen., 3 March 1758.
176 JWP, 4/11.26, J. Watt to J. Watt sen., March 1758.
177 JWP, 4/47, W. Withering to J. Watt, March 1795.
178 JWP, 4/11.27, J. Watt to J. Watt sen., 11 March 1758.
179 B. & W. Col., MIII, 5/12, Map of the Firth of Clyde.
180 B. & W. Col., MIII, 5/9, Map of the West Coast of Scotland.
181 JWP, 4/11.28, J. Watt to J. Watt sen., 31 March 1758.
182 N.L.S., Rennie Papers 19824.16, J. Watt to J. Rennie, 2 Dec. 1794.
183 JWP, 4/11.29, J. Watt to J. Watt sen., 14 April 1758.
184 JWP, 4/11.46, J. Watt to J. Watt sen., 4 Dec. 1758.
185 This map may be in Royal Society, Smeaton, Designs, Vol VI, f. 54.
186 JWP, 4/11.45, J. Watt to J. Watt sen., 13 Dec. 1758.
187 JWP, 4/11.48, J. Watt to J. Watt sen., 15 Dec. 1758.
188 JWP, 4/11.50, And. Bell & Co. to J. Watt, 14 Sept. 1759.
189 Ibid.
190 Royal Society, Smeaton, Designs, Vol VI, ff. 52 v & 53.
191 Labarre, (1952), p. 260.
192 Balston, (1957), pp. 157 - 8.
193 Williamson, (1856), p. 55, note 14.
194 JWP, 4/11.54, J. Watt to J. Watt sen., 17 Oct. 1760.
195 See JWP, 4/11.85, 'Acct. of John Willison', 17 Feb. & 25 March 1761, also C1/1, 'Accounts', 23 June 1761.
196 B. & W. Col., MIII, 3/1, 'Waste Book', 27 Dec. 1761.
197 Ibid, 4 April 1760 and 25 March 1761.
198 JWP, 4/11.84, J. Watt to J. Watt sen., 6 April 1762.
199 N.L.S., Rennie Papers 19824.16, J. Watt to J. Rennie, 2 Dec. 1794.
200 JWP, 6/32.1, J. Rennie to J. Watt, 26 Nov. 1794.
201 JWP, Letter Book 2, p. 187, J. Watt to Andrew Bell, 1 Dec. 1794.
202 Ibid, p. 190, J. Watt to G. Hamilton, 7 Jan. 1795.
203 JWP, C6/9.16, Alex. Wilson to J. Watt, 21 Oct. 1817.
204 JWP, C6/9.37, J. Smith to J. Watt, 24 Feb. 1819.
205 I wish to thank the staff of the Map Library at the British Library for their help in finding this watermark.

THE PARTNERSHIP WITH JOHN CRAIG, 1759 – 1765

John Craig, ? - 1765

Watt's success in establishing his instrument business became known to some of the citizens of Glasgow among whom was John Craig, a merchant and architect. (1) Robison described him as 'a very intimate Acquaintance of Mr. Watt's'. (2) Craig went with Watt to Edinburgh when Watt visited the printers over his Uncle Watt's map. But even before that, Watt had borrowed money from him as early as October 1757. (3) On that occasion, the sum was £2. There was a further £2 in September 1758 and another £3 in May 1759. (4) These entries may not show the total indebtedness of Watt to Craig and the position may have been formalised when they became partners in October 1759.

Among Watt's accounts is 'An Inventory of the Tools Goods etc belonging to us James Watt & John Craig each one half taken Oct. 7th 1759 at Glasgow' (5) which must mark the beginning of their partnership which lasted for just over six years. Craig may have been willing to invest more of his money in Watt's business because he recognised that here was somebody with an attractive personality, with ambition, not afraid to work hard and above all was enterprising in his efforts to tackle new tasks and venture into new spheres. In only two years of trading, Watt had developed a small concern in an important city. He had forged strong links with members of the University who had been able to put much work in his way. An instance of this was the partnership with Wilson and Black, two influential people. He was manufacturing instruments which were finding a market among the nautical fraternity and he had developed outlets for them beyond Glasgow at Greenock and elsewhere. His production techniques and specialised tools had given him a competitive advantage. Recently he had expanded his business to include a retail shop which provided a further outlet. As Bryden has pointed out, here was no dedicated craftsman beavering away in a small workshop but a business man with vision and ambition. (6) Craig would help Watt to fulfil part of his ambition and enable him to change from craftsman into managing partner. But this was at the expense of changing policy from that of a small concern slowly growing by ploughing back its profits into one with quicker development through borrowing capital.

While Watt provided the expertise and contacts in the new partnership, Craig provided capital and kept the books, thus helping Watt to run the business. (7) The Inventory of October 1759 showed a valuation of £91.19s.3½d. for tools, materials and stocks of goods for sale, with £108.0s.8½d. as 'cash on hand', making a total of £200. In 1755, Watt had drawn up a list of the tools he wanted to purchase in London which amounted to £23.12s.6d. (8) We do not know how much was spent but, at the beginning of January 1757 which may be the date he considered that he started business on his own, the valuation was £21.4s.2d, including £5 'ready money'. (9) An undated Inventory, probably of the move to Glasgow in August 1757, showed a valuation of £26.5s.1½d. (10) An increase in the following couple of years from £26 to £91 confirms Watt's competence and ability. Presumably he contributed this £91 and possibly a little cash to the partnership while Craig undertook to make up the other £100. This would have given them equal shares.

It is not until 15 May 1760 that we find the next injection by Craig of £32. There continues a series of mostly small contributions by him of between £2 and £5, sometimes £10. More notable were additions of £91 on 24 October 1761; £30 on 10 February 1762; another £30 on 26 October 1762; £42 on 19 August and £112.5s. on 22 August 1763; £143, on 6 March and £40 on 13 November 1764. (11). Possible reasons for the high investment in October 1761 and August 1763 will be examined later. (See pp111 & 115) It may be that the smaller more regular sums were to pay wages. Annual totals were 1760, £88; 1761, £150;

1762, £108; 1763, £183. This money was lent on interest towards which £49.2s. was paid in March 1764 and £57.4s.7d. on 2 June 1766. The total indebtedness on Craig's account was £757.4s.7d, no mean sum for those days! (12) From Watt's ledgers, we can see that Craig supported Watt's shops. For instance he purchased a microscope in May 1759. (13) When Watt, Craig and Black later opened a shop in the Trongate, Craig's personal account listed knives and forks, cards, buckles, snuffers, wax taper, a pair of 'sizers', spectacles, stick of wax and wafers, bread basket, plyers, files, corkscrew, watch hook, powder horn and so on. (14) Craig died in early December 1765 when Watt senior sent his condolences. (15)

Watt's Personal Life, 1759 - 1765

Watt's salary from the partnership with Craig should have been £35 per annum. He received his first year's 'cellary' on 4 October 1760. The next payment of £70 was on 23 October 1762 for two years but, after that, the accounts are silent about further amounts. (16) By February 1760, Watt had begun to call himself 'Merchant', presumably reflecting his enhanced status with a shop. (17) We can only speculate about any influence that might have been derived from his father as Watt began to move away from his instrument business into that of being a merchant and shopkeeper.

Watt's personal life and affairs changed during this period also. We have seen already that he furnished his own room somewhere in Glasgow during October 1760. (See p81) Apparently in 1761 he tried to obtain a 19-year lease of a garden from the University but this came to nothing. When his talents became better recognised, he was invited to attend meetings of the Anderston Club where his superiors in education and social position welcomed him. These, according to Smiles, included 'Mr. (later Professor) Millar, Dr. Robert Simson, the mathematician, Dr. Adam Smith, Dr. Black and Dr. Cullen'. (18) As Cullen left Glasgow in January 1756, (See p44) it is doubtful if he were a member at the same time as Watt. The group met on Saturdays at Anderston to the west of Glasgow where they walked from the University, dined, discussed and returned late in the evening.

In addition, there was also the circle of his own friends who formed an informal club. Watt named them as 'Wm Morehead of Herbertshire, James Muirheid of Stonehouse, John Allan of [?] your father [John Craig] Mr Millar and myself'. (19) Of these, Millar was the only one with strong university connections to whom they looked up as 'the oracle of the company'. Their discussions 'Turned principally upon literary subjects, religion, morality, Belles lettres, etc. & to these conversations my mind owned its first ?bias? to such subjects'. Watt felt himself to be very much inferior to the others through his lack of any college education. (20) However, Watt's ties with the University began to loosen during the period of the Craig partnership as he developed more connections with people in the city of Glasgow and beyond. He joined the Free and Accepted Masonic Lodge in June 1763. (21) In the same year, he moved his workshop out of the College precincts. (22) Adam Smith was deputed to negotiate with him on the recovery of some of this space. (23) Watt senior continued to address letters to the College until 2 May, then to the Saltmarket in November 1763, followed by the Trongate in February 1764. (24)

Up to the autumn of 1761, Watt's health was good, but then he became unwell, (25) possibly through the headaches that were to plague him for the rest of his working life. The first definite evidence we have of them is when he wrote to his father on 1 January 1762, 'I would write you more fully but am badly with a headach wch have troubled me very much of late. I am wishing you a good new year'. (26) It sounds as if these headaches were fairly recent. A few days later, he wrote, 'I am pretty well today & in general have not the headach above one a week & find nothing so good for it as fasting till it grow better as it generally proceeds from some Disorder in the stomach'. (27) From that time onwards, they are men-

Watt's House, Delftfield Lane.

tioned in letters and recorded in his journals quite frequently. Although in 1762 it does not appear that he was under great pressure or strain at work, later his headaches often occurred at times of high stress or anxiety and almost certainly were some form of migraine. He might be incapacitated for days on end, so severe were his attacks sometimes.

Another change in his life took place when he married Margaret Millar on 16 July 1764. (28) She was a cousin for whom he had 'an early and constant attachment' and was affectionately called Peggy. (29) They are said to have moved into a house, possibly the manager's, belonging to the Delftfield Pottery Company (30) but no contemporary evidence for this has survived. While Watt had links with this Company in 1768, there is nothing to show any connection earlier. A drawing of the building reveals that, far from being a poor place on a 'humble scale', as Muirhead asserted, it was more a 'plain but comfortable style' as Smiles pointed out. (31) It had two storeys and was more commodious than the College rooms he had occupied earlier as well as being on the edge of the country. We shall see that they changed their place of residence in 1766 or 1767. Peggy became a great support and helped in the shop as well as looking after the house and later their children.

On 18 July 1765, Watt wrote to his father announcing the birth of their first child. Although Peggy had been ill that April, (32) the birth seems to have gone smoothly and Peggy was 'as well as can be supposed'. (33) As they had an early opportunity to have the child baptized by a Mr. Thom, who must have performed the ceremony at their house because the other ministers would do it only in church, they did not consult with Watt senior about the name beforehand, which would have delayed the baptism for a week. They decided to call this child, their first son, John in memory of 'poor Jockey'. (34) Young Jockey suffered from some illness that November but recovered, (35) only to die a little later that month or early December from 'an overgrown liver'. (36)

There are few surviving financial papers so that it is difficult to judge how successful Watt had been. If his comments over a proposed marriage settlement are anything to go by, and knowing also that he was borrowing money to finance his enterprises, then it would seem that he had not made much progress. Towards the end of the summer of 1766, Uncle Muirheid wanted Peggy to have £100 and thought that a settlement of £25 per annum should be arranged for her in case Watt himself should die. Watt senior might provide part of this, Watt thought £10 at most. It is Watt's comments in a letter to his father which suggest that he had not done as well as he had hoped.

> I ought before now to have been in a way of giving to you rather than asking from you but want of health and other mislucks have prevented me my health has been much better lately & at same time I think I have laid up a stock of Experience that will soon pay me for the trouble it has cost me. (37)

We do not know whether the marriage settlement was ever arranged but once again we see Watt's optimism for he continued to drop deeper into debt.

Employees

The effects of Craig's capital can be seen most clearly in the increasing numbers of journeymen employees. The earliest list we have is in Watt's partnership accounts when wages were paid at the beginning of October 1760. (38) Then follow two 'Journey Mens Books', the first running from October 1760 to April 1764 (39) and the second from the beginning of January 1764 till June 1771. (40) However these may not reveal the whole picture, particularly after 1765. In all, twenty-eight names are listed but only one, Gardner, was employed for the whole of this time. Thomas Millar was employed for only forty-seven days before October 1760. Ten names appear in the 1760 list and sixteen during 1763 but, as their employment did not overlap, the maximum number on Watt's books at once was fourteen in June 1763. While there were eight in April 1765, only two remained after the end of the summer, Gardner and James Sym, the latter described as 'Mr Watt Shop keeper' in May 1766. These figures may be compared with those of William Robb at Montrose who was active between about 1750 until 1804. He had a workforce of fourteen apprentices and journeymen working for him in the 1750s. (41)

The summaries of wages are:- October 1760, £55.7s.6d; October 1762, £214.6s.9$^1/_2$d; January 1764, £201.2s.11$^1/_2$d. (42) Few, if any except Gardner, of these people were employed full time. In October 1760, Robert Allan received £10.4s.9d. for 175 $^1/_2$ days at 14d. per day, George Taylor £4.19s. for 91 days at 1s. and Andrew Perry £4.2s.11d. for 99 $^1/_2$ days at 10d. McMillan and Gardner must have been apprentices in 1760 as they only received subsistence at 3s. per week. This raises a puzzle over Gardner because his birth has been given as 1734 which would make him older than Watt himself and certainly very old to be apprenticed. (43) Lack of a Burgess ticket neither stopped Watt trading nor engaging apprentices. For example, James Dunshie was 'to learn the art and business of instrument-making for five years'. (44) Presumably Watt employed these other people on either a daily basis or some form of contract to carry out particular tasks or to supply particular parts. This is supported by Muirhead when he wrote, 'One journeyman throughout the year, and three or four others, from time to time as occasion required, were all that Mr. Watt at first found it necessary to employ'. (45) But the records do not agree with the number of sixteen which Muirhead says Watt employed in 1764, for they show that the numbers had already begun to decline in that year.

Where the account books give any details of the work for which these men were paid, we can see it could be quite varied. Allan had made some unspecified parts for 42 quadrants before October 1760 but then in 1761 and 1762 was making fiddles and fiddle bows, guitars and a couple of base guitars. James Couper was mostly engaged on quadrants including an 'ibionie' one, but he also made a set of organ keys. James Lyme was paid for three dozen quadrants and Alex Gardner for glass grinding. John Picken and John McLean apparently specialised in flutes. Gardner was making guitar plates and keys for flutes in 1762 as well as other parts for these instruments right through into 1765. By that time he was out of his apprenticeship and helped Watt with his steam engine experiments. He became Watt's chief technician and was responsible for Murray Osborne, an apprentice, on whose behalf he was charging Watt 2s.6d. a week for providing 'bed, bord and washen'. These two worked on the engine at Kinneil as well as some of Watt's atmospheric engines. Gardner took over the scientific instrument business and was definitely running it independently by 1773 as shown by an advertisement in the *Glasgow Journal*. (46) Gardner's family continued with the business until well after 1900, which indicates that Watt must have developed a viable concern in this part at least of his activities. Osborne also set himself up as a mathematical instrument maker. James Sym junior, whose father had been employed as a journeyman by Watt, was in 1801 a third instrument maker in Glasgow so that three out of the five then practising there had been trained either directly or indirectly by Watt. (47)

The Workshop

At the beginning of the partnership with Craig, Watt maintained his workshop in the College precincts but must have moved it elsewhere when he had to leave in 1763. In his Waste Book beginning in November 1765 and continuing into March 1769, (48) Watt divided his accounts into two parts, 'Shop in Trongate' and 'Workhouse'. The Workhouse is thought to have been in or close to King Street, 'in a little court [at the] north end of the beef market'. (49) Tools mentioned in the 1759 Inventory include the great turning lathe, a hand lathe and a large and small turn-bench. As well as the usual hand tools such as hammers, files and pliers, there were screw plates and dies, a 'milling tool', draw plates, diamond for cutting glass, a round dividing plate, a fluting mill, as well as printing plates for compass cards and his book on the Hadley's quadrant.

Watt's reputation had spread beyond Glasgow and reached Moffat by August 1765, when he received an inquiry from James Tait.

> Tho' I have not the pleasure of being acquainted, yet I am no stranger to your reputation for ingenuity: as I have heard many Gentlemen make mention of you in the most genteel manner, on that Head - I have seen some of your Drawing Machines the workmanship of which I am satisfy'd with -
> The intention of this letter, Sir, is to know whether you work in all the Different branches of the Mathematical Instrument maker way - my reason for desiring this information is that I make use of several Instruments of various kinds in my business... - getting small things from London is frequently attended with inconveniency... whereas less time would be lost in that way to Glasgow. I must in particular beg to know whether or not you

> divide (I don't mean if you can, that I am persuaded of) but whether you have the proper apparatus for dividing upon any radius that might be wanted, as I have a pretty curious instrument to be made shortly and should be glad to have it done by you.
> I also want one of yr drawing machines. (50)

Amongs the instruments which Tait wanted was a 9 ins. circular protractor, (51) so we can be sure that Watt's workshop was equipped with some form of dividing apparatus for circles which has not survived. The workshop was reasonably well fitted out for general jobbing tasks as well as more specialised work on small instruments. Supplies of materials were obtained through contacts across the country. For example, barometer tubes came from Bristol (52) and Newcastle upon Tyne, lignum vitae from his father, boxwood from London (53), and sheet brass from his brother in Bristol. (54)

The Products

At the beginning of the partnership, production and sales continued along the same lines as previously. The 1759 Inventory lists quadrants, Gunter's scales, a wide variety of straight and folding rules, sectors, compasses, pencils, box and ring dials, globes, telescopes, prospects, microscopes, burning and reading glasses and magic lantern glasses. (55) This was still the stock-in-trade of a scientific instrument seller and obviously much of it must have been purchased by Watt on his recent trip to London. But he may have begun to realise that he could not make every type of instrument himself. For example, in October 1760, he wrote to his father that he had no telescopes but would commission some. (56) While he might have had telescopes made by one of his lads, in April 1761 Watt intended to order compasses from London (57) and in August he bought dividers, compasses and Gunter's scales from one of his regular London suppliers, Stedman. (58)

Production and repair of quadrants continued, as we have seen above. Some were made from mahogany (59) and others from green or black ebony. (60) The only indication we have of the scale of their production, or indeed any of Watt's lines, is in a letter to his brother at Bristol in 1761. Jockey had a couple of quadrants which were samples to be sold at 25s. for the standard model and 26s. for the green ebony one.

> But would not chuse to sell at that price unless to those that would take 4 dzn in a year At least if you can get any body to sell them in Bristol at 5 pr ct commission & sell a little cheaper than others in the place I think it would do but would not chuse that way if you can sell them in wholesale. I think the whole quantity I shall be able to make this year will be about 14 Doz of which I can sell about 4 Doz here at a higher price. (61)

Not only do we see Watt being willing to sell at a lower price for bulk orders but also that he was willing to undercut his competitors. Equally interesting is the figure of fourteen dozen quadrants which he could produce in a year. Of course we do not know whether he achieved this in either that year or any others, yet it does show the potential scale of his business.

Outlets for quadrant sales widened. There was still his father at Greenock who had around eighteen from 1760 to 1762 as well as rules, dividers, compasses and Gunter's scales. (62) While McDowall took some in 1760, this outlet ceased by the beginning of 1762 but then resumed again a year later. (63) Also mentioned as receiving quadrants were Capt. J. Campbell and John Forsyth, both of Greenock, the latter taking six of the Davis model. (64) Other customers were Willison at Port Glasgow, (65) Jockey and Lean at Bristol (66) and Messrs. Sibbale and Houlston at Liverpool. (67) In 1762, a dozen quadrants were consigned on account for Charles Cunningham to sell in Jamaica (68) as well as Simple in Plymouth. (See p67) There were other sales where the names were not recorded. We are unable to determine when production ceased from the remaining account books. On the first of January 1771, Watt noted in his journal that he had finished a quadrant (69) and that on 26 August 1773 he divided another. (70) These must have been special instances because, by that time, he was well launched on his career as a civil engineer.

Work for the University continued since Watt was paid £5.5s. in June 1760 for repairs to instruments which could have been in conjunction with the appointment of Wilson as Professor of Practical Astronomy and the opening of the observatory. (71) We will investigate later Watt's repair of the Newcomen engine model for Anderson but here merely note that Watt's experiments on it were carried out during his partnership with Craig.

The Perspective Machine

During this partnership, Watt's noted ingenuity began to expand from adapting tools to improve production into improving instruments themselves. The first of these of which we have any knowledge was his 'perspective machine'. (see p109) Lind brought from India a machine invented there probably by a Mr. Hurst. (72) It consisted of a vertical board mounted on a tripod. At the end of an arm was an eyepiece through which could be observed the object or scene to be drawn as well as the tip of a pointer. The operator moved the pointer to cover the outline of what he wanted to draw. The pointer was connected to a series of arms on which was mounted a pencil to make the actual drawing on paper placed on the board. Hurst's machine was made mostly of brass and so was very heavy. Watt

> wished to make a machine more portable and easier in its use; and, at the suggestion of my friend, Mr. John Robison, I turned my thoughts to the double parallel ruler, an instrument then very little known, and not at all used that I know of. After some meditation, I contrived the means of applying it to this purpose, and of making the machine extremely light and portable. (73)

Among other improvements, he made the legs of the tripod from tinned iron, tapered so that they slid into each other for carrying and turned into a walking-stick. The board became a box which could be opened to form the mount for the paper and the various arms. These arms, which were wooden, folded away inside the box when not in use so the whole could be put into a great-coat pocket. Surviving pieces of these machines in the Garret Workshop indicate that some form of batch production was employed. This is confirmed in a letter to Lind in which Watt said they would be put 'out of sorts' if he had to make a bigger one. (74) Robison's suggestion of the double parallel ruler should be noted because this perspective machine could have been the source of inspiration for Watt's later parallel motion on his rotative steam engines.

Watt said he made these improvements around 1765, but it must have during the autumn of 1764 because he sold a 'Drawing machine' to Black in October 1764. (75) A bill for 14s. sent by Urie for printing directions for perspective machines is dated 19 December 1764. (76) The leaflet consisted of four pages and 100 were printed. Black was charged £2 for his machine and, in 1765, George Clerk Maxwell £2.2s.2d. Lind himself asked for one in April 1765. (77) Watt says that he sold between fifty and eighty but it was soon copied in London by Adams (78) who could have developed batch production techniques to manufacture them more cheaply. When he was in London in the spring of 1767, he wrote to Peggy that he had 'got a person to take all my perspective machines which I think a piece of good luck'. (79)

Watt's perspective machine.

Presumably he meant all those which were finished at that time because, to meet what must have been a later order, twelve were sent away in August 1768. (80) Sketches survive which Watt made with one during his survey of the Crinan Canal in July 1771. Just after that he tried some improvements by using a diagonal mirror and a telescope (81) and a little later a horizontal board (82) but these were not successful although he continued with modifications until the end of 1773. (83) He sold some to Alexander Cumming in 1775 (84) and he received £2.12s.6d. from a Mr. Wallin in 1779 (85) which is where the story really ends. However in 1814, Brewster invited Watt to contribute a description of it for the *Edinburgh Encyclopaedia*. (86)

Musical Instruments

Another new business venture which occurred during the Craig partnership involved the production and sale of a wide variety of musical instruments. This move on Watt's part was quite extraordinary because he 'did not know one musical note from another' and 'had an absolute deficiency of any musical ear'. (87) Muirhead gave a colourful description:

> Succeeding beyond expectation in his first attempts in that novel line, it is wonderful how many dumb flutes and gouty harps, dislocated violins and fractured guitars, nervous viol-di-gambas, hysterical mandolins, and thorough basses suffering from hoarseness, came thenceforward to be cured by him of their complaints, and restored to health and harmony. (88)

While we have not discovered a harp, fifes and even a bagpipe could be added to this list. Such instruments would have needed the same manufacturing techniques as the common scientific instruments of those days, most being principally of wood with a few metal parts. Wooden quadrants have been mentioned; cases for barometers were wood; rules would have been made from wood and in the Garret Workshop there is a wooden tube for a telescope.

Watt may have diversified into musical instruments after being asked to repair some and then realising he could manufacture them. In October 1757, there is an enigmatic entry in his Waste Book when he charged the Rev. Mr. James Graham for perpetual screws, keys and adjusting screws. (89) Our earliest definite reference to musical instruments is on 12 April 1760 for mending a couple of flutes, a bagpipe and a fife. (90) That August, another flute was repaired and a fife sold. (91) In December 1760, Watt supplied four screws 'to Double Bass for Musical Society' through John Campbell (92) for whom a year later he mended two flutes. (93) In the meantime, during the spring of 1761, Allan was paid for mending a flute and making three fiddles. In the following summer and autumn, he made ten guitars. (94) By that time, the musical instrument business was firmly launched, possibly through further investment by Craig of £91, some of which was used to purchase brass wire, varnish, brushes, harpsichord wire and wood for guitars. (95)

Fiddles

Although Cumming would ask in December 1761, 'Pray goes fiddle making on', (96) fiddles were probably never a large section of Watt's production. Some were sold in 1762 (97) but in 1763 Watt bought a batch in London (98) as well as 28 different fiddles and 36 bows costing £135.12s. from Molewater & Enslie in Rotterdam. (99) We have a record of one being sold in 1765 (100) and another in November 1766 to James Ferguson, who did not have the forty shillings at the time but offered to pay double that in May 1791 when he made the only complaint we have about any of Watt's instruments, that 'She did not turn out a good one, which made me part with her'. (101)

Flutes

In contrast, sales of flutes are located on many pages of Watt's books and he must have developed a considerable trade. This justified the construction of special tools such as the rotary gouge for cutting out three swells together. (See p85/Plate 7) Also among the small tools in the Garret Workshop there is a stamp with the letters T. LOT. This is the name of a well-known Parisian flute maker of the mid-eighteenth century and it casts a serious question mark over some of the ethics of Watt's marketing policies. (102) There are also stamps for J WATT and GLASGOW.

The first intimation we have that Watt was considering manufacturing flutes comes in a letter to Jockey at Bristol in March 1760 when he asked,

> If you have time shall be obliged to you to make enquiry if I could sell some flutes at Bristoll. I can afford the best box flutes at 7/- the 2nd sort at 6/- & shall uphold them good. (103)

This sounds as if Watt were just considering launching out into this field. He sent his brother five fifes in April (104) but it was not until over a year later that we have a mention of

> a Dozn German Flutes to sell as markets rule in the place not under 5/- pr ea at that price unless in Quantity you may try to get somebody to sell them on Comm[ission] & speak to the masters that teach it to recommend them for which you may pay them 2/6 or 3/- for each they can procure the sale of not under 9s & you can get them sold for exportation or otherwise I can supply any demand You'll observe that these are all ready money prices. (105)

This letter is significant as it shows Watt's attempts to introduce a new product. He must have already planned his production methods and was confident of meeting the demand.

Towards the end of the summer in 1761, Watt made a trip through the western side of England which is the only occasion of which we know when he was definitely trying to sell his products rather than purchase goods for his shop. He set off in the middle of August possibly for London; at least his Cash Memorandum Book notes a Mr. B. Duke of 'Holburn' ordering 18 flutes. (106) He returned by Bristol, Bath, Gloucester, Worcester, Birmingham, Wolverhampton, Shrewsbury, Chester and Liverpool. He saw Jockey in Bristol but then near Gloucester his horse broke down and he had to buy another. (See p38) Payment for this new mare left him without sufficient money to get home so he had to draw a bill on his brother which he asked to be kept secret from their acquaintances. (107)

Watt took orders for 48 flutes at Bristol, around 62 at Bath, 18 at Worcester, 56 at Shrewsbury and some quadrants at Liverpool. Some of these orders were despatched in April 1762 on the ship *Concord*. (108) Of these contacts, Charles Crump at Bristol at least placed further orders. (109) Other people who purchased Watt's flutes were Cumming in London, (110) John Campbell in Glasgow, (111) Robert Home in Edinburgh, (112) Charles Claggett in Dublin (113) and even Boulton in Birmingham who took a dozen in 1768. (114)

A major customer was Neil Stewart, a music seller in Edinburgh, who purchased 126 between September 1762 and March 1766 at a cost of £28.7s. (115) He seems to have been an important contact, having outlets also in Dublin. He supplied Watt with suitable wood for fiddles and guitars. (116) A letter sent to him in May 1766 suggests when Watt ceased production. 'It is not in my power to supply you with any more flutes than those at Edinr. & had it not been for our former dealings I could not have sent these having few enough for my own retail'. (117) While some sales of other flutes are mentioned, the accounts are too fragmented to draw any conclusions either about the scale of production or its profitability. This business may have been drawn to a final conclusion in April and May 1774 when Watt sold J. Aird 'flutes to the value of £22.17s.6d'. (118) As this deal included many parts for making flutes, he was probably clearing off the remainder of his stock.

Guitars

A similar pattern can be seen with guitars. In November 1760, Watt charged for mending a guitar (119) and the following May sold a couple. (120) Allan was paid for making 24 guitars between July 1761 and December 1762. (121) Some sales are recorded to Home in Edinburgh who traded with Watt in various other lines. (122) But the more remarkable sales were to Claggett in Dublin who was invoiced first on 19 July 1761 for three guitars and a viola de gamba. (123) From then through to September 1762, Claggett's bill rose to £133.14s.6d. (124) with a further £30.10s. up to April 1764. (125) Watt sent some guitars to his father for despatch to Dublin with instructions that, if there were no boat sailing immediately, they were to be sent on to Irvine. (126) Claggett ran a theatre, pleasure gardens, a band and some sort of music school in Dublin. He entered into partnership with a Mr. Gibson who was capable of supplying all the necessary instruments and keeping them in repair. But this partnership had broken down a year before Claggett wrote to Watt in September 1765 offering to form a new partnership with him. He proposed that Watt should move to Dublin where he could continue with his instruments and his other trades for there was no other regular shop there in these lines. He also offered to explain to Watt all the improvements which Gibson had made to guitars. (127) Unfortunately, Watt's reply has not survived. This was just after he had invented his separate condenser for the steam engine which could have influenced his decision to turn down Claggett's offer. However he remained friendly with Claggett for many years. In 1776, Claggett patented a form of finger-board for the violin which would assist beginners to play such instruments more easily. (128) There are no references in the archive collections to Watt's involvement with this but in the Garret Workshop are several examples of these fingerboards, possibly made by Watt or one of his journeymen. Claggett asked Watt for help again in 1788 for a further patent on this (129) and also in 1793 about a new type of piano and a mechanical organ which he had invented. (130)

Organs

Watt had made organs himself. Robison described how Watt first became involved in this new venture.

> A mason-lodge in Glasgow wanted an organ. The office-bearers were acquaintances of Mr. Watt. We imagined that Mr. Watt could do anything; and, though we all knew that he did not know one musical note from another, he was asked if he could build this organ. He had repaired one, and it had amused him. He said, 'Yes'; but he began by building a very small one for his intimate friend Dr. Black, which is now in my possession. In doing this, a thousand things occurred to him which no organ-builder ever dreamed of, - nice indicators of the strength of the blast, regulators of it, etc. etc. He began to the great one. He then began to study the philosophical theory of music. Fortunately for me, no book was at hand but the most refined of all, and the only one that can be said to contain any theory at all, - Smith's Harmonics. Before Mr. Watt had half finished this organ, he and I were completely masters of that most refined and beautiful theory of the beats of imperfect consonances. He found that by these beats it would be possible for him, totally ignorant of music, to tune this organ according to any system of temperament; and he did so, to the delight and astonishment of our best performers. In prosecution of this, he invented a real monochord of continued tone; and, in playing with this, he made an observation which, had it then been known,

would have terminated a dispute between the first mathematicians in Europe, - Euler and D'Alembert. (131)

Robison's account is wrong in many details but clearly shows that Watt tackled his organ building in the same way that he would later tackle his experiments on the steam engine. Having gained experience by repairing an organ, he was asked to build one so he studied the background theory by reading the available literature. He became quite proficient in this subject, to the surprise of his acquaintances who looked on him as a sort of 'oracle'. (132) He went on to build a small organ which was an improvement on others.

If Robison and Watt studied the theory of music together, this must have been after March 1762 when the *Merlin* had returned from Jamaica with Robison on board following the trial of Harrison's chronometer, or even later, possibly not until 1764. But the first record we have of Watt being involved with organs is a letter from Cumming dated 27 December 1761. (133) Cumming had been employed by the Duke of Argyll for some years after 1752 to construct models of scientific instruments, steam engines and so on. The Duke was an influential person known as 'The King of Scotland' through his unparalled political power, and was also the leading patron of the University of Glasgow. (134) Cumming, assisted by his brother John, had assembled an organ at Inveraray Castle in 1757. (135) We have no idea how Watt came to know Cumming but they were to become life-long friends since Cumming helped Watt later, first with the patent specification in 1769 and secondly with the patent trials in 1794 when he was a watchmaker in London.

Watt turned to Cumming who replied, 'As I know the situation of a one projecting new things I went directly to an organ Builder to enquire of Him the Prices of different pipes'. (136) This suggests that Watt was only just beginning to build an organ. Cumming gave descriptions and prices of different pipes which may indicate that Watt did not build these himself. He thought that Watt's scheme for a portable organ was very good and considered it an advantage that the organ could be lifted from its frame, confirming that Watt showed his ingenuity in its construction. Cumming said that, as there was a shortage of journeymen organ builders, Watt's commissions might take time to be finished and wished him success. Cumming wrote in September 1762 that he was 'glad you succeed so well in Organ building'. (137) Perhaps the organ had been finished but for whom we do not know.

There is no mention in the accounts of this period about Black paying for an organ. But, on 10 January 1768, Black wrote from Edinburgh, 'Pray when will the Organ be ready - Unless I get it soon I shall be disappointed because I expect soon to have one of my nieces in the house who can play upon it'. (138) Possibly Robison was mistaken about Watt building his first one for Black. In December 1762, there are two entries concerning organs in the accounts, 'To Tuning Mr. Dunlops Organ, £ 3.3.0' and 'Mending the Mason's organ, £12.0.0'. (139) We know nothing more about Dunlop's organ but we find that Watt received £20 on 10 June 1763 'from the Royall Arch Mason Lodge in part of their organ £20'. (140) There was a further £10 in June 1765. (141) These payments must have been for the organ mentioned by Robison and the keys for it may have been made by one of Watt's journeymen, Couper. (142) In March 1764, a Mr. Lockart paid £24 for an organ (143) and Watt may have sold another one at the same time. (144) Watt's journals contain references to what must have been his last involvement with organs at the end of 1770 when he 'looked out sundries of organ'. (145) He gave parts to 'Morison' who had got 'my little organ almost compleated'. Morison had made some new pipes of which Watt did not approve and neither did he approve of springs being substituted for pendulum weights, which were possibly connected with the blowing apparatus. (146) Watt obviously thought his own designs superior.

The Finish of the Musical Instrument Business

In contrast to the scientific instrument business, nobody took over the complete musical instrument business from Watt. The sales trip into England was not repeated and demand in the Glasgow region is unlikely to have proved sufficient to sustain the levels of production initiated by him. The business was drawn to a close as the lads were laid off and the shop closed. We have no clues about whether it was profitable and few indications about the quality of the instruments. It certainly seems far removed from Watt's main interests and this could be one reason for its abandonment.

The Shop

The final venture of the partners was perhaps the most surprising of all, which was their expansion during 1763 into a large shop selling a wide variety of goods. Watt may have relinquished his shop in the College precincts when he was asked to move his workshop and found temporary accommodation in the Saltmarket where his father wrote in November 1763. (147) But then appeared an advertisement in the *Glasgow Journal* for 1 December 1763, 'James Watt has removed his shop from the Sautmercat to Mr. Buchanan's land on the Trongate where he sells all sorts of mathematical and musical instruments, with variety of Toys and other goods'. (148) This was on the north side of the Trongate near the Exchange. (149) The growth in previous years of Watt's businesses must have seemed dramatic, revealing his entrepreneurial skill, with the result that both Craig and Black were willing to make large investments in this shop which branched out into goods in which none of them had ever dealt before. We do not know whether they competed with Carlisle who was selling similar lines in 1755. (150 See p46) Perhaps his shop had closed and Watt followed this earlier precedent for selling scientific instruments alongside a wide range of other items.

To stock their shop, Watt made another journey to London and returned by Birmingham and Sheffield during July and August 1763. He took with him £382.8s. which included £4.15s. for a lottery ticket for Black. Of this money, Black supplied £ 150, James Watt senior £42.10s, John Craig £142.10s. and J. Colville £34.13s. (151) His father's and Colville's money may have been to purchase special commissions for them but Watt had the rest available to spend. His own expenses for the trip were £34.5s.1d. and he took the opportunity to purchase new clothes for himself at £9.9s.6d. (152) In London he spent £92.18s, mostly as might be expected on scientific instruments. His purchases included fiddles, a spinet, barrel organ, clarinet, reflecting telescopes, prospects, reading glasses, reflecting microscope in brass, Dollond's opera glasses, scales, compasses, universal ring dials, electric machines, barometers, thermometers, pencils, Gunter's scales, Napier's bones, chess boards, shagreen cases for knives, battle boards, shuttlecocks, inkstands, pocket ink pots, leather cases, memorandum books, sticks and canes, Chinese crackers and fire wheels, pick tooth cases [sic], straw boxes, fishing rods, spangles, necklaces and earings, etc. At Birmingham he spent £74.2s. on plated steel spurs, corkscrews and snuffers, tweezer cases, steel pencil cases, ladies and men's steel watch chains, trinkets, silver seals, thimbles, nutcrackers and nippers, snuff boxes, buckles, smelling bottles, brass scales and steelyards, brass and iron bag locks, coffee mills, table bells, pistol tinder boxes, guns, pistols, letter punches, spectacles, etc. Sheffield came last with only £27 being spent on best white and stained ivory, wood and box table knives, dessert and carving knives, razors and lancets. (153) We have the names of these suppliers but Boulton is not among them. This list shows the broad range of goods which Watt and his partners offered for sale but it in no way covers everything! Watt reported to his father in October 1763, 'My shop does very well'. (154)

NOTES: THE PARTNERSHIP WITH CRAIG

1 Smith, (1994), Bryden, p. 12 lft. and Dickinson, (1935), p. 28.
2 Robinson & Musson, (1969), p. 28, 'Professor Robison's Narrative of Mr. Watt's Invention of the improved Engine', 1796, see JWP 3/36.
3 B. & W. Col., MIII, 3/1, 'Waste Book', 24 Oct. 1757.
4 Ibid, 30 Sept. 1758 and 18 May 1759.
5 JWP, C1/1, 'Accounts', Inventory, 7 Oct. 1759.
6 Smith, (1994), Bryden, p. 12 lft.
7 Muirhead, (1858), p. 45. 'Craig appearing to have taken no share in the manufacturing part of the business, but only, (as is shown by a memorandum in the Journal), to have been book-keeper to the concern'.
8 JWP, 6/46.6, List of Tools, J. Watt to J. Watt sen., 19 June 1756.
9 B. & W. Col., MIII, 3/1, 'Waste Book', 3 Jan. 1757.
10 B. & W. Col., MIII, 3/3, Ledger, Jan. - April 1757.
11 B. & W. Col., MIII, 3/4, 'Personal Ledger of James Watt, Oct. 1759 - Dec. 1764', f. 14, J. Craig's Account and JWP, 3/21, 'Household Expenses'.
12 B. & W. Col., 3/4, 'Personal Ledger', f. 14, J. Craig's Account and Smith, (1994), Bryden, p. 12 lft. and JWP, 3/21, 'Household Expenses, 1757 - 64'.
13 B. & W. Col., MIII, 3/1, 'Waste Book', 6 May 1759.
14 B. & W. Col., MIII, 3/6, 'Ledger of Personal Accounts', f. 37, J. Craig's Account.
15 JWP, 4/32.31, J. Watt to J. Roebuck, 5 Dec. 1765 and JWP, 4/7.46, J. Watt sen. to J. Watt, 6 Dec. 1765.
16 B. & W. Col., MIII, 3/4, 'Personal Ledger', f. 2, 4 Oct. 1760 and f. 4, 23 Oct. 1762.
17 Ibid, f. 11, Feb. 1760.
18 Dickinson & Jenkins, (1927), p. 19 and Smiles, (1878), p. 33.
19 JWP, Letter Book 4, J. Watt to J. Craig, 30 April 1805.
20 Ibid.
21 Smith, (1994), Bryden, p. 16, Note 72 and Dickinson & Jenkins, (1927), p. 19.
22 Muirhead, (1858), p. 44.
23 Crowther, (1962), p. 111.
24 JWP, 4/7.64, J. Watt sen. to J. Watt, 2 May 1763; 63, 18 Nov. 1763 and 62, 11 Feb. 1764.
25 JWP, C1/36, A. Cumming to J. Watt, 27 Dec. 1761.
26 JWP, 4/11.75, J. Watt to J. Watt sen., 1 Jan. 1762.
27 JWP, 4/11.78, J. Watt to J. Watt sen., 12 Jan. 1762.
28 See JWP, 4/44, note on wrapper by J. Watt jun. Also Bryden, (1972), p. 25, Scottish Record Society, *Burgess and Guild Brethren of Glasgow, 1751 - 1846*, Edinburgh, 1935, p. 59, 'James Watt, merchant, Burgess and Guild brother, as married Margaret, lawful daughter to Daniel Miller, wright, Burgess and Guild Brother, 17.3.68'.
29 Muirhead, (1858), p. 50 and Dickinson & Jenkins, (1927), p. 19.
30 Dickinson & Jenkins, (1927), p. 21 and Dickinson, (1935), p. 31.
31 Muirhead, (1854), Vol I, p. cvii and Smiles, (1878), p. 77.
32 JWP, 4/7.52, J. Watt sen. to J. Watt, 25 April 1765.

33 JWP, 4/11.127, J. Watt to J. Watt sen., 18 July 1765.
34 Ibid.
35 JWP, 4/7.47, J. Watt sen. to J. Watt, 15 Nov. 1765.
36 JWP, 4/32.32, J. Watt to J. Roebuck, see Muirhead, (1854), Vol I, p. 12, Letter 13, gives date as 4 Dec. 1765 but it is probably later.
37 JWP, 4/11.137, J. Watt to J. Watt sen., 8 Sept. 1766.
38 JWP, C1/1, 'Accounts', f. 11, 4 Oct. 1760.
39 B. & W. Col., MIII, 3/7, 'Journey Mens Book, Oct. 1760 - April 1764'.
40 B. & W. Col., MIII, 3/8, 'Journey Mens Book, 5 Jan. 1764 - June 1771'.
41 Ibid, see under John Gardner and A. Morrison Low, 'Scottish Barometers', a talk prepared for a meeting of the Royal Meteorological Society Historical Group, 8 March 2000.
42 JWP, C1/1, 'Accounts', f. 11, 4 Oct. 1760; f. 27, 23 Oct. 1762; f. 69, 1 Jan. 1764.
43 Clarke, (1989), p. 164 and D.C. Soutter, 'Additional information relating to John Gardner, senior (1734 - 1822) Maltman, scientific instruments maker, town surveyor to the city of Glasgow', typescript Glasgow 1976.
44 Bryden, (1972), p. 27.
45 Muirhead, (1858), p. 45.
46 Clarke, (1989), p. 170.
47 Bryden, (1972), p. 30.
48 B. & W. Col., MIII, 3/2, 'Waste Book', various entries.
49 Muirhead, (1858), p. 50.
50 JWP, 3/41.2, J. Tait to J. Watt, 6 Aug. 1765.
51 JWP, 3/41, J. Tait to J. Watt, 18 Aug. 1765.
52 JWP, C1/1, 'Accounts', f. 19, 24 Oct. 1761.
53 Smith, (1994), Bryden, p. 12 lft. and B. & W. Col., 3/4, 'Private Ledger', f. 12, 'Hunter & Miller, merchants in London', 8 May 1760, Boxwood, £ 19.19s.11d.
54 JWP, 4/11.57, J. Watt to John Watt, 3 March 1760.
55 JWP, C1/1, 'Accounts', various entries.
56 JWP, 4/11.53, J. Watt to J. Watt sen., 24 Oct. 1760.
57 JWP, 4/11.58, J. Watt to John Watt, April 1761.
58 JWP, 4/11.71, C. Stedman to J. Watt, 31 Aug. 1761.
59 JWP, 4/11.79, J. Watt to J. Watt sen., 13 Jan. 1762.
60 Ibid and JWP, 4/7.66, J. Watt sen. to J. Watt, 9 March 1763.
61 JWP, 4/11.59 J. Watt to John Watt, 15 May 1761.
62 JWP. 4/11.82, J. Watt to J. Watt sen., 13 Feb. 1760 to 1 Oct. 1762.
63 JWP, C1/1, 'Accounts', f. 7, 21 Feb. 1760 and JWP, 4/11.79, J. Watt to J. Watt sen., 13 Jan. 1762 and B. & W. Col., MIII, 3/4, 'Private ledger'.
64 B. & W. Col., MIII, 3/4, 'Private Ledger', various entries.
65 JWP, C1/1, 'Accounts', f. 7, 23 June 1760.
66 Ibid, f. 19, 24 Oct. 1761.
67 B. & W. Col., MI, 2/1, 'Cash Memorandum of J. Watt, Suppose 1761'.
68 JWP, C1/1, 'Accounts', f. 20, 1 Jan. 1762.
69 B. & W. Col., MI, 1/16, 'Journal', 1 Jan. 1771.
70 B. & W. Col., MI, 1/20, 'Journal', 26 Aug. 1773.
71 Dickinson & Jenkins, (1927). p. 19 and Coutts, (1909), p. 265.
72 Robinson & McKie, (1970), p. 433, J. Watt's Note Book, JWP, W/14.

73 Muirhead, (1854), Vol I, p. cxv and pp. cxi - cxviii and Muirhead, (1858), pp. 51 - 6.
74 Muirhead, (1854), Vol I, p. 3, Letter 1, J. Watt to J. Lind, 29 April 1765, JWP, C1/15.
75 B. & W. Col., MIII, 3/6, 'Ledger of Personal Accounts', f. 5, J. Black's Account, Oct. 1764.
76 B. & W. Col., MII, 8/7.1, Invoices from R. Urie.
77 B. & W. Col., MIII, 3/6, 'Ledger of Personal Accounts', f. 42, G. Maxwell's account and JWP, C1/15, J. Lind to J. Watt, 29 April 1765.
78 Muirhead, (1854), Vol I, p. cxvii.
79 JWP, 4/4.20, J. Watt to Margaret Watt, 11 April 1767.
80 JWP, 4/4.16, Margaret Watt to J. Watt, 9 Aug. 1768.
81 B. & W. Col., MI, 1/17, 'Journal', 8 Aug. 1771.
82 B. & W. Col., MI, 1/18, 'Journal', 4 Feb. 1772.
83 Muirhead, (1854), Vol II, p. 67, Letter 127, J. Watt to W. Small, 11 Dec. 1773; Boulton 348.38.
84 B. & W. Col., MI, 2/6, 'Cash Memorandum', 7 Dec. 1775.
85 B. & W. Col., MI, 2/11, 'Cash Book, Cornwall', 3 Jan. 1779.
86 JWP, C6/8.112, D. Brewster to J. Watt, 12 June 1814.
87 Muirhead, (1858), p. 46.
88 Ibid, p. 48.
89 B. & W. Col., MIII, 3/1, 'Waste Book', 8 Oct. 1757.
90 JWP, C1/1, 'Accounts', f. 8, 12 April 1760.
91 Ibid, f. 8, 7 Aug. 1760.
92 B. & W. Col., MIII, 3/4, 'Private Ledger', 13 Dec. 1760.
93 Ibid, 13 Nov. 1761.
94 B. & W. Col., MIII, 3/7, 'Journey Mens Book', f. 11, April - July 1761.
95 JWP, C1/1, 'Accounts', f. 18, 24 Oct. 1761.
96 JWP, C1/36, A. Cumming to J. Watt, 27 Dec. 1761.
97 B. & W. Col., MIII, 3/4, 'Private Ledger', f. 4, 28 July & 12 Aug. 1762.
98 B. & W. Col., MI, 2/7, 'Cash Memorandum of J. Watt, Suppose 1763'.
99 C1/1, 'Accounts', f. 57, 1 Nov. 1763.
100 JWP, 6/28.64, Mr. Stanford to J. Watt, 19 Sept. 1765.
101 JWP, 6/28.60, J. Ferguson, 'Bought of J. Watt', 10 Nov. 1766 and JWP, 4/13.28, J. Ferguson to J. Watt, 2 May 1791.
102 I wish to thank Michael Wright for pointing this out to me and explaining its significance.
103 JWP, 4/11.57, J. Watt to John Watt, 3 March 1760.
104 JWP, 4/11.53, J. Watt to John Watt, 22 April 1760.
105 JWP, 4/11.59, J. Watt to John Watt, 15 May, 1761.
106 B. & W. Col., MI, 2/1, 'Cash Memorandum'.
107 JWP, 4/11.62, J. Watt to John Watt, 1 Oct. 1761.
108 JWP, C1/1, 'Accounts', f. 20, 8 Apr. 1762.
109 Ibid, f. 20, 8 April 1762, 16 July 1762 and B. & W. Col., MI, 2/2, 'Cash Memorandum of J. Watt, 1763 - 5', 13 Oct. 1763.
110 B. & W. Col., MIII, 3/4, 'Private Ledger', f. 23, 28 Aug. 1762.
111 Ibid, f. 12, 12 Jan. 1763.

112 B. & W. Col., MII, 5/3, R. Home to J. Watt, 7 June 1766.
113 JWP, C1/1, 'Accounts', f. 25, 20 Aug. 1762.
114 Boulton 348.2, J. Watt to M. Boulton, 12 Dec. 1768.
115 JWP, 6/28.86, J. Watt to N. Stewart, 15 May 1766.
116 JWP, 6/28.100, N. Stewart to J. Watt, 29 Dec. 1762.
117 JWP, 6/28.86, J. Watt to N. Stewart, 15 May 1766.
118 B. & W. Col., MI, 2/6, 'Cash Memorandum', 2 May 1774.
119 JWP, C1/1, 'Accounts', f. 12, 3 Nov. 1760.
120 Ibid, f. 12, 25 May 1761.
121 B. & W. Col., MIII, 3/7, 'Journey Mens Book', f. 11.
122 B. & W. Col., MIII, 3/4, 'Private Ledger', f. 17, 2 Aug. 1762.
123 JWP, C1/1, 'Accounts', f. 12, 19 July 1761.
124 B. & W. Col., MIII, 3/4, 'Private Ledger', f. 15.
125 Ibid, f. 6, 4 April 1764.
126 JWP, 4/11.87 & 88, J. Watt to J. Watt sen., 4 & 10 May 1762.
127 JWP, C1/38, C. Claggett to J. Watt, 30 Sept. 1765.
128 Patent 1140, 7 Dec. 1776.
129 JWP, 4/64.2 - 9, C. Claggett to J. Watt, Jan. - Aug. 1788.
130 JWP, 6/23.19, C. Claggett to J. Watt, 15 Aug. 1793.
131 Muirhead, (1858), p. 47 and Robinson & McKie, (1970), 'Memorial by Dr. J. Robison', pp. 258 - 60.
132 Muirhead, (1958), p. 48.
133 Sobel, (1996), p. 121 and JWP, C1/36, A. Cumming to J. Watt, 27 Dec. 1761.
134 Donovan, (1975), p. 96 and *Country Life*, 12 June 1969, M. Cosh, 'Clockmaker Extraordinary, The Career of Alexander Cumming'.
135 Lindsay & Cosh, (1973), p. 88.
136 JWP, C1/36, A. Cumming to J. Watt, 27 Dec. 1761.
137 JWP, 4/87.2, A. Cumming to J. Watt, 8 Sept. 1762.
138 Robinson & McKie, (1970), p. 8, Letter 1, J. Black to J. Watt, 10 Jan. 1768.
139 JWP, C1/1, 'Accounts', f. 28, 10 Dec. 1762.
140 Ibid, f. 33, 8 June 1763, see also B. & W. Col., MIII, 3/4, "Private Ledger", f. 32.
141 B. & W. Col., MI, 2/3, 'Cash Memorandum', 7 June 1765.
142 B. & W. Col., MIII, 3/7, 'Journey Mens Book', f. 8, Sept. 1763.
143 B. & W. Col., MIII, 3/4, 'Private Ledger', f. 6, 7 March 1764.
144 B. & W. Col., MI, 2/2, 'Cash Memorandum', 10 March 1764, from ?&ie? £24.10.0.
145 B. & W. Col., MI, 1/16, 'Journal', 19 Dec. 1770.
146 Ibid, 25 Feb. 1771,
147 JWP, 4/7.63, J. Watt sen. to J. Watt, 18 Nov. 1763.
148 Dickinson & Jenkins, (1927), p. 28.
149 B. & W. Col., MII, 5/2, 'Receipts', 6 Nov. 1764, J. Buchanan received £10.10.0 for half year's rent.
150 Bryden, (1972), p. 19.
151 JWP, 3/21, 'Household Expenses 1757 - 64'.
152 Ibid.
153 B. & W. Col., MI, 2/7, 'Cash Memorandum'.
154 JWP, 4/11. 105, J. Watt to J. Watt sen., 13 Oct. 1763.

FROM MERCHANT TO ENGINEER, 1765 – 1774

Watt's Life and Personality, 1765 - 1774

In the period after the death of Craig in 1765 until he moved to Birmingham in May 1774, Watt changed the direction of his career from that of instrument maker, production engineer and merchant to being a chemical, civil and mechanical engineer, thus broadening his knowledge and interests even further. Of course he would not have segregated his activities into these separate compartments since they overlap each other but, in order to understand his ambitions and achievements, it is convenient to investigate them individually. While his merchanting interests seem to have been abandoned completely, his experiences in the management of men and money must have been useful to him later in Birmingham. Watt had to assume financial control during a crisis in 1782 which almost ruined his partnership with Boulton. We will assess Watt's own financial position during the last years of his residence in Scotland when we examine later the arrangements between Watt, Roebuck and Boulton concerning the steam engine patent of 1769. (See p446)

Watt had already shown mechanical ingenuity in his production engineering devices which broadened into inventions in other spheres, culminating with the separate condenser in 1765. Until that time, he was involved only with small scale engineering associated with mathematical, scientific and musical instruments. He retained an interest in the first two for the rest of his life, continuing to make minor improvements to them, but experiments on steam engines led him to pioneer investigations into new fields of scientific research. His model Newcomen engines were the largest objects he constructed up to 1765 and he must have quickly realised the necessity of erecting full-sized engines to exploit his separate condenser and reap any financial rewards. He formed a partnership with another Scottish engineer, Robert Mackell, to build Newcomen pumping engines and it is most likely that Mackell helped to launch Watt into his brief career as a civil engineer.

According to Smiles, Watt set up as a surveyor in the summer of 1766 and took an office in King Street (1) but this cannot be confirmed. However he did employ one Adams as a clerk at Christmas 1771 (2) and regularly worked with another surveyor, James Morison, an Englishman. This practice as surveyor or civil engineer brought Watt into a different circle of people with higher social status. He had to meet the projectors of the various canal or similar schemes who might include the Lord Provost of Glasgow and other worthies of that city, Cathcart and other Commissioners of Police or the Forfeited Estates as well as of course other engineers like Smeaton or John Golborne of Chester. He became a Burgess and Guild Brother in March 1768, claiming his right through his father-in-law. (3) His enhanced social position was matched by the purchase of a chaise in 1773. (4) All this is certainly at variance with Muirhead's picture of him leading a 'retired course of life, which from choice as well as necessity he appears to have followed'. (5) Watt was definitely ambitious, at least at this period of his life.

Yet, apparently for the first time, we find a weakness appearing in Watt's character, which was his inability to deal with men and face up to opposition from them or arguments with them. This will be examined in greater depth when looking at his period as resident engineer constructing the Monkland Canal. It would turn him from being a directing into a consulting engineer and would also influence the way he developed his steam engine business. It is surprising that there were no signs of similar problems with the men he employed in his instrument business with one exception of James in the shop. Robison described Watt as 'modest, timid, easily frightened by rubs and misgivings and too apt to despond'. (6) From that assessment, we can understand why Watt was afraid of tackling people. How much this was caused by his continuing ill-health, and how much this fear augmented the ill-health, is difficult to determine. Darwin in his first letter to Watt after meeting him in 1767 wrote, 'Now my dear new Friend, I first hope you are well, and less hypochondriacal'. (7) Darwin had quickly summed up his new acquaintance.

Assessment of Watt's Health

In 1771, Watt declined the offer of a position in Russia to supervise an iron foundry. In his letter of refusal to their Ambassador, he pleaded ill-health,

> A bad constitution which was rendered worse by a sedentary life has hitherto prevented me from making the proper advantage of that knowledge I had purchased at a great expense of time and money... Besides my ignorance of the business required, I am when afflicted by bad health, (which occurs but too often) timid & irresolute & perfectly incapable of struggling with the prejudices which all mankind have for their own opinions. (8)

Here Watt anticipates Robison's assessment of himself as timid and irresolute. His ill-health may have been caused not only through his childhood weaknesses but also by spending too much time indoors working on scientific instruments and chemical experiments. We have seen how he was interested in mechanics and chemistry from his youth (See p41) and he told the Ambassador that 'I have dabbled too much in Chemistry to imagine that I could immediately succeed in so difficult a process [as iron founding]'. (9) Watt had his own workshop in the garrets of his various houses. His ill-health could have been caused partly through experimenting in these places where working conditions would have been poor to say the least. He might have been breathing noxious fumes from the chemical reactions but his symptoms were not those of mercurial poinsoning. Muirhead commented, 'He was gradually led more frequently to foresake the solitary vigils of his work-shop in the city for the active labours of his profession of a civil engineer'. (10) In addition, Watt was extremely busy at this time, with a young family, his shop, the remnant of his scientific and musical instrument businesses, involvement with the Delftfield pottery and all the experiments that this entailed, experiments on the separate condenser as well as launching out into a new career of civil engineer.

His career as civil engineer brought both disadvantages and benefits regarding his health. Some of his surveying commissions had to be carried out in adverse Scottish weather conditions. He told Roebuck in September 1769 that 'I got only two days' surveying before I was drove home by the rainy weather, which has also somewhat hurt my health'. (11) He continued to suffer from headaches which often prevented him from working and delayed production of his reports. For example, while helping to survey the River Clyde for Golborne and starting as Resident Engineer to build the Monkland Canal in 1770, he had headaches on 15 June, 5 and 6 July, a violent headache on 17 July, was 'much fatigued' on the 19th. and on the 25th. had another 'headache, did very little all day'. (12) There is little doubt that these may have been caused partly by overwork and strain through trying to do too much, possibly to reduce his debts.

Yet early in 1770, he reported to Small that, 'The time has not been thrown away, for the vaguing [sic] about the country, and bodily fatigue, have given me health and spirits beyond what I commonly enjoy at this dreary season'. (13) He informed the Russian Ambassador, "The open air & exercise have rather mended my health but it is still at times very bad". (14) By the end of 1771 he recognised that the change to being a civil engineer had brought about considerable improvement in spite of the problems he had to face.

> I must say that my present life is a life of much vexation, besides bodily fatigue, of hunger, cold, wet feet, etc., which I could not endure had I the least of gout, the gravel, or many other diseases. I don't know how it is, but I think my health rather better in these gloomy months of December and November than it was in summer. (15)

While Watt's constitution endured all these privations, the headaches persisted, causing him to be frequently incapacitated throughout this period. Without any workshop or shop to worry about and away from the problems of constructing the Monkland Canal, Watt enjoyed time spent surveying in the field so that his health began to improve. Watt junior commented after his father's death, 'The organic defect appeared to be in his digestion and its effects were intensely severe sick headachs, but by continual temperance and good management of his constitution, which he treated with much medical skill, it improved as he advanced in years'. (16)

Watt's Need for Moral Support

It is through his correspondence with Small that we learn more about Watt's private views during this period of his life than any other. Their first meeting occurred when Watt was returning from London in the summer of 1767 after the failure to secure an Act for a canal between the firths of Forth and Clyde. Small was born in 1734 at Carmylie, Angus, where his father was the minister. He presumably trained as a physician in Scotland before taking up the post of professor of mathematics and natural philosophy in the college of Williamsburgh, Virginia. He settled there for some years, practising also as a physician, but the climate did not suit him and he had to return home. He moved to Birmingham in 1765 where he became Boulton's doctor on strong recommendations from Benjamin Franklin. James Keir, a Midlands chemical manufacturer, described him as,

> A gentleman of very uncommon merit... who, to the most extensive, various, and accurate knowledge in the sciences, in literature, and in life, joined engaging manners, a most exact conduct, a liberality of sentiment, and an enlightened humanity; being a great master in the exact sciences, and seeming to carry their regularity and precision into his reasonings and opinions on all other subjects. (17)

Small and Watt were nearly the same age and shared many similar interests particularly in natural philosophy so that they struck up an instant friendship.

Throughout his life, Watt needed people to whom he could turn for moral support. One of the reasons he gave their Ambassador for declining the offer of a job in Russia was 'that I am now in my native country in the middle of my friends and connections'. (18) Jockey was possibly the first person in whom he could confide. Robison was the next deep friend, but he was often away from Glasgow. Peggy was obviously another but, even though she was his wife, she may not have been accepted quite as a confidante. Miss Jane Campbell wrote of the help he received from Peggy as

> having added to his enjoyment of life, and as having had the most beneficial influence on his character. Even his powerful mind sank occasionally into misanthropic gloom and from the pressure of nervous head-aches, and repeated disappointments in his hopes of success in life, Mrs. Watt from her sweetness of temper, and lively, cheerful disposition, had power to win him from every wayward fancy; to rouse and animate him to active exertion. She drew out all his gentle virtues, his native benevolence, and warm affections. (19)

William Small,
1734 - 1775.

But it was Small to whom he poured out his innermost feelings in long letters and, after Peggy's death in 1773 followed by that of Small in 1775, it was Boulton who became the perfect supportive partner.

Watt's Ambitions

The death of Craig in 1765 and the need to repay his loans may have brought home to Watt his perilous financial position and lack of real success in his business enterprises. The misanthropic gloom mentioned by Jane Campbell may have been compounded by ill-health and headaches, caused partly by overwork and stress brought on by his business problems. Any ambitions to become an instrument maker of the calibre of Short or Ramsden were never fulfilled and it is impossible to tell when he recognised that he would never achieve great success in this profession. His venture into the merchant's shop does not seem to have been a great success either and at this period his first really important invention, the separate condenser, had to languish at Kinneil waiting for someone to provide the finance necessary to bring it to a successful conclusion. In these circumstances, Watt turned to civil engineering because the financial rewards appeared much greater, and possibly more easily gained. In

1771, he held out great hopes for his new career, provided that he had good health, when he wrote to Small,

> The remainder of my time is taken up partly by headaches and other bad health and partly by consultations on various subjects, of which I can have more than I am able to answer, and people pay me pretty well. In short, I want little but health and vigour to make money as fast as is fit...
> Now doctor, if you and your friend Hygeia can impart to me these blessings, I may be a rich and happy man. (20)

His new career was proving a success even though Muirhead commented that 'the toils of his business were severe and the profitable returns small'. (21)

This success may explain why Watt turned down the offer of the job in Russia which had come through Robison. Cathcart was British ambassador in Russia at that time and may have had some influence in the background. Robison had moved there at the request of Admiral Sir Charles Knowles who had accepted a post as president of the Russian Board of Admiralty. (22) Russia was at war with Turkey and needed to develop their armament industries. Robison wrote to Watt in the spring of 1771,

> ... in order to see how you would relish the Scheme of coming here in quality of Master Founder of Iron Ordnance to Her Imperial Majesty. I made no hesitation in recommending my friend, because I was well acquainted with your knowledge in Metallurgy and Mechanics, and your intimate acquaintance with all the process as carry'd on at Carron... I am not as yet able to tell you what may be your encouragement, but I am sure her Majesty's Generosity will make it very suitable to Her sense of the importance of the Employment. (23)

Watt was surveying a canal when the Ambassador's letter with its offer arrived in July so he could not reply until early August.

> In this case I think he [Robison] is mistaken as to myself the only virtue I pretend to is honesty, that, leads me at present to trouble your Excellency with a more particular account of myself & of my reasons for refusing an offer that to many people in my station would be very ['tempting' crossed out] pleasing. (24)

We do not have any idea of the salary Watt was offered but obviously he found it was 'tempting'. He also told the Ambassador

> Add to all this... that I have as much employment as I can properly perform that the Public works in this country are daily encreasing & that I have a rational prospect while I behave myself properly, of making as much money here as I could reasonably ask tho I was willing to leave my country, I crave your Excellencys pardon. (25)

He declined a similar offer a few years later.

Although he turned down these offers, Watt hoped to make his fortune from his inventions and career so that he could enjoy some form of retirement but this always eluded him while in Scotland. He must have discussed this with Roebuck and Roebuck's secretary, George Jardine. Jardine was trying to rouse Watt to action over his steam engine in 1768.

> You are surely very near to something that will be much to your advantage: the happiness, or at least the interest of your family, - your own ease and amusement, - that life of ingenious indolence which you have figured out for yourself, are all within prospect: - not to mention the honour of a discovery of so much importance which few would think so moderately of as yourself. (26)

In the autumn of 1771, Watt must have had another period of depression so that Small had to encourage him with the prospect of a better fortune if he stuck to his career.

> I beg you may reconcile yourself to engineering in the vulgar manner; it will by degrees give you both reputation and fortune, and even sell your fire-engines. I admire your delicacy about being paid for your genius and industry: pray, what do lawyers and doctors do? (27)

The desire for 'ingenious indolence' appears again when Small was encouraging Watt to move to Birmingham in 1773,

> I have no doubt that after a little time you might have good business in England as an engineer, could you once find an opportunity of giving a specimen of your talents. You might live by inventing only an hour in a week for mathematical-instrument makers. (28)

But Watt constantly complained about his indolence and decline in his inventive powers and told Small at the beginning of 1770,

> I really think that on the whole I grow less inventive as I grow older, but I have the consolation that what I do contrive generally answers better. Today I entered the thirty-fifth year of my life, and I think I have hardly done thirty-five pence worth of good in the world; but I cannot help it. I do as much as indolence and *ennui* permit, and I think myself less indolent than I once was. (29)

Two years later, he carried on similarly, 'I am still much troubled with headaches, laziness, and confusion of ideas; but am not so much addicted to low spirits and despair as I used to be'. (30) This was Watt's opinion of himself but during these years he was involved in a wide variety of enterprises which may have added to his confusion and headaches but certainly not to his laziness. Many of his important inventions for the steam engine still lay in the future.

Watt consistently underrated himself. He told Small, 'I cannot sufficiently thank Mr. B. and you for your good wishes towards me; you have deceived yourselves into a better opinion of me than I deserve; the only virtue I can lay claim to is honesty; that inclines at present to tell you my failings'. (31) He also told the Russian Ambassador,

> I am a person of no great learning... It is now some years since I began to take upon me the profession of a *civil* engineer. The prejudices of my friends have prevailed upon the public to view me in a more favorable light than I really deserve. I have in consequence been entrusted with public business. (32)

We will discover further instances of this modesty in his civil engineering career when he failed to settle the terms of his employment and relied, for instance, on Glasgow Council paying him fairly.

Possibly through this modesty and more certainly through this claimed indolence, most of Watt's inventions were never developed beyond the experimental stage. He admitted this himself when he wrote to Brewster in 1814. 'Another imputation I cannot, however, repel; - with so many new ideas, why did I not complete more of them? *The Spirit was willing, but the flesh was weak.* I was always averse to labour, and never was a mathematician'. (33) In addition, there were so many other claims on his time as well as so many bright ideas that he could scarcely have been able to bring them all to fruition on his own. But, from the very first letter to Lind in April 1765, he seems to have regarded his invention of the separate condenser in a different light and wrote, 'I have now almost a certainty of the *facturum* of the fire-engine'. (34) Later that summer, he told Roebuck, 'I have tried my new engine with good success'. (35) At the beginning of 1768, Watt again reported to Roebuck that the engine, 'doing twice as well as the common, is, I think, absolutely certain from what Mr. Jardine saw; and there is little reason to fear but what it will do all we expected'. (36) After the engine had stood neglected at Kinneil for over two years, Watt still hoped to succeed with it and told Small, 'Notwithstanding my natural despondence, I am convinced that the machine may be made to answer in a very considerable degree, and in more forms than one, but I am by no means a proper person to carry it into execution'. (37) But it was Small who once more roused Watt to carry on with the engine after the death of Peggy and the collapse of his civil engineering career in 1773. Watt had 'lost much of my attachment to the world... I am heart-sick of this country... I am indolent to excess'. (38) Small remained 'solicitous you should convert your discoveries into fame and money as fast as possible' (39) and encouraged Watt to join him and Boulton in Birmingham.

Watt's Family

The years from 1765 to 1774 saw many changes in Watt's personal life. He moved house at least a couple of times. Peggy wrote to him in November 1766 saying, 'I have taken James Martine the upholsterers house... the Garrat is very small and am afraid it will not [seal] you for there is no fireplace there is some to sell in the next close but I will not take any till you come home as you will be best judge'. (40) From the lack of a fireplace we may deduce that Watt carried out some sort of work in the garret. Probably they did not move at that time and she may have been pregnant which could be why Watt wrote in the following vein in May 1767.

> I am very sorry to be absent from you in your present situation... I beg my dearest you would keep a good heart & believe tho I am absent I have not forgot you I am very uneasy about that cursed flittin I wish you could prevail on Mrs Martin to lett you in before the term as to my Garret I beg you would speak to Mr Macintosh to lett the things stay in it for a month or so till I come down, as they cannot be removed by anybody but myself rather offer him a years rent of it for that time nothing could be more unlucky than this affair at this time. (41)

We should note Watt's affection for Peggy as well as his concern about the things in his garret. They moved again in May 1770 when Watt was busy packing up the shop goods and clearing an immense amount from his garrets. (42) There is also a note in June 1772, 'came to Glasgow and found that Mrs Watt had flitted to the Caltons'. (43) We have no knowledge of where any of these places may have been.

The birth of their first child, John, has been mentioned already. (See p106) Two years later their first daughter, Margaret, was born sometime in 1767. She was their only child who married in 1791 (44) and had children but died just before 5 June 1796. (45) Then followed James Watt junior born on 5 February 1769. (46) Watt told Roebuck, 'Mrs Watt has not recovered so fast as we at first expected but is getting better the young ones are well'. (47) Watt junior was the only child to outlive their father, dying on 2 June 1848 after taking over the steam engine business. Agnes was born on 24 December 1770 (48) but died on 15 February 1772. (49)

Peggy was pregnant again in 1773 when Watt had to go to Inverness for a survey of a canal through the Great Glen. He must have been very uneasy about her, especially when he found no letters from her waiting for him there. He wrote telling her not to worry about him. (50) He had started south and reached Fort Augustus where he received a letter. This was almost certainly her last to him; she said that she had been very ill for the last four days but 'Thank God I am now able to sit up'. (51) Dr. Hill was doing all he could to help and she hoped she would be strong enough to bear the child and that Watt would be home by then. Robert Muirheid wrote on 24 September that Peggy had been taken dangerously ill and was almost despaired of. (52) Watt received this two days later, a Sunday, and set out for Glasgow immediately. He spent the night at Fort William and travelled south all day in the rain. When he reached Dumbarton, he could go no further, 'the certainty of my loss bore so strongly upon me'. (53) Hamilton arrived there in a chaise the next morning,

> by his black coat & his countenance I saw I had nothing to hope - We met without speaking ... he informed me I had lost my dear friend on Friday morning [Sept. 24]... I did what I could to force grief from my mind but feared to come home where I had lost my kind welcomer. (54)

Watt learnt that Peggy had been ill and had recovered only for the child to be delivered stillborn. She appeared easy after that but then became restless and later died without a groan.

> In her I lost the comfort of my life, a dear friend and a faithfull wife, may she enjoy that happiness in another state she wanted in this life of sorrow... She bore me 4 living children 2 sons & 2 daughters & died of the 3 son who was buried with her of these children a son & daughter are left me whom God long preserve. (55)

His grief was profound. There can be little doubt but that the loss of Peggy was one reason why Watt decided to leave Glasgow and go to Birmingham eighteen months later to erect his experimental steam engine with Boulton's help.

The Shop

From Watt's comment in October, 1763, 'My shop does very well', (56) and some figures for the first sales, the partners must have been satisfied at first with their adventure. Muirhead said that the 'ready money' sales gradually 'reached £50 per month or £600 per annum'. (57) The following totals from sales at the Trongate 'Toy Shop' in the first months were not as high as that.

13 Dec. 1763 to 16 April 1764,	£158.3s.7½d.
16 Apr to 22 Aug	£157.14s.2d.
22 Aug to 20 Nov	£118.13s.3½d.
21 Nov to 16 March 1765	£196.17s.10½d. (58)

But we have no way of relating this to stock held, expenditure on wages, rent, interest on capital, etc. Therefore we cannot tell whether it was profitable. The bulk of the sales was derived from the 'toys' with musical and scientific instruments accounting for about fifteen and thirty percent respectively. (59) Because this shop existed for around six years, presumably Watt thought it must have been paying its way. However, there was no refunding of Craig's loans although some may have been repaid to Black. (60)

From various surviving invoices, we can see that stocks were maintained during the initial years after 1763. Contacts continued with suppliers in London, Birmingham, Bristol, Edinburgh and Sheffield. John Leviston in Birmingham undertook to 'take the liberty to send you anything that is new in the toy way' and, besides needles, tongs and pencil cases, sent a fruit basket and ink stand 'of new fashion' in 1765. (61) Startin and Moody, also in Birmingham, invoiced shoes and buckles, 'models of Boy & Girl, Gold Finches, large Squirrells', etc. (62) Watt bought goods to the value of £74.19s.2d. from this company before they went bankrupt in 1768 when he refused to pay part of their bill because they had sent items for which he had not asked. (63)

When Watt went to London in the spring of 1767 to assist with applying for an Act for the Forth and Clyde Canal, he took the opportunity of ordering more stock from his suppliers. (64) From Charles Lincoln, with whom he became friendly, he bought telescopes in brass, burning and reading glasses, in other words many of the same sort of things as before. Lincoln wrote later, hoping that Watt 'had a pleasant journey, got home in good health & found your family well to whom tho' unknown you'll please to make my respects acceptable'. (65) Watt also took the opportunity to stock up with clothes, spending at least £8.16s.5d. (66) After finishing in London, he returned by way of Birmingham where he first met Small who showed him round the Soho Manufactory while Boulton was absent. (67) Watt was duly impressed but once again did not order anything from Boulton. (See p115) Yet various invoices show that he placed orders with other suppliers in Birmingham and afterwards in Sheffield.

Watt was in London once more in August 1768 to submit his patent application for steam engines with his separate condenser. Although there are invoices for his shop dated to the early part of this year, there are none for the latter. We know that he stayed for a while with Small and Boulton in Birmingham so he had the opportunity to place more orders both there and in London. One explanation may be that no archives have been preserved from this period but another may be that Watt had already decided to close his shop which had not been without its difficulties. In May 1765, his father had written to him about leaving some money with 'James your Shop Keeper', showing that Watt had some assistance in running it. (68) We cannot tell if this was James Sym who appears in the Journeymen's books from January 1764 to October 1769 and is referred to as Shop Keeper by Gardner. (see p106) In January 1766, Watt was proposing to discharge 'James' from the shop but he did not say what the problem was for he had only suspicions and no proof. His father recommended someone called Pate as a replacement but James was not to be dismissed immediately, there being a two month overlap to ease Pate into his new job. Pate remained with Watt until February 1769. (69) Peggy was also in the shop and proposed to attend regularly. (70) This was when Watt was beginning to be away from home on his steam engine and surveying business. It is possible that the shop ran into difficulties without Craig's watchful eye.

During the summer of 1766, Urie printed 10 quires (240 copies) of shop 'catalogues' which may suggest that sales were diminishing. (71) There may have been further problems at the beginning of 1767 for his father hoped that Peggy would 'not sit long at a time in ye shop or be out during ye severe weather'. (72) He also gave his son a warning that going to London over the Canal bill might be 'a furder loss to your shop or town business'. (73) Watt had told his father in January 1766 that he intended 'dropping most of his lads' (74) but, according to his Journeymen's Book he had already reduced his labour force much earlier. Yet he retained his 'Workhouse' certainly into early 1769, probably as a place where parts could be made or assembled for his engineering business. (75) We have seen that flute manufacture may have ceased around 1766 (See p113) and a change from scientific instruments into heavier engineering products could have commenced at this time with supplies for the shop being purchased wholesale from London and elsewhere. The instrument business may have been sold to Gardner in December 1770 when Watt's journal records that he spent three days 'sorting John Gardners account', finishing on Christmas Day. (76)

As for the shop, Watt's father wrote in March 1769, 'I suppose you are now (with the assistance of some person) inventoring your Shop goods so that you can be selling off parcels at a time to prevent your having to many on hand when you quit the shop'. (77) Entries in the accounts for individual customers die out around May 1769. (78) We can be certain that this venture ceased before Watt accepted the post of resident engineer for the construction of the Monkland Canal in June 1770 (79) because he noted in his Journal for 24 May 1770, 'Finished packing up of shop goods'. (80) These goods were included in the discussions over the settlement of the transfer of the 1769 patent to Boulton (81) and remained stored until after Watt had moved to Birmingham in 1774. In November, Hamilton sold Watt's household furniture but took the shop goods to his own apartments, saying, 'I intend to have all the shop things cleaned up & if the Hardware people will not give a tolerable offer for them to sell them by auction'. (82) Thus ended the career of James Watt, merchant.

Ingenious Devices

While Watt may have abandoned manufacturing and selling scientific instruments, this did not mean that he abandoned his interest in them which he would retain for the rest of his life. In fact his move into surveying and practical civil engineering gave him fresh opportunities for adaptation and invention of ingenious devices. There have been attempts to claim Watt's priority for some of them, especially his 'micrometers' for measuring distances. But, like so many of his ideas, he does not seem to have brought them into general use other than for himself or a few of his friends at the time and so other people reinvented them later, claiming them as theirs. (83) Watt's secrecy about his inventions has certainly led to his being denied recognition subsequently. He was a person who could not use an instrument or see a machine without seeking to improve it. Muirhead wrote,

> He ardently seized every opportunity of extending his acquaintance with the various branches of physical philosophy, and of investigating the principles of its phenomena, as if prophetically conscious that to his untaught but earnest apprehension might be revealed those secrets which hitherto had been hidden even from the wise and learned; endeavouring, - to use an expression of his own, - "to find out the weak side of Nature; and to vanquish her". (84)

The surveying instruments he used came in for this scrutiny. This was in addition to keeping his own and other people's instruments in repair. For example, in September 1770, he was dividing theodolites and that October 'Centred glass for Mr. Golbornes level'. (85) Then, as he travelled through other parts of Scotland, he took the opportunity to visit industrial sites. While papermaking and letterpress printing did not become subjects for his inventive genius, the textile and pottery industries certainly did.

Surveying Instruments

In September 1767, Watt was pleased to announce a new 'trifle' to his friends which was so simple that it could be applied to 'any telescope whatever'. (86) He used a double glass prism, with one part attached to the indexing arm of a quadrant. In one position, the two prism glasses would be parallel and there would be only one image of the object. The letter ends at this point, but he must have intended to focus the telescope which moved one prism and then, by moving the indexing arm to restore the image, the distance of the object could be indicated. He gave an example of this to Anderson. (87) An adaptation of a quadrant was tried again in 1773. Here he used the 'principle, that of Bird's octant, in which the objects are only once reflected. In this I am making, the fixed glass stands at 45 deg. to the first radius; and by shifting the place of the eye, the head is never in the way'. (88) In the Garret Workshop, there are incomplete frames for two surveying quadrants cast in brass with provision for a compass dial in the centre. It is likely that they date from this period and may be prototypes for this device.

In 1769, Watt contrived

> a most excellent method of measuring distances by means of a tele[scope]: *a.b.* is a telescope with two parallel crossed hairs in the focus of the eye glass, *c.d.* is a pole divided into feet and inches. If the hairs comprehend one foot at one chain distance they will comprehend ten feet at ten chains etc. I have tryed this experimentally & can measure half a mile to a yard; don't publish this unless it be printed already. (89)

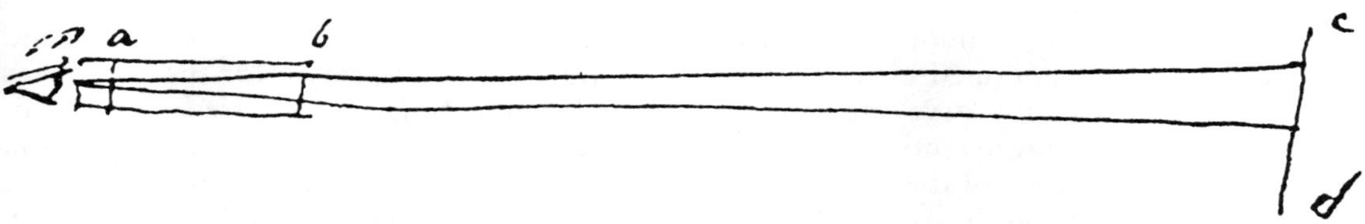

Watt's sketch of his telescope for measuring distances. 'a - b' is the telescope with cross hairs in the eye glass and 'c - d' the pole divided into feet and inches.

Here we see Watt's secrecy over his first attempt at measuring distances in a way that would be easier than dragging a surveyor's chain across the ground. It would have been used with a pole which an assistant took to the spot where the distance was required to be measured. Judging by his later method, the hairs would have been horizontal. The lower one would have been lined up to the zero at the bottom of the pole and the distance measured where the upper one crossed the scale on the pole. He improved this in 1771 by placing on the pole two white discs, preserved in the Garret Workshop, about eight inch diameter with a one inch red stripe across them. The lower one was fixed at the bottom of the pole and aligned through the telescope with the lower cross hair. The assistant moved the other disc up the pole until the surveyor aligned it with the upper hair. The distance was then read off on the pole which was was 12 ft. long. It was possible to measure up to 30 chains. These discs would have been much easier to see than lines on a pole and Watt used them during his surveys at Crinan and Tarbert in 1772 and in the Great Glen the following year. He also showed the arrangement to Smeaton. (90) A similar device invented by a Mr. Green was awarded a premium by the Society of Arts in 1778, although Watt had informed the Society of his prior invention. Green used a more powerful telescope which could measure greater distances and may be the reason why he was given the premium. (91)

However, Watt found that this method left him too much in the power of his assistants where the distances were too great for him to be able to read the number of chains on the pole. In 1772, he realised that distances might be measured when focusing a telescope because the distance between lenses had to be altered. (92) He returned to this idea in October 1773. (93) Accordingly he made a rough wooden telescope which still exists in the Garret Workshop with a slot in one side nearly as long as its focus. The object glass was placed in a tube which could be moved from end to end of this slot by means of a short protruding piece of metal. He added a field glass and had two indexes on the pole which had to be brought into focus by moving the object glass. This would show the comparative distance on a scale. Small pointed out that it would require tables to be drawn up for optical corrections and that Watt ought to use a different term from 'micrometer', finding one which implied that it measured distances. (94) However, owing to the termination of his surveying career, he never completed this instrument and, although he described it to several other people including Ramsden, it does not seem to have been taken up at that time. (95)

Watt criticised the surveyor's levels he used, telling Small in November 1769 that he had

> ...discovered a fault but all levels I have seen are subject to & which gave me much plague before I observed it. it is this you know levels are now adjusted by looking thro the telescope and observing when the hair cuts some object then turning [hole] half round on its axis & observing if the hair cuts the same mark, that being done the bubble is brought round to its mark & the telescope lifted up & turned end for end and if the bubble then remains at the same place the instrument is supposed to be adjusted. (96)

Watt realised this did not allow for any inaccuracy in the bubble or spirit level or whether the axis of the telescope be properly aligned, either horizontally or vertically, for the spirit levels were fixed to the body of the telescopes. In addition, the mounting itself might be disturbed when the telescope was being lifted in or out. Therefore Watt suggested placing the level on the mounting itself which would reveal if the mounting had moved, a position where it is today. (97)

A month later, Watt again complained to Small,

> If I was to follow the business of Levelling I should certainly make some more alterations on Levels, for, from different causes while you are looking thro' the Telescope it is possibly altering its situation this you cannot know till you remove your head & examine the level & if it blows or the ground is soft you can never be certain take what time you will & I assure you this circumstance kills not a little time not much to the comfort of the observer whose cold fingers & toes urge him to be gone. (98)

He proposed making a parallelogram of thin glass tube which could be partly filled with 'liquor' to act as a spirit level. He thought water would be preferable to mercury. (99) This level would be

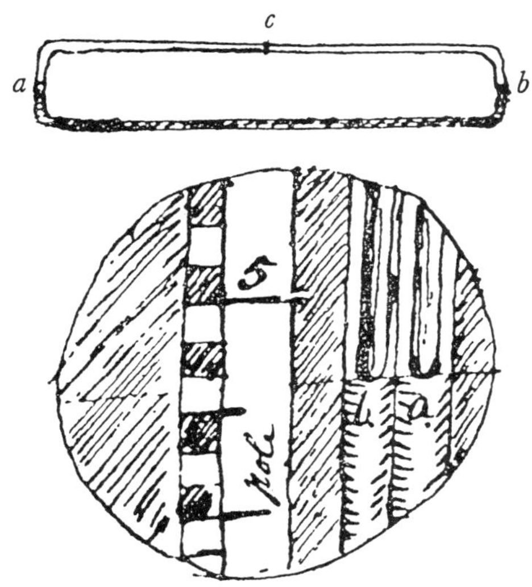

Watt's method of levelling a telescope with a liquid in a glass tube and how this could be checked through the eyeglass.

inserted into the body of a telescope with suitable lenses so that the height of the 'liquor' could be observed in the two vertical legs of the tube as well as the object or surveyor's pole through the telescope. As long as the two surfaces of the liquor remained aligned, the level itself could not have moved. The sighting lenses were made from different types of glass, the convex being crown glass and the concave flint. This may have been an early attempt at correcting chromatic aberration. (100)

Dividing Scales on Instruments

From his youth, Watt had been aware of the need to engrave or divide scales on mathematical instruments. Morgan taught him some methods and, on his return to Greenock, he quickly made himself a dividing plate for quadrants. This must have been some form of master template which would have been an easy method for mass-producing standard instruments. (101. See p60) Tait enquired in 1765 whether Watt had 'the proper apparatus for dividing upon any radius that might be wanted', possibly because this was the time when such machines were being developed. (102. See p107-8) At the beginning of Watt's career, the more accurate instruments would have been divided manually, using geometric methods. Around 1725, Graham introduced important improvements in the division of arcs and turned the beam compass into an accurate instrument used directly for engraving. Bird and Sisson continued to perfect the technique of dividing instruments by a geometric method which attained its height in the middle of the century. Bird's method, published in 1767, achieved greatest perfection possible for such work. (103)

But already between 1765 and 1768, the Duc de Chaulnes had constructed the first machines for dividing straight lines and circles and Ramsden had begun his attempts to construct a dividing engine which he completed in 1773. (104) These were based on turning a screw so that the thread either moved a nut a predetermined distance for each revolution for linear scales, or the screw became in effect a worm gear meshing into the circumference of a circle for circular scales. The Duc de Chaulnes published his book in 1768 but his machines do not seem to have been taken up by craftsmen and remained purely of historic interest. (105) Watt's endeavours may have been stimulated by knowledge of the work of these others.

At that time, it was very difficult to cut the screw of what was in effect a micrometer with a regular thread but this was improved through the need for more accurate astronomical instruments. Today we consider the micrometer as an engineer's tool for taking linear measurements, but the same principle of turning a screw to move some other part was applied to focus a telescope as well as to elevate it. Smeaton was interested in astronomy and was well acquainted with making accurate measurements. His knowledge of micrometers on such instruments may have prompted him to use one to measure thermal expansion. (106) Watt used the term 'micrometer' for a variety of measuring or sighting instruments. One has been mentioned above in connection with a telescope. In 1772, he 'contrived a new micrometer' which consisted of two converging lines drawn on glass intersected by a scale of equal parts. (107) In the collections of the Science Museum is what today we call a micrometer for accurately measuring the length of pieces of metal. It came from the Soho Foundry (108) and has one fixed head and another which can be moved by a screw with an indexing wheel attached to the screw. The date of construction is unknown but it is an early example of such a tool.

During the latter part of 1772, Watt was experimenting with a linear dividing machine that had a rack wheel moved by a fine screw of 38 threads to the inch. (109) He described it to Small,

> It has a wheel fixed upon it with 150 teeth and only 1 $^{1}/_{4}$ inch diameter; it is moved any portion of a turn or number of turns by a straight-line rack, the teeth of which fix it, without shake, and is moved by the hand or foot. It divides distinctly an inch into 400 equal parts. (110)

He continued working on this in the new year and in March 1773 proudly told Small, 'My dividing-screw can divide an inch into 1000 tolerably equal and distinct parts on glass'. (111) Such accuracy was a remarkable achievement at that time. Black asked Watt to bring a 'divided plate for graduating Thermometers - Your machine with the Screw is admirable for its exactness but the plate will be exact enough for some Operations I wish to have done - and it saves calculation'. (112)

Watt strove to improve his machine further. He had a new one in February 1774 and used it to divide the scales for a barometer. '332 teeth of the leading screw correspond to $^{1}/_{20}$th inch exactly & 241 to one revolution of the screw this screw seems now to be very exact it is 9 inches long and is made of steel'. (113) He told Small a little later,

> I had occasion to use my last dividing screw for the first time the other day. It divided 9 inches into 20ths, and did not err the $^{1}/_{200}$th of an inch in the whole 9 inches. I did not find that there was the least inequality among the divisions, though I subjected them to the most severe trial, and I have found a way by which I can divide a foot into thousandths of an inch without erring above $^{1}/_{200}$th of an inch in the whole length, and the divisions shall be equal among themselves; so that I reckon that machine exceeding near perfect, and find it very useful, as it saves much needless compass work, and, moreover, can divide lines into the ordinates of any curve whatsoever. (114)

Watt would appear to have been well in advance of his contemporaries with this machine for it was not until after 1797 that Henry Maudslay developed his screw cutting lathe and micrometer which have always been acclaimed as the first to achieve great accuracy on this principle. Once again Watt does not appear to have turned it into a commercial proposition.

Textile Interests

Watt's practice as a civil engineer of course involved him, as we would expect, in surveying canals, harbours, roads and water supply. But, just as Smeaton also built waterwheels and other machinery, so Watt followed his example, giving advice on both prime movers such as waterwheels and steam engines and the machinery driven by them such as slit mills and forges. Another interest was textile machines. He could not have known that, at the same time as he was becoming a civil engineer, the inventions in Lancashire by James Hargreaves of the spinning jenny around 1764 (115) and Richard Arkwright of the waterframe patented in 1769 (116) would cause a dramatic expansion of the cotton industry, creating an enormous increase in the demand for power. Indeed, by far the largest number of Watt's rotative steam engines built before 1800 were erected in textile mills.

The Linen Industry

During the years that Watt lived in Scotland, linen manufacture became unquestionably the most important industry there. (117) The quantity of cloth produced increased by leaps and bounds.

1750	7,752,500 yards valued at	£361,736
1760	11,747,729	£523,153
1770	13,049,535	£634,411
1780	13,410,934	£622,187
1790	18,092,249	£722,544
1800	24,235,633	£1,047,598 (118)

The dramatic increase in production for the last ten years may reflect the introduction of chlorine as a bleach which Watt helped to develop. A large market for coarse linen cloths was found among the colonists and, in 1771, two million yards, valued at £452,000, were exported from the Clyde. (119) This trade was linked with the tobacco industry (120) and both suffered a prolonged depression which began with the failure of the Ayr Bank in 1772. Bleaching was necessary to whiten the linen but bleachfields were expensive to set up. Luncarty Mill, beside the Tay just north of Perth and the largest in Scotland, was begun in 1752 and needed a capital investment of just over £2,500 in the following ten years. (121) While finance on this scale might come from individual merchants or landowners, co-partnerships became a more common source. (122)

Watt's first connection with bleaching linen possibly came through the person who became his second father-in-law, James McGrigor, a Glasgow bleacher who discounted bills from Claggett in 1763 and 1765 for Watt. (123) McGrigor told Watt in April 1769 that English oil of vitriol was being offered to Scottish bleachers as cheaply as Roebuck's acid (124) and Watt warned Roebuck that either he would have to show why his product was superior or else reduce his price. William Sandeman described McGrigor as 'our mutual friend' (125) and it was no doubt through McGrigor's good offices that Watt visited Sandeman's Luncarty Mill while he was surveying the line of a canal up the Vale of Strathmore in April 1770. (126) Smeaton had already visited this mill in the previous year because in his *Designs* there is a plan of Sandeman's beetling engine. (127)

Watt was asked to advise about waterwheels and machinery connected with bleaching. To give a polished finish to bleached linen cloth, it was passed through a beetling mill consisting of a heavy horizontal roller against which the cloth was beaten by wooden hammers lifted by a camshaft. The cloth might be treated for up to four days in these machines which were normally driven by waterpower. In 1754, Sandeman had paid £102.15s.11¾d. for a beetling machine and in 1761 an additional one was installed, possibly to Mackell's designs. (128) Sandeman was considering replacing one with a 'mangle', which sounds as if it was some form of calender with rolls, perhaps following the example at Saltoun where one was installed in 1760. (129). As well as recommending improvements to Sandeman's beetling mill where the wooden roller was at some time replaced with a cast-iron one, Watt carried out experiments on the calender by wrapping the cloth round the middle roller very tightly. While this flattened the cloth, it did not give a glossy finish like the beetles. (130) At the same time, he suggested alterations to keep the cloth stretched out sideways and later suggested placing pairs of felt-covered rollers at an angle at the sides of the cloth to keep it taut in a manner similar to later automatic temples on looms. (131) Watt continued to advise Sandeman for some time. Overheated bearings were one problem, (132) so Watt recommended a bell metal, 2oz. tin to 1 lb. copper, which he considered the best material for bushes. (133) Sandeman wished that he had consulted Watt about his new waterwheel when the old one failed while Watt enquired from Sandeman about boat builders for barges on the Monkland Canal. (134)

Watt recorded two visits to McGrigor, one on 12 September 1771 and another on 20 May 1772. (135) The second concerned a beetling mill when he drew a plan, perhaps incorporating his experience gained at Luncarty. (136) Later he gave directions about puddling McGrigor's mill dam and some gearing, (137) followed by alterations to a traversing machine and its gearing, (138) as well as trying the waterwheel at McGrigor's new mill. (139) Among the James Watt Papers is a document, not in Watt's hand, titled 'The Dementions of a Beetling Machine and Value of Materials and Work' (140) which shows that the main roller was 13 ft. long, 13 ins. diameter with 36 beetles or hammers and that the whole machine cost £41.4s.4d, including £13 for workmanship. This document also has estimates for washing mills and other gearing and shafting. Another one has an estimate for a snuff mill. (141) While at Perth in May 1770, Watt visited the Ruthven paper mill and sketched its machinery. (142) All this was considerably different in variety and scale from the work of a scientific instrument maker.

Watt developed an ink which would not lose its colour when used to mark cloth being bleached. Joseph Fry, a Quaker typefounder and chocolate merchant at Bristol, wrote to Watt in 1782,

> At... W. Sandemans at Perth I saw a liquid which I suppose is a solution of Silver, with which the Bleachers mark their cloth with a Pen, with a fine neat stroke, which is said to remain as long as the cloth, & that it is of thy Invention. If thou keeps the Process a Secret, I should be glad to know where to get it for the use of my Friends & customers the Bleachers and Callico Printers in the South - Wm Sandeman could not tell me where 't was sold, but said he thought at Glasgow. (143)

Watt must have invented this ink before December 1771 when he started to send invoices to Alex Monro who was buying bottles of silver solution from him. (144) However there seems to be no record of when he made the actual discovery. Gardner continued with its production after Watt had left Scotland. (145)

Other Textile Machines

Notes in Watt's *Journals* illustrate his ability to examine textile machines which he had never seen before and quickly grasp how they operated. Although there is a set of hand cards in the Garret Workshop he does not seem to have ever spun. He sorted out the basic principles and, typically, proceeded to suggest improvements where he could have had no previous practical experience. An example is the occasion when John Mackenzie and his millwright, Thomas Bell, visited Watt in 1770 either over their waterwheel or because they had heard that Watt had advised Sandeman at Luncarty about his beetling mill and cylinder calender machine. In his *Journal*, Watt commented on the copper rollers or cylinders Bell had devised for his continuous calico printing machine which is the earliest reference to this invention. (146) The machines were driven by a waterwheel and the cylinders must have been engraved like intaglio copper printing plates where the ink to be transferred onto the cloth was left in the hollows. Mackenzie and Bell may have been having trouble with the doctor blades for scraping off the excess ink because Watt thought that lantern horn or tortoise shell might work very well. (147) He commented, 'They keep their whole process secret', and gave no further details.

In October 1770, Watt observed a demonstration of a knitting machine which had been improved by a G. Jardine and recorded the performance, noting how quickly it could be operated. (148) The original knitting machine had been invented by William Lee in 1589 and was the most complex of any machine which had been developed up to that time. An improvement had been added soon after 1610 but from then until the middle of the eighteenth century there were few further changes. (149) In 1759, Jedediah Strutt patented an attachment for ribbed knitting, (150) making the fortune which enabled him to finance Arkwright. His development unleashed a series of further additions by others, although the basic principle of operation remained the same. Henson, writing in 1831, listed over 2,000 parts on a frame with 360 needles which needed eleven operations to make a single row of stitches. (151)

Watt commented that the frame which he saw was much lighter and easier to move than the common one (152) and must have observed it closely for he soon 'conceived an idea of a new stocking frame'. He may have tried some experiments because there are a few needles in his Garret Workshop. (153) On Lee's frame, all the needles were pushed backward and forward together and their looped or 'bearded' heads were closed by a pusher bar so that the last row of stitches in the knitted fabric could be pushed over the new loops held in the needle heads in a combined movement. Watt proposed to draw the needles back sequentially as the slurcock was driven across the machine making the loops but realised that then the pusher bar could not be used to close the beards and suggested individual pulley wheels for each needle instead. This might have speeded up the knitting but it is doubtful if it would have worked in practice.

Notes in his Journal are too brief to enable us to discover why Mr. Robertson of the bank took tea with Watt and gave him particulars about working for him 'at cotton' in 'the Joland of Caricow'. (154) In February 1774 Watt visited the demonstration of a machine for twisting thread in Dr. Poteau's new mill at Paisley. From his description, this was based on the circular silk throwing mills developed in Italy in the fourteenth century and first brought to Britain by the Lombe brothers who erected some at Derby in 1718. (155) Watt made full notes on the Poteau machine which was ten feet in diameter and had 60 spindles. Just as he remarked that the knitting machine he had seen could be moved more easily than common ones, so, on this twisting machine, he recorded that this 'machine goes very easily & may be driven by one hand'. (156) As well as the mechanism, he was always interested in the power required to operate machines.

In the last sections of this chapter, we have seen how, in the years after 1765, Watt left behind his careers of mathematical instrument maker and merchant and became involved in other activities. We will next examine some of the other enterprises in which Watt was engaged at this time. He was involved as a chemical engineer in experiments to make alkali from salt with Black and Roebuck, and acted as technical adviser to the Delftfield Pottery. He launched his practice as civil engineer which covered surveying canals, roads and harbours, becoming resident engineer for the construction of the Monkland Canal. Finally there was, of course, the development of his career as a mechanical engineer through his experiments on the Newcomen model atmospheric steam engine which led to his invention of the separate condenser and the erection of several Newcomen engines to drain coal mines.

NOTES: FROM MERCHANT TO ENGINEER

1 Smiles, (1878), p. 92 and Dickinson, (1935), p. 44.
2 B. & W. Col., MI, 1/18, 'Journal', 23 Dec. 1771.
3 Bryden, (1972), p. 25.
4 B. & W. Col., MI, 1/20, 'Journal', 18 July & 10 Aug. 1773 and MI, 2/6, 'Cash Memorandum', 6 March 1774.
5 Muirhead, (1858), p. 45.
6 Ibid, p. 96.
7 King-Hele, (1981), p. 46, Letter 67D, E. Darwin to J. Watt, 18 Aug. 1767.
8 Royal Society, J. Watt Letters, PH 13, Letter 71, J. Watt to Russian Ambassador, 7 Aug. 1771.
9 Ibid.
10 Muirhead, (1858), p. 202.
11 JWP, 4/32.15, J. Watt to Dr. J. Roebuck, 24 Sept. 1769.
12 B. & W. col., MI, 1/15, 'Journal', various dates.
13 Muirhead, (1854), Vol I, p. 85, Letter 72, J. Watt to W. Small, 3 Jan. 1770; Boulton 348.15.
14 Royal Society, J. Watt Letters, PH 13, Letter 71, J. Watt to Russian Ambassador, 7 Aug. 1771.
15 Muirhead, (1854), Vol II, p. 20, Letter 93, J. Watt to W. Small, 24 Dec. 1771, Boulton 348.23.
16 JWP, 3/36, J. Watt jun., 'Memoir of James Watt', p. 21.
17 Muirhead, (1858), pp. 249 - 50, quoting from J. Keir, 'Account of the Life and Writings of Thomas Day, Esq.', 1791, p. 111, note.
18 Royal Society, J. Watt Letters, PH 13, Letter 71, J. Watt to Russian Ambassador, 7 Aug. 1771.
19 Muirhead, (1858), p. 50.
20 Muirhead, (1854), Vol II, p. 20, Letter 93, J. Watt to W. Small, 24 Dec. 1771; Boulton 348.23.
21 Muirhead, (1858), p. 46.
22 Dickinson & Jenkins, (1927), p. 35.
23 Robinson & McKie, (1970), p. 24, Letter 20, J. Robison to J. Watt, 22 April 1771.
24 Royal Society, J. Watt Letters, PH 13, Letter 71, J. Watt to Russian Ambassador, 7 Aug. 1771.
25 Ibid.
26 Muirhead, (1854), Vol I, p. 25, Letter 31, G. Jardine to J. Watt, Monday 1768; JWP 4/57.4.
27 Ibid, Vol II, p. 19, Letter 91, W. Small to J. Watt, 19 Oct. 1771; Boulton 340.17.
28 Ibid, Vol II, p. 72, Letter 129, W. Small to J. Watt, Dec. 1773; Boulton 340.38.
29 Ibid, Vol II, p. 87, Letter 75, J. Watt to W. Small, 31 Jan. 1770; Boulton 348.18.

30 Ibid, Vol II, p. 20, Letter 93, J. Watt to W. Small, 24 Dec. 1771; Boulton 348.23.
31 Ibid, Vol II, p. 9, Letter 85, J. Watt to W. Small, 30 Sept. 1770.
32 Royal Society, J. Watt Letters, PH 13, Letter 71, J. Watt to Russian Ambassador, 7 Aug. 1771.
33 Muirhead, (1854), Vol II, p. 357, Letter 395, J. Watt to D. Brewster, 7 June 1814.
34 Ibid, Vol I, p. 3, Letter 1, J. Watt to J. Lind, 29 April 1765; JWP, C1/15.
35 Ibid, Vol I, p. 4, Letter 2, J. Watt to J. Roebuck, 23 Aug.; JWP, 4/32.38.
36 Ibid, Vol I, p. 44, Letter 44, J. Watt to J. Roebuck, 10 Feb. 1768; JWP, 4/32.20.
37 Ibid, Vol II, p. 24, Letter 97, J. Watt to W. Small, 30 Aug. 1772; JWP, 4/59.4.
38 Ibid, Vol II, p. 69, Letter 127, J. Watt to W. Small, 11 Dec. 1773; Boulton 348.38.
39 Ibid, Vol II, p. 76, Letter 133, W. Small to J. Watt, 29 March 1774; Boulton 340.40.
40 JWP, 4/4.22, Margaret Watt to J. Watt, 15 Nov. 1766.
41 JWP, 4/4.18, J. Watt to Margaret Watt, 4 May 1767.
42 Muirhead, (1854), Vol II, p. 95, Letter 79, J. Watt to W. Small, 10 March 1770 and B. & W. Col., MI, 1/15, 'Journal', 23 - 9 May 1770.
43 B. & W. Col., MI, 1/18, 'Journal', 19 June 1772, see also 29 May 1772, 'Flitted my Garret to Calton'.
44 JWP, W/6.53, 26 May & 52, 10 June 1791, J. Watt jun to J. Watt.
45 JWP, 4/42.44, G. Hamilton to J. Watt, 5 June 1796.
46 Dickinson & Jenkins, (1927), p. 30, J. Watt jun. died 2 June 1848.
47 JWP, 4/32, J. Watt to J. Roebuck, 10 Feb. 1769.
48 B. & W. Col., MI, 1/16, 'Journal', 24 Dec. 1770.
49 B. & W. Col., MI, 1/18, 'Journal', 15 Feb. 1772.
50 JWP, 4/4.4, 3 Sept. & 5, 9 Sept. 1773, J. Watt to Margaret Watt.
51 JWP, 4/4.2, Margaret Watt to J. Watt, 14 Sept. 1773.
52 JWP, 4/4.1, R. Muirheid to J. Watt, 24 Sept. 1773.
53 B. & W. Col., 1/20, 'Journal', 28 Sept. 1773.
54 Ibid, 29 Sept. 1773.
55 Ibid.
56 JWP, 4/11.105, J. Watt to J. Watt sen., 13 Oct. 1763.
57 Muirhead, (1858), p. 45.
58 JWP, C1/1, 'Accounts', f. 69, 13 Dec. - 16 April 1764 and f. 72, 16 Apr. 1764 - 16 March 1765.
59 Smith, (1994), Bryden, p. 12 lft.
60 B. & W. Col., MIII, 3/4, 'Private Ledger', f. 31.
61 JWP, 4/58.16, J. Leviston to J. Watt, 24 Aug. 1765.
62 JWP, 4/46.11, Startin & Moody to J. Watt, 4 May 1767.
63 JWP, 6/28.37, from Startin & Moody, 8 June 1768.
64 JWP, 4/17, J. Watt to Margaret Watt, 14 May 1767, '& shall sett out for you as soon as I have finished buying a few things'.

65 JWP, 4/46.30, C. Lincoln to J. Watt, 18 June 1767.
66 JWP, 4/46.16, A. Lee to J. Watt, 19 May 1767, see also JWP, 4/46.25, J. Donaldson to J. Watt, 29 Apr. 1767.
67 Muirhead, (1854), Vol I, p. cxvii.
68 JWP, 4/7.50, J. Watt sen. to J. Watt, 3 May 1765.
69 JWP, 4/11.148, J. Watt to J. Watt sen., 21 Feb. 1769.
70 JWP, 4/11.134, J. Watt to J. Watt sen., 23 Jan. 1766.
71 B. & W. Col., MII, 8/7.1, Invoice from R. Urie, 21 June 1766.
72 JWP, 4/7.29, J. Watt sen. to J. Watt, 21 Jan. 1767.
73 JWP, 4/7.28, J. Watt sen. to J. Watt, 9 Feb. 1767.
74 JWP, 4/11.134, J. Watt to J. Watt sen., 23 Jan. 1766.
75 JWP, C1/1 'Accounts' and MIII, 3/2, 'Waste Book, 1765 - 9'.
76 B. & W. Col., MI, 1/16, 'Journal', 22 & 25 Dec. 1770.
77 JWP, 4/7.11, J. Watt sen. to J. Watt, 23 March 1769.
78 See B, & W. Col., MIII, 3/6, 'Ledger of Personal Accounts'.
79 B. & W. Col., MI, 1/15, 'Journal', 27 June 1770.
80 Ibid, 24 May and 23 - 29 May 1770.
81 B. & W. Col., MI, 1/19, 'Journal', 13 April 1773.
82 JWP, 4/20.21, G. Hamilton to J. Watt, 15 Nov. 1775.
83 Muirhead, (1858), pp. 229 - 39.
84 Ibid, p. 45.
85 B. & W. Col., MI, 1/15, 'Journal', 28 Sept. & 13 Oct. 1770.
86 JWP, C1/14, Letter from J. Watt, 9 Sept. 1767.
87 Muirhead, (1854), Vol I, p. cxxxv.
88 Ibid, Vol II, p. 37, Letter 104, J. Watt to W. Small, 17 Jan. 1773; Boulton 348.26.
89 Boulton 348.4, J. Watt to W. Small, 28 Jan. 1769. See Dickinson, (1935), p. 56 for illustration.
90 Dickinson, (1935), p. 75 and Muirhead, (1854), Vol I, p. cxxxi.
91 Muirhead, (1854), Vol I, p. cxxxv.
92 B. & W. Col., MI, 1/18, 'Journal', 19 Feb. & 4 March 1772.
93 B. & W. Col., MI, 1/20, 'Journal', 8 - 24 Oct. 1773.
94 Muirhead, (1854), Vol I, p. 64, Letter 125, W. Small to J. Watt, 16 Oct. 1773; Boulton 340.36.
95 Muirhead, (1854), Vol I, p. cxxxvii.
96 Boulton 348.13, J. Watt to W. Small, 3 Nov. 1769.
97 B. & W. Col., MI, 1/17, 'Journal', 23 Sept. 1771.
98 Boulton 348.14, J. Watt to W. Small, 12 Dec. 1769.
99 B. & W. Col., MI, 2/8, 'Note Book of J. Watt, 1770 - 81'.
100 B. & W. Col., MI, 1/16, 'Journal', 25 Nov. 1770.
101 JWP, C1/1, 'Accounts', Oct. 1759.
102 JWP, 3/41.2, J. Tait to J. Watt, 6 Aug. 1765.
103 Daumas, (1989), p. 194.

104 Ibid, p. 194.
105 Ibid, p. 199
106 A.J. Pacey, 'Vis Viva and Power', typescript in Dept. of History of Science and Technology, UMIST, p. 192.
107 Muirhead, (1854), Vol II, p. 35, Letter 102, J. Watt to W. Small, 24 Nov. 1772; Boulton 348.25 and B. & W. Col., MI, 1/19, 'Journal', 18 Nov. 1772.
108 Dickinson & Jenkins, (1927), p. 37.
109 B. & W. Col., MI, 1/19, 'Journal', 8 Nov. 1772.
110 Muirhead, (1854), Vol II, p. 35, Letter 102, J. Watt to W. Small, 24 Nov. 1772; Boulton 348.25.
111 Ibid, Vol II, p. 39, Letter 106, J. Watt to W. Small, 7 March 1773; Boulton 348.27.
112 Robinson & McKie, (1970), p. 39, Letter 33, J. Black to J. Watt, 22 May 1773, with illustration.
113 B. & W. Col., MI, 1/20, 'Journal', 25 Feb. 1774.
114 Muirhead, (1854), Vol II, p. 75, Letter 132, J. Watt to W. Small, 3 March 1774; Boulton 348.41.
115 Aspin & Chapman, (1964), p. 13.
116 Fitton, (1989), p. 26.
117 Durie, (1979), p. 95.
118 Hamilton, (1932), p. 90.
119 Ibid, p. 120.
120 Durie, (1979), p. 161.
121 Shaw, (1984), pp. 237 - 8.
122 Ibid, p. 245.
123 JWP, 3/21, 'Household Expenses', Aug. 1763 and B. & W. Col., MI, 2/3, 'Cash Memorandum', 11 July 1765.
124 JWP, 4/32.27, J. Watt to J. Roebuck, 28 April 1769.
125 JWP, 4/80.1, W. Sandeman to J. Watt, 11 Jan. 1770 [?1].
126 B. & W. Col., MI, 1/14, 'Journal', 9 April 1770.
127 Smeaton, (1950), p. 14, see Royal Society, Smeaton, Designs,Vol I, f. 64, Feb. 1769.
128 Shaw, (1984), p. 224.
129 Ibid, p. 226.
130 B. & W. Col., MI, 1/14, 'Journal', 26 April 1770.
131 B. & W. Col., MI, 1/16, 'Journal', 1 Feb. 1771.
132 Ibid, 14 Dec. 1770.
133 JWP, 4/80.3, J. Watt to H. Turnbull, 23 Feb. 1771. A modern metal for bearings would have additional small proportions of zinc and lead.
134 JWP, 4/80.4, H. Turnbull to J.Watt, 2 March 1771.
135 B. & W. Col., MI, 2/42, Note Book on Surveying Expenses.
136 B. & W. Col., MI, 1/18, 'Journal', 20 May 1772.

137 B. & W. Col., MI, 1/19, 'Journal', 25 July 1772.
138 Ibid, 9 Oct. & 9 Dec. 1772.
139 B. & W. Col., MI, 1/20, 'Journal', 27 May 1773.
140 JWP, 4/77.7, no date, 'Dementions of a Beetling Machine'.
141 JWP, 4/77.8, no date, 'Estimate for a snuff mill'.
142 B. & W. Col., MI, 1/15, 'Journal', 10 May 1770. See British Association of Paper Historians, *The Quarterly*, Issue 7, July 1993, p. 16, R.L. Hills & J. Reece, 'British Paper Mills: Ruthven and Methven'.
143 JWP. JW/23.6, J. Fry to J. Watt, 1 Dec. 1782.
144 JWP, 6/28.55, A. Monro to J. Watt, 11 Dec. 1771.
145 Moore, (1996), p. 15 and Bryden, (1972), p. 33.
146 Day & McNeil, (1996), p. 55, 'Thomas Bell'.
147 B. & W. Col., MI, 1/16, 'Journal', 7 Nov. 1770.
148 Perhaps this was George Jarden, member of the Guild of Hammermen, who helped Watt earlier.
149 *Journal of the Textile Institute*, Vol 80, No. 2, 1989, R.L. Hills, 'William Lee and His Knitting Machine', pp. 169 - 84.
150 Henson, (1831), p. 264.
151 Ibid, pp. 65 - 7.
152 B. & W. Col., MI, 1/16, 'Journal', 20 Oct. 1770.
153 Ibid, 27 Oct. 1770.
154 B. & W. Col., MI, 1/18, 'Journal', 14 & 18 March 1772.
155 Istituto Internazionale di Storia Economica 'F. Datini', Serie II, Vol 24, *La Seta in Europa, Secc. XIII - XX*, 1993, R.L. Hills, 'From Cocoon to Cloth', p. 70.
156 B. & W. Col., MI, 1/20, 'Journal', 14 Feb. 1774.

3
Chemical Engineer

THE STATE OF CHEMISTRY BEFORE 1760

Watt said that he had been interested in chemistry as a youth but we have no way of telling either what roused this interest or what direction it took. This interest was further stimulated by Black and remained an important occupation all through his life. It was shared by other close friends such as Roebuck, Hamilton, Boulton and other members of the Lunar Society. It was a period when many new discoveries were made and there were rapid developments, but, at that time, the term chemistry might imply something very different from what we understand today, and it changed in meaning during Watt's own life. This has been described recently:-

> At one time, chemistry might have been called a theory of life, at another time a department of metallurgy; at one time a study of combustion, and at another an aid to medicine; at one time an attempt to define a single word, the word *element*, and at another the quest for the unchanging basis of all phenomena. (1)

Towards the end of Watt's life, John Dalton put forward his atomic theory with its classification of distinct elements following the Chemical Revolution of A.L. Lavoisier while at the beginning of the eighteenth century, there was still belief in Aristotle's four elements.

Aristotle's Elements

Watt wrote to Smeaton in April 1783, 'By the help of Dr. Priestley's experiments I have attempted to demolish 2 of the most ancient elements (air & Water) a third has been destroyed for some time; but in return we have made 2 or 3 more'. (2) As a youth, Watt probably followed the age-old theory that there were four elements from which everything was constituted. Although this originated long before, Aristotle taught that the world was composed of the basic 'elements' of earth, air, fire and water, which, in their turn, contained the four 'qualities' of heat, cold, dryness and moisture. This doctrine remained almost unquestioned until the seventeenth century but was investigated and disputed more and more as the eighteenth progressed. (3)

From being actual physical elements themselves, Aristotle's elements were gradually changed into theoretical concepts as true chemical elements were identified during the eighteenth century and the basic constituents of the air and matter around us were discovered. Hermann Boerhaave, the famous Dutch medical doctor and chemist, retained the four perfect forms or concepts of Aristotle's elements when he wrote his *New Method of Chemistry*, first translated into English in 1727. Cullen, Black and Watt were well aware of his work. (4) Boerhaave said,

> All the instruments of chemistry may be reduced to six; the four first whereof, are the four vulgar elements.
> The 1°, is *fire*, by which all motion is produced.
> The 2°, is *air*, by means whereof motion is preserved.
> The 3°, is *water*, by which motion is communicated.
> The 4°, is *earth*, whereof all vessels are made.
> The 5°, are *menstrums*, whereby bodies are intimately dissolved and mov'd.
> The 6°, are *vessels*, wherein the bodies to be changed by fire, water, *etc.*, are contained. (5)

Chemists who came after Boerhaave, particularly those in Scotland, constantly advanced their subject. By the time Black was lecturing in Edinburgh after 1766, he doubted the truth of Aristotle's elements. When speaking about the Greek philosophers, he said,

> They assigned four elements, or elementary kinds of matter, viz. fire, or the matter of *fire*, *air*, *water*, and *earth*; and by the different combination of those elements, they supposed that all the different bodies in nature were formed.
> These four supposed elements were accordingly considered until very lately, as the primary elements of natural bodies; in so much, that Mr. Macquor, in his System of Chemistry, thought it proper to begin by treating these four kinds as elementary. (6)

While dismissing Aristotle's elements, Black was still forced to admit that, 'The truth is that our knowledge of nature is not yet brought to that state of advancement, which might enable us to follow with propriety an order of this kind [the new chemistry]'. (7) A few of the steps leading to the start of the change in perception will be briefly summarised.

The Start of New Understanding

As a better understanding of chemistry developed towards the end of the seventeenth and at the beginning of the eighteenth centuries, so it was realised that alchemy was a pseudo-science and that transmutation of one substance into another was technically impossible. Alchemy became history as it became clearer that there were more basic elements than the four described by Aristotle. One of the people who contributed towards this was Robert Boyle who has been called the founder of modern chemistry for three reasons. He devoted his life to bringing chemistry to the attention of natural philosophers as a subject worthy of attention for its own sake and not as an adjunct to medicine or alchemy. (8) He introduced rigorous experimental methods and also gave a clear definition of an element. Through his experiments, he showed that the four elements of Aristotle and the three principles of the alchemists (mercury, sulphur and salt) did not deserve to be called elements or principles at all because 'Elements... [are] certain Primitive and Simple, or perfectly unmingled bodies; which not being made of any other bodies, or of one another, are the ingredients of which all those call'd perfectly mixt Bodies are immediately compounded'. But nowhere did Boyle say what substances he considered to be elements. (9)

Boyle's other important contribution to natural philosophy and Watt's development of the steam engine was his investigation of air, made possible through the air pump, one of which he built in 1658. Boyle never thought of air as a chemical entity. It was a peculiar elastic fluid in which floated various reactive particles responsible for the phenomena of respiration, the rusting of iron, deliquescence, the greening of copper and so on. The air pump became a common piece of equipment in most laboratories and its use would lead to a better understanding of the atmosphere and gases in general. Pneumatic chemistry became a subject in which Watt was involved in many experiments both with Priestley and later on his own for Beddoes and the Pneumatic Institution at Clifton.

Transmutation was also dealt a blow by Newton who recognised that laws of physics applied not only in the world but also throughout the universe. 'With Newton, the Universe acquired an independent rationality, and the whole cosmology of Aristotle, of Galen and of the Middle Ages lay in the dust'. (10) It was the beginning of the establishment of a new order where law prevailed. There was a mathematical certainty in the motion of the planets and much else. Experiments could be repeated with the assurance, that if conditions were the same, results would be the same. For example, the value of the latent heat of steam would be approximately the same in experiments carried out by both Watt and Black.

In addition to certain laws which were universally applicable, there began to emerge identification of the basic chemical elements which would eventually form the atomic theory. Black considered that certain elements of substances were exempted from change and decay and were 'capable of being variously compounded together, and separated again, so as to give origin to the productions we see daily arise, and to the perpetual changes of one form or state of matter into another'. (11) The constituent elements of bodies were permanent and unchangeable. In their work, he, and later H. Cavendish, Priestley and C.W. Scheele, assumed that matter was 'conserved', that is to say, it was neither created nor destroyed in the course of their experiments. They also assumed that weight was a measure of the amount of matter. (12) They began the quantitative study of chemistry and showed that there could be a high degree of exactness possible in chemical operations with definite proportions. However they did not begin to advance the theory of the conservation of heat.

The Nature of Air

Although Boyle carried out some experiments on air, the main investigations undertaken by chemists concentrated on solids and liquids through the difficulty of collecting gases. This was partially solved by Stephen Hales in the 1720s by collecting the gas in a vessel suspended over water. He discovered that plants or minerals contained or held within their pores large quantities of gases. He heated solids and liquids in a gun barrel and collected the air that was given off. He attempted to determine the amount of air but, once this was measured, it was thrown away. He made no attempt to investigate the quality of the air, which in some cases was coal gas from coal, oxygen from red lead and saltpetre, hydrogen from iron filings and dilute sulphuric acid, so he missed identifying these different gases. In some experiments on breathing, he found he could breathe a fixed volume of air much longer if it were passed through rags soaked with potash between exhalation and inhalation but he did not pursue this experiment 'fearing that I might thereby some way injure my lungs'. (13)

Hales's breathing this air shows the reliance of experimenters on their own senses to distinguish different chemicals. The senses of sight, smell, and feeling were employed extensively even by Watt. It is surprising that these people did not cause themselves great bodily harm, which indeed Watt and members of his family nearly did when experimenting with chlorine for bleaching as well as carbon monoxide for Dr. Beddoes. The use of the senses was paramount in Boerhaave's tests to identify chemicals for he wrote,

> Now we call *body* whatever is extended, solid, hard, impenetrable, moveable, figurable, ductile, etc....
> The *sensible* [bodies] are such as affect our senses; or such as have a determinate magnitude, figure, and distance, with regard to our organs of sense, as to produce changes therein, whose effects come under the notice of the mind.
> *Insensible* are such as do not fall under our senses; that is such as are either so small, or so remote, that their action on the organ works no notable change. Thus the air, wherewith we are everyway surrounded, is full of an infinite number of heterogeneous corpuscles, which have indeed an effect on our bodies; but 'tis such an one as our senses take no cognizance of.
> Now, chemistry only regards *insensible bodies* as its objects, so far as they *are capable of being rendered sensible*; whence, in effect, it considers none but sensible bodies. (14)

With only such methods of determining different substances, it must have been particularly difficult to distinguish different gases in the atmosphere around us. Boerhaave would have agreed for he wrote,

> To define the nature of air, would be extremely difficult; in as much as its intimate affections are unknown to us. All we know, is, 1^0 That air is naturally an homogeneous similar body, 2^0, Fluid, 3^0, Heavy, 4^0, Elastic, 5^0, That it rarifies by fire, and contracts by cold, 6^0, That it is compressible by a weight laid thereon, and rises and restores itself upon removal of the same. All which circumstances, should incline it to coalesce into a solid, were fire wanting. (15)

J.T. Desaguliers thought that Air was

> ... a Fluid, whose Parts are endow'd with a centrifugal Force, whereby they repel each other without touching...
> The Atmosphere or Air which incloses our Earth, is a Compound of several kinds of Vapours and Exhalations that come out of several solid and fluid Bodies, and continue to be suspended and float about in the Air, no longer than the centrifugal Force of their Parts continues; for every compressible Fluid that loses the centrifugal Force of its Parts, (and consequently its Elasticity) ceases to be a Fluid, and becomes a solid. (16)

Here at least we have the possibility of the atmosphere consisting of different vapours. It was one of the remarkable achievements of Thomas Newcomen that he recognised the difference between steam and ordinary atmospheric air. For his engine to run continuously, it was essential to snift or blow any air out of the cylinder in order to achieve sufficient vacuum while steam had to be retained which could be condensed to create that vacuum. Watt must also have been aware of this difference when carrying out trials on the University model Newcomen engine.

In 1754, Black reported to Cullen about his experiments for his doctoral thesis,

> It was, indeed, an experiment I was trying that amused me, in which I mixed together some chalk and vitriolic acid at the bottom of a large cylindrical glass jar; the strong effervescence produced an air or vapour, which, flowing out at the top of the glass, extinguished a candle that stood close to it and a piece of burning paper immersed in it, was put out as effectually as if it had been dipped in water, yet the smell of it was not disagreeable. (17)

We should note that Black must have tested this air by breathing it. He had created 'fixed air' or carbon dioxide but he was not interested in its properties and, by not following it up, did not advance the study of what would become pneumatic chemistry. Concerning air and steam he wrote,

> As heat can thus produce from water an elastic vapour, which resembles air in so many of its qualities, it is not improbable that air itself is produced in the same manner from some other matter... And although we are not able to cool it so much as by that way to reduce it alone to a dense unelastic body, we find, however, that the more it is cooled, its elasticity is the more diminished. (18)

He recognised there was a difference between the vapour of water and air itself but was more interested in the phenomenon of heat and its effects than in the study of gases.

The Importance of Chemistry

Boerhaave wrote,

> Chemistry is an Art, whereby sensible bodies contained in vessels,... are so chang'd, by means of certain instruments, and especially fire, that their several powers, and virtues are thereby discovered; with a view to the uses of medicine, natural philosophy, and other arts and occasions of life. (19)

Cullen, following Boerhaave, made chemistry a popular study because he saw its relevance to the 'arts' or what we would call industrial applications. Black was keenly aware of the differences between this approach and that held by the medical faculty when he moved to Edinburgh as a student in 1752. Black continued Cullen's example when he returned to Glasgow in 1756 and this may have roused Watt's interest in the application of chemistry to industrial problems. Many of Watt's chemical investigations were concerned with improving industrial processes and here again we see the practical rather than the theoretical personality appearing. Also Watt may have been hoping to make some important discovery which would have created his fortune in the same way as Roebuck and others.

But, for Watt's later career, the importance of chemistry lay more in the methods which Black employed. Black was a cautious investigator and his success was due to the accuracy of his measurements using the balance. In the chemical sphere, his triumph consisted in showing that the reactions of quicklime discussed in his thesis were reversible, and that the same amount could always be extracted as was put in. His work was of great importance, being the first detailed quantitative study of a chemical reaction and its reversal. Following Black's example, it would be the exactness of his measurements, not only with the balance but also with the thermometer, that gave Watt the key to understanding what was happening to the steam as it passed through his model Newcomen engines.

Watt's Early Chemical Interests

Our surviving archives contain virtually no references to Watt's interests in chemistry before Black left Glasgow for Edinburgh in 1766. The explanation may be quite simply that they worked together and therefore had no need to record any of their experiments. Yet, when Watt was left on his own, he was able to develop his own ideas which we will follow in two main areas of research, that of making alkali from salt and that of improving the products of the Delftfield Pottery. But first, we will examine a few of his more general enquiries.

The only indication about chemical experiments before 1765 is in a letter from a M. Turner of Liverpool in 1761. (20) He sent Watt a box containing some ten dozen one ounce phials, four dozen half ounce phials of 'ather' and 180 pamphlets which he hoped Watt could find a shopkeeper to sell. Turner also enclosed another bottle for Watt and one for Black which he described,

> I had collected a Bladder full of foul Air which is produced in the process for making Ather, with design to send it to Dr Black, but an ingenious Friend of mine accidently laying his Hands on it, consumed it all in Experimts. The result of which was that the Electrical spark readily setts it on fire - But if you think any quantity of this air wou'd be a curiosity, I can readily collect more in my next process if it would be acceptable to the Doctor. (21)

Although Watt does not seem to have been directly involved with the attempts to produce the 'Ather', this letter suggests that he was well aware of the process and may have had close links with Black in similar experiments.

Watt's interest in textiles included dyeing and bleaching. In 1769, he was helped to read a book in German about the mines in the Upper Hartz by Chaillet, a Swiss dyer, who had come to Glasgow to learn about dyeing linen and cotton red. Watt sent Small some samples of Chaillet's red yarn and must have tried to discover Chaillet's method of replacing vitriolic acid with some other solvent for alum but failed. (22) Later he told Small about Chaillet's discoveries in producing soap and in making carmine. (23) Watt found out Chaillet's secret of his glue recipe before Chaillet left Glasgow towards the end of 1770. (24) In the meantime, Watt had noted that Mackenzie and Bell were dyeing cloth with indigo to produce a china blue colour and washing it afterwards in several hot liquors of which a solution of copperas was one. (25) Later Irvine passed to Watt a method of dyeing silk by dipping it first in some solution and then boiling it in cochineal, alum and tartar. (26)

As might be expected with someone engaged in civil engineering, Watt noted the different types of rock in the country through which he happened to pass. Near Ayr, he spotted a granite which he thought might make a good 'water cement'; did this mean one which would set under water? (27) Just before leaving for Birmingham in 1774, he sent Baron Mure a recipe for making a waterproof cement to act as a damp course in his house 'to prevent moisture from rising in the wall'. (28) He described to Small some experiments with alum and arsenic, (29) preparing cobalt, experiments on glass for Dollond, (30) Irvine's experiments with 'flint acid', (31) the action of quick lime in decomposing a solution of tartar, (32) and so on. These experiments on acids were mentioned to Black. (33) There can be little doubt that this list contains only a small fraction of Watt's interests in and experiments on chemical substances.

Notes: Chemical Engineer

1 Partington, (1957), p. 62.
2 JWP, Letter Book 1, p. 29, J. Watt to J. Smeaton, 27 April 1783.
3 Singer, (1941), p. 48.
4 Cardwell, (1971), pp. 34 - 35 and Brock, (1992), p. 77.
5 Boerhaave, (1727), p. 218.
6 Robison, (1803), Vol I, p. 343.
7 Ibid.
8 Brock, (1992), p. 42 and Partington, (1957), p. 67.
9 Partington, (1957), pp. 70 - 71.
10 Singer & Underwood, (1962), p. 146.
11 Robison, (1803), Vol I, pp. 341 - 342.
12 Singer, (1941), p. 288.
13 Brock, 1992), p. 78 and Partington, (1957), p. 91.
14 Boerhaave, (1727), p. 53.
15 Ibid, p. 290.
16 Desaguliers, (1763), pp. 339 - 340.
17 Donovan, (1975). p. 200.
18 Robison, (1803), Vol I, p. 201.
19 Boerhaave, (1727), p. 51.
20 Dr. F.G. Page has kindly pointed out that this would almost certainly be Dr. Matthew Turner, a surgeon, chemist and scholar and the person who at tended Wedgwood's injured knee while in Liverpool and in so doing introduced Thomas Bentley, later Wedgwood's partner.
21 JWP, 3/41.10, M. Turner to J. Watt, 12 Nov. 1761. See also ibid, 3/41.7, M. Turner to J. Watt, 7 Sept. 1762. The 'ingenious Friend' was probably Joseph Priestley who was invited to go to the Warrington Academy on 25 June 1761. See Gibbs, (1965), p. 16.
22 Muirhead, (1854), Vol I, p. 63, Letter 58, J. Watt to W. Small, 28 May 1769; Boulton 348.9.
23 Boulton 348.14, J. Watt to W. Small, 12 Dec. 1769.
24 Boulton 348.22, J. Watt to W. Small, 21 Dec. 1770.
25 B. & W. Col., MI, 1/16, 'Journal', 7 Nov. 1770.
26 Ibid, MI, 1.18, 'Journal', 7 Dec. 1771.
27 Ibid, MI, 1/17, 'Journal', 16 June 1771.
28 Ibid, MI, 1/21, 'Journal', 1 May 1774.
29 Boulton 348.9, J. Watt to W. Small, 28 May 1769.
30 Boulton 348.16, J. Watt to W. Small, 31 Jan. 1770.
31 Boulton 348.26, J. Watt to W. Small, 17 Jan. 1773.
32 Boulton 348.27, J. Watt to W. Small, 7 March 1773.
33 Robinson & McKie, (1970), p. 36, Letter 30, J. Watt to J. Black, 26 Dec. 1772.

ROEBUCK, BLACK AND SALT

John Roebuck, 1718 - 1794

Between 1765 and at least 1770, the person who had the greatest influence on Watt's life was Dr. John Roebuck. In 1765, Roebuck was approaching the summit of his industrial career. He was eighteen years older than Watt, having been born in Sheffield in 1718. He went to Edinburgh in 1737 to study medicine and chemistry where he distinguished himself and was impressed with the society there - 'a circumstance which produced in his mind, a partiality ever afterwards in favour of Scotland, and contributed not a little to his making choice of it for the chief field of his exertions and industry'. (1) He moved to Leyden to continue his medical studies where he received his M.D. in February 1743. (2) After travelling around Europe, he obtained a practice in Birmingham where he also set up a small laboratory. Here he was successful both as a medical doctor and a chemist. His friendship with Boulton and Small originated at this time and would be crucial later for Watt's final development of his steam engine but it was his chemical investigations which took him back to Scotland.

At Birmingham, there was a demand for oil of vitriol, sulphuric acid, to clean the many metal objects produced there. In 1747, Roebuck replaced the glass retorts used in making this acid with lead chambers, which enabled more acid to be produced more cheaply. To exploit this, and probably to avoid English patents taken out by a Dr. Ward, Roebuck formed a partnership with another Birmingham person, Samuel Garbett, who later became known to Watt. Roebuck and Garbett set up a manufactory at Prestonpans near Edinburgh in 1749. They were able to keep their process secret for a time and are believed to have made a large profit from it, partly because sulphuric acid was in great demand in Scotland as a bleach for linen. 'The success of the establishment at Prestonpans, which had far exceeded their expectation, enabled the Doctor and his partner Mr GARBET, to plan and execute other works of still greater benefit and public utility'. (3)

At that time, there was no iron foundry in Scotland. Roebuck had been experimenting with pit coal to replace charcoal to produce iron. He, Garbett and William Cadell, a Scot, established the Carron Foundry where the first furnace was blown on 1 January 1760. (4) To Roebuck was due the original conception since his were the plans for the novel production processes on a scale unrivalled even in the south. (5) Later, Carron supplied Watt with castings for his steam engines and other engineering applications. 'But though Roebuck was full of ideas, and brimfull of confidence, he lacked the organising ability so necessary in such an enterprise'. (6) These qualities were supplied at first by Garbett, but he returned to Birmingham and a manager, Charles Gascoigne, was employed. Once Carron was up and running, Roebuck started to look for fresh enterprises. His chief failing, not unusual in men of like temperament, was an inability to remain in one enterprise for long, so that he constantly became involved in financial obligations beyond his means. (7) George Jardine, (see p124) wrote, "When the business at Carron sunk by degrees into a matter of ordinary detail, and afforded less scope for the Doctor's peculiar talents, he was unfortunately tempted to engage in a new and different undertaking'. (8) In order to ensure adequate coal supplies for the voracious appetites of the furnaces at Carron, Roebuck became lessor of the Duke of Hamilton's extensive coal and saltworks at Bo'ness, where the coal was reputed to be of superior quality. The lease included the ancient Kinneil house overlooking the Firth of Forth where Roebuck moved with his family in about 1764. (9) He failed to persuade Boulton to join him in this venture so he had to seek alternative financial sources which came partly from the Carron Company itself. Gascoigne wanted control of the Company and no doubt encouraged Roebuck to become involved in other enterprises. Roebuck borrowed £8,000 on security of his $^{2}/_{13}$ of Carron stock in 1764. (10)

Roebuck hoped that the Bo'ness Coal Company, worked in conjunction with the saltpans, would be a profitable unit but, as happened with the iron works, the collieries involved much greater expense than had been anticipated. Although he opened up new seams of coal and increased production, the mines were very wet and, in 1765, the volume of water in them proved to be beyond the capacity of the only available steam engines, the Newcomen atmospheric type, to remove economically. (11) Roebuck borrowed from his friends and acquaintances as well as taking large sums from his other industrial concerns so that by 1766, these withdrawals came to about £22,000. (12) However, in 1768, the Bo'ness Coal Company began to show some signs of success and was supposed to be yielding a profit of about £80 a week. At his zenith, in his coalmines around Kinneil and Fife as well as in agriculture, Roebuck was employing around one thousand people. (13) Since his debts to the Carron Company showed no signs of decreasing, on 5 December 1768 Roebuck surrendered his last share, ending the formal connection. (14) However, the prosperity of the coalmines did not continue, leading to his first financial failure in 1770.

The account Jardine gave of his employer's achievements and character helps us to understand Roebuck's relationship with Watt. Roebuck was said to have introduced

> a spirit of enterprise and industry, before that time little known in Scotland, which soon pervaded many other departments of labour, and gave birth to many other useful projects. He brought from England, then much farther advanced in arts and industry, many ingenious and industrious workmen, at great expence, who, by their instructions and example, communicated and diffused skill and knowledge to others. At all times Dr ROEBUCK held out liberal encouragement to rising genius, and industrious merit; and spared no expence in making trials of improvements and discoveries, which were connected with the different projects and works which he was carrying on.
>
> Such was the active and useful life of Dr ROEBUCK, a man of no common cast, who united, in a very high degree, a great number of solid and brilliant talents, which, even separately, fall to the lot of but few individuals. Distinguished by an ardent and inventive mind, delighting in pursuit and investigation, always aspiring at something beyond the present state of science and art, and eagerly pressing forward to something better or more perfect, he thus united energies the most powerful, with the most unwearied and persevering industry. To that peculiarity of imagination, so fitted for scientific pursuit, which readily combines and unites, which steadily preserves its combinations before the eye of the mind, and quickly discovers relations, results and consequences, was added, in his character, great promptitude and firmness in decision. (15)

Watt and Roebuck

Owing to the need to drain his collieries, we can understand why Roebuck was interested in Watt's improvements to the steam engine and was willing to help him through his desire to support new endeavours. The first indication we have that Roebuck and Watt knew each other is a letter to him from Watt written on 23 August 1765 when Watt was experimenting with models of his separate condenser. (16) From the remains of this damaged letter, it is evident that there had been earlier communications. First Watt mentioned a model of some waggon he had made for Roebuck which could have been a truck for carrying coal on a waggonway. Then there was a reference to a couple of tin pans which were probably for experiments on the heat of steam in which Watt and Roebuck were involved later that autumn. It is only at this point that Watt came to the models of his new engine. It is evident that by this time these two men were involved with much more than Watt's separate condenser and engine. In Roebuck's first surviving letter to Watt in September 1765, we can see that he had a great respect for Watt's scientific knowledge and ability after Watt had told him about his discoveries on steam, Watt's Law and his co-operation in these enquiries with Black. Roebuck hoped 'to see you at Kinneil before the end of this month, and hope Dr. Black will accompany you'. (17) Here Black and Watt are associated in Roebuck's mind. His friendship with Watt is implicit when he invited Watt to bring his wife and family to stay at Kinneil in April 1769. (18)

Watt incurred considerable expenditure on the trials of his model engines. According to tradition, he was at first helped financially by Black.

> Dr. Black continued to take a lively interest in Watt's experiments, and lent him occasional sums of money from time to time to enable him to prosecute them to an issue. But the Doctor's means were too limited to permit him to do more than supply Watt's pressing necessities. Meanwhile, the debts which the latter had already incurred, small though they were in amount, hung like a millstone round his neck. (19)

Once again, Smiles has contracted many facts to give a dramatic, if somewhat misleading, impression. Our records do not show Black lending any money to Watt in 1765, but there was £50 in February and £22.10s. in June 1766. (20) Soon after that, Black moved to Edinburgh and was not in a position to lend any more to help with the patent fees, as has been suggested by Muirhead. (21) Muirhead may have been confused with proposals to patent Black's process of making alkali from salt which was being discussed at the same time.

We do not know when Roebuck first offered Watt financial assistance. He was in serious difficulties with his flooded coalmines when Black told him about Watt's new engine. (22) Muirhead asserts that Roebuck was quick to appreciate Watt's inventive genius and to realise the great profit that might accrue through its successful application. (23) Although it is most likely to have been the steam engine which first interested Roebuck with the alkali from salt project not far behind, it seems that he may not have become involved financially until the end of 1765 with the first engine to be built at Kinneil or even in 1767 over Watt's indebtedness to Craig.

Watt found himself faced with a financial crisis through the death of his partner, Craig, who died on or just before 5 December 1765. (See p104) He must have realised that Craig's loans for the shop would soon be called in by Craig's brother, but he told Roebuck,

> Mr. William Craig who is Brother & Heir to my friend Mr. John Craig has assured me that I may in everything depend on his friendship which I never had any reason to doubt of as in everything in his power he has always been my friend & he is one of those people on whom I would not suspect an Increase of fortune to make an alteration. (24)

It may not have been until sometime in 1767 when some agreement was reached. (See p446) Roebuck took over these loans and shouldered Watt's indebtedness to the amount of £757.4s.7d. (25) Roebuck paid William Craig £200 and £300 in January and October 1769 respectively. (26)

Yet, when recollecting events later during the trials of Watt's patent in 1796, both Robison and Black thought that Watt had been introduced to Roebuck to obtain financial backing for Watt's experiments. Robison was unknown to Roebuck in 1765 (27) but said at a patent trial in 1796,

> [Watt's] situation in life made it imprudent to engage in great expenses, and he was obliged to look out for an assistant. Most fortunately there was in the neighbourhood such a person as he wished, - Dr. Roebuck, a gentleman of very uncommon knowledge in all the branches of civil engineering, familiarly acquainted with the steam-engine, of which he employed several on his collieries, and deeply interested in its improvement. He was also well accustomed to great enterprises, of an undaunted spirit, not scared by difficulties, nor a niggard of expense. Such a man was indispensibly necessary to one of Mr. Watt's character, - modest, timid, easily frightened by rules and misgivings, and too apt to despond. I do not know who pointed him out to Mr. Watt. (28)

Later in the course of the same speech, Robison said, 'I believe that Dr. Black was the chief means of forming the connection between Mr. Watt and Dr. Roebuck'. (29)

Black himself laid no claim to having introduced the two but did stress the financial involvement in connection with constructing a larger engine.

> [Watt's] mind became now very much employed in contriving the machinery by which this improvement might be reduced to practice; and he soon planned it to such a degree, that he thought he was ready to make an experiment on a large scale. But here he was stopped by the want of funds; and he found it necessary to associate himself with some person who had money and spirit for such an undertaking, and to participate with him the advantages which might be derived from this invention. He addressed himself to the late Dr. Roebuck, whose spirit for enterprise and improvement in the arts was very well known, and the Doctor accordingly received with zeal the opportunity offered to him. A small engine was soon built in one of the offices of Kinneil House, near Borrowstoness, where various trials were made and some difficulties surmounted, so as to give satisfaction. (30)

In 1768, Roebuck told Boulton that a very small engine had been erected at Kinneil some years earlier which had convinced him of the 'justness of his principles' and encouraged him to help Watt financially. (31. See p372 & 419)

But financial assistance may not have been the only way in which Roebuck helped. Black said that Roebuck, in addition to his knowledge of medicine and chemistry, 'had no small degree of mechanical knowledge and ingenuity; and was well qualified to perceive and value the talents of Mr. Watt. He had also much experience of the use of common steam-engines, which he employed in working his colliery'. (32) Robison also suggested that Roebuck's assistance was much greater than moral and financial support.

> Dr. Roebuck knew Mr. Watt's talents and most liberally praised them. His timidity, his disposition to despond under unforeseen difficulties, and his painful anxiety and diffidence in himself were the subjects of frequent merriment at the Doctor's fireside; and I have often heard him say, that without his help, and even his instruction, Mr. Watt could never have gone on. I have heard him mention some important, but subordinate, parts of the engine which were of his contrivance. (33)

We can see that, if Black had introduced Roebuck to Watt because he thought that Roebuck could use Watt's engines in his coalmines, Watt must have been equally pleased to meet Roebuck in view of the other practical help which he could give. Trials on Newcomen engine models had shown their great consumption of steam, while experiments with the first model of the separate condenser had shown the probability of great savings. Watt needed to test a full-size Newcomen engine to confirm their performance. Roebuck had coalmines with such engines on one of which Watt carried out trials in the autumn of 1765. Then Watt would have realised that nobody would buy his engine unless they were convinced it would perform as economically as he claimed, so he needed to build one on a large scale. He needed to do this in secret where it could not be seen by other people who might copy it. His workshop in Glasgow was too small, but Roebuck had space and privacy at Kinneil House. It is more difficult to judge whether Watt hoped for Roebuck's financial support at the beginning but we can understand why he saw many other practical advantages in sharing his exciting discovery of the separate condenser with such an influential and important industrialist as Roebuck. Watt overcame his reticence and 'the connection was very soon formed. Dr. Black and all Mr. Watt's friends were happy at seeing so fair a commencement'. (34)

But a person of Roebuck's at times impetuous and impulsive character would find difficulty in keeping patience with the timid and irresolute Watt. Roebuck commented about Watt's engine in his first letter, 'It is certainly best to press it forward with all speed, whether you pursue it as a philosopher or as a man of business'. (35) This not only shows Roebuck's drive and enthusiasm but also points to the crux of their relationship. Watt, the philosopher and engineer, wanted to achieve perfection in all parts of his engine which meant endless experiments, modifications and hence delays. Roebuck, the man of business, wanted to see it running as soon as possible to ensure a quick return on the investment. These differences may explain the tensions underlying two of Roebuck's letters; the first in April 1768 when he wrote, 'I should have rejoiced to have received more favourable accounts of your trial; but I hope in a few days you will have better success'. (36) Was his patience wearing a little thin? In the second sent during the autumn, he wrote

> I want much effectually to try the machine at large. You are letting the most active part of your life insensibly glide away. A day, a moment, ought not to be lost. And you should not suffer your thoughts to be diverted by any other object, or even improvement of this, but only the speediest and most effectual manner of executing one of a proper size, according to your present ideas. (37)

Roebuck wanted results and recognised the need to concentrate effort and application. Watt, at this time, had many other irons in the fire, probably too many to be able to deal with any one properly or thoroughly. Many years later, Black summed up the relationship between the two.

> [Roebuck] was withal ardent and sanguine in the pursuit of his undertakings, and was therefore a fortunate associate for Mr. Watt. Mr Watt was a valetudinarian, more or less, ever since I knew him; and his mind was liable to be too much depressed by little cross accidents, or by the necessity of a greater expense than he had foreseen; whereas the Doctor was undaunted on such occasions and roused Mr. Watt to disregard expense, and to double his exertions, until the difficulty was overcome. (38)

Salt for Curing Herrings

From the earliest days of their friendship, Roebuck and Watt were involved in trading in sea salt for curing herrings. Roebuck was producing salt and very soon he was supplying it to Watt senior through Watt forming the link between them. We have seen how Watt acted as his father's agent while in Glasgow and how his Uncle Muirheid and his father were trading partners. Their commerce in barrels of herrings, which came from Fort William and the Western Isles, can be traced back to March 1751 (39) To preserve herrings, salt was needed and this was another of their commodities. One account mentions 70 bolls of Spanish salt, (40) which confirms that Scottish salt was unsuitable for preserving fish. For centuries, the Scots had produced salt by boiling sea water with the coal that was readily available around the Firth of Forth. Thanks to a preferential tax, this industry survived English competition after the Union. But Scottish, and to a lesser extent, English salt, contained so many impurities that only imported salt could be used to preserve fish, especially if they were to be exported to foreign markets. (41) This was why Cullen had investigated ways of purifying the Scottish salt but the method he recommended to remove the bitter calcareous contaminants used expensive potash. (42) This may be one reason why Black studied the chemical properties of calcareous substances following his professor's interest in the problem of purifying salt. It is not known whether Black and Roebuck worked together on this but during the autumn of 1765 Roebuck was trading with the Watts and Uncle Muirheid in salt.

There may have been some dealings in August 1765; (43) in a letter to his father at the beginning of September, we find Watt saying that Roebuck's salt at 12s. per boll seemed a good price. Roebuck was intending to make salt 'proper for curing exportation herrings' and, as his father had plenty of space, Watt recommended stocking a cellar and selling it wholesale. Should any be left at the end of the season, he knew somebody who could dispose of it to people in Fife. (44) Three weeks later, he told his father that 'everybody is vastly pleased with [Roebuck's] salt' and that he would send a barrel as soon as one arrived to cure a few herrings for themselves and their friends. (45) But no salt had arrived early in October when needed at the height of the preserving season. (46) Perhaps this was not Roebuck's fault for Watt wrote to say he was sorry that Roebuck was being 'molested with the excise men but hope you will gett the better of them soon'. (47) In November, Watt complained to Roebuck that only three casks of salt had arrived and that, although they had sold one and part of another, 'it was considerably dirtied by the cask' so that two bushels were unsaleable. (48) It seems that some barrels were not properly filled either. (49) Some salt was sent by sea from Bo'ness around the north of Scotland because the Forth and Clyde Canal did not exist. The *Jean & Peggy* of Borrowstowness loaded 2,576 bushels of Scots salt in March 1766 bound for Greenock but unfortunately struck rocks in the Orkneys and some of the salt had to be off-loaded to save her. (50) This trade in salt between Watt senior and Roebuck may have continued for some time or else Watt senior was very slow in paying his bills. Towards the end of 1768, there were two letters from Roebuck asking Watt if he could nudge his father into settling the salt accounts. (51) Watt explained to Roebuck that his father had trouble receiving payment from his Highland customers which Watt explained to Roebuck. His father sent £60 in December 1768 and a further £40 in the following February so, if this was all to pay for salt, then their trade must have been quite extensive. (52)

Alkali from Salt

Alkalis were important during the eighteenth century because they were used in glass-making, soap manufacture, the production of alum and saltpetre, as a fertilizer and in bleaching. All of these industries expanded but the natural sources from which alkali was made, such as potash from wood or domestic kelp and foreign barilla, became scarcer and more expensive. (53) The method of bleaching was to boil the cloth with ashes, an alkali, and then with a sour, an acid, followed by exposure to sun light. These processes were repeated until the requisite degree of whiteness was attained. In a country such as Scotland where strong sunshine was frequently lacking, bleaching was a bottle-neck in the expansion of the linen industry because the process took a long time. The acid used at first was sour milk, but, in

1756, Dr. Francis Home recommended in his book, *Experiments on Bleaching*, that sulphuric acid be used as a substitute. In fact, Roebuck had already tried it and frequently recommended it to bleachers before Home's publication. (54) Bleachers formed an important market for Roebuck's sulphuric acid.

While the replacement of sour milk by sulphuric acid helped to lessen the time needed in bleaching, these acids still had to be neutralised by an alkali. It seemed that lime might be an alternative to alkalis but it was liable to damage the cloth and its use was prohibited for many years. Cullen failed to discover a way to make inexpensive alkali from natural substances but the potential of lime interested Black who wrote to Cullen in 1754, 'I long vastly to hear what improvement you have made on bleaching ashes. Do you make a cheaper ley from pearl-ashes and lime than that commonly employed?' (55) This subject continued to attract Black's attention and in 1763 he informed his father that 'I have at last wrote a paper upon the use of lime in bleaching'. (56) This may be the essay which was published in 1771 as 'An Explanation of the Effect of Lime upon Alkaline Salts; and a Method pointed out whereby it may be used with Safety and Advantage in Bleaching'. (57)

Black had long been interested in the classification of alkalis and the nature of causticity . It was presumably this which caused him to join forces with Roebuck and Watt in the second half of 1765 to attempt production of alkali from a mixture of salt and lime. Possibly Roebuck's production of salt was one motivation. They probably hoped to be as successful with alkali manufacture as Roebuck had been with sulphuric acid. Roebuck and Black must have become acquainted through their shared interests in chemistry from the time when Black was first at Edinburgh before taking up the Glasgow professorship in 1756. Jardine said that Roebuck lived with his friend Dr. Black 'till his [Roebuck's] death, in close habits of intimacy; and he often acknowledged, with much frankness, the advantages which he derived, in his various pursuits, from a free and unreserved communication with that eminent chemist'. (58)

The idea of making alkali from salt and lime originated with Black. (59) Watt wrote to Small in October 1769, 'You know Dr B. first invented the theory of the Alcali which he communicated to me & I tryed Experiment till I gave a probability of its succeeding in practice. Doctor R. was taken into the scheme soon after me'. (60) However in the rough draft of this letter, Watt said that Roebuck was 'taken into the scheme at the beginning of it'. (61) Today we have no way of telling which is correct but these experiments show Watt being accepted by Black and Roebuck as an equal partner in the venture and that it was Watt who devised the equipment and supervised the practical trials. Perhaps this reflected a change in Black's attitude towards him after his invention of the separate condenser because he had shown that he had the expertise to become a first-rate natural philosopher. The accounts of these trials are the first we have of Watt's experiments in chemistry.

When these trials started and where they were carried out remain unknown but, because Watt was supervising them, Glasgow was the most likely location. There is an entry in an account book in December 1766 'for caraige to the Gorbals of salt, 3d.' (62) Uncle Muirheid had property there so that could have been the place. Our first definite record of these trials is found in Watt's letter to Roebuck of 19 February 1766, which reveals that something had been going on for five months, so stretching back to the previous September. Watt said, 'I would have wrote you sooner but delayed expecting some of the salt experiments to come to a favourable result', and then he gave a summary of progress.

> 1st Lime was mixed with about its own bulk of Salt & sett in a cool Cellar it rarified in Del;[iquescence] the liquor being a brine of Salt. owthing [nothing] being expected from it it was neglected for 5 months it was then found Covered with Crystals of mild alcali

> 2d a flat plate was filled with Lime made into a paste with strong brine drops of caustic alcali appeared on its surface which soon shot into Crystals during its Crystalization a considerable quantity of brine ran from it having somewhat of an alcaline taste.
> 3d three flat plates were filled one with a mixture of Coal ashes & Lime equal parts moistened with Strong pickle the second with lime moistened with pickle of $^1/_2$ strength the third with Lime moistened with pickle of $^1/_3$ strength. in about 4 days the ashes & Lime produced drops of alcali on the surface which in about 2 weeks were Cristal: & a quantity of brine ran from it considerably alcaline the second also had produced a small Quantity of Cristals like wool.
> the third produced scarce any.
> The Lime (still wet) of the second experiment was mixed with the alcali produced by it the mixture lixiviated filtrated & evaporated the alcali was Lost & the Com[mo]n salt reproduced. (63)

We learn about these experiments only from Watt's letters to Roebuck. There is no mention of Black in any of them but presumably Watt must have also kept Black informed and sought his advice.

We can see that Watt expected alkali or soda (sodium carbonate) to be produced from lime (calcium hydroxide) and salt (sodium chloride). This should have appeared on the surface of the blocks of the mixture in the form of crystals which could be scraped off by hand, rather similar to the way saltpetre was produced at that time. The two basic constituents were available cheaply so the process might have been profitable. However modern chemistry shows that this reaction does not work but yet Watt produced some alkali. Some recent experiments suggest that the answer must lie in the use of sea salt in which there was a sulphate impurity. A lime-sulphate process for soda is quite feasible and became the basis of the process later used by Keir. Coal ashes added a further variable because they could contain both carbonate and sulphate. (64) These small varying amounts of sulphate in the salt used by Watt gave him some alkali and a false optimism which encouraged him to go on experimenting.

Watt was puzzled at his lack of success and wrote again to Roebuck at the end of February 1766 to say that he had 'not yett gott any certain footing about the alcaline process'. (65) He had filled two earthenware pots with equal parts lime and salt and left one covered and the other uncovered. All that had happened was that the brine soaked through the earthenware! But a few days later he was more hopeful. He thought that the reaction occurred on the surface of the mixture and therefore had made moulds for producing blocks like bricks. When dry, these blocks could be built up into walls 3 ft. thick, 9 ft. high and 30 ft. long. He estimated that three of these walls would have an acre of surface and cost £80 in materials and construction. He expected a yield of sixty pounds of alkali a day. (66) Watt at times could certainly be ambitious and have grand schemes!

He was more hopeful that April for, in his opinion, 'the Alcali has made great progress'. (67) He had obtained some samples and hoped to send Roebuck an account of its success. Then he must have tried floating the lime in the brine but found that this method was not as successful. (68) At this point silence descends until November, probably because Watt was trying to launch his surveying career and was engaged in erecting an atmospheric engine at Kennet. Another reason may have been Black's departure to Edinburgh. There is an undated document describing the process and showing a mould for the blocks and framing on which to stack them. (69) Trials must have continued during the rest of the summer and into the autumn when Watt reported to Roebuck,

> I would sooner have wrote you account of the produce of the lime had it been equal to what I though it should be it has turned out only one fourth of what the Lime on the plate produced but it is 4 times more than what this performed in the experiment before so if goes on in that ratio it will soon come up to what the plate did I have always observed that it was some time before the Lime began to produce properly probably owing to the want of a proper Quantity of fixd air or surface. I am now going on with another experiment of which shall write the success. Small as this produce is it would amount to £270 - pr annum in a house 10 houses would clear about £2000 so it would still be a good trade but I expect better. (70)

Here again we see Watt's great optimism, going straight from meagre experiments into ambitious plans for large scale production. But at that point, this series of experiments seems to have been drawn to a close.

There the matter might well have ended through Watt becoming more and more involved with his steam engines and surveying. But, as he returned from London in September 1768 after completing the initial legal formalities for his patent on the separate condenser, he stayed with Small in Birmingham. He later commented to Lind, 'When at Birmingham, I spent a fortnight in the same house with your ingenious friend Captain [James] Keir, who is a mighty chemist before the Lord, and a very agreeable man'. (71) Keir was involved in the manufacture of glass and other industries which used alkali and he was also investigating the possibility of making alkali from salt. Small must have found himself in a difficult situation because he knew that Keir and others were experimenting along these lines and then Watt also confided his experiments. On 7 January 1769, Small warned Watt, 'One experiment was shewed to me, and two were described to me, since you left me, which will certainly soon lead the operators, altho they were not great cymists to your alcali secret. This is what I alluded to and I beg you will attend to it'. (72) When Watt became aware of Keir's interests, he worried that he might have revealed too much of his own secrets when he was staying with Small. (73)

It may have been Small's letter which stirred up the Scottish partners to renew their interest in the alkali project. Watt must have suggested to Black that they take out a patent. Black agreed, provided the experiments worked and Watt and Roebuck joined him in sharing the expenses. (74) Should Roebuck not be willing to join them, Black was sure that he could find other partners. (75) Roebuck must have expressed interest so Watt wrote to Small, 'The Alcaline affair will be put in agitation immediately I have many things I would talk with you about much contrived and little executed how much would health & spirits be worth to me!' (76) Watt's comments about his health should be noted.

Roebuck suggested that Kinneil would be a suitable place for further experiments but Watt disagreed, perhaps because he could not supervise them properly there. Also he told Black, 'Indeed if my health does not mend I can not at all [be concerned in the affair] as I can be of no service to any body. In any case we would be much the better of an Industrious partner and the scheme if successfull will answer us all'. (77) Watt began to realise at last that he needed to delegate some of his work but also that the process needed to be improved. His first experiments had indicated that in a house '18 feet wide 36 long and 15 high 9 tun of Alcali might be made in a year'. (78) Watt then sent Black an account of his experiments which suggests that he had developed his own process independently without Black's advice; (79) the pupil was beginning to equal the master. Black suggested a meeting of himself, Roebuck and Watt at Kinneil and was 'getting a pair of boots of sufficient capacity to contain two legs and six stockings' to keep himself warm on the journey in the bitter January weather. (80)

At that point, Watt may have carried on with a few more experiments (81) but his steam engine was occupying more and more of his time so it was left to Black to continue. While Small was urging them to make haste, Watt told him, 'Doctor Black has it at present under tuition he proceeds slowly according to the way of Philosophers. whether it be lost or not I can do no more than I do'. (82) But in July 1769, he was assisting Black once more and asked Small to tell Roebuck about schemes any rival might have for a patent. (83) However, in September, Watt reported to Small, 'As to the Alkali we have been trying some experiments which are not yet come [hole] issue & indeed are not operating so fast as we wish them [hole] the 2 Drs therefore seem willing to defer compleating the pat.t till they are more au fait & I cannot urge it'. (84)

That is almost the end of the story as far as Watt was concerned over experiments to make alkali from salt. Although Roebuck claimed to have secured a patent, (85) this was never in fact done. In the autumn of 1769, John Roebuck junior reported that he had seen Black 'who was going very lazily about the cakes & seems rather to despair of them'. (86) At this point, the Scottish partners seem to abandon the race. Black may have realised that the reaction would not work. Watt was nearing success with his steam engine and so was concentrating on that. Roebuck, on the other hand, must have realised that he was becoming embarrassed financially and so could not help. In addition, Small had pointed out that a book had been published earlier in which Black's method of producing alkali had been described. (87) This would have negated any patent. In 1771, Keir explained to Watt how he had achieved success in related experiments which were based on a true lime-sulphate reaction. (88) But for a long time, Small thought Watt's method would succeed (89) and even in the spring of 1773 asked Watt, 'Once more why do you neglect the fossil Alkali?' (90) But Watt did not take up the subject again while he remained in Scotland.

Notes: Roebuck, Black and Salt

1 *Transactions of the Royal Society of Edinburgh*, Vol 4, Pt. 1, 1796, G. Jardine, 'Account of John Roebuck, M.D., F.R.S. Edin.', p. 66.
2 Ibid, p. 67.
3 Ibid, p. 71.
4 Ibid, p. 73.
5 Campbell, (1961), p. 9.
6 Ibid, p. 10.
7 Ibid, p. 11.
8 Jardine, (1796), p. 74.
9 *Dictionary of National Biography*, S. Lee Ed., Smith Elder & Co., London, 1897, 'Roebuck, John, M.D., 1718 - 1794', p. 93 ff.
10 Campbell, (1961), p. 12.
11 *Dictionary of National Biography*, 'Roebuck', p. 94 and Schofield, (1963), p. 67.
12 Campbell, (1961), p. 12.
13 Jardine, (1796), p. 83.
14 Campbell, (1961), p. 13.
15 Jardine, (1796), p. 84.
16 Muirhead, (1854), Vol I, p. 4, Letter 2, J. Watt to J. Roebuck, 23 Aug. 1765; JWP, 4/32.28.

17 Muirhead, (1854), Vol I, p. 6, Letter 5, J. Roebuck to J. Watt, 13 Sept. 1765; JWP, 4/57.1.
18 JWP, 4/32.116, Dr. J. Roebuck to J. Watt, 8 April 1769.
19 Smiles, (1878), p. 90.
20 B. & W. Col., MI, 2/3, 'Cash Memorandum', Feb. 1766 and Robinson & McKie, (1970), p. 40, Letter 34, J. Black to J. West, 23 Feb. 1774.
21 Muirhead, (1854), Vol I, p. ci.
22 Schofield, (1963), p. 67.
23 Muirhead, (1854), Vol I, p. ci.
24 JWP, 4/32.32, J. Watt to J. Roebuck, no date (Muirhead gives a date as 4 Dec. 1765 which is certainly wrong).
25 B. & W. Col., MIII, 3/4, 'Personal Ledger', f. 14. See also Muirhead, (1854), Vol II, pp. 47 - 48, 'Discharge by Mr. Watt to Dr. Roebuck, 17 May, 1773'; JWP, 4/5.21.
26 JWP, 4/32.108, J. Roebuck jun. to J. Watt, 23 Oct. 1769 and 119, J. Roebuck to J. Watt, 31 Jan. 1769.
27 Muirhead, (1854), Vol I, p. ix.
28 Ibid, Vol I, p. liv.
29 Ibid, Vol I, p. ix.
30 Ibid, Vol I, pp. xxxvii - xxxviii.
31 Boulton 334.5, J. Roebuck to M. Boulton, 12 Dec. 1768. See Robinson & Musson, (1969), p. 42, 'Memoranda of James Watt Esquire, from the Messrs. Harts. Communicated by John Smith, 19 March 1845' for a 9 inch cylinder at Kinneil.
32 Muirhead, (1854), Vol I, p. xxxviii.
33 Ibid, Vol I, p. lxi.
34 Ibid.
35 Muirhead, (1854), Vol I, p. 6, Letter 5, J. Roebuck to J. Watt, 13 Sept. 1765, JWP, 4/57.1.
36 Ibid, Vol I, p. 19, Letter 24, J. Roebuck to J. Watt, 18 April 1768; JWP, 4/57.3.
37 Ibid, Vol I, p. 33, Letter 36, J. Roebuck to J. Watt, 30 Oct. 1768; JWP, 4/57.7.
38 Muirhead, (1854), Vol I, p. xxxviii.
39 JWP, C4/A11 Letter Book of J. Watt sen, 1754 - 1758 and JW/19.57, J. Muirheid to J. Watt sen., 20 March 1754.
40 Ibid, JW/19.41, J. Muirheid to J. Watt sen., 16 Dec. 1765.
41 Donovan, (1975), p. 83.
42 Ibid, p. 86.
43 B. & W. Col., MII, 5/4, 'Accounts', 23 Aug. and 2 & 9 Nov. 1765.
44 JWP, 4/11.128, J. Watt to J. Watt sen., 5 Sept. 1765.
45 Ibid, 4/11.130, J. Watt to J. Watt sen., 28 Sept. 1765.
46 Ibid, 4/32.35, J. Watt to J. Roebuck, 11 Oct. 176?5.
47 Ibid, 4/32.11, J. Watt to J. Roebuck, 18 Oct. 1765.
48 Ibid, 4/32.7, J. Watt to J. Roebuck, 13 Nov. 1765.
49 Ibid, 4/32.33, J. Watt to J. Roebuck, 18 Nov. 1765.
50 Ibid, JW/19.34, 'Miscellaneous Salt & Other Accounts', 25 March 1766.
51 Ibid, 4/57.8, J. Roebuck to J. Watt, 9 Nov. and 9 & 22 Dec. 1768.
52 See JWP, 4/7.18, J. Watt sen. to J. Watt, 9 Dec. 1768 and 4/7.15, 2, 12 & 20 Feb.

1769; 4/11.146, J. Watt to J. Watt sen., 9 Dec. 1768, 4/11.147, 10 Dec. 1768 and 4/11.148, 21 Feb. 1769.

53 Donovan, (1975), p. 79.
54 Clow, (1952), p. 136.
55 Donovan, (1975), pp. 77 & 79 - 80.
56 Ramsay, (1918), p. 36.
57 Ibid, p. 80, contained in F. Home, *Experiments on Bleaching*, 2 edn., 1771. See also F.G. Page, 'Chemical and Analytical Aspects of the Early Alkali and Bleach ing Industries in Britain', Ph.D. thesis, University of Leicester, 1999.
58 Jardine, (1796), p. 86 and Ramsay, (1918), p. 54.
59 See Musson & Robinson, (1969), Chap X, 'James Watt and early Experiments in Alkali Manufacture', pp. 352 - 371.
60 Boulton 348.12, J. Watt to W. Small, 21 Oct. 1769.
61 JWP, 4/59.7, J. Watt to W. Small, 20 Oct. 1769.
62 B. & W. Col., MIII, 3/2, 'Waste Book 1765 - 1769', 11 Dec. 1766.
63 JWP, 4/32.30, J. Watt to J. Roebuck, 19 Feb. 1766.
64 See Musson & Robinson, (1969), pp. 369 - 371 for appendix by Drs. W.V. & K.R. Farrar.
65 JWP, 4/32.29, J. Watt to J. Roebuck, 29 Feb. 1766.
66 JWP, 4/32.6, J. Watt to J. Roebuck, 3 March 1766.
67 JWP, 4/32.28, J. Watt to J. Roebuck, 20 April 1766.
68 JWP, 4/32.27, J. Watt to J. Roebuck, 14 May 1766.
69 JWP, 4/32.4, '1766 or 1767'.
70 JWP, 4/32.5, J. Watt to J. Roebuck, 1 Nov. 1766.
71 JWP, C1/15, J. Watt to J. Lind, 25 Oct. 1768.
72 Boulton 340.1, W. Small to J. Watt, 7 Jan. 1768 but must be 1769.
73 Boulton 348.12, J. Watt to W. Small, 21 Oct. 1769.
74 Robinson & McKie, (1970), p. 16, Letter 9, J. Black to J. Watt, 23 Jan. 1769.
75 Ibid, p. 17, Letter 10, J. Black to J. Watt, 26 Jan. 1769.
76 Boulton 348.4, J. Watt to W. Small, 28 Jan. 1769.
77 Robinson & McKie, (1970), p. 18, Letter 11, J. Watt to J. Black, 27 Jan. 1769.
78 Ibid.
79 Ibid, p. 18, Letter 12, J. Watt to J. Black, 28 Jan. 1769.
80 Ibid, p. 19, Letter 13, J. Black to J. Watt, 28 Jan. 1769.
81 Ibid, p. 20, Letter 14, J. Watt to J. Black, 14 March 1769.
82 Boulton 348.8, J. Watt to W. Small, 28 April 1769.
83 Boulton 348.10, J. Watt to W. Small, 5 July 1769.
84 Boulton 348.11, J. Watt to W. Small, 20 Sept. 1769.
85 Boulton 340.8, W. Small to J. Watt, 10 Oct. 1769.
86 JWP, 4/32.105, J. Roebuck jun. to J. Watt, Nov. 1769.
87 Musson & Robinson, (1969), p. 3.
88 Robinson & McKie, (1970), p. 25, Letter 21, J. Keir to J. Watt, 14 Dec. 1771 and p. 26, Letter 22, J. Keir to J. Watt, 27 Feb. 1772.
89 Boulton 340.24, W. Small to J. Watt, 16 Nov. 1772.
90 Boulton 340.29, W. Small to J. Watt, 15 March 1773.

Watt and the Delftfield Pottery

The Origins (1)

> 'I have now a small share in the Delft work along with a new company. We intend carrying on a Stone manufactory likewise, and are getting proper people for that purpose' (2)

Watt had borrowed money from his relatives to invest in the Delftfield Pottery in Glasgow and this letter of 15 February 1768 to Black is the earliest reference that has survived confirming Watt's involvement. He was then acting as a sort of technical adviser as well as a partner. In 1786, he claimed that he had laid the foundations of this Company's prosperity in a letter to Robert Muirhead, his cousin, when it had been suggested that he should surrender his partnership.

> I certainly would not willingly relinquish my share in the Delftfield Company but cannot think of retaining it against the opinion of a majority of the company. I am sorry it has not been in my power to be of more service to the concern lately but on the other hand it should be considered that I took much more trouble upon me in laying the foundations of their present prosperity than the share I had required me to do and that I have hitherto run all risks along with them nevertheless if it is their will that I should resign, I shall comply without repining. (3)

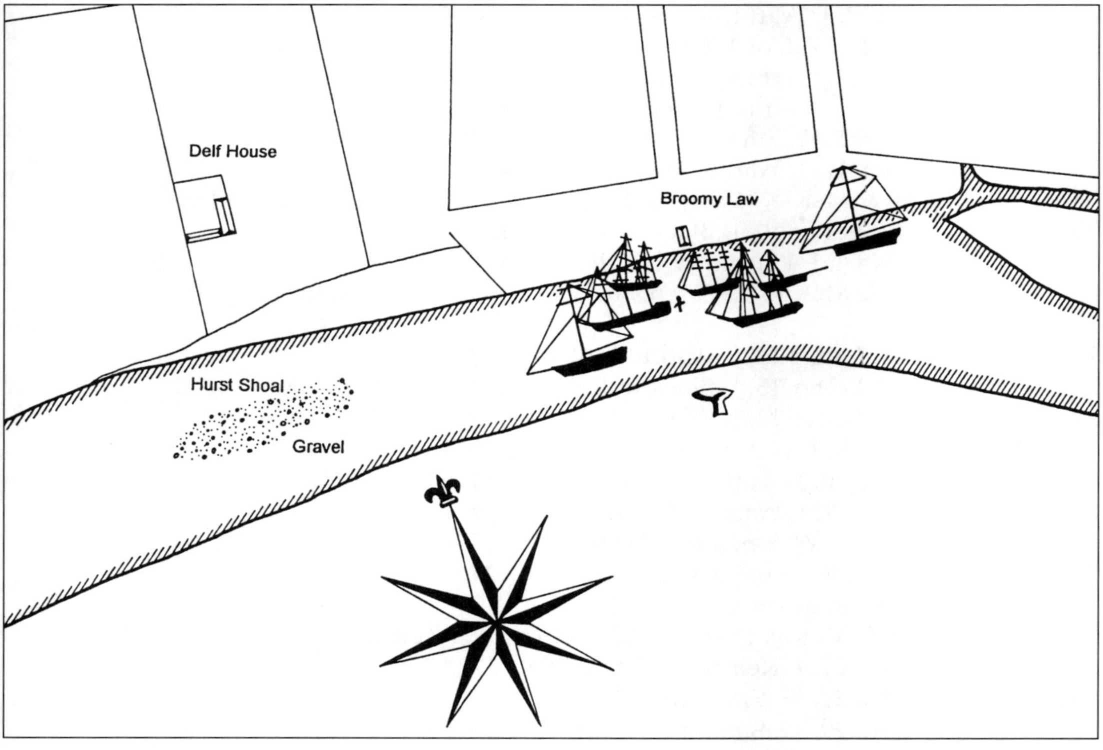

A Redrawn Portion of a Map prepared for Smeaton by Barrie and Wilson in 1758 showing the Delftfield Pottery.

He wrote in a similar vein to Hamilton. (4) The important point to notice here is Watt's claim of setting the Delftfield Pottery on the road to prosperity, which he achieved through his knowledge of chemistry. His involvement in this partnership was different from those with Black, Boulton and Roebuck, because it never developed into any close friendships. The Delftfield partnership remained on a business level, but nevertheless was important to the story of the steam engine because it provided Watt with a regular income at a time when he was running down his scientific and merchanting businesses to start his career as a civil engineer, as well as after he had moved to Birmingham and was struggling to launch his steam engine.

The Delftfield Pottery (often referred to by Watt as Delfthouse) was established in 1748 when Laurence Dinwiddie, who was one of Glasgow's leading tobacco merchants, invited his elder brother, Robert, to join him. Robert had been trading as a merchant in London until 1751 when he became Governor and Commander in Chief of His Majesty's Colony and Dominion of Virginia. He retired in 1757 through ill-health and went to live at Clifton, near Bristol. Robert Finlay, a cordwainer (shoemaker), tanner and merchant, was the third partner; Patrick Nisbet, of whom little is known, the fourth. (5) In the original agreement, Laurence and Robert Dinwiddie between them had three fifths of the land on which the pottery was built and Robert Finlay and Patrick Nisbet one fifth each. This reflected their financial investment in the concern. (6)

None of these investors had any knowledge of pottery manufacture and, throughout its history, the partners, who provided the bulk of the capital were business men whose main interests lay in other enterprises. Therefore they had to employ a manager who had some knowledge of this industry to run the company for them. This person might be offered a share in the partnership but was never the leading figure and probably had little interest in chemical experiments. This may be a significant point when their success is compared with that of Josiah Wedgwood, who was trained as a potter and was passionately interested in the development of his works.

Laurence Dinwiddie owned land at Germiston to the east of Glasgow where there was clay and there was more clay on the site where they decided to erect their pottery. The intention was to produce tin-glazed delftwares from the local clay deposits in the English and Dutch manner which, at that time, were in demand both at home and also overseas. Delftware held an intermediate position between expensive porcelain and cheap earthenware so that the growing prosperity of a middle-class in Glasgow might be expected to provide one market while Laurence Dinwiddie's trading links across the Atlantic would have provided another.

The site chosen was situated on the west side of Glasgow close to the River Clyde, just down-stream from the main quay, the Broomielaw. The land was purchased from the Merchants' House of Glasgow. A point, that was stressed in the sale, was that 'they have no right or interest by virtue hereof to the waterside ground or Grass thereof upon the South of the lands but only free access to their said lands by the waterside ground'. Between their land and the river itself, there was to be left a strip free to act as a highway. (7) The map of the Clyde, drawn by Barrie and Wilson in 1758 for Smeaton, shows the pottery as an 'L' shaped building at the southern end of the plot. There is a space between it and the river bank which appears to slope down to the river. Opposite the 'Delf House' in the river is the Hirst shoal of hard gravel while, upstream, ships are anchored off the Broomielaw quay. The quay itself stops short before it reaches the Delftfield plot. (8) Therefore, while the pottery was situated in a good position close to the head of the navigable part of the Clyde, it did not have direct access to the water so that goods would still have to have been transhipped from the Broomielaw.

The partners appointed a young man, John Bird, who was a painter at the Lambeth pottery, to act as manager. He drew up plans, but apparently the buildings were not constructed according to them. The local clay proved to be unsatisfactory so more had to be imported from Carrickfergus in Ireland. There were many disputes. The initial firing of the kiln went badly wrong which was only one of several misfortunes. Bird was soon replaced which led to a lengthy lawsuit. However, after this, the Company seems to have settled down to reasonably successful production.

On the death of Laurence Dinwiddie on 3 May 1764, his share passed to his son Robert. It has been suggested that he might not have had the same interest in the concern as his father and, both with his Uncle Robert living in Bristol and perhaps through trading difficulties, the pottery was advertised for sale in the August and September of 1765. (9) As there was no response, there was a further advertisement in May 1766. (10) But we shall see that Robert Dinwiddie, the son, was still a partner in 1783 so it seems more likely that the proposed sale was caused through Nisbet who seems to have got into financial difficulties in 1765 and transferred his share in the business to four merchants, James Buchanan, James Coulter, John Shortridge and David Niven (or Nivien). There is no mention of Watt. (11) The new partners were Freemasons, members of the Glasgow Kilwinning Lodge, to which Robert Finlay's brother John and the lawyer who handled the proposed sale of the works also belonged. Watt was already a Mason in a different lodge. James Coulter's brother was Robert Dinwiddie's uncle so these people had close links.

The Product

When it was established, the Delftfield Pottery was in competition both with other rival manufacturers particularly in London, Bristol and Liverpool as well as with the development of newer types such as cream-coloured earthenware which appealed to a wider market. (12) In England, the name 'delftware' came to be given to earthenware glazed with a mixture of tin (from 25% down to 5%) and lead through its similarity to the products of the Dutch town of Delft which were at the height of their popularity up to the middle of the eighteenth century. English delftware was made in much the same way as earlier Italian maiolica or faience. The clay was first thrown on a wheel or pressed into a mould to produce the required shape. This was then fired in a kiln for the first time. When it was cool, the resulting 'biscuit' body was dipped into the glaze which was allowed to dry. This absorbent surface was then ready to be painted, taking care not to make any mistakes as these were impossible to correct. The piece was fired a second time in order to fuse the glaze to the body and to fix the decoration. However delftware had a soft but thick and heavy body which was quite fragile and so easily chipped or broken. The glaze did not fit closely and might craze. Often it had to be thick to be sufficiently opaque to conceal the colour of the body because frequently coloured clays were used which otherwise might show through the glaze. Although delftware was easily cracked with boiling water, it cost much less than imported porcelain and hence was very popular, large quantities being made.

But the great majority of people sought after porcelain through its brighter colours, thin, hard, translucent body and tough clear glaze yet they could not afford it because it was imported from China. So it was the dream of every ambitious potter to develop either true porcelain itself or a fine white earthenware that could imitate its appearance without incurring highly expensive methods and destructive losses in production. True or hard-paste porcelain was first made in China in the ninth century AD. It was composed of a mixture of china clay (*kaolin*, a pure white clay formed by the decay of felspar, the chief constituent of granite) with china stone (*petuntse*, less fully decomposed felspar). After the vessel had been formed, these constituents were fused together by firing in a kiln, first at about 900°C, then dipped in glaze and refired at about 1300°C. The blue decoration of cobalt was particularly admired.

Many Europeans tried to discover the secrets of Chinese porcelain. Hard-paste porcelain was first produced at Dresden, Saxony, about 1709 by Johann Friedrich Bottger and from there attempts to copy porcelain spread slowly across Europe. Investigations into methods of making porcelain in England began in the 1740s. These used a mixture of china clay and ground glass which resulted in a soft-paste variety. This type had a white translucent body but easily warped in the kiln; hot water was likely to make it crack; the glaze surface was easily scratched and absorbed stains. At Bow, the addition of bone-ash improved the ability of the porcelain to withstand the heat of the kiln and this was adopted in a number of factories. The next major discovery was that Cornish stone and china clay would form a true hard-paste porcelain body. William Cookworthy, a native of Kingsbridge in Devon, after training in London, set up as a chemist and druggist in Plymouth. He started experimenting around 1747 but it was not until 1768 that he took out a patent for his porcelain formula and set up a manufactory in Plymouth. Sometime before he gained his patent, he sent to Black in Edinburgh samples of kaolin and petuntse which he had found in Cornwall in 'immense quantities'. (14) Black carried out tests and found that the kaolin burnt 'to a very pretty stone porcelain'.

In the meantime during the early 1700s, while European potteries concentrated on developing formulas for porcelain, most British counterparts improved existing ceramic types to produce stoneware. Clay was the raw material which most influenced the appearance of the pottery and it was this which underwent large changes before 1760. Improvements which resulted in pottery having paler, stronger bodies were associated with the introduction of white 'ball' clay together with ground flint. Ball clay must be distinguished from china clay with similar properties but also important differences. The principle advantage of ball clay was that it fired at temperatures up to 1300°C to produce white pottery, which meant that ware made with it could compete with expensive imported porcelain. It could also challenge the soft white-coloured delftware. (15) Ball clay was used in Staffordshire from about 1710 and flint about ten years later. Ground flint was mixed with clay to make the body stronger and more able to withstand the high temperatures necessary in the firing of stoneware. Sometimes as much as half the clay mixture for the body was ground flint but more usually about one-third. Flint was also mixed with lead in the preparation of liquid glazes at least by 1760. (16)

These newly imported materials of flint and ball clay allowed the increased manufacture of a completely cream-coloured body and, by 1740, this type was much in demand for popular domestic tableware. It is from this period that the growth of the Staffordshire potteries really began. Wedgwood's contribution was to discover reliable formulas for both body and glaze after conducting a series of experiments on a scientific basis and suffering many disappointments and heavy losses before control of the firing was mastered. His reward came in 1766 when Queen Charlotte bought a service of Wedgwood's refined cream-coloured earthenware and allowed it to be called 'Queen's Ware' in 1765. (17) The Queen's acceptance of this improved earthenware or stoneware helped to persuade the greatest people in the land that, although much cheaper to produce, it was of such quality that it could be used instead of porcelain and so it became fashionable. (18) This cream-coloured earthenware would extinguish the manufacture of tin-glazed earthenware or delftware across the Western world. This outlines the competition faced by Delftfield.

Watt Joins the Delftfield Pottery

We have no clues over when or why Watt started to take an interest in the manufacture of pottery. The most likely reason was his interest in chemistry. We have seen how the startling advances in pottery techniques which occurred during the years up to 1760 were linked with the discovery of new raw materials and their chemical analysis. If others could make improvements through scientific experiments, perhaps he, Watt, could as well and create the fortune for himself which he so much coveted. His knowledge about some of these changes may have come through seeing the latest examples of ceramic design when he travelled to London, Bristol and elsewhere on behalf of his scientific instrument business. Perhaps it was the brilliant colours in the best porcelain which prompted him to try and emulate them through experimenting with different chemical compositions. We have no indication that he was involved at this early period with the design of either shape or style of the products although later he did criticise some of the output.

We have seen how Cookworthy consulted Black about his discovery of kaolin and petuntse in Cornwall which may show that Black's interest in the pottery industry was widely known. So perhaps Black could have influenced Watt. Black ventured to define chemistry in the following words:-

> CHEMISTRY is the science or study of those effects and qualities of matter which are discovered by mixing bodies variously together or by applying them to one another with a view to mixture, and by exposing them to different degrees of heat, alone, or in mixture with one another, in order to enlarge our knowledge of nature, and to promote the useful arts. (19)

Surely it must have been unintentional that Black should have given such a fine description of the potter's art. Like Cullen, he was interested in the application of chemistry to improve industrial processes; a lesson quickly learnt by Watt.

Although the topic of porcelain may not have been included in his lectures at Glasgow, Black later covered its manufacture. 'Porcelain is another production of art which has been invented upon principle by making chemical experiments upon the earthy and stony substances, and which derives to be admired for the elegance and usefulness of its productions'. (20) He outlined its history and described its production. He also commented that nobody had made so many experiments in the study of earthenware and porcelain as 'Mr. Wedgewood'. He must have kept himself well informed about progress in the pottery industry and knew the importance of scientific experiments in its improvement.

Black studied the quantitative science of heat which was at the heart of so many industries for he realised that heat was the chief means of promoting chemical changes. Watt had manufactured Black's design of a furnace, which may have given useful background knowledge for the pottery industry and its need for kilns. The earliest letters that have survived between these two scientists concern kilns and experiments with them on ceramics. It looks as if Watt was seeking advice from Black because he knew Black had had experience not only of furnaces but of ceramic experiments as well.

At this point, the question must be raised of Watt's involvement in the Delftfield Pottery before he wrote to Black on 15 February 1768 telling him that he had purchased a small share. It has been suggested that he was a partner much earlier. G. Quail gives no reference other than Dickinson for his statement, 'Watt's association with the Delftfield pottery possibly started about 1760. At this time he may have sold some of their wares at his shops in the Saltmarket and Buchanans Land. He possibly did odd experiments for the company'. (21)

There are no entries in Watt's accounts for his shops of purchasing anything from Delftfield but in December 1762, his uncle John Muirhead sent to 'Greenoak Eight Hodgets of Delphware to go to Jamica'. (22) Likewise no evidence has been found to support Dickinson's claim that Watt acquired an interest in the Company in 1763. (23) J. Kinghorn and Quail say that Watt was a partner in 1764 with £244.4s.2d. invested. (24) This figure corresponds with one quoted in Watt's Journal for 1772 so may have been given an earlier date. When the Company being offered for sale around 1765, Watt's name was not included among the partners. (See p164)

In Watt's association with the company, the Masonic links may have been important. (25) With the Company up for sale in 1765, it is possible that production had ceased which, together with the Masonic links, may explain how tradition claims that Watt occupied the manager's house after his marriage to Peggy. This must have been the building which William Martin occupied in 1760 when he became Managing Partner. (see p105) Martin was still involved with the concern in 1772 and died in January 1779 'at the Delftfield in the neighbourhood of this city'. (26) Presumably he had vacated the building to allow Watt to move in.

In the 1760s, delftware was being eclipsed by other types of earthenware so that, if Delftfield Company were to survive, it had to change its product. At this period, Delftfield was experimenting with other types of pottery and in 1766 introduced white stoneware which was harder in the 'paste' than delftware. (27) The *Glasgow Journal* for May 1766, (28) which carried an advertisement for a new ware from the Delftfield Company, claimed that "by some new compositions luckily discovered... [they had] brought it to a still greater degree of perfection as it now holds boiling water and is made stronger than formerly'. We do not know whether Watt was involved with this improvement so it remains speculation that he may have helped Delftfield at this time which led to the partnership a couple of years later.

Watt as Technical Adviser

To produce stoneware would involve changing nearly all of the processes at the Delftfield Pottery with the exception of the potter's wheel. We have seen that experiments must have been carried out, but we have no details of anything before 1768. It is not clear whether Watt tried to improve delftware or whether his trials were for new types of pottery. The science of chemical analysis was in its infancy and all progress in the development of a ceramic body, or glaze or any other technique, was, by necessity, empirical. Potters tried to conceal their secrets from their rivals so each person had to discover anew the latest advances. Identification and isolation of materials was the first necessity before any work could begin on observing their behaviour. We will examine Watt's investigations through the various production stages.

Clay

Different types of clay were essential to make the change from delftware into stoneware. Watt must have experimented with clays soon after joining the Company in 1768 using some which his father had sent him that February. His father mentioned various types which he had seen or heard of in Scotland and where they were to be found. (29) A little later, Watt told Black,

> I shall be extremely happy to see you, and can communicate a great many particulars relating to earthenware. From an accidental experiment I found that p[ipe] clay may be burnt as white and hard as any porcelain Nothing hitherto discovered can be wrought at

> the wheel with double the trouble it can. If, therefore, this could be made to succeed in practice, nothing would be necessary for china but stone body. (30)

Perhaps Watt was trying to copy porcelain.

Pigott, an agent in Liverpool, had some clay, possibly from Carrickfergus, ready for shipping in June 1768 and could send thirty tons. (31) Watt's father was forwarding a small cask from Liverpool for Delftfield a year later but failed to mention the contents. (32) Clay is not mentioned again in Watt's correspondence while he lived in Scotland so presumably he must have established a satisfactory mixture. His interest in clays was renewed later through erecting steam engines to drain the tin and copper mines in Cornwall. This county and Devon were the main sources for ball clay, kaolin, and petuntse, all crucial for cream-coloured earthenwares and porcelain. He continued to help both Delftfield and Wedgwood by advising on types of Cornish clays.

Grinding Flints and Colours

Part of the toughness of stoneware was achieved through the addition of ground flint to the clay. The Delftfield Company had taken possession of the North Woodside Mills on the River Kelvin in June 1758 when they were offered for sale. (33) At this period, the mills were used to prepare colours. Then an advertisement appeared in the *Glasgow Journal* in June 1768,

> A MILL TO BE SET
> That the Delft Colour Mill at Woodside which is fit for a Snuff or a Waulk Mill is presently to be set. For further particulars enquire at the Delph House or at Mr. Watts shop in Buchanans land. (34)

The Company still retained possession much later when another business was threatening to build another mill on the same fall in the river and Watt was asked for his advice in his capacity as a civil engineer together with Martin their manager. (35) At a meeting of the Company at Woodside in September 1772 to see the situation of this other mill, Watt's opinion was 'They should be allowed to proceed upon binding themselves to do nothing to hinder our milne'. (36)

The Delftfield Company appears to have retained the Woodside Mills in order to grind flint there. In 1726, Thomas Benson had patented a method of grinding calcined flints immersed in water in a large iron pan in which iron balls were driven round by arms attached to a central rotating shaft. In a further patent of 1732, he proposed using granite balls on a similar bed which avoided iron particles discolouring the pottery. (37) This would have been the type of mill installed by Watt for Delftfield but just when is unknown. In 1783, Hamilton was hoping to drive his pans with a steam engine so that he could bring all the processes under one roof instead of having them three miles apart. (38) Later Watt described what he had done.

> The small round tubs for grinding flints were originally used in Staffordshire but are now entirely laid aside, & they have even got to tubs of 20 or more feet in diam. I erected one of 8 feet in diam. at Delft mill, but proof was brought by Martin and his men that the small tubs did infinitely better & it was laid aside, & I am afraid condemned unheard all I can say is that it holds with reason that the large tubs should do best, as having superfluous friction... they have commonly 4, 6, or 8 large stones and they take out or put in according to the quantity of water they have to drive them. (39)

Glazes

On joining the Delftfield Company, Watt experimented with glazes. He told Black,

> I have contrived a glazing for stone-ware that will bear the ware to be burnt with it on it; for if that were not the case the trouble of putting the glazing on a hard-burnt close body would raise the price too high, but this can be sucked in upon a bisket and, when burnt, is very little if any inferior in beauty to your china, which appears to me to be made of pipe-clay and petuntse, and glazed with an opaque white glazing. What that glazing is I have not discovered; my glazing is a glass of lead and tin without salts, which in great heats are apt to render the tin transparent, and at same time they prevent the glazing from running smooth on the ware without it be made very fusible, in which case it runs about. But I have by no means acquired all the knowledge I must have in this article, having just now little time to spare on it". (40)

While salt glazed earthenware had been popular during the earlier part of the eighteenth century, towards the 1760s, lead again became significant as a glaze. Lead ore, mixed with flint or ball clay and water, was normally applied in liquid form to the already fired 'biscuit' body. But the lead tended to give the glaze a yellow colour, varying in shade according to its thickness and the whiteness of the underlying fired clay; (41) hence possibly Wedgwood's cream ware. After about 1770, Wedgwood was able to make it appear whiter by adding minute quantities of cobalt, so he called this 'pearl ware'. One of the challenges was to produce a true white glaze in imitation of porcelain, something to which Watt would return later.

Cobalt

The work of painters on delftware had been dominated by one thing above all else and that was the blue and white porcelain imported from China. Cobalt was the most satisfactory blue colouring agent because the colour produced was much less susceptible to changes in firing conditions in the kilns and it could be counted on to produce true blue colours varying in intensity with the concentration of cobalt in the glaze. But the high price of cobalt was a continuing problem. It was imported from Saxony which kept supplies small and prices unnaturally high through regulation of its export. Smuggling was punishable by death. It was claimed that around 1760, '£200,000 Sterling goes out annually for that Article of Commerce'. (42) Hence cobalt was a much sought-after commodity.

Just before 1759, in a glen above Alva on the Ochils, a vein of cobalt was discovered. However the veins of cobalt at Alva proved to be elusive and mining may have ceased in 1766 although a stockpile of ore remained available and samples were sent to interested people. Sir Charles Erskine sent samples of ore to a number of people for testing and appears to have received encouraging reports. Black was one of these and said, 'I shall compare some trials to be made by painting some delftware and comparing the colour with saffer used here'. (43) Saffer or zaffer was an impure oxide of cobalt. In our first surviving letter from Black to Watt written on 10 January 1768, Black asked Watt to send the 'Manganese and Cobalt'. (44) Manganese could be used to produce black and, in conjunction with cobalt, mulberry ceramic colours. This again suggests Black's interest in the pottery industry.

The Delftfield Company received samples at the beginning of 1769 when Watt was able to carry out experiments. He reported his results in a letter to Lord Barjarg.

> I took out at random 12 lb of Cobalt. This I separated by picking into two parcells of which the best weigh'd 6 lb. 10 1/2 oz. The other 5 lb. 5 1/2 oz. These were carefully

> roasted & pounded. The best then weigh'd 5 lb. 5 $^3/_4$ oz. The second weigh'd 4. 11 oz. both together 10 lb. 10 lb. of Ground Flints was added to them & they were levegated seperately by the Mill the best gave on the ware a very good blue but not so deep as the Dutch Zaffre tho superior in Colour. The second was very light & not so good a Colour. From this Experiment it appears that the best roasted Cobalt will take no more than $^3/_4$ of its weight of Flint the other none at all. On the whole 18 oz. to the pound will make it about as deep as the lightest Dutch Zaffre. 12 lb of Cobalt unroasted then is equal to 15 lb. of Zaffre These 15 lb. of Zaffre wou'd cost us about 30s. (45)

Watt may well have visited this mine because it was situated quite close to where he was surveying to make the River Devon navigable. However production had declined to nothing in 1768 and the lease was forfeited. (46)

But this did not end Watt's involvement with cobalt because another source was discovered while he was in Cornwall at the end of 1781. At the back of his Journal for this period is a note on chemical tests he carried out on 'Mr. Beauchamps Cobalt' (47) and probably about the same time he entered into his Common Place Book notes on how to make royal blue from impure cobalt based on the method of Gerhard Abbé Rozier. (48) In 1782, he investigated a cobalt mine in Cornwall but this turned out a failure.

Other Glazes

In his Journal he wrote in August 1770, 'Visited the Delfthouse in the evening think they are improved in manufacturing their plates but still deficient in glazing the man Austin is come but not begun work gave directions for building a new house for part side of square'. (49) This also shows other ways in which Watt was involved and we will look at the personnel aspects later. On the first of September, Watt went again to 'the Delfthouse saw Austin's glaze which is very bad - I imagine the greeness of our glazing is owing to the want of manganese". (50) This was followed by another visit on the seventeenth,

> Went to Delfthouse - their last kiln had been pretty hard burnt & was of a tolerable colour the glazing is however still deficient in not being uniformly smooth having misty spots upon it, being too transparent & of a greenish colour. I imagine that the addition of a little mang.e & arsenic might remedy these. (51)

He must have continued with experiments to produce a white glaze for in the autumn he tried mixing bone ashes with clay or window glass, bone ashes and something else (word illegible) and heating them violently but discovered that they did not produce the same whitening effect that they once had. (52)

In spite of some failures, these notes show how Watt's knowledge of the chemistry helped put the Company's products on a firm footing. In 1772, the concern, advertising as Dinwiddie, Martin and Company, claimed that 'having brought the STONE and DELFT Ware to the greatest perfection are determined to serve the Public on as low terms as possible and have already reduced the price of several of their articles'. (53) Part of their success was due to Watt's investigations and part due to recruiting men from a rival company in Glasgow which had imported men from Staffordshire (54) so that in 1773 they placed an advertisement, probably written by Watt, in the *Glasgow Journal*, (55)

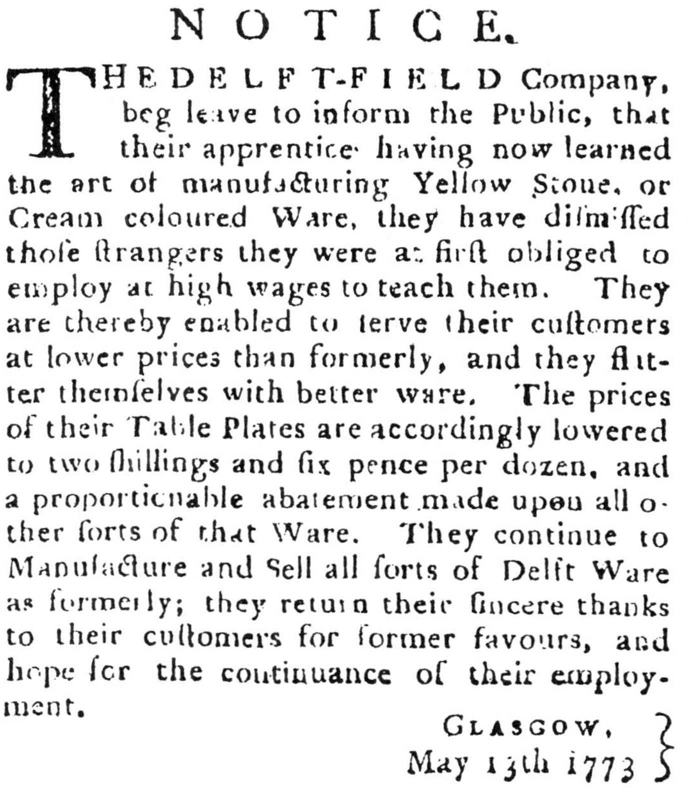
NOTICE.

THE DELFT-FIELD Company, beg leave to inform the Public, that their apprentice having now learned the art of manufacturing Yellow Stone, or Cream coloured Ware, they have dismissed those strangers they were at first obliged to employ at high wages to teach them. They are thereby enabled to serve their customers at lower prices than formerly, and they flatter themselves with better ware. The prices of their Table Plates are accordingly lowered to two shillings and six pence per dozen, and a proportionable abatement made upon all other sorts of that Ware. They continue to Manufacture and Sell all sorts of Delft Ware as formerly; they return their sincere thanks to their customers for former favours, and hope for the continuance of their employment.

GLASGOW, May 13th 1773

Advertisement for the Delftfield Pottery probably prepared by Watt in 1773.

> THE DELFTFIELD Company beg leave to inform the Public, that their apprentice having now learned the art of manufacturing Yellow Stone or Cream Coloured Ware, they have dismissed those strangers they were at first obliged to employ at high wages to teach them. They are thereby enabled to serve their customers at lower prices than formerly, and they flatter themselves with better ware. (56)

We know that Watt had improved their cream ware through a letter from Robert Muirhead in 1775,

> At the time you was paying so much attention to the bettering of the cream colour'd ware you found out a method of giving it as fine a glaze as China & perfectly white We think as our buildings are sufficient for such an undertaking that upon your describing to us that secret it will be worth our applying for a patent for making such Ware, and that you shall have such annuity during the continuance of such patent for your trouble & Expence and ingenuity as we can agree upon with you - If you like this proposal let me know as also when you'll be down in Scotland. (57)

Watt could not have answered because Muirhead asked again about it in March 1778 when he described it as 'perfectly white... and an enamel as fine as any China'. Watt had apparently lost the recipe but then found it again so Muirhead asked whether Watt was willing to reveal the secret only to himself and Dinwiddie and what would be Watt's terms as he supposed that this was 'an entire new art and has not been discovered & practised by any other persons in Britain'. (58) This leads to the speculation about what Watt could have achieved if he had concentrated fully on the pottery industry. Again there could have been no reply because Muirhead wrote once more in February 1779 wondering whether Watt could do some more experiments (59) and, in July Hamilton passed on a request from Duncan Nivien to Watt to see if he could discover the components of Wedgwood's galzes. (60)

At last Watt replied and it is from this that we derive most of our knowledge about his work on glazes.

> I know not the composition of the cream Coloured body as used in Staffordshire but was formerly told they used 3 parts clay and one of flints for it and for the white stone ware 2 ps clay & one flint - The flints give whiteness but encrease the tendency to ring crack in the ?...? I should imagine the first composition a good one - all whitish or opake whitish flints should be rejected they contain calcarious crust which is highly prejudicial, all calcarious crust should be rubbed off the flints before calcination - The utmost care should be used to prevent dirt entering the clay or flints, or *smoking the biscuit* as to the Glazing I cannot readily find my memorandums but think it was 12 lbs white lead *from Bristol* 20 lb flints and 10 lb clay The clay gives hardness & prevents crazing, but if too much is used, causes dryness & want of glass. If the glazing is wanted to be paler may add to the whole ?mass? of the above composition 10 lb best flint glass ground to very fine powder but let this be tryed only on a parcel of ware dispersed in various parts of the kiln - If the growan clay has not reached you from Cornwall I should expect it to be a good succedareum in place of part of the flints in the body - neutral salt in small proportions whiten clay Alum or vitriolated tartar are I believe the best the latter can be had here very cheap may try in small batches - Above all things a strict attention to cleanliness adds the most to the beauty of the ware but could never be obtained at the Delftwork - no ware which is not clean & white in the biscuit will look well in the glaze Care should be taken to eradicate all ?dran? clay out of the sagar clay the smell of it makes brown our sagar clay not being white should be slipt with white clay ?...? or have some gypsum mixed in their glazing'. (61)

Improvements to Kilns

Of much greater importance were Watt's improvements to kiln design. (62) Soon after he set up his scientific instrument business in Glasgow in 1757, he made some parts for Black's furnaces and around 1764 he tried to improve the design of small steam boilers. In view of Watt's later patenting in 1780 a steam engine boiler with fire grate for consuming smoke, his improvements to pottery kilns must have given him good experience. Watt's letter to Black in February 1768 shows that the Delftfield Pottery was still employing the traditional square type of kiln fired with wood. These needed a great deal of skill to achieve successful results, partly because different types of wood had to be used at different stages and care had to be taken to prevent ash settling on the pieces being fired, causing black spots on the glazing. Watt pointed out that 'This way of firing is not only expensive, but inconvenient' and that a great deal of the ware was spoilt. (63) He commented on the route of the gases passing through the kiln, realising that some parts heated more than others. It is surprising that coal, readily available in Glasgow, does not seem to have been tried because Henry Delamain, a delftware potter in Dublin, had spent more than £1,000 in the 1750s on designing new kilns for coal. (64) Perhaps the Delftfield partners were unwilling to invest such capital without being certain of success.

Illustration of the later circular type of kiln which was installed at Delftfield while Watt was there.

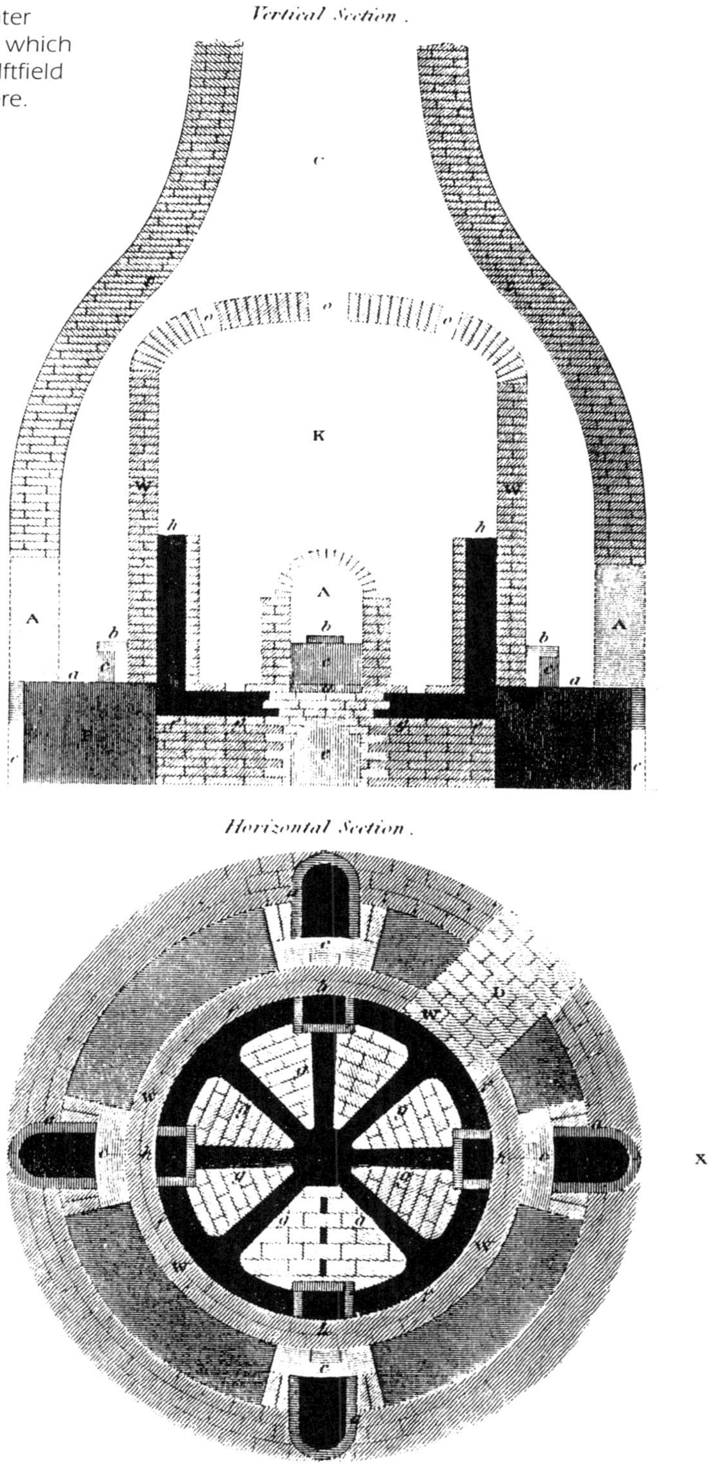

However Watt was willing to carry out experiments, but no mention is made at whose expense, probably his own. He sent Black a description of a small new kiln he intended to construct and fire with coke and sought Black's advice on its suitability. It was cylindrical, four feet diameter and four feet high, closed at the top with an arch, terminating in a two foot high chimney. (65) So his experiments were on quite a large scale. He must have been pleased with his results for he later told Black, 'I have tried the cokes for burning our Delft; they answer very well'. (66) But it was only after alterations to the grate that he made the kiln heat equally, although ash still settled on the ware, which he proposed to prevent by closing the chimney when the fire was stoked and sealing the saggars better. These small scale experiments were carried out after Watt had been successful with the model Newcomen engine of Glasgow University. Much later, he had models constructed of his different inventions for the steam engine so he frequently employed models for his experiments.

Watt also described how the type of kiln for stoneware was like that of common potters. Pit coal was used and the firing lasted for forty-eight hours. The advantage for the Delftfield Company would have been cheaper fuel than wood. Watt experimented in a different little furnace that he had but found difficulty in raising more than 'a bold red heat'. He then tried coke, which soon had the furnace up to a 'bright white' heat. (67) The delftware fired in it was equal in colour to any. Watt was surprised at this difference in heat from the two types of fuel and so once again sought advice from Black. Watt doubted whether the heat from coke would spread as far as that from flames. Black replied, 'A kiln which is heated with flaming fuel may be said to be heated by means of a torrent of *liquid fire* which flows through it, but I am persuaded that cokes act mostly by a radiation like that of the sun'. (68) Watt carried on with his experiments and told Black, 'I have strange and incredible things to tell you about stone kilns, which, if true, cokes will beat pit-coal even as to quantity or cheapness'. (69) There was still much to be discovered about the nature of heat.

Watt was experimenting for over a year from February 1768 until at least May 1769. It may have been of great significance that he announced to his father that December, 'I have fired a kiln with coals which has succeeded', (70) because coals would be the fuel generally used. He must have been proud of this for that May he recommended Matthew Boulton to consider a china factory as a good trade and that he, Watt, thought 'I almost understand how to make a furnace for it'. (71) We may note that bottle ovens were first recorded at the Delftfield on MacArthur's map of Glasgow, published in 1773, but they must have been built earlier, possibly in connection with the introduction of stoneware. (72)

Personnel

We do not have any idea of the extent to which Watt was responsible for employment of personnel at Delftfield. When he went to London in the summer of 1768 to apply for his steam engine patent, he took with him £30 in cash from 'Delfhouse' and returned through Lichfield, Newcastle and Stone as well as Sandon. He paid 10s.9d. for 'pottery & horse' and £1.11s.6d. for 'Cream Coloured Ware'. Then he 'agreed John Hinton for the Delft work at £50 pr annum gave him 7 guinneas to engage other hands'. (72) Later he noted that the 'man Austin will not answer' (73) and in 1771 recorded, 'Called at Delfthouse with Mr Dinwidder & desired Mr. Martin to provide for a proper assistant for himself found the business there going on very well'. (74) More than ten years after he had moved to Birmingham, he was still actively involved in the Delftfield management and wrote in 1784 to a Mr. John Widows about possible employment at Glasgow.

The Partnership

While Watt remained in Scotland, we have few details about the partnership because he could see his colleagues and discuss the situation with them. At first it would seem that he did not hold a high opinion of his partners and wrote to Small in 1769,

> Our pottery is doing tolerably, tho' not as I wish. I am sick of the people I have to do with, tho' not of the business, which I expect will turn out a very good one. I have a fine scheme for doing it all by fire or water mills, but not in this country, nor with the present people. I have tryed no chemical experiments this winter. (76)

He was right about the concern doing well for he recorded in his Journal for 10 June 1771,

> Dined with the Delftfield company, found on Ballc. our books and throwing out the bad debts that our stock had increased £1700 = since we entered into partnership. We agreed that the interest of our capital should be paid up from the time we begun viz £100 to each full share. (77)

Watt's later comments to Small show that, while he was pleased with the financial result, he was not satisfied with the products. 'Our pottery does very well tho we make damned bad ware & my 8th part worth £70 p annum clear to me. I am still f[seal] the china scheme & am certain it may be made to answer'. (78) His investment of apparently an eighth part of one of the shares was paying a good return in 1771.

In 1772, Delftfield did even better, and this is the possible source of the quotation used wrongly by Dickinson and others about Watt's capital and involvement in 1764.

> 13 June - At meeting of Delftfield Compy found the encrease of our stock since last year to be £833 from which deducting bad and dubious debts £195.1 remains our true profit £637.13.7. and we ordained Mr Martine to place the same to credit in proportion having paid to each one years interest of our capital being our proportion of the whole interest of £3320.14.8. or £41.10 to each full share.
> 14 June - My capital in the Delfthouse is now £474.4.2. besides £20 of interest to be speedily paid me I owe of that to Mr Moreheid £120 & to Nancy Millar £110 both with one years interest my free capital there being £244.4.2. (79)

So it looks as if Watt borrowed £230 from his relations to finance the purchase of an eighth of a share. The interest he paid them may have been 5 per cent so, with returns of around £70 per annum, he must have made a handsome profit. The £3320.14s.8d appears to have been the total capital invested in the Company at this date.

Business continued to do well and there was a possible payment of £20 on 2 March 1773. (80) Then we find another entry in Watt's Journal for July 1773.

> Delfthouse Company our clear profit this year are much the same as last year but we ordered only so much to be carried to our capital as made it £4000 in all & ordered payt. of interest at £40 to each full share. I recd. my £20 and paid Miss Millar interest upon bill £5.10. (81)

From this we may deduce that Watt had increased his holding to a half of one of the shares. Profits continued to come in because, that August, he wrote, 'Received from Delfthouse (of the £72 lent them) £32'. (82) As well as his capital investment in the part share, Watt lent money on interest to the Company at various times. Fortune continued to smile and, in April 1774, Watt dined 'with the Delftfield Company', presumably to discuss the results before the annual meeting which was held on 5 May. (83) He wrote about that meeting, 'dined with the Delftfield Company when we resolved to continue the partnership for one year more'. (84) For Watt, this was to be the last of such dinners for many a year because, at the end of that month, he arrived in Birmingham to erect his new type of steam engine at Boulton's Soho Manufactory.

The extent to which Watt's own discoveries contributed any improvement to pottery is impossible to determine, but it is reasonable to assume that his knowledge of the chemical processes enabled him to assess developments by others and to recommend to his partners whether such innovations were worth copying. What is surprising is how Watt found the time to carry out all the experiments which must have been necessary for him to explore the constitution of glazes, methods of kiln firing and much more. However this was never his full time employment as it was with Wedgwood. During his civil engineering career, he was often not paid promptly for his surveys so it would be interesting to know how much he relied on this income from Delftfield to support his other ventures.

Smiles's picture of the Broomielaw in around 1760. The cone on the left is at a glass works. The Delftfield pottery was out of the picture to the left.

NOTES: JAMES WATT AND THE DELFTFIELD POTTERY

1 A full account of James Watt's involvement with the Delftfield Pottery is being published by the Society of Antiquaries of Scotland.
2 Robinson & McKie, (1970), p. 8, Letter 2, J. Watt to J. Black, 15 Feb. 1768.
3 JWP, Letter Book 1, p. 176, J. Watt to R. Muirhead, 17 June 1786.
4 Ibid, p. 174, J. Watt to G. Hamilton, 18 June 1786.
5 Kinghorn & Quail, (1986), p. 11.
6 *Scottish Pottery*, 15th. Historical Review, (1993), H. Kelly, 'Documents relating to Delftfield', p. 44.
7 Ibid, pp. 43 - 44.
8 Royal Society, Smeaton, *Designs*, Vol VI, f. 54.
9 See *The Glasgow Journal*, 29 Aug. - 5 Sept., 5 - 12 Sept. amd 26 Sept. - 3 Oct. 1765.
10 Ibid, 1 - 8 May 1766.
11 SRA, B10/15/7148, 'Disposition Patrick Nisbet to Trustees for the Creditors', 25 March 1765.
12 *Scottish Pottery Historical Review*, No. 6, 1981, G. Quail, 'James Watt at Delftfield;the work of James Watt as a potter at Delftfield, Glasgow, and the development of cream-coloured earthenware', p. 45.
13 Weatherill, (1971), p. 14n.
14 Robinson & McKie, (1970), p. 10, Letter 3, J. Black to J. Watt, 19 Feb. 1768.
15 Weatherill, (1971), pp. 10 - 14.
16 Ibid, p. 23.
17 Copeland, (1972), p. 6 and Reilly, (1992), p. 42.
18 Reilly, (1992), p. 41.
19 Robison, (1803), Vol I, p. 11.
20 Ibid, Vol II, pp. 191 - 196.
21 Quail, (1981), p. 47.
22 JWP, JW/19.44, 'Miscellaneous Salt & Other Accounts, 1735 - 1771', from J. Muirhead, 30 Dec. 1762.
23 Dickinson, (1935), p. 29.
24 Kinghorn & Quail, (1986), p. 39.
25 Smith, (1994), Bryden, p. 16 lft.
26 Kinghorn & Quail, (1986), pp. 21 & 22.
27 Dickinson & Jenkins, (1927), p. 20. See also J.A. Fleming, *Scottish Pottery*, 1923.
28 *The Glasgow Journal*, 1 - 8 May, 1766.
29 JWP,4/7.11, J. Watt sen. to J. Watt, 4 Feb. 1768.
30 Robinson & McKie, (1970), p. 12, Letter 5, J. Watt to J. Black, 20 April 1768.
31 JWP, 3/41.3, Pigott to J. Watt, 6 June 1768.
32 Ibid, 4/7.8, J. Watt sen. to J. Watt, 6 May 1768.

33 Quail, (1981), p. 48.
34 *The Glasgow Journal*, 9 - 16 June 1768.
35 B. & W. Col., MI, 1/19, 'Journal', 22 Aug. 1772.
36 Ibid, 5 Sept. 1772.
37 Hills, (1994), pp. 178 - 9.
38 JWP, 4/43.24, G. Hamilton to J. Watt, 21 Aug. 1786.
39 Ibid, Letter Book 1, p. 159, J. Watt to G. Hamilton, 8 Jan. 1786.
40 Robinson & McKie, (1970), p. 11, Letter 4, J. Watt to J. Black, 1 April 1768.
41 Reilly, (1992), p. 17.
42 *Transactions of the English Ceramic Circle*, Vol 16, Pt. 2, 1997, J. Turnbull, 'Scottish Cobalt and Nicholas Crisp', p. 144.
43 Ibid, p. 144, and NLS, Ms. 5098, f. 49, Erskine Murray Papers.
44 Robinson & McKie, (1970), p. 8, Letter 1, J. Black to J. Watt, 10 Jan. 1768.
45 NLS, Ms. 5099, Erskine Murray Papers, J. Watt to Ld. Barjarg, 14 March 1769. I wish to thank Mrs. Jean Jones for drawing my attention to this letter.
46 Turnbull, (1977), p. 150.
47 JWP, C3/6, 'Diary, June 1780 - 1782'.
48 Ibid, C3/10, 'Common Place Book', p. 30.
49 B. & W. Col., MI, 1/15, 'Journal', 8 Aug. 1770.
50 Ibid, 1 Sept. 1770.
51 Ibid, 17 Sept. 1770.
52 Ibid, MI, 1/17, 'Journal', 31 Oct. 1771.
53 *The Glasgow Journal*, 13 - 20 Feb. 1772. See Kinghorn & Quail, (1986), p. 52 for full text.
54 Kinghorn & Quail, (1986), p. 23.
55 B. & W. Col., MI, 1/19, 'Journal', 19 April 1773.
56 *The Glasgow Journal*, 6 - 13 May 1773. See Quail, (1981), p. 52 for full text.
57 JWP, 4/20.14, R. Muirhead to J. Watt, 3 May 1775.
58 Ibid, 4/15.5, R. Muirhead to J. Watt, 10 March 1778.
59 Ibid, 4/15.13, R. Muirhead to J. Watt, 25 Feb. 1779.
60 Ibid, 4/15.20, G. Hamilton to J. Watt, 8 July 1779.
61 Ibid, W/5, J. Watt to G. Hamilton, 15 July 1779.
62 Fuller information can be found in the Watt/Black correspondence in Robinson & McKie, (1970), and Quail, (1981).
63 Robinson & McKie, (1970), p. 8, Letter 2, J. Watt to J. Black, 15 Feb. 1768.
64 I am indebted to Dr. J.N. Black for this information.
65 Robinson & McKie, (1970), p. 12, Letter 4, J. Watt to J. Black, 7 April 1768.
66 Ibid, p. 11.
67 Ibid, p. 9, Letter 2, J. Watt to J. Black, 15 Feb. 1768.
68 Ibid, p. 10, Letter 3, J. Black to J. Watt, 19 Feb. 1768.
69 Ibid, p. 13, Letter 5, J. Watt to J. Black, 20 April 1768.

70 JWP, 4/11.146, J. Watt to J. Watt sen., 9 Dec. 1768.
71 Boulton 348.9, J. Watt to W. Small, 28 May 1769.
72 Kinghorn & Quail, (1986), p. 15.
73 B. & W. Col., MI, 1/14, 'Journal', 24 Sept. 1768.
74 Ibid, MI, 1/15, 'Journal', 23 Aug. 1770.
75 Ibid, MI, 1/16, 'Journal', 8 April 1771.
76 Boulton 348.4, J. Watt to W. Small, 28 Jan. 1769.
77 B. & W. Col., MI, 1/17, 'Journal', 10 June 1771.
78 Boulton 348.23, J. Watt to W. Small, 24 Dec. 1771.
79 B. & W. Col., MI, 1/18, 'Journal', 13 & 14 June 1772.
80 JWP, Letter Book 1, p. 25, J. Watt to G. Hamilton, 18 Feb. 1783.
81 B. & W. Col., MI, 1/20, 'Journal', 16 July 1773.
82 Ibid, 28 Aug. 1773.
83 Ibid, MI, 1/21, 'Journal', 19 April 1774.
84 Ibid, 5 May 1774.

4
Civil Engineer

The Background to Watt's Career as Civil Engineer

The eighteenth century was a time of general development in Scotland. 1723 saw the founding of the Honourable the Society of Improvers in the Knowledge of Agriculture. (1) This was followed in 1735 by the Edinburgh Society for Encouraging Arts, Sciences, and Agriculture, which offered prizes to farmers who planted the largest amounts of timber. (2) After 1745, Scotland made astonishing headway in every department of economic activity. Contemporary authors were full of the new spirit of enterprise. A writer to the *Scots Magazine* said,

> Since 1746, the whole system of trade, husbandry, and manufactures began to advance... It was not merely the fashion for landlords to become improvers, they engaged in every useful project, they are the chief adventurers in our fisheries, manufactures and trading companies. (3)

Scottish landlords looked south and saw the great improvements to agriculture in England and wished to follow that example on their own poor estates. This 'improving movement' became extensive during the second half of the century when, almost everywhere except in the Western Highlands, turnips and potatoes were being sown, land was being cleared and drained, agricultural implements were being improved and small holdings were being thrown into large farms. Farming for profit was taking the place of subsistence farming. (4) The Duke of Argyll, Campbell of Shawfield and proprietor of Islay, was a well-known agriculturalist who had doubled the revenues of his estates through improvements before his death in 1761. (5)

Other parts of these fundamental changes were based on the energy supplied by coal. Down to the close of the eighteenth century, coalmining in Scotland was concentrated along the shores of Fifeshire and Ayrshire and in the immediate vicinity of Glasgow. This was owing to the high cost of transport over land. But the coalfields extended right across the lowlands, starting in the west with the small field at Drumlemble on the Kintyre peninsula and stretching through Ayr, past Glasgow and along either side of the Firth of Forth, reaching to the east of Edinburgh. In Ayrshire, the mines supplied coal for saltpans, while another important market was Ireland. The collieries in the neighbourhood of Glasgow met the demands from both domestic and industrial consumption as well as the distilleries and sugar houses of Greenock and the glass-works of Dumbarton. (6) Up to around 1750, this demand was relatively stable but then, particularly after 1760, a combination of agricultural, industrial and urban development increased its level. (7) Figures of coal production are very incomplete but it would appear that production for the whole of Scotland rose from half a million tons in 1707 to probably two million tons by 1800. (8) Complaints were made about the coal-masters near Glasgow who, through their monopoly of the trade, raised the price from about 1s.1d. or 1s.3d. per cart to 2s.1d. or 2s.6d. by 1760. (9) It was too expensive to bring coal by road from further afield and consequently improved transport communications became essential for continuing industrial expansion and economic prosperity.

Watt's Background and Training as a Civil Engineer

Adam Smith wrote in 1776 that 'Good roads, canals and navigable rivers are the greatest of all improvements. It is... along the banks of navigable rivers, that industry of every kind begins to subdivide and improve itself'. (10) These were the areas on which Watt concentrated in his career as a civil engineer. From his surveys and reports, we can see the primitive state of communications in the Scotland of his day and we can follow the techniques of civil engineering which he used as well as discover his methods of surveying and preparing his reports.

We have seen that there was a tradition of surveying in the Watt family, starting with Watt's grandfather Thomas (See p30) and his Uncle Watt who had made his career as a land surveyor. (See p31) Although both had died before Watt could have known them, he was well aware of their attainments. He may have gained his early knowledge of mathematics and engineering from their books and manuscripts (11) from which he may have copied his 'Notes on Surveying of Land'. (12) From an early age, he studied how to undertake a land survey, work out the areas of land with odd-shaped boundaries, carry out a level between places as well as how to use some surveyor's instruments. (See p41) Through his father's ship chandlering concern and his own scientific instrument business, he was familiar with the necessary surveying instruments. He also acquired the basic competence in mathematics which would enable him to complete the necessary calculations. Once again, we see his study of the literature preceding practical application.

We can trace other people who gave Watt practical assistance and advice which helped to establish him in his career as civil engineer. There was Wilson but Watt's participation with Wilson in the Clyde survey of 1757 (See p74) seems to have been exceptional because he did not carry out any other surveys until 1766. Even the resurveying of parts of Uncle Watt's 'Map of the Clyde' was undertaken by his father and brother in the summer of 1758. (See p95) However the Clyde survey must have involved another person who helped Watt, James Barrie, because he and Wilson both put their names on the map which they prepared. (13)

James Barrie, c. 1703 - 1792

James Barrie moved to Glasgow in 1734 when he was already experienced in surveying. (14) He would have been thirty-one at that time for he died in 1792, aged eighty-nine. (15) Since Uncle Watt died in 1737, they could have known each other and this may have been another reason for young Watt being taken along to assist with the Clyde survey. In addition, Barrie may have known about Uncle Watt's earlier surveys of the Clyde and his maps which must have been stored at Greenock. Barrie acted as surveyor for Glasgow unofficially for a number of years and may have assumed that position on Uncle Watt's death. He became Glasgow's first paid city surveyor in March 1773 when the Council agreed that he should receive the sum of £15 sterling annually. (16) Therefore he was in a good position to be able to advance Watt's surveying career.

Watt must have learnt a great deal from the practical work on the Clyde survey and it probably spurred him to have printed the 'Map of the Clyde' from his uncle's originals. He sold Barrie a dozen copies in 1761. (17) It is with another survey of the Clyde that our records next show Barrie helping Watt. Golborne had gone to Glasgow for the first time in September 1768 to investigate how to improve the river. He may have asked that another survey should be carried out to check his soundings and possibly recommended Watt who undertook this in September and October 1769. (18) Watt opened his report with the words, 'Agreeably to the desire of the Magistrates of Glasgow, I have, accompanied by Mr. James Berry, examined the declivity of the bed of the River, from the Broomielaw Quay to Dumbuck-ford according to Mr Golborne's directions'. (19) Watt had experienced difficulty with his levels and possibly Barrie was consulted to confirm that these were correct. Barrie also assisted with the soundings. (20)

A similar situation seems to have arisen with Watt's survey of the Monkland Canal which he carried out soon after that for the Clyde. He wrote a draft report in December 1769 which made no mention of Barrie, (21) but, in the one published at the end of January 1770, he said, 'I have the satisfaction of Mr. Barrys agreeing with me in the propriety of the above alterations and in the practicability of the whole scheme'. (22) In his notebook on 'Expenses on Surveys, etc.' Watt recorded for 27 January 1770, 'To three days with Mr Barry' (23) which suggests that he and Barrie went over the ground before finalizing that report. We do not know what part Barrie's influence played in the background, but Watt's plans were accepted in Glasgow, funds were raised and he became the Resident Engineer.

There is a further instance where Barrie may have used his influence to secure work for Watt. When Barrie was appointed as Glasgow's first official surveyor in 1773, part of his duty was 'to levell the new streets to be made in the said City as also the old streets when they are to be recausewayed'. (24) The Council passed a resolution in June for Clyde Street to be causewayed 'from the Water port to the Broomielaw' which had to take precedence over other works. (25) Watt received instructions from a Robert Finlay in a letter dated 20 July 1773 (26) and quickly responded for he looked at the street on the following day and levelled it on 26 and 27 July. (27) We do not know the full details of this commission but Barrie may have been able to exert some influence on Watt's behalf on this and other jobs such as repairs to the harbour at Port Glasgow.

Robert Mackell, ? - 1779

If James Barrie was able to help Watt with the surveying side of his career, Robert Mackell (or Meikle) helped him to develop the engineering side. The date of Mackell's birth is unknown but he may have been related to the more famous millwright, Andrew Meikle, of Phantasie near Haddington who was born in 1719. (28) The first we hear of Mackell is that he made a model for Glasgow Council in 1734 of an engine for chipping and slitting iron and rolling iron hoops. (29) He undertook this work as a 'stranger millwright' but in September 1739 he was admitted a burgess of Glasgow. He worked for the Duke of Argyll at Inveraray during the 1740s making models of wind and watermills for grinding corn, slitting iron, as well as preparing flax and thread. He was also involved in the construction of wind and watermills for various industries and inspected the coal mines near Campbeltown in 1750. Presumably building watermills gave him experience with levelling and constructing leats to bring water to them which would not have been dissimilar from making canals for navigation.

Mackell was asked in 1762 to make a survey for a possible canal route joining the Forth to the Clyde from the River Carron at Abbotshaugh to the Clyde at Yoker Burn below Glasgow. (30) This may have changed the direction of Mackell's career more towards civil engineering for in October 1764 he bought from Watt's shop some items for £2.13s.0d. (31) and in November a levelling instrument for £3.12s.0d. and a small theodolite for £1.1s.0d. (32) We have no record of earlier transactions. At about this time, he and Watt joined forces in some joint engineering practice. They entered into an agreement about building Newcomen type atmospheric steam engines during the latter part of 1765 (See p354) and we find Watt assisting Mackell to survey a new route for the Forth and Clyde Canal in the early part of 1767.

Smeaton's route for the Forth and Clyde Canal took it far down the river Clyde to the west of Glasgow so that that city would have been effectively by-passed. The Glasgow tobacco merchants, who wanted a canal to facilitate the re-export of American tobacco to the continent, were indignant and, in December 1766, commissioned Mackell to examine whether another route could be found which would take the course of any canal nearer to or through Glasgow itself. Mackell sought Watt's assistance and together they made their report, 'An Account of the Navigable Canal Proposed to be cut from the River CLYDE to the River CARRON, as surveyed by ROBERT MACKELL and JAMES WATT', which was printed in

London. (33) This survey in the early part of 1767 must mark Watt's serious entry into the surveying and civil engineering profession. He was invited to go to London with Mackell to help pass the bill through Parliament. Although their bill was withdrawn, Watt returned to Scotland with a much better understanding of engineering which we will see he adapted to meet Scottish requirements.

It must have been during the summer of 1767 that Mackell and Watt surveyed the line of a canal to carry coal from the hinterland of Alloa to the Forth possibly for Cathcart. Then Watt branched off on his own but Mackell's help enabling him to do this would have been invaluable. Soon after, Mackell became fully occupied with building the Forth and Clyde Canal because, at the first General Meeting in London on 14 March 1768, Smeaton was appointed 'Head Engineer' at £500 per annum and Robert Mackell 'Sub-Engineer' at £315 per annum. (34) Mackell's position was what we today would call Resident Engineer so he became responsible for supervising the whole organisation to construct the canal.

Once he had taken up his appointment, Mackell became too busy to undertake any other surveying work. The Commissioners of the Board of the Annexed or Forfeited Estates (See p224) wanted a survey of a canal up the Vale of Strathmore. Smeaton declined to do it and recommended a Mr. William Keir to carry out a brief survey. (35) A more thorough survey was necessary which again Smeaton declined but suggested either Mackell or James Brindley. (36) But neither of them could undertake the task and it was recommended that the Commissioners should employ 'Mr Watt from Glasgow, as there is reason to believe he is sufficiently qualified to execute the same'. (37) Was the suggestion of Watt's employment put forward by Mackell? If so, it brought Watt to the attention of the Board of the Forfeited Estates which was comprised of some of the important landowners and politicians in Scotland. Watt was surveying the Monkland Canal at this time and it shows how quickly he had risen to both competence and recognition in his new profession, no doubt through Mackell's tuition. Mackell continued supervising the construction of the Forth and Clyde Canal until near his death on 8 November 1779. (38)

John Smeaton, 1724 - 1792

While Barrie and Mackell helped to train Watt at the start of his career, Smeaton gave Watt professional advice later on. Watt would have been aware of Smeaton's reputation at least from the survey of the Clyde in 1758 and had a great respect for him. Watt's knowledge of the principles of waterwheels shows that he had read the papers Smeaton gave to the Royal Society (39) but we do not know when the two first became acquainted. Smeaton must have been at the Houses of Parliament in the spring of 1767 opposing the bill for Mackell and Watt's Forth and Clyde Canal in favour of his own more northerly route but we have no evidence that they met. Our earliest record of any contact between the two starts in January 1770 concerning Watt's report on the Monkland Canal. Presumably Glasgow Council had sought Smeaton's advice on the feasibility of Watt's scheme for a dead level canal and waggon way to connect it with Glasgow. Smeaton replied, 'I think then, it will not admit of a doubt, whether this canal ought to be pursued upon the dead level'. (40) This opened the way for Watt to reply and to ask for advice about certain points. He also apologized for errors in his levels because part of the survey had been carried out in mist. He tried to excuse himself, 'You will no doubt blame me for committing so considerable an error as four feet in nine miles But upon making allowance for bad weather & a bad headach I had that day it is possible I may not appear so blameable'. (41)

Now Smeaton might well have seen Watt as a potential rival, for after all Watt was involved with opposition to his schemes for the Forth and Clyde Canal, both with Mackell as we have seen and then later by surveying a more northerly route through Loch Lomond at

John Smeaton, F.R.S., 1724 - 1792.

the request of Cathcart. In 1772, Smeaton, Golborne and Watt submitted rival schemes for the improvement of Ayr harbour. (42) Yet Smeaton was willing to help the younger man. When Watt pointed out that his Monkland Canal would not take any water from Smeaton's Forth and Clyde Canal, Smeaton replied,

> I shall be glad if any hints of mine be of service to you in the affair of the Monkland Canal or otherwise. I have no apprehensions myself of any interfering in point of water with the great canal; but in such cases, it is always best to be clear; that ill founded jealousies may not interrupt these publick undertakings; for which I hope the clause you mention will remove all foundation'. (43)

Watt and Smeaton became well known to each other and their friendship lasted many years.

Not content with starting to supervise the construction of the Monkland Canal in the summer of 1770, Watt also undertook to prepare a design for a new bridge across the Clyde at Hamilton on behalf of the tutors of the Duke of Hamilton who was a minor at that time. His tutors wanted Smeaton's advice but not to have to pay for it (44) and obtained it in a roundabout way because Watt submitted his schemes to Smeaton for approval who replied in a long letter commenting on Watt's proposals and suggesting improvements. (45) As a quid pro quo, Watt sent Smeaton details of the tests he had carried out on Newcomen atmospheric steam engines in Scotland and Smeaton replied, 'I am obliged to you for the proportions you have given me of different engines. I have not yet had time to compare their effects with that of the New River [engine]'. (46) The subject of steam engines was raised again when Watt and Smeaton met in the summer of 1773 (47) and they continued to correspond after Watt had moved to Birmingham.

Charles Shaw, ninth Baron Cathcart, 1721 - 1776

This was an age of patronage and Watt was helped in his surveying career in a different way by Charles Shaw, ninth Baron Cathcart, who knew Watt senior. Lord Cathcart was born at Edinburgh in 1721. He succeeded his father in 1740 and, on the death of his maternal grandfather, Sir John Shaw, inherited his estates which included Schaw Park, the estate of Sauchie in Clackmannanshire, as well as the Schaw estates in Renfrewshire where the town of Greenock was situated. (48) Cathcart was an important political figure, being a Lord of the Bedchamber and aide-de-camp to H.R.H. the Duke of Cumberland at the battle of Fontenoy in 1745. As well as continuing his military career, he held many high positions in Scotland such as that of Lord Commissioner of Police between 1764 and 1768. In the latter year, he became Ambassador to the Court of St. Petersburg until 1771 when his wife, Jane, died there. (49) While in Russia, he secured important orders for Wedgwood with Empress Catherine II as well as helping Boulton. (50) He joined the Board of the Forfeited Estates in 1773 (51) and died at London after a long illness in August 1776.

Greenock formed the crucial link between Cathcart and Watt. He provided some land for the church which Watt senior built as well as more for other public works. (See p36) Cathcart's position as Commissioner of Police and his friendship with Watt senior probably helped to advance Watt's career. The Board of Police was established in 1714 with a First Commissioner, five Lords, three Gentlemen, a Secretary, cashier, solicitor and doorkeeper, all paid by the English Treasury. It was intended that the Board should help to improve the government in Scotland and at first had the duty of making proposals on various topics such as the exercise of Crown patronage in the Kirk, reducing the Highlands to tranquillity, assisting the poor, repairing the highways, encouraging the export of naval stores from Scotland, etc. However, membership became a comfortable sinecure and the Commission was finally disbanded in 1782 when its records were lost. (52) Yet it seems that Cathcart may have tried to revive it because he used it to support at least three of the surveys carried out by Watt. When he sent Watt instructions about the first survey, Cathcart concluded with, 'I beg you will thank your father from me for his dispatch relating to Tarbert & Creinan Passages and tell him these things will be examined'. (53) This was just one of the projects in which Cathcart and Watt senior cooperated.

Cathcart was involved with the promotion of Scottish canals certainly as early as 1764 when he asked Smeaton for further information about parts of his report on the project for making a navigable canal between the firths of Forth and Clyde. (54) Cathcart was particularly interested in making the Forth navigable past Stirling and up to Aberfoyle. Through the Commissioners of Police, he asked Morison, who lived at Alva and later Alloa, (55) to carry out a quick survey.

James Morison

James Morison, an English land surveyor, was able to help Watt in a different way because he acted as professional assistant on many of Watt's canal surveys and even carried out parts of them himself. Watt may have been introduced to Morison through Cathcart when the latter asked them to survey a possible northern route of the Forth and Clyde Canal. As a result of Mackell and Murray's survey in 1762, in the following year, the Board of Trustees for the Encouragement of Fisheries, Manufacturers and Improvements in Scotland asked Smeaton 'to give them a design for the sort of canal, that in my judgement would best suit the country, when taken in every point of view'. (56) The northerly route proposed by Smeaton ran up the Forth beyond Stirling nearly to Aberfoyle where it turned south westerly to cross the Pass of Bollat, descended by the Endrick River into Loch Lomond and thence down the Leven River to the Clyde at Dumbarton, but his survey of this line was very brief.

Once it became clear that the Forth and Clyde Canal would be built along the more southerly route, Cathcart tried to renew interest in the Lomond Passage. Morison had surveyed routes for canals through the Great Glen and River Gudie but for whom is not known. He was probably asked by Cathcart to look again at Smeaton's route for the Lomond Passage because he delivered his report to the 'Lords of Police' in July 1767. (57) On the seventh of that month, Cathcart sent Watt instructions to carry out a further survey.

> Instructions for Mr Watt
> Mr Watt will herewith receive a copy of James Morrisons report of the Passage from Alloa to Dumbarton through Loch Lomond.
> James Morrison will meet Mr Watt as soon as he can go to Dumbarton from whence they will proceed in the Tract described in the Report to Alloa; they will examine the same and give their joint opinion upon it specifying any observations and amendment that may occur to them illustrated with proper Plans & sections of such alterations'. (58)

Morison's report contained no estimates of costings which was one of the details Cathcart wanted to know. It would seem that, as a land surveyor, Morison had no experience of civil engineering nor of determining costs of constructing locks or other canal works. This may have been one reason why Watt was asked to prepare a detailed survey and estimates which shows Cathcart's confidence in his attainments as a civil engineer. Another may have been an instance of the Scots wanting the work of an Englishman to be checked by a person from their own country whom they could trust for Watt was the person in charge.

Morison assisted Watt on most of his subsequent major surveys of canals. In April and May 1770, they went to Perth to prepare a report on a canal up the Vale of Strathmore for the Board of the Forfeited Estates. Also for the same Board they went to Crinan and Tarbert in July 1771. On this survey, while Watt was investigating one area, Morison might be either supervising the men or surveying another part so that they worked as a team but once again under Watt's supervision. They set out from Cathcart's Schaw Park near Alloa in September 1773 to survey the Great Glen. (59) This was to be a quick survey on behalf of the Commissioners of Police to ascertain the nature of the country through which any canal might pass and determine whether one would be feasible. The two of them reached Inverness and started to make their survey towards Fort William. However when they had reached Fort Augustus, Watt received a message that his Peggy was critically ill. So he set off for Glasgow, leaving Morison to finish. (60) Later Morison also helped prepare the maps and plans for the report. We will see how Morison prepared a rival report to Watt's after they had surveyed the River Devon for Cathcart late in 1773 which was the final task in Watt's surveying career. (See p286)

Assistants

When carrying out a thorough survey such as that for the Crinan route, Watt took with him men who could undertake the manual labour. On the earlier survey of the Lomond Passage, he was still employing people in his workshop and so took some of his 'lads', including Gardner. (61) Osborne is mentioned not only in connection with the Crinan survey but also on a survey of the waggon way which was intended to connect the end of the Monkland Canal with Glasgow itself. (62) We do not know whether Watt had any assistance with the 'office' work in addition to Morison.

From the Crinan commission, we learn some details about what was required to make a detailed survey both of the route and of the geological conditions through which it would pass. Watt took three assistants, John and Robert Johnston, at 2s.6d. per diem, and Osborne. (63) Morison arrived in Glasgow on 5 July with his assistant and helped with the preparations. Surveying instruments, measuring chains and stakes would have been necessary for the actual survey. In addition, to help ascertain the nature of the soil, John Johnston took,

> Thy Borring Roads [rods], nine fathoms, two ?Peckys? [?picks], two spades, four shovels, two Pinches, four quarry wedges, thy 6 jumpers [for making holes in rocks], and hammer, charger, and pricker, your Big Coat, your spatter dashes, your spurs and dog bag. (64)

The boring rods were repaired and the chisels sharpened at a cost of 1s.6d. (65) This equipment was transported from Glasgow to Greenock in a cart. (66) In addition quarrymen and other labourers were hired when required during the actual survey for digging holes, driving the boring rods and so on.

NOTES: THE BACKGROUND TO WATT'S CAREER AS A CIVIL ENGINEER

1 Hamilton, (1932), p. 43.
2 Ibid, p. 45 and Thomson, (1945), p. 1.
3 Hamilton, (1932), p. 36.
4 Ibid, p. 5.
5 This information was supplied by Roger Emerson.
6 Hamilton, (1932), p. 169.
7 Lythe & Butt, (1975), p. 162.
8 Ibid.
9 Hamilton, (1932), p. 233.
10 Cited by Reilly, (1992), p. 57.
11 JWP, C6/10, J. Watt jun. to J. Rennie, 14 Oct. 1819.
12 JWP, C4/B27, Notes on 'Surveying of Land', by James Watt.
13 Royal Society, Smeaton, *Designs*, Vol VI, f. 57 - 58.
14 Moore, (1996), p. 27.
15 Ibid, p. 27, note 59.
16 Ibid, p. 14 and SRA, C1.1.35, Glasgow Council Minute Book, March 1772 - Sept. 1777, p. 157, 29 March 1773.
17 B. & W. Col., MIII, 3/1, 'Waste Book', 27 Dec. 1761.
18 Riddell, (1979), p. 35 and Cleland, (1816), p. 290.
19 JWP, 3/55, 'River Clyde, Reports from 1752 to 1834' and B. & W. Col., MI, 3/49, 'Report of J. Watt... on the River Clyde, Sept. Oct. 1769'.
20 B. & W. Col., MI, 2/54, 'Report on the Level Monkland Canal', Glasgow, 29 Jan. 1770.
21 Ibid, MI, 2/80, 'A Scheme for Making a Navigable Canal from the City of Glasgow to the Monkland Coaleries', by James Watt, [Dec. 1769]".
22 Ibid, MI, 2/54, 'Report on the Level Monkland Canal'.
23 Ibid, MI, 2/42, 'Expenses on Surveys', 27 Jan. 1770.
24 SRA, C1.1.35, Glasgow Council Minute Book, March 1772 - Sept. 1777, p. 157, 29 March 1773.
25 Ibid, p. 174, 22 June 1773.
26 JWP, 4/50.30, R. Finlay to J. Watt, 20 July 1773.
27 B. & W. Col, MI, 1/20, 'Journal', 26 - 27 July 1773 and MI, 2/42, 'Expenses on Surveys'.
28 Day & McNeil, (1996), p. 478 for biography on Andrew Meikle.
29 Lindsay & Cosh, (1973), p. 423 and Shaw, (1984), pp. 102 - 103 for brief biography of Mackell.
30 Lindsay, (1968), p. 15.
31 B. & W. Col., MIII, 3/4, 'Personal Ledger', f. 6, 29 Oct. 1764.
32 JWP, C1/1, 'Accounts', f. 70, 30 Nov. 1764 and bill for £3.16s.6d for another list of instruments sold to Mackell, B. & W. Col., MI, 2/44, 'Mr. Rt. Mackell, Bought of James Watt', 20 Dec. 1773.

33 NLS, 5.3651 (19), R. Mackell & J. Watt, 'An Account of the Navigable Canal Proposed to be cut from the River Clyde to the River CARRON, as surveyed by ROBERT MACKELL and JAMES WATT', London 1767.
34 Lindsay, (1968), p. 20.
35 SRO, Forfeited Estates, E 727/37/1(2), G. Young to My Lord Coupar Angus, 4 Sept. 1767.
36 Ibid, E 727/37/2, G. Young to My Lord Coupar Angus, 14 Nov. 1769.
37 Ibid, E 727/37/3, G. Young to My Lord Coupar Angus, no date.
38 *Transport History*, Vol I, 1968, J. Lindsay, 'Robert Mackell and the Forth & Clyde Canal', p. 291.
39 See, for example, Watt's comment in the Culross Tide Mill report, B. & W. Col., MI, 3/67, no date.
40 B. & W. Col., MI, 2/60, J. Smeaton to J. Watt, 31 Jan. 1770.
41 Ibid, MI, 2/54, J. Watt to J. Smeaton, no date.
42 Ibid, MI, 1/18, 'Journal', 5 June 1772.
43 JWP, 4/19.2, J. Smeaton to J. Watt, 21 Feb. 1770.
44 Ibid, 3/43, J. Naismith to J. Watt, 7 July 1770.
45 B. & W. Col., MI, 3/60, J. Smeaton to J. Watt, 8 Sept. 1770.
46 Ibid.
47 Ibid, MI, 1/20, 'Journal', 27 & 29 Aug. 1773.
48 Paul, (1905), Vol II, p. 520.
49 Reilly, (1992), p. 80.
50 Ibid, pp. 80 & 123.
51 SRO, Forfeited Estates, E 721.11, Minute Books General, p. 116, 26 July 1773.
52 Riley, (1964), pp. 185 - 187 and Murdock, (1980), p. 20.
53 B. & W. Col., MI, 3/36, Ld. Cathcart to J. Watt, 7 July 1767.
54 Smeaton, (1812), Vol II, p. 56, 'Queries proposed by the Right Honorable Lord CATHCART'.
55 B. & W. Col., MI, 3/2, 'Waste Book', note at back.
56 Skempton, (1981), p. 113 and Smeaton, (1812), Vol II, pp. 31 - 65, 'The FORTH and CLYDE NAVIGATION'.
57 B. & W. Col., MI, 3/10, J. Morison's Report on the Lomond Passage, Sept. 1767.
58 Ibid, MI, 3/36, Ld. Cathcart to J. Watt, 7 July 1767.
59 JWP, 4/4.5, J. Watt to Margaret Watt, 3 Sept. 1773.
60 B. & W. Col., MI, 1/20, 'Journal', 26 Sept. 1773.
61 Ibid, MI, 1/14, 'Journal', 5 Oct. 1767.
62 JWP, 3/8, J. Watt to W. Craig, 4 Feb. 1781.
63 B. & W. Col., MI, 1/17, 'Journal', 10 July 1771.
64 JWP, 4/82, A. Harley, Langton, to J. Watt, 9 July 1771.
65 JWP, 6/28.11, 'Misc. Accounts', J. Watt dr. to J. Wilson.
66 JWP, 4/62.20, 'General Account of Charges'.

The Start of Watt's Career as Surveyor

Dickinson and Jenkins give the summer of 1766 when Watt closed his shop and started his career as surveyor or engineer by setting up an office in King Street (1) but there is no archival evidence to support this. Smiles wrote,

> It would appear that he succeeded in obtaining a fair share of business in his new vocation. He already possessed a sufficient knowledge of surveying from the study of the instruments which it had been his business to make, and application and industry did the rest. His first jobs were in surveying lands, defining boundaries, and surveyor's work of the ordinary sort; from which he gradually proceeded to surveys of a more important character. (2)

The comments by Smiles about Watt's application and industry are certainly correct for he must have been extremely busy with a young family, his shop, the remnant of his scientific instrument business, involvement with the Delftfield pottery and all the experiments that entailed, experiments on the separate condenser for his steam engine as well as launching out into his new career. But the records are silent about his starting as an ordinary land surveyor. It looks as if he commenced and continued with works of an engineering character, and in fact was soon calling himself 'Engineer' and no longer 'merchant'. One aspect of this change and training for it has not been discovered. We will see how he learnt about costing of structures and estimating for construction but we have no signs how he learnt about the geology of the area through which his canals were to pass. Presumably he picked up what he could from books and then observed the ground and lands around him.

Navigations between the Forth and Clyde

Watt's early career as an engineer was linked with attempts to give a better navigable route from the Firth of Clyde to Glasgow as well as one from that city across the isthmus to the Firth of Forth. The suggestion for a canal across this part of Scotland can be traced back at least to the autumn of 1726 when Alexander Gordon carried out a survey and was impressed with the magnitude of the undertaking. (3) Uncle Watt may have surveyed the Clyde up to Glasgow in 1734 and suggested deepening it. (4. See p31) After it was realised that Port Glasgow was silting up, Glasgow Council asked James Stirling to survey their river which he did in the summer of 1752 and proposed a series of weirs and locks which would contain sufficient depth to float coastal vessels at all times as far as the Broomielaw. (5. See p74) This resulted in Smeaton's survey in 1755 followed by the one in which Watt was involved a couple of years later with Wilson and Barrie. (See p74) The collapse of the walls of the lock which Smeaton had tried to build at Marlin Ford in 1761 and then again in 1762 led to the abandonment of his schemes so that the River Clyde remained in a state navigable only for small boats. This may be why, in 1762, Mackell and Murray recommended that their line of the Forth and Clyde Canal should terminate at Yoker Burn, thus avoiding about five miles of the poor navigation of the Clyde up to Glasgow.

In 1764, Smeaton was asked to make a survey for this canal. In his first report, he looked at two lines. The northerly one went up the Forth, over the Bollat Pass and entered the Clyde at Dumbarton. He thought that this would cost more than the direct route which left the Forth to the south of the Carron river, followed the Bonny River to the watershed in Dullatur bog and then down the course of the River Kelvin. However the Kelvin entered a narrow rocky valley with steep sides just to the west of Glasgow where there were many mill sites.

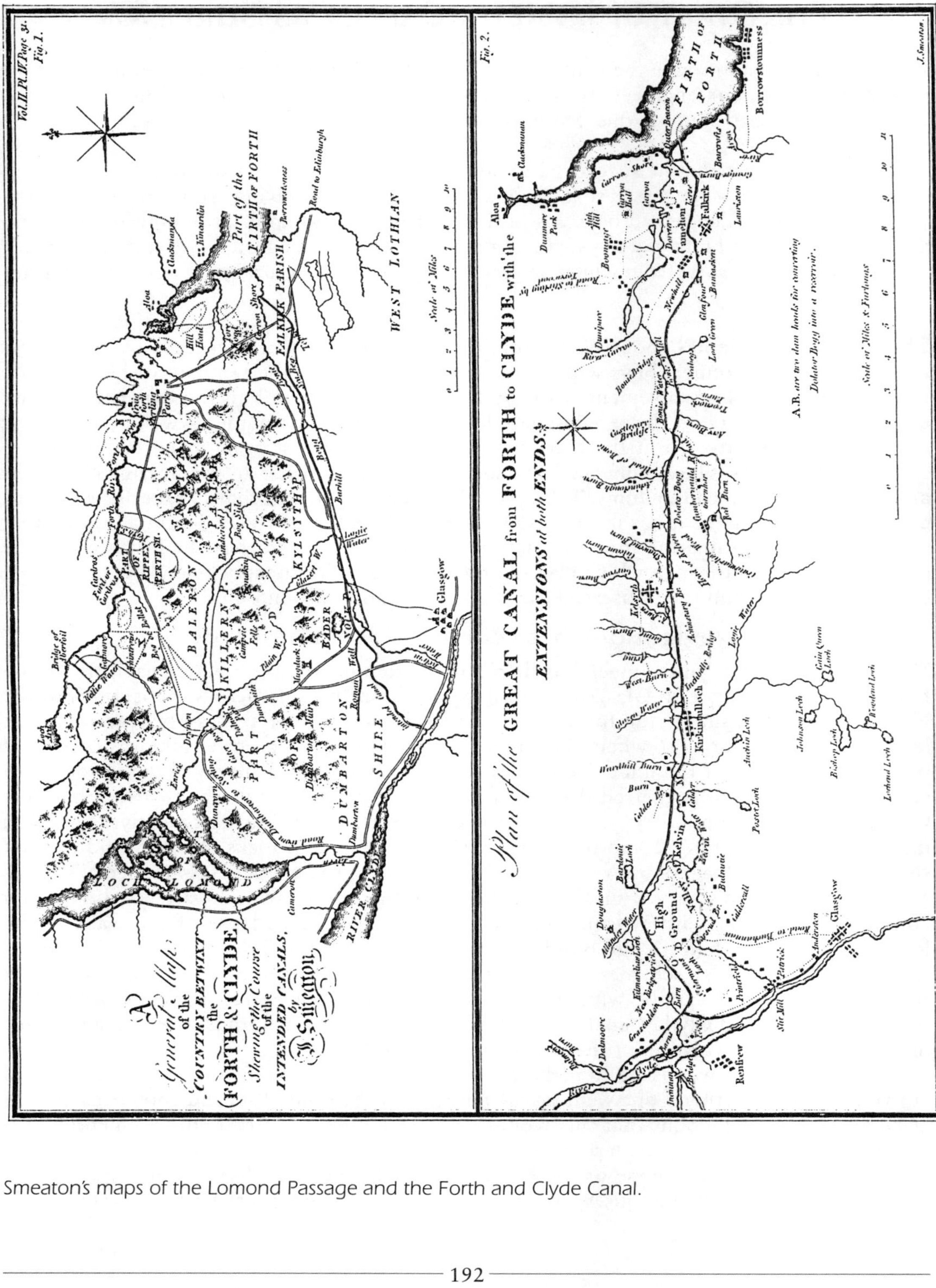

Smeaton's maps of the Lomond Passage and the Forth and Clyde Canal.

To avoid these, he took his route further north at this point, by-passing Glasgow, and his first proposal was to reach the Clyde at the mouth of the Yoker Burn, thus avoiding part of the unnavigable Clyde. Smeaton had 'found that a middling ga[bert] boat of fifty-six feet length, stem and stern, 17 $^1/_2$ wide, and drawing four feet water, will carry at least forty tons: and this I look upon to be the largest size that will be convenient for an artificial navigation'. (6) These boats were capable of navigating in both firths and carrying some coastal traffic. He estimated the cost of a canal 5 ft. deep along this southern route would be £78,970. (7)

Arguments raged over his proposals for a couple of years. There was Cathcart interested in the northerly route. Glasgow people wanted the canal to pass nearer to their town while another faction, probably mostly from Edinburgh, wanted a larger canal to take coastal shipping. We do not know when Watt joined the controversy but he wrote a report which has no date criticising Smeaton. He pointed out that gaberts were 'unfitt to go to sea for the Common purposes of commerce. But the Forth Coasting vessels go to sea in all weathers and if it is intended that such vessels should pass this canal, it must be 9 foot deep'. (8) Watt produced revised estimates for such a canal as well as for a branch to Glasgow which he thought the city deserved and arrived at the estimate of £195,472.8s.9d. He also pointed out that it was uncertain whether the Clyde would be deepened enough to take these boats with a 9 ft. draught so that the canal might have to be extended to Dumbarton - 'what that would cost may be difficult to determine the ground being very Rocky'. (9)

It must have been in response to such criticisms that Smeaton, once the bill for the Forth and Clyde Canal had been passed, asked Mackell to re-survey the line (10) and prepared a second report which he finished in October 1767. In it, he recommended extending the canal further westwards to the 'Dalmoore Burn' and building a branch from the Yoker Burn eastwards to Glasgow itself. One advantage of this would be that ships could pass along this section between Glasgow and Port Glasgow when the Clyde was unnavigable. (11) He prepared new estimates for canals of different sizes. One with a depth of 10 ft. would cost £226,503. The one which was built was for a depth of 7 ft. The height at the watershed was 147 ft. which meant that 42 locks with an 8 ft. fall in each would be needed. This estimate was £147,337. (12)

Two factors may have caused people in Glasgow to investigate schemes for alternative small canals. One was that Smeaton's route would have by-passed the city and the other was the high cost of both Smeaton's scheme for a 7 ft. canal and Watt's estimates for a ship canal. In December 1766, the tobacco merchants commissioned Mackell 'to examine whether there might not be found some track which, though it was higher, might be shorter, and which should fall into the Clyde nearer Glasgow'. (13) Mackell drew up proposals for a canal 7 miles shorter than Smeaton's which would pass through the North Lanarkshire coalfield and have a 240 ft. high summit near the Bishop Loch. This scheme would have used narrow boats but was apparently rejected as too expensive. (14) It was at this point that Mackell and Watt set out to make a more thorough survey with firm estimates.

The 'Ditch' Canal

It may be that it was this survey to which Watt senior referred when he wrote to his son on 21 January 1767. 'My complements to Peggy & hope she will not sitt long at a time in ye shop or be out during ye sever[e] weather and I hope you will not jurnay in bad weather or night time in this uncarten cleamate'. (15) About a year earlier Peggy had helped out in the shop when Watt had to dismiss his shop manager. (16) While Watt was away surveying, presumably she kept an eye on business affairs until the shop was closed. The first definite reference we have to this survey comes in another letter dated 9 February 1767 in which we see that his father had doubts about the project.

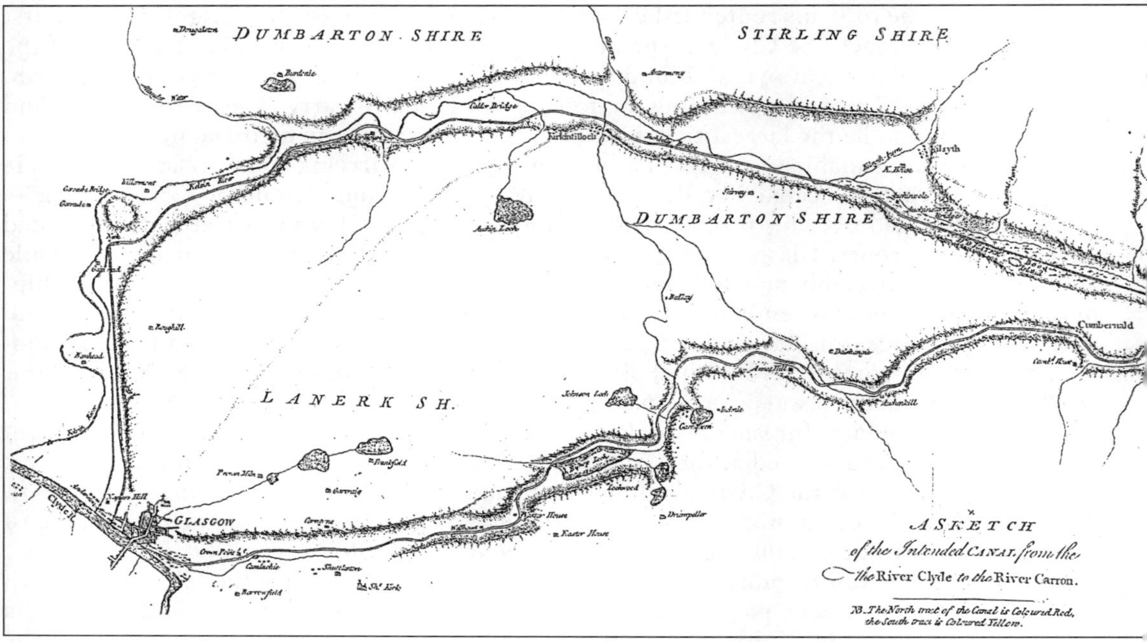

> Its unserten how fare yt small canall will give content to ye London merchts I think you should keep yourself as mutch disingeged as you cane from atendence on ye executing of it as your helth may be impared with such weather as we have of ordinary [next part illegible]... tho you get £100 for going to London & atending their sometime it may be a furder loss to your shop or town bussiness, how ever you [?...?] or drew for them or have jurnays about yt work mack positive agreements & keep short acounts with them as some of their Generositys will not be great. (17)

The prospect of earning £100 for going to London for a short time to help support the bill for this canal through Parliament must have seemed very attractive to Watt with all his debts. But he failed to take note of his father's warning because he made no proper financial agreements for carrying out some of his later surveys and also the promoters of many schemes were late in paying him.

Mackell and Watt were engaged with surveying and attending committees during the whole of February and early March 1767. (18) Their report suggested two routes but both were for much smaller boats than Smeaton's, only 50 ft. long by 9 ft. wide, drawing 3 ft. water. These could have navigated the Firths. Either canal would have been 24 ft. wide and 4 ft. deep, hence its opponents called it a 'ditch, a gutter, a mere puddle'. (19) One line would have terminated a little to the west of the Broomielaw and followed the east side of the Kelvin until it joined Smeaton's earlier route. There would be problems cutting through the free-stone and millstone grit in this area but, with 48 locks of 6 ft. fall, it would cost £44,510.6s.8d. The other would have started to the east of the Broomielaw and passed through the coal mining areas around Camlachie and Shettleston, then north of Coatbridge, past Bishop Loch, which would have been at the summit, before dropping down past Cumbernauld and joining Smeaton's route at Redburn Bridge. This would have necessitated some deep cutting around

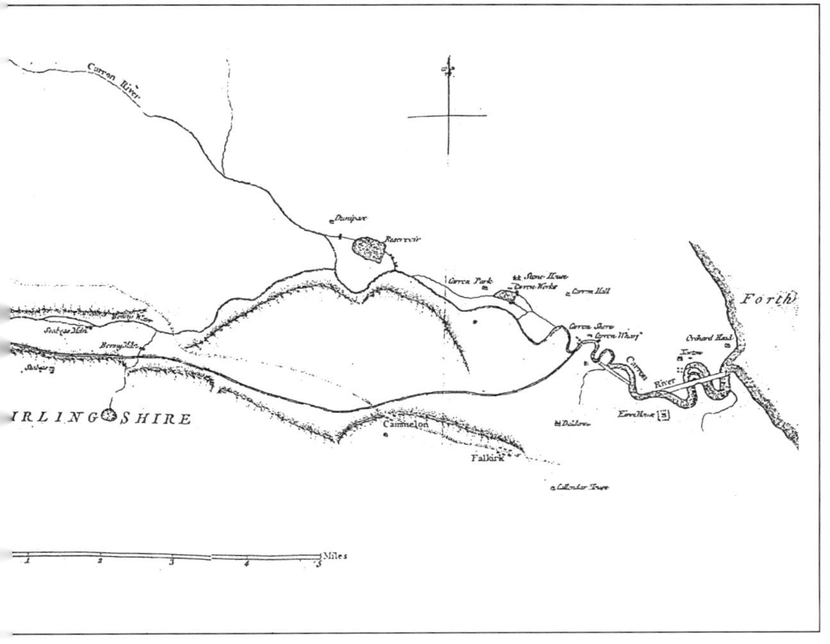

Mackell and Watt's map of the routes for their narrow canals between Glasgow and the Firth of Forth.

Bishop Loch as well as 82 locks to surmount the climb of 240 ft. The estimate was £55,563.11s.8d. (20) With such dramatically lower costs than Smeaton's proposals, it is hardly surprising that there was support among Glasgow merchants who would have a direct line of communication to the Firth of Forth. In addition, it would pass through a coal mining area and so generate an important traffic along its route.

Watt was asked to go to London to support a bill which was introduced by Lord Frederick Campbell. (21) He set out on 10 March passing through Edinburgh. He took the opportunity to visit other canal works on his journey south and arrived at Brighouse three days later where he looked at the Calder Navigation which was being built to Smeaton's designs. To improve the river, a large number of locks was needed which Smeaton 'proposed to make... better than most of the Locks he had seen on any River in England'. (22) This was most likely the first time Watt had closely inspected any canal under construction and it would leave a profound impression on his later schemes. The drawings of these locks preserved at the Royal Society in London show well piled foundations and massive stonework for the walls. Watt made notes on the construction and costs as well as the time it took for a boat to pass through. He wrote, 'A boat took 12 minutes to pass a lock but was only 8 minutes in the lock. the boats travel 30 miles a day'. (23) This visit brought home to him the great expense of constructing a locked canal, with its need for skilled masons to build the locks. Once opened, there was the requirement for adequate water supplies to operate it as well as the extra costs of maintenance and the time spent in passing barges through the locks.

Watt's next visit showed him a very different type of canal. He crossed the Pennines to Manchester where he set out for Worsley. 'Left Brighouse at 4 o'clock in the morning came to Manchester about 1 o'clock set out in a chaise for Worsely mill, the horses knocked up about 4 miles from Manchester we walked the rest on foot saw the canal & came back in a boat to Manchester'. (24) At Worsley Delph he would have seen how coal was brought out of the mines in containers on small boats. A crane transhipped the containers onto larger barges which took them along a level canal to Manchester where another crane unloaded the containers and placed them on horse-drawn carts for distribution in the town. During his return to Manchester by canal, he would have passed over Barton aqueduct, a major engineering feat of the time. Between Worsley and Manchester there were no locks, and indeed none for many miles, as he saw on the next day when he followed its route south to Altrincham. While bridges and aqueducts were still necessary, such a level canal avoided the need for reservoirs and major supplies of water as well as expensive lock construction. The Bridgewater Canal could have been the source of inspiration for many of Watt's later canal schemes in Scotland, including some for Cathcart.

Watt left Manchester and went south through Birmingham probably to deliver a letter from Roebuck to Garbett but Garbett was not at home. So Watt left there, apparently without meeting any other members of the Lunar Society, and proceeded to Oxford where he again missed Garbett. Only after his arrival in London on 17 March was he finally able to call on Garbett. For the rest of that month, he was drawing plans, having the report printed and attending committees. (25) In his first letter to Peggy written on 5 April, we find that he was carrying out commissions for his father but was not impressed with Parliament.

> I have yett seen nobody being close confined attending this confounded Committee of Parliament Barclay [who seems to have been in charge] is extremely fashous & I have great difficulty to keep peace between Mc [Mackell] & him but I agree very well with them both I think I shall not long to have any thing to do with the house of Commons again I never saw so many wrong headed people on all sides gathered together as Mc says I believe the deevil has possession of them I am much afraid that after all we will lose our bill but you must not say so B. keeps us constantly employed or hanging on about something or other but I declare I do not know what we do & yett we do all that is done & if we had not been here I shall be bold to say it would have gone before now as he has not the surest sense of anything about it but this to yourself - My coming home is as you see uncertain I would write you oftener but the pen is never almost out of my hand & you know how I like that... you see I am very angry but I keep myself very quiet. (26)

This visit to London made Watt very wary about being involved with obtaining Acts of Parliament again. It is also evident that he did not like clerical work which may account for delays later in publishing the reports on his surveys. But it was not all work because three days before he wrote this letter, he had been to the opera at a cost of ten shillings. (27)

A few days later, Watt wrote that their opponents had caused delays to the passage of their bill until after the holidays which would give a fortnight's respite when he might go into the country. He thought their bill hung in the balance. The disputes between Mackell and Barclay continued. Yet, while they were not on speaking terms with each other, Watt remained friendly with both. (28) At the beginning of May, Watt wrote that he was sorry that he was still not on the road home. The people in Edinburgh had given a great deal of trouble but the third reading of their bill was scheduled for the following Monday and he still had hopes that they would carry it. Even if the Edinburgh people did stop their canal, Watt thought that there would then be another canal on which he stood a good chance of being employed. (29)

In the meantime, this bill had roused the latent hostilities between Edinburgh and Glasgow. In opposition to the 'little despicable scheme of the Glasgow merchants' for their ditch, (30) the Edinburgh faction drew support from many other parts of Scotland for a 'great canal' which would be capable of accommodating sea-going ships. A subscription for such a canal was opened in London on 1 May 1767 and over £100,000 was raised quickly. The canal bill in which Watt was concerned should have been read on 14 May but it was withdrawn in return for £1,200 compensation to the Glasgow and Carron subscribers together with the promise of collateral cuts to Glasgow and Bo'ness. Watt wrote to Peggy,

> The subscribers to the great canal being people of Great Interest etc. our managers here thought it best to make up the matter with them last night & have given up our Canal on conditions. I am now at my own liberty & shall sett out for you as soon as I have finished buying a few things... even tho we have lost the day there is a great burthen off my spirits but I am very concerned that I am so unlucky as to be absent from you at such a critical time on every account. (31)

Peggy was pregnant and also had to move house. (see p126) The Act for the rival Forth and Clyde Canal received the Royal Assent on 8 March 1768. (32) While in London, Watt bought items to restock his shop and found 'a person to take all my perspective machines which I think a piece of good luck'. (33) He visited some of his friends such as Ramsden. This was probably how he learnt about the latest developments in scientific instruments. He then set out for home. (See p128)

This trip to London greatly enlarged his experience of surveying and civil engineering and gave him the idea of a narrow, level canal with no locks but with some method of transhipping facilities at both ends. He based this concept partly on what he had seen at Manchester on the Bridgewater Canal. This would save the expense of digging broader ones, save the expense of locks and reservoirs, need less costly boats and so be suitable for his native country which was still comparatively undeveloped. He could have discussed civil engineering works with his colleagues while he was sitting around in the Houses of Parliament. Even if he did not meet Brindley, one likely topic of conversation would have been Brindley's narrow canals. They would require far less soil to be dug out. In the case of their own canal, Mackell and Watt allowed for removing soil £11,919.6s.8d. on their northern route while Smeaton estimated £42,328 for his longer route including the branch to Glasgow. Figures for building the necessary locks were £9,600 and £37,800 respectively. Mackell and Watt pointed out that their deeper locks would require only 4,320 cubic feet of water for each boat to pass while Smeaton's would use 4,608. (34) On narrow canals such as those built by Brindley, it would be possible to haul boats in tandem. For a poor country like Scotland, where the larger 40-ton boats envisaged by Smeaton were, according to Watt and Mackell, rarely used, there were obvious advantages in building the narrow canals even though their capacity was for boats of only 20 tons. There were importance consequences for Watt in the failure of this ditch canal. The way opened for renewed interest to improve navigation in the Clyde itself because the Forth and Clyde Canal by-passed Glasgow. But more important for him, the practicality had been shown of making a canal to tap the coal fields in the high land to the east of Glasgow which would result in his only major civil engineering work, the Monkland Canal. All this was still in the future.

NOTES: THE START OF WATT'S CAREER AS SURVEYOR

1 Dickinson & Jenkins, (1927), p. 26.
2 Smiles, (1878), p. 92.
3 Lindsay, (1968), p. 15.
4 B. & W. Col., MIII, 5/6,7 & 8 for Uncle John Watt's plans and sections.
5 Riddell, (1979), pp. 8 - 23.
6 Smeaton, (1812), Vol II, p. 69, 'MINUTES concerning Forth and Clyde'.
7 Ibid, Vol II, p. 51, 'ESTIMATE of the EXPENSE'.
8 JWP, C7/7.4, Notes and Costings by J. Watt on canal proposed by J. Smeaton.
9 Ibid.
10 Smeaton, (1812), Vol II, p. 71, 'Instructions for Mr. Mackell relative to the proposed Canal from Forth to Clyde'.
11 Ibid, Vol II, p. 81, 'The Second REPORT of John Smeaton, engineer, F.R.S., touching the practicability and expense of making a navigable canal from the river Forth to the river Clyde'.
12 Ibid, p. 97, 'Third ESTIMATE', 8 Oct. 1767.
13 Lindsay, (1968), p. 16.
14 Ibid.
15 JWP, 4/7.29, J. Watt sen. to J. Watt, 21 Jan. 1767.
16 JWP, 4/11.134, J. Watt to J. Watt sen., 23 Jan. 1766.
17 JWP, 4/7.28, J. Watt sen. to J. Watt, 9 Feb. 1767.
18 B. & W. Col., MI, 2/4, 'Cash Memorandum', various dates in February and March 1767.
19 Lindsay, (1968), p. 18.
20 Mackell & Watt, (1767), pp. 13 & 15.
21 Cleland, (1816), p. 298.
22 Skempton, (1981), p. 105.
23 B. & W. Col., MI, 2/4, 'Cash Memorandum', 13 March 1767 and MI, 2/5, 'Cash Memorandum'.
24 Ibid, MI, 2/4, 'Cash Memorandum', 14 & 16 - 18 March 1767.
25 Ibid, various entries.
26 JWP, 4/4.25, J. Watt to Margaret Watt, 5 April 1767.
27 B. & W. Col., MI, 2/4, 'Cash Memorandum', 2 April 1767.
28 JWP, 4/4.19, J. Watt to Margaret Watt, 18 April 1767.
29 Ibid, 4/4.18, J. Watt to Margaret Watt, 4 May 1767.
30 Lindsay, (1968), p. 18.
31 JWP, 4/4.17, J. Watt to Margaret Watt, 14 May 1767.
32 Priestley, (1969), p. 266; 8 Geo. III, C. 63, Royal Assent 8 March 1768.
33 JWP, 4/4.20, J. Watt to Margaret Watt, 11 April 1767.
34 Mackell & Watt, (1767), p. 5 and Smeaton, (1812), Vol II, p. 95, 'Third ESTIMATE'.

Early River and Canal Surveys

The River Devon (1)

Soon after Watt had returned from London in May 1767, he and Mackell must have surveyed the River Devon and the area around Alloa for a canal. Watt's report has not survived and the only reference occurs in Smeaton's report to Cathcart that autumn. The River Devon rises to the north of the range of the Ochil hills but eventually flows south through the Crook of Devon. Then, in a sudden bend, it heads more or less west towards Stirling, following the foot of the Ochils, but at Menstrie twists again due south to reach the River Forth at Cambus, a couple of miles west of Alloa. While there was exploitation of mineral deposits in these hills, such as a vein of cobalt in 1759 and the Airthrey silver mine in 1761, (2) they proved to be disappointing in their extent and alone would not have justified making the River Devon navigable.

In its course along the foot of the Ochils, the River Devon follows a geological fault and beneath the relatively flat lands to the south, stretching down to Alloa and the Forth, were large deposits of coal at quite shallow depths. This is the western end of the coalfield running through Culross and eastwards into Fife. (3) Coal had been mined here from at least 1519 and, by 1640, there were extensive workings in the Clackmannan and Sauchie districts. Much was shipped both around the Firth of Forth and further afield to the Netherlands and the Baltic, principally through the port of Alloa where a Custom House was provided in 1710. (4) This coal was described as 'the finest... in all this country... and has always been held in the highest estimation, on account of its superior quality'. (5)

The principal landowners in the neighbourhood of Alloa in the 1760s were the Erskines of Mar and Lords Cathcart and Kennet. Around 1730, Sir John Erskine had canalised part of the Devon in order to carry his coal to the banks of the Forth at Cambus from where it could be more easily despatched to market. (6) He achieved this partly by improving the channel of the Devon and partly by installing sluices at 'Tilli Body Bridge', presumably a form of flash lock. (7) This canal could not have been functioning in the 1760s when Cathcart wanted to extend the exploitation of the deposits at his Sauchie colliery. This mine was very ancient and, at this period, was drained by an engine which took its water from the River Devon, probably a waterwheel operating lever pumps.

Morison had surveyed the higher reaches of the Devon above Sauchie in 1765 (8) and, possibly as a result of this, Cathcart sought Smeaton's advice in February 1766. (9) Smeaton was at Cambus quay on 20 November 1766 when he looked at the tidal Forth as well. However his report on the Devon is dated 14 September 1767, nearly a year later, which may be why Cathcart asked Watt to make another survey, or perhaps he wanted a second opinion. Smeaton's report mentioned a scheme by Mackell and Watt

> ...of carrying a canal upon a dead level from *Tillicoutery* to a point opposite the schoolmaster's house at *Alloa*, following the foot of the rising grounds the whole way, and which will be about 30 feet higher than the sea, proposing to continue the same from thence between banks across the flat ground to the waterside (which is not far), or making a cut through the flat grounds to the dead level termination of the canal, which plan would save locks and waste of water, I can only say, that, from my view of the country, I believe such a project practicable; but, as I have neither plans nor sections of the course that it would take, I am in no capacity to judge of the expence.
>
> It is obvious that this canal would be defective in not suffering the vessels to pass into the *Forth*, and that either the banking across the valley, which would be a considerable

> depth, would be a considerable article of expence; and that in any view of the affair, if a communication with the *Forth* is dispensed with, a considerable saving may be made, but I apprehend not above $^1/_3$d. (10)

Presumably it was meant to carry coal from Cathcart's mine at Sauchie to the port of Alloa. It would have taken a shorter route than following the Devon, but it is impossible to compare the saving made by having no locks with the greater expense of higher embankments or deeper cuts.

This was Watt's first scheme for a dead level canal on which goods would be transhipped into and out of small barges. Hence it would have been a narrow canal, cheap to construct with no locks and needing no great reservoirs to supply it. But it would have been too small for sea-going vessels even if it had been connected to the harbour at Alloa. Watt's suggestion was clearly based on the Bridgewater Canal and he would propose this concept again for some of his other schemes in Scotland. This canal may have been made partly redundant in 1766 or 1768 when one of the earliest Scottish wooden waggon ways was built by John Earl of Mar from the Waterwheel Pit to the docks at Alloa. (11) Although this may not have reached Cathcart's pits initially, later it was extended further into the coalfield.

Cathcart's correspondence with Smeaton on 5 November 1767 (12) mentions a survey of the River Devon carried out by Watt. This must have been made that August or September, possibly in conjunction with a survey of the tidal Forth.

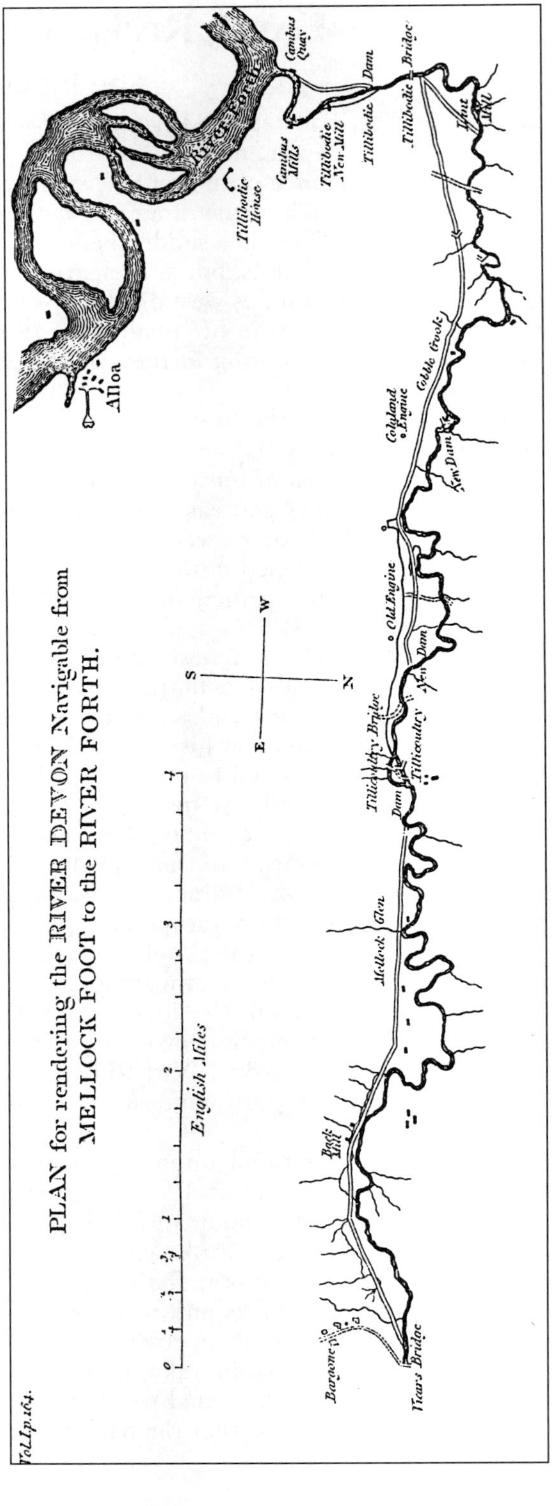

Smeaton's route for his canal to improve the navigation of the River Devon.

Opposite page: Sketch map of the area around Alloa and the lower reaches of the River Forth.

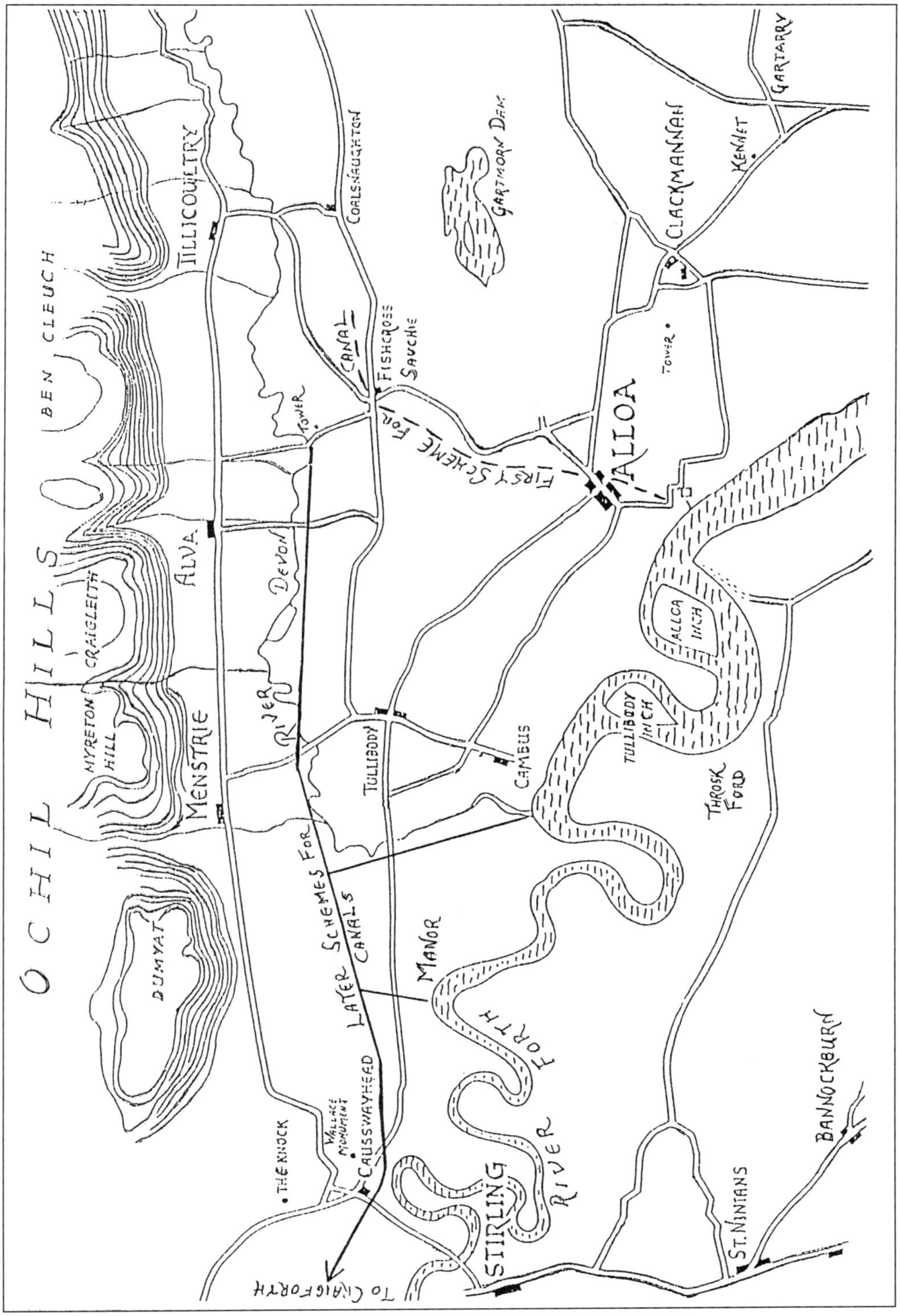
OCHIL HILLS
BEN CLEUCH
TILLICOULTRY
COALSNAUGHTON
GARTMORN DAM
CLACKMANNAN
KENNET
GARTARRY
CANAL
FISHCROSS
SAUCHIE
TOWER
FIRST SCHEME FOR
ALLOA
TOWER
ALVA
CRAIGLEITH
RIVER DEVON
ALLOA INCH
TULLIBODY INCH
TULLIBODY
CAMBUS
MYRETON HILL
MENSTRIE
THROSK FORD
DUMYAT
LATER SCHEMES FOR CANALS
MANOR
RIVER FORTH
BANNOCKBURN
THE KNOCK
WALLACE MONUMENT
CAUSSWAYHEAD
STIRLING
ST. NINIANS
TO CRAIGFORTH

Cathcart may have been trying to extend the sale of his coal up the Forth itself into the region beyond Stirling, and, if so, this could account for Watt's suggested line of a canal here. Watt realised that the mill dam at Menstrie, where the Devon made its last southerly turn to fall into the Forth at Cambus, was on the same level as the weir at Craigforth above Stirling. Therefore he proposed making a canal to carry the line of the River Devon further westwards towards Stirling, passing round the foot of the hill where the Wallace Memorial now stands, crossing the River Allan either by damming that river or by an aqueduct, and meeting the Forth just above Craigforth weir. This scheme would have level water almost from Alva, where Cathcart's colliery lay to the south, as far as Frew, about two-thirds of the way up the Forth towards Offering and Aberfoyle. In essence, this was another of Watt's dead level canals, which would give a good route for Cathcart's coal into wider markets because it would avoid the delays of boats locking into and out of the tidal Forth. This might have provided a cheaper and safer alternative than improving the Forth itself.

In another new idea, Watt proposed a wooden aqueduct in order to make a low level crossing of the River Allan. Cathcart realised the difficulty of constructing a masonry aqueduct with little clearance for floods and for which expensive foundations would be necessary, rendering the cost enormous. Smeaton commented on Watt's idea,

> It is executed in wood, upon the plan of *Julius Caesar's* bridge, which is well adapted to resist ice, or other floating bodies, or upon any better more modern construction, it is apprehended it would come much cheaper, would stand against the force of the stream, and of the ice in the winter, and would not be affected by the land floods, which cannot rise to the sole of the aqueduct, but must be liable to the expence of repairs to which a stone bridge would not be subject. (13)

Smeaton was very guarded in his reply about the aqueduct and canal, pointing out that, if the purpose of this canal were only to link the Devon collieries with the upper Forth, then Watt's scheme 'seems very well adapted for that purpose, and in that case would merit serious consideration'. (14) However, if coal were to be sent down the Firth of Forth, he could not see ships passing up the tidal Forth and then back along the canal. All these schemes must have been temporarily abandoned when Cathcart left to take up his post as Russian Ambassador in 1768.

The Lomond Passage

In between carrying out these surveys of the River Devon, Watt was asked by Cathcart to look at Smeaton's northerly route for the Forth and Clyde Canal, the Lomond Passage. With the withdrawal of the bill for the 'ditch' canal, Cathcart may have hoped to revive interest in this alternative which might have served his collieries better than the Carron route because it would have provided good transport up the River Forth itself. From Aberfoyle on the edge of the Highlands, the River Forth still follows a winding course across flat peaty lands until it reaches the weir at Craigforth about a mile to the north of Stirling, where the lower, tidal stretch ends. Smeaton reckoned that, from where his route would join the Forth at Keltie about four miles below Aberfoyle, the distance to Stirling was about twelve miles as the crow flies but more than twenty by the river. (15) Sir John Sinclair pointed out in 1799, 'The fall of the river, in its whole extremity of this parish, through the great moss, to Stirling, is very inconsiderable; and the ground consisting almost wholly of moss and clay, is easily wrought'. (16) In fact, the river at Aberfoyle is only about sixty feet above sea level and the fall from Gartmore to Craigforth about thirty feet. The river was described as deep and slow moving but there were shoals where the depth was only a couple of feet.

Between Stirling and Alloa, the tidal stretch of the Forth is also extremely twisting and, in places, shallow. Smeaton commented that it had 'so winding a passage as is almost without parallel, it being accounted four miles by land, and twenty-four by water'. (17) There was a quay at Stirling which coastal vessels could reach, but often only with great difficulty. At spring tides, those which could lower their masts to pass under Stirling bridge, could get as far as the weir at Craigforth where all goods had to be transhipped into very small boats which could make their way up the Forth, being dragged over the fords and shoals.

The region to the north-west of Stirling seemed to present great potential for development. There was slate at Aberfoyle and limestone had been found at Gartmore. (18) Agricultural improvement was all the rage at this time, although the various reports about developing the navigation make no mention of it. Treating the soil with lime was one method of increasing agricultural yields. If navigation up the river was better, then coal could be brought up more cheaply from the collieries around Alloa both for making lime and for household fuel. Schemes for improving the Forth could be linked to improving the River Devon for the transport of coal. An improved river would also have assisted with drainage of the moss or peat lands. Lord Kames, who entered into the possession of the Blair-Drummond estate in 1766, improved part of the area near Monteith and straightened the Goodie or Gudie Water. (19)

Because Smeaton's survey of the Lomond Passage was very brief, Cathcart wanted a more detailed one and first asked Morison. In his report, (20) Morison gave the length of what would become the Forth and Clyde Canal as 28 $^1/_2$ miles of artificial cuts and the distance from its westerly end to Dumbarton as eight, making a total of 36 $^1/_2$ miles. On the other hand, the Lomond Passage, while climbing 40 ft. higher and longer at 47 miles, had fewer artificial cuts, so Morison wrote,

> Upon these principles in point of time there cannot be great difference betwixt the two passages. The one being shorter and more uniformly certain at all times, the other being readier by having fewer locks; but the Lomond passage must cost less than the Carron passage and for that reason as well as from the addition of the natural commerce of the rivers Forth and Leven which would be very considerable may in all probability afford an equal profit to the undertakers at a much lower toll. (21)

But Morison's report contained no proper costings and he had not ascertained the nature of the soil through which the canal was to be cut. Therefore Cathcart instructed Watt to go over the ground again with Morison. This was Watt's first survey where he was in charge although he was no doubt helped by Morison's existing knowledge. They left Dumbarton on 20 July 1767 and passed through Drummond, Bucklivie, Alloa and over to Falkirk. Watt also took some of his 'lads' who were employed not only in helping with the surveying apparatus but also in ascertaining the nature of the ground through which the canal would pass by digging holes and using boring rods. They returned to Dumbarton on the 29th. They covered much the same ground between 7 to 11 August. On 5 October, Watt received 'from the police £35, & my Expenses £7.1.0, the lads £3.12.0'. (22)

One place which needed special investigation was the summit level over the Bollat. Smeaton's route from Loch Lomond lay north-eastwards up the valley of the Endrick Water before turning off and climbing steeply up through difficult country to the Bollat Bog or Pass which was 222 ft. above high water of neap tides in the Clyde. One of Smeaton's maps shows that water would have to have been brought from quite long distances to keep the summit level supplied. Watt employed his lads to make borings because Cathcart thought a tunnel might be better than a cutting. He had heard about the tunnel at Harecastle for the

Trent and Mersey Canal and, having asked Smeaton about this in 1764, (23) also wrote to Watt, 'It has been found in England that it is better to Pierce than to *surmount*, & in the great Staffordshire Navigation they conduct it 2 $^1/_2$ miles under a mountain, estimating the Drift at £5,000 per mile'. (24) Watt thought that the canal could be lowered 20 ft. through this pass but that a tunnel would be impracticable owing to the hardness of the rock. To supply the summit level with water, he followed Smeaton's proposal of constructing a reservoir of 60 acres at Burghmore and taking water from the higher reaches of the Endrick Water through an aqueduct or leat 4 $^1/_2$ miles long. Here we discover the first mention of Watt's use of wood for aqueducts. He proposed to carry the line of this channel over narrow gullies in wooden troughs, similar to his slightly later proposal for crossing the River Allan. (25) This idea may have originated with troughs or launders which supplied water to waterwheels. In another example at the beginning of the seventeenth century, Hugh Middleton installed wooden troughs on his New River which supplied water to London. (26)

Cathcart later sent Watt instructions about writing his report. Watt was to follow Morison's, commenting on it section by section. Watt was also to compare his suggestions with Smeaton's original report, except that his canal would be eight feet deep. His calculations were to be 'founded on real Particulars & on principles which you will explain'. (27) Watt's report was much longer than Morison's and covered three main sections; from the Clyde up the Leven to Loch Lomond, from Loch Lomond over the Bollat Pass to the Forth at Offering and finally the Forth itself as far as Craigforth mill and dam. To make the River Leven navigable would require three locks to overcome a difference in height of just over 19 ft. (28) This section, a distance of 5 $^1/_2$ miles, would cost £6,575.18s.11d. (29) To surmount the climb up to the Bollat Pass, 19 locks would be needed, followed by a similar descent of 17 locks. These would be 18 ft. wide, 64 ft. long, 10 ft. fall, with 7 ft. 6 ins. water for boats of 60 to 70 tons. This section was to be 13 miles long, costing £60,386.10s.7d. (30) The River Forth itself was to be shortened from 30 to 22 miles by means of cuts, although Watt doubted whether some of those proposed by Morison were necessary and he recommended leaving more of the route in the existing river bed. Three locks were to be built at the shallows or fords. This section was estimated to cost £12,766.13s.7$^3/_4$d, giving a total for a canal from the Firth of Forth to the Firth of Clyde of £79,729.3s.1$^3/_4$d. (31) The extra expense of the lockage over the Bollat Pass should be noted as it may have confirmed Watt's views about the advantages of level canals.

The Tidal Forth

If the Lomond Passage were to finish at Craigforth weir, it would not have been much used unless the tidal Forth were improved as well. Watt made more journeys to the Stirling area in August and September 1767 to investigate the lower reaches of this river. His survey of the section below the weir at Craigforth appears to have been undertaken to check on Morison's work and complete the route of the Lomond Passage. A glance at a modern map will show just how twisting this river still is and how navigation could have been improved by straightening it. There was also the problem of a shoal called the 'Thrask' between Alloa and Cambus, where the River Devon joined the Forth. Watt took soundings in the river on 21 September, 'Being Two days before the full moon'. (32) If cuts were made and the river deepened, vessels of 100 tons might go always up to Stirling every spring tide without fear of lying on the shoals. To reach Craigforth, he recommended cutting a canal to by-pass Stirling bridge so that masts need not be lowered. A swing bridge at the northern end of the existing bridge would accommodate passenger and road traffic. The total estimate for the work was £5,333.17s.3$^1/_2$d, while shortening the river and reclaiming the old bed would gain 237 acres valued at £5,925. (33) With the reports on the Lomond Passage and the River Devon

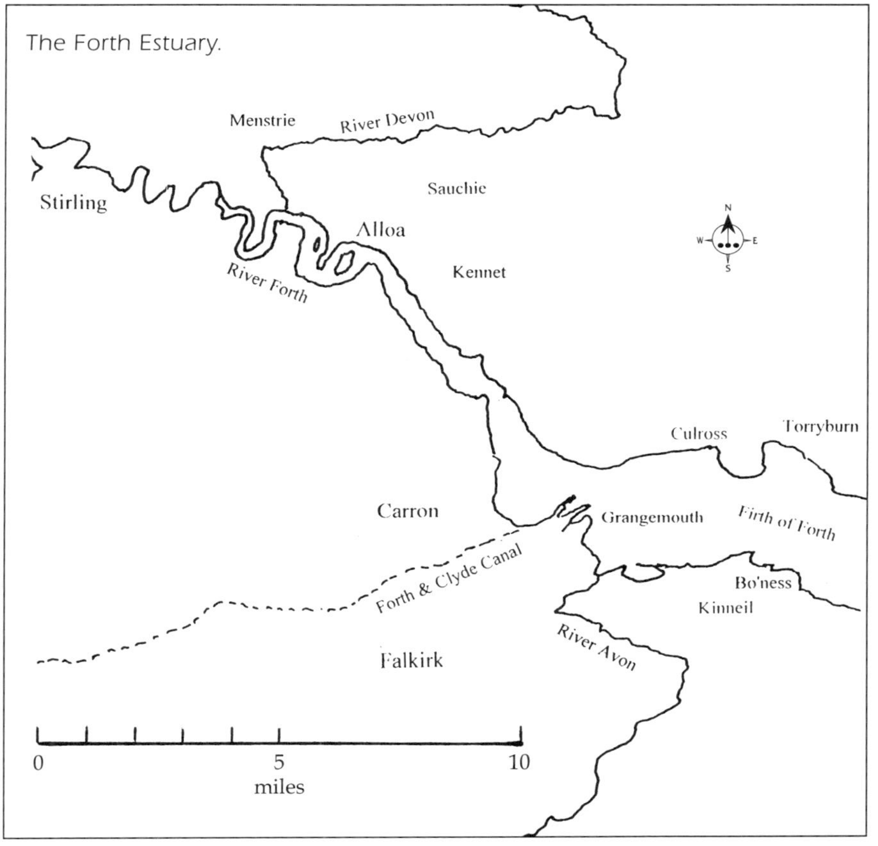

submitted to Cathcart, Watt's surveys of rivers and canals drew to a close for a couple of years. This may have been due either to Cathcart moving to Russia and so not finding work for Watt or Watt wishing to concentrate on developing his separate condenser for the steam engine.

The River Clyde

After the failure of Smeaton's schemes in 1762, it was not until 1768 that further progress was made on the improvement of the River Clyde. Another Act was passed for improvements to the city of Glasgow and the river navigation but it included no specific plans relating to the latter. (34) In September, Golborne came to Glasgow for the first time together with Brindley to discuss with the Glasgow magistrates plans for a ship canal across central Scotland. But Golborne also looked at the Clyde itself and suggested jetties or dykes to restrict the channel of the river so that it would scour out by itself a deeper channel through a more confined bed. The greatest obstacle, he declared, was the Dumbuck Ford, about twelve miles below Glasgow Bridge. Here the river divided in two and Golborne recommended the construction of a jetty or obstruction over the south channel to confine the current in the other and to remove the hard crust in the river bed by dredging. By these means, he was confident that a depth of water of between four and five feet would be obtained at the Broomielaw at low tide. (35) Before leaving for England with his fee of £85, he may have suggested that the magistrates should commission another survey of the river to check his soundings and that this should be carried out by James Watt, (36) or it is possible that the magistrates were dissatisfied with his report and called on Watt to check it.

That Watt should have been invited was slightly ironic because his father actively opposed improvements to the river, possibly fearing the effects this would have on the prosperity of Greenock. He thought that

> As it is said they want to make the river considerably deeper than 7 foot by wh they will be under the ordinery Levell of Low water in which sand & Rubage will constently lay as the landwater will be continually caring more than can be cared out of it without ane uncommon expense I wish they could be deverted from that plan. (37)

He thought that a three foot depth would be quite adequate because then the existing lighters could carry goods just as cheaply and with as great despatch as any larger boats. (38) Watt senior's worst fears were realised once the navigation up to Glasgow was opened for ocean-going ships. Jean Cochrane told Watt in 1782, 'Trade in this place [Greenock] is almost gone - consequently rents of houses much lower, many standing empty... [We] were obliged to lower the rent of this House, otherways could not let them'. (39)

Apparently the local people wanted someone they could trust so they turned to Watt who carried out his survey in the latter part of the summer of 1769. He was assisted by Barrie to whom he was 'much indebted to [his] Knowledge of the River and fords, and the observations formerly made, and also for his assistance in the present observations'. (40) However it was Watt who wrote the report and it was Watt who carried out most of the survey so it does not seem that this was a joint commission as has been suggested. (41) Golborne was not so impressed with Barrie and complained to presumably the Lord Provost of Glasgow that August.

> When Mr. Barry left us he went to fetch his instrument & said he would call on me at the Bull in half an hour, this was about 12 o'clock I waited till past four, this is such usage as I never met with... Upon the whole I would wish your Lordship would trust intirely to Mr Watts whose abilities may be confided in & who I doubt not will do it with great accuracy. (42)

In spite of Golborne's opinion of Watt's ability, relationships between the two never developed in the same close way as those with Smeaton.

Watt told Small,

> I recd. yours on Tuesday night & would have answered it sooner but had one of my headachs when it came & the day after. yesterday I was busy writing & jumbling my Brains reconciling discordant Observations of my Survey of the River which I have accomplished to an inch. If you can come nearer in 13 miles levelling I lay down my instrument to you. (43)

Poor Watt suffered these headaches probably through worry about this survey which gave him great difficulty. At one time, he had an error of 3 $^1/_2$ ins. in his levelling. (44) He asked Wilson and Irvine to check with double stations. (45) Another problem was measuring the depth of water at the shallows in the short period of low tides and he feared that he may not have carried out Golborne's directions precisely so he sought Barrie's assistance. (46) He compared his readings and soundings with both those of Smeaton in 1755 and those taken by Wilson as far as the Marlin Ford in 1758. Then at one period he was able to carry out only two days surveying before he 'was drove home by the rainy weather, which has also somewhat hurt my health'. (47) However he was pleased with his results even though they gave him a great deal of trouble.

Watt took as his base a fixed point which was a stone in the Broomielaw quay. He worked out the levels of all the other shoals and fords from this, beginning with the Hirst opposite the Delftfield pottery. Wilson had started at a different place so their figures did not always agree. Watt found at low water that the depth at the Hirst was only 1 ft. 2 ins., on the Windhole, 1 ft. 5 ins. and on Pointhouse, 1 ft. 4 ins. At high water on the same tide there was 5 ft. 3 ins. on the Marlin Ford which shows how improvements to the river were essential for ocean-going ships.

Watt opened his report with,

> The General wish that Mr Golborne had been more particular in Explaining the methods by which he proposed to deepen the river Clyde & that he had shown the methods by which he estimated the expence have produced this attempt to show the probability of succeeding by the methods he proposes & att the same time to give some idea how he may have formed his estimate. (48)

He continued with outlining the different ways of clearing the shoals. Where the river was barely eighteen inches deep at low water, the largest lumps of rock might be removed by hand or special tongs. Then a horse and plough might loosen the hard layers so that the spate of water might wash away the mud and gravel.

> If the crust under that depth still continues to be solid the machine call the porcupine may be used instead of the plough. these machines are made of different forms the most common consists of several wheels of wood of 3, 4, or more feet diameter fixed on an axis so as to be about a foot distant from one another in the edges of these wheels are fixed spikes of iron a little crooked on each end of the axis is a gudgeon by means of which the machine is fixed in a frame of wood so it can turn round like a rolling stone for land. It is used by drawing it up and down the river by a boat or otherwise the sharp spikes being pressed into the ground by the weight of the wheels loosen & tear it up in coming out again.
>
> The ground loosened by the porcupine may be lifted by a dredging boat which is a flatt bottomed vessell having on each side a large shovel or spoon with a long handle by which a man guides it & presses it against the bottom in a proper direction while other men pull it for[war]d by means of a rope or chain fastened to the spoon & wound round a small windlass. when it is full it is pulled up & emptyed either into the big boat herself or which is better into a small boat with a false bottom that carrys it away & emptys it in some proper place. The harbour of Toulon was deepened by this method. (49)

Watt loved mechanical devices. He costed Golborne's method of building jetties to confine the river to a narrower bed and so scour out its channel and concluded,

> It is by no means meant by the above to alledge that Mr Golborne has overcharged his estimate but to show that works of this kind he proposes may probably be excuted [sic] at the expence & it must be considered that a very great latitude must be assumed in Estimating every work where water is concerned as spates must always lay the men idle besides sometimes undoing part of what has been done. (50)

A new Act was obtained which received the Royal Assent on 12 April 1770. (51) Watt attended a meeting of the Glasgow Town Council sometime in May and was asked to write to Golborne because, having started to charge higher rates on shipping, they felt obliged to make some improvements immediately. They hoped that Golborne would soon visit Glasgow. (52) In the meantime, Watt inquired what works might be started under his supervision and asked Golborne to 'transmit to my Lord provost your directions concerning the above which I shall endeavour to execute in the manner you direct'. He asked for full instructions so that 'no mistakes may happen'. (53) Clearly he was hoping to undertake this work himself and even went so far as to suggest that he might join Golborne in a partnership, probably to give himself some experience of actually handling a contract. Possibly this spurred Golborne to hasten to Glasgow where he summoned Watt to view the river on 13 June. (54) Golborne was given the contract for deepening the Clyde on the 28th (55) and then returned to Chester. He told Watt that a partnership 'would [not] be adviseable in a Work of so precarious a Nature'. (56) By then, Watt had agreed to undertake the Monkland Canal. Work started in the Clyde on dredging the various shallows and confining the course of the river by a series of jetties so that it would scour out a deeper bed. In this way, there would be no need for any locks or dams. So successful were the improvements that Golborne, in his report of 1781, was able to write that the channel, which he had dredged to a depth of 6 ft. 10 ins., was now, by the action of the river, 14 ft. and in some places as much as 22 ft. deep. (57) He was paid £578.14s.9d. in January 1771. (58)

Notes: Early River and Canal Surveys

1 A fuller version of Watt's surveys in the Stirling and Alloa areas was published in the *Forth Naturalist and Historian,* Vol 21, 1998, pp. 71 - 84, R.L. Hills, 'James Watt's Surveys Around Stirling'.

2 Dickie & Foster, (1974), p. 5 and *English Ceramic Circle,* Vol 16, pt 2., 1997, J. Turnbull, 'Scottish cobalt and Nicholas Crisp', pp. 144 - 151.

3 *Forth Naturalist & Historian,* Vol 7, 1984, I. Bauman, 'Coalmining at Culross in the 16th and 17th Centuries', pp. 84 - 125.

4 Drummond, (1953), pp. 15 - 16.

5 Sinclair, (1978), p. 160.

6 Ibid, p. 161.

7 Smeaton, (1812), Vol I, p. 142, 'River Devon Navigation', 20 April 1765.

8 JWP, 3/22.2, J. Watt, 'Report on Rivers Forth & Devon, 18 Jan. 1774'.

9 Smeaton, (1812), Vol I, pp. 154 - 161, 'The REPORT of John Smeaton, engineer, concerning the practicability and expence of making navigable the River *Devon...*',14 Sept. 1767.

10 Ibid, Vol I, p. 161.

11 Adamson, (1988), p. 5 and Lewis, (1970), p. 255.

12 Smeaton, (1812), Vol I, p. 147, 'QUERIES &. communicated by Lord Cathcart, relative to a navigation on the *Forth,* November 5, 1767'.

13 Ibid, Vol I, p. 148.

14 Ibid, Vol I, p. 149, 'Answer to the five Queries concerning the improvement of the Navigation of the river *Forth* from the River *Devon* upwards...', 11 Nov. 1767.

15 Ibid, Vol II, p. 37, 'The FORTH and CLYDE NAVIGATION... GENERAL DESCRIPTION of the Passage by way of Loch Lomond', 22 Dec. 1764.

16 Sinclair, (1977), p. 16.

17 Smeaton, (1812), Vol II, p. 37, 'Forth & Clyde Navigation'.
18 See B. & W. Col., MI, 3/7, 'J. Watt's Report on Lomond Passage, Lower Forth to Craigforth'.
19 *The Farmer's Magazine*, Monday, 4 August 1817, pp. 261 & 263.
20 B. & W. Col., MI, 2/40, 'Mr. Morison's Report upon a Canal from the Forth to the Clyde by the Lomond Passage, July 1767'.
21 Ibid, see also his report as MI, 3/10.
22 Ibid, MI, 1/14, 'Journal', 5 Oct. 1767.
23 Smeaton, (1812), Vol II, pp. 56 - 57, 'Queries proposed by the Right Honorable Lord Cathcart'.
24 B. & W. Col., MI, 3/36, Ld. Cathcart to J. Watt, 7 July 1767.
25 Ibid, MI, 3/8a, 'J. Watt's Report on the Lomond Passage over the Bollat Pass'.
26 Smiles, (1874), *Early Engineering*, p. 69.
27 Ibid, 3/43, Ld. Cathcart to J. Watt, 24 ?Aug. 1767.
28 Ibid, MI, 3/9, 'J. Watt's Report on making the River Leven navigable'.
29 Ibid, MI, 3/8b, Report on Bollat Pass.
30 Ibid, MI, 3/8a, Report on Bollat Pass.
31 Ibid.
32 Ibid, MI, 3/7a, Report on Lower Forth.
33 Ibid, MI, 3/7c, Report on section from Craigforth to Offering.
34 Priestley, (1969), p. 157, 8 Geo. III, C. 16, Royal Assent 24 Feb. 1768.
35 Hamilton, (1932), p.p. 241 - 242.
36 Riddell, (1979), p. 35.
37 JWP, 4/7.10, J. Watt sen. to J. Watt, 1 April 1769.
38 Ibid, 4/3.26, J. Watt sen. to J. Watt, 4 April 1770.
39 Ibid, JW/22.38, Miss Cochrane to J. Watt, 7 Feb. 1782.
40 Ibid, 3/55, 'J. Watt's Report on River Clyde, 20 Oct. 1769'.
41 Smith, (1994), Bryden, p. 14 lft.
42 B. & W. Col., MI, 3/4, J. Golborne to 'My Lord', 2 Aug. 1769.
43 JWP, 4/59.7, J. Watt to W. Small, 20 Oct. 1769.
44 B. & W. Col., MI, 3/14, 'Survey Notes of Clyde'.
45 Ibid, MI, 3/16, 'Survey Notes of Clyde, Fifth Survey'.
46 Ibid, MI, 3/54, J. Watt to J. Golborne, 3 Nov. 1769.
47 JWP, 4/32.15, J. Watt to J. Roebuck, 24 Sept. 1769.
48 B. & W. Col., MI, 2/54, 'Book of Incomplete Draft Reports'.
49 Ibid.
50 Ibid.
51 Priestley, (1969), p. 158, 10 Geo. III, C. 104, Royal Assent 12 April 1770.
52 SRA, C1.1.34, Glasgow Council Act Book, Jan. 1768 - March 1772, p. 284, 8 May 1770.
53 B. & W. Col., MI, 2/54, J. Watt 'to Mr. Golborne'.
54 Ibid, MI, 1/15, 'Journal', 13 - 19 June 1770.
55 SRA, C1.1.34, Glasgow Council Act Book, Jan. 1768 - March 1772, p. 292, 28 June 1770.
56 JWP, 4/50.5, J. Golborne to J. Watt, 19 July 1770.
57 Hamilton, (1932), p. 242.
58 SRA, C1.1.34, Glasgow Council Act Book, Jan. 1768 - March 1772, p. 374, 8 Jan. 1770 [1771].

The Monkland Canal

After the Clyde, the next survey Watt undertook was a quick one for the Monkland Canal. He pointed out in his initial report that,

> The encreasing price, and scarcity of coals in the city of Glasgow at some seasons, has induced the Magistracy to consider of the most probable means of remedying these evils: This appears to be the opening a communication with some distant coalierys by water. The Monkland coalierys are, above all others, those which bid the fairest to reduce the price of fuel. (1)

Watt wrote to Small, 'though coal is everywhere here about in plenty and the very town stands upon it, yet measures have been taken by industrious people to monopolise it and raise its price 50 per cent within these ten years'. (2) Watt reckoned that a cartload weighed 7 cwt. which sold at Monkland for 6d. Yet the price of land transport made it uneconomic to fetch this coal from the collieries at Monkland to Glasgow where it fetched 2s.6d. (3)

The Lord Provost at the time was James Buchanan of Drumpellier who was one of the main supporters, being present at the annual meeting of the committee of the Monkland Canal in February 1771. (4) He would have been aware of Watt's recent survey of the Clyde but it seems it may have been Watt who put himself forward for the Monkland survey. He told Small,

> Somehow or other [I] got into the good graces of the present magistracy who have employed me in engineering for them (as Mr Smeaton terms it)... a canal to bring coals into the town... Vanity also bade me tell Glasgow people they might be served as well at home as by strangers. (5)

A little later, he wrote, 'Many people here had conceived a much higher idea of my abilities than they merit; - they had resolved to encourage a man that lived among them rather than a stranger'. (6)

Watt may have been very anxious to secure this survey because he finished it without having made any agreement over payment. He wrote to an unidentified person at the end of March 1770 about his expenses.

> I have not yett taken upon me to determine in what Rank I stand as an Engineer & shall only observe that what knowledge I have in that way has not been acquired without great loss of time & money & that it is not everyday that a man is fitt either for the Business of the Field or Closet. These things being laid before them I reffer the price of my labour to the gentlemen concerned as I do not wish them to pay me in such a manner as may deter them from employing me afterwards & trust they will do it so that I shall remain willing to serve them. (7)

Watt would have been much cheaper than a Smeaton or a Brindley, which is why he may have been employed on some later surveys. But, for Watt, his Monkland survey and subsequent construction of the canal marked a double turning-point in his career as civil engineer. Not only was he inundated with commissions for the next three years until the economic depression following the collapse of the Ayr Bank in June 1772, but also his experiences on site turned him from being a supervising or executive engineer into a consulting engineer which had profound significance for the development of his steam engine.

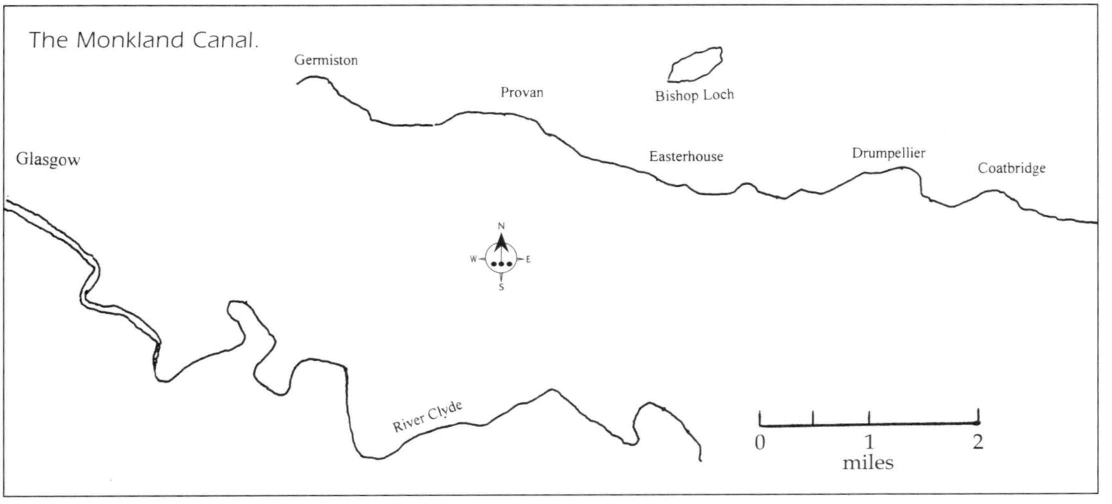

Monkland Surveys and Reports

Watt and two assistants started surveying the Monkland Canal on 26 October 1769. One day he 'viewed the country on Horseback'. (8) Weather conditions must have been bad again because this was the occasion when he made a serious mistake in his levels of four feet through not being able to see through the mist for which he apologised to Smeaton. (9. See p184) His fear of being shown incompetent may have been one reason for his later accepting the post of resident engineer because he told Small,

> I now had a choice... to carry into execution a canal projected by myself with much trouble, or to leave it to some other person that might not have entered into my views and might have had an interest to expose my errors; (for everybody commits them in those cases). (10)

Perhaps this reflects the sort of mental stress which Watt incurred at such times and which may partly account for his headaches. The survey was finished by 5 December although they did not go out every day. Nevertheless, part of the time was spent making a thorough investigation of soil conditions and Watt reported, 'Every part of this tract that was suspicious, has been examined by boring; and no rock has been found, but what was visible at the surface in one or two places, where there is no extra cutting; the deep cuts are tilly clay, except that at Cotes bridge, which is sand'. (11) Perhaps Watt would not have been quite so optimistic had he known what lay ahead.

His initial report must have been written and printed quickly because a meeting was arranged for 3 January 1770 to open a subscription to launch the project. (12) The Monkland collieries lay in the higher land about ten miles east of Glasgow. Watt proposed two routes. One would have followed a valley from Glasgow itself through Camlachie, Shettlestone, Drumpellier and Cotes bridge but it had 'a rise of 266 feet perpendicular above our river'. (13) This line would be 11 miles long and require twenty five locks to surmount the climb which alone were estimated at £12,500. For purchasing the land, £1,560 was allowed and cutting the channel of the canal £3,333. The total, with no allowances for obtaining the Act of Parliament or payment for damages to mills, etc., came to £20,317.11s. (14) so over half the expense was for the locks. He envisaged boats 50 ft. long, 9 ft. wide, drawing 30 ins., which carried 20 tons or 60 carts. (15)

Watt then outlined his scheme for a level canal without any locks, 'that is to say, a canal, whose end next Glasgow should be on a level with the coalieries: and I find that it can be brought within a mile of the town, and that a waggon way, or good causey, can be made for the remainder of the road'. (16) He had made a thorough investigation of the line for this canal which would take a more northerly route and, like the other, it was to be 16 ft. wide at the bottom, 22 ft. at the top and 3 ft. deep with passing places as he envisaged the same size of boats. His estimates for the land were £1,420 and digging £3,030. Extra costs were included for cutting at the summit and elsewhere as well as for tunnels or culverts, road bridges, etc. so that the total came to £9,653.10s. including £1,000 for the Act of Parliament and unforeseen expenses. (17) Most of such a canal would be cheap to construct particularly as there would be no need for skilled masons to build locks. He made no allowances for reservoirs as he thought that leakage would soon be inconsiderable because mud would fill any porous parts and evaporation would be replaced by springs or burns along the route. He pointed out another advantage, that, while it would take a boat ten minutes to pass each lock or 10 hours 50 minutes for a single journey on the locked canal, a journey on the level canal took only 6 hours 40 minutes. But he failed to point out the extra costs and time of transhipping cargoes for the final stretch into Glasgow. The level of this canal would be 246 ft. higher than the Clyde and would finish at Germiston near Glasgow where there would be a harbour. This was where the waggon way or causey would commence to take coal and goods to and from the city itself and the canal. Here we see Watt's final concept of the level canal; one where contents of the barges would be transhipped onto waggons and moved up or down inclined railways to overcome differences in height. The causey must have been a paved road but it was the waggon way or railway which Watt preferred. His proposals were challenged by someone suggesting an alternative canal route or a waggon way constructed for the whole distance but these schemes were rejected. (18)

Watt's Background Knowledge of Railways

We have seen that Watt was introduced to the concept of a level canal in England and, again, there were few railways in Scotland before the 1760s from which he could learn about their construction. The earliest is generally given as that opened in 1722 by the York Buildings company between their mines at Tranent and the harbour at Port Seton near Prestonpans, East Lothian. (19) Soon after 1750, John Dixon laid a line from his Gartnavel and Knightswood pits, north-west of Glasgow, to Yoker on the Clyde. (20) This railway must have been seen by Watt during his Clyde survey in 1769. (21) Another line, but one which Watt is unlikely to have seen, was that from Fordell colliery to St. David's harbour in Inverkeithing Bay. This could have been opened as early as 1752. (22) It would seem that the only other railway opened before 1766 was that taking coal from the collieries at Carronhall and Kinnaird to the Carron Ironworks. (23) Watt may have been looking at waggons for this line and suggesting improvements to them in his earliest surviving letter to Roebuck in August 1765. Unfortunately part of the letter has been torn off but, in the remaining fragment, Watt mentions a model of a waggon and fore and hind wheels on an axle. (24) Then there was the waggon way built in either 1766 or 1768 by John, Earl of Mar, from his Waterwheel pit to the docks at Alloa. (25. See p200) Watt must have been aware of this line through his surveys for canals in that area as well as erecting a Newcomen engine nearby for Lord Kennet. (26)

When Watt travelled to London in 1755, he would have crossed over the routes of various waggon ways in the Newcastle area but there is no record that he ever stopped to look at them. Also when he visited the west of England on his sales trip in 1761, we know that he called on a Mr. Underwood at Bath but again there is no mention that he inspected Ralph Allen's railway which was still functioning, bringing down stone from the quarries at Combe Down. (27) The trucks for this railway are illustrated in Desaguliers's *A Course of Experimental Philosophy* (28) which we know Watt read in order to learn about Newcomen engines and so he may have studied this account as well. J. Leupold's *Theatrum Machinarum* of 1724 has a drawing of a rope-worked balance incline with presumably loaded waggons hauling up empty ones. (29) He knew that F. de Belidor's volumes contained nothing about railways because he had bought a copy in 1767. (30) But this would have helped his other civil engineering activities because, for example, he read it on the subject of a new pier for Greenock in February 1773. (31)

In his initial report, Watt gave a brief description of the waggon way he proposed to connect Germiston with Glasgow.

> I think the waggon-way may be so conducted, as to have a regular descent to the town, and if the high street were levelled at the bell of the wyndhead, as I am told is proposed, I do not see any considerable obstructions to the horse drawing back a half loaded waggon; particularly as about 60 feet of the perpendicular height may be cut off, by making a double waggon-way down a steep bank at the end of the canal, where, by means of a rope and pulley, the loaded waggons going down, may draw the half loaded ones up.
>
> I have not stated the expence of this waggon-way in the estimate, as the ordinary tolls on coals, which in this case ought to be paid to the waggon way, will be sufficient to make and keep it in order.
>
> I propose the waggons to carry three carts of coals, and to be so constructed, that they can be drawn on the streets as well as on the waggon-way, and to prevent their damaging the pavement, their wheels should be 6 inches broad in the rim. (32)

In this instance, he gave a figure of 8 cwt. for Glasgow carts so the capacity of the waggons would have been 24 cwt. (33)

His proposal to use waggons with broad and presumably flangeless wheels may have been derived from Smeaton who used such trucks at Plymouth to move heavy stones during the construction of the Eddystone lighthouse which was finished in 1759. (34) Smeaton thought that Watt's 'waggons so constructed as to pass the streets, which though unusual, is very feasible'. (35) We have seen that Smeaton supported Watt's scheme for this level canal even though he must have realised that it would involve heavy costs in transhipping cargoes. He had tried an experiment on the Calder navigation at Brighouse which was supplied with coals from a colliery about five miles away but only half a mile from the river.

> Some cargoes of coals were brought up by water, and though the tolls upon coals are but one quarter of merchant's goods, yet they could not deliver them cheaper than by land-carriage, for the loading and carriage down to the river side, and the loading and carriage from the river to people's houses, together with freight and tolls, made the coals rather dearer than if fetched by land carriage from the pits'. (36)

Watt ought to have taken warning.

Watt envisaged that part of his waggon way would be a rope-worked balance incline in which the loaded coal waggons would haul up empties or partly loaded ones. But he never mentioned how transhipment of coal or other goods was to take place between canal and railway in this and indeed any of his schemes. Cranes, such as those he saw in Manchester, are never described but he told Smeaton, 'If the coals are to be shipt in boxes according to Mr. Brindleys way these boxes will weigh about one fourth of the coals they contain & will besides hinder stowage I apprehend one of these boxes would be cheap at 10/-'. (37) He may have looked into other methods of transhipping goods or even complete waggons in a special boat because in December 1771, he recorded that he

> ...gave william directions about Bevels & model of new boat. I propose she shall be 60 feet long overall & four feet deep that she shall be floored at 2 feet from the bottom which floor shall be caulked tight so that if she should be overset she may be easily got up. (38)

The advantage of being able to raise the boat easily was soon put to the test after its launch on 2 May 1772. The boat, which cost £60, was found to be leaking and Watt was employed 'lifting Boat' for a couple of days in the middle of the month. (39) Then problems were found with the steering which were overcome by making a large oar so that the first trial was not held until 8 June when 50 carts of coal were loaded. Watt expected that it would carry 75 carts. (40) The higher deck made loading and unloading much easier so Watt was satisfied with it. (41)

We should note Watt's optimism that the waggon way would be self-financing, both in its construction and maintenance, but he does not say how the money would be raised to build it. His initial report contained no estimates for this waggon way and he must have realised, or it was pointed out, that it would be necessary to determine this quickly. Therefore he spent three days at the end of February 1770 'measuring and estimating the waggon way' (42) when he may have produced provisional estimates. These matters seem to have been left until he noted in his journal for 28 September 1771, 'Viewing ground for a waggon way'. (43) But it was only in the New Year that he again returned to this task when he looked for a 'tract for the waggon way' on 17 January 1772 and 'began survey of waggon way at Germiston' on the 31st of that month. He continued on February first, third and fourth (the second being a Sunday). Possibly this was to make preparations for a meeting of the canal committee on the fifth. (44) He spent a further day in February on the waggon way just before one of his children, Agnes, died which no doubt caused further delay. (45)

During this period, Watt corresponded about the cost of materials with people who were accustomed to supply equipment for such railways. One of these was John Hunter junior from Kilmarnock who, with his partner Mr. Staggart, gave costs of wooden rails and sleepers. They had been supplying places in the Lake District and Ayr. (46) There was another quotation of 'Newcastle prices of Wagon way' which showed that rails of beech, birch or fir cost less than oak, oak being used for the main line carrying loaded waggons. The main line would have two sleepers each yard, but the return 'by-way' only one. (47) Scotland was still following English practice and importing new technology from her neighbour.

Among 'Queries for Mr. Golborne', there is an estimate for a waggon way at £4,335 which, as it was to be twelve miles long, could have been an alternative to replace the canal but no further details are given. (48) The surviving estimate for the line to link the Monkland canal with Glasgow was more expensive per mile.

Estimate of Waggon Way from Jermiston to Broomylaw being 2 1/4 miles.		
2 1/4 miles railway leveling included at £ 300	£	675
Byway 2 1/2 miles at £ 200		500
Ground for whole 10 ells wide being in all about 7 acres at £60		420
	£	1665
3/4 mile causeway to join Kilsyth road		300
		1965
Tolls on the above 20,000 tons at 3d.		250
additional to do. 1 1/2		125
		375
Maintenance of 2 1/4 miles Wag way yearly		200
	£	175 (49)

From this, we can see that Watt estimated that this waggon way would cost £1,965 to construct and that there ought to be a profit of £175 annually, but it was never built.

Construction of the Monkland Canal

After Watt's scheme had been laid before a 'very numerous meeting' on 3 January 1770, a subscription list was opened on the eleventh. (50) The City Council agreed to contribute £500 on the understanding that the collieries would send 30,000 tons of coal annually by the canal. (51) This was later changed to the proprietors of the 'ground on the tract' subscribing £5,000. (52) Support came from other public bodies such as the Trades House, the University, the Maltmen, Wrights, Bakers, Masons and Fleshers, since the canal was regarded as an undertaking for the common good.

Watt's scheme for a level canal and waggon way had been accepted partly after a second survey carried out at the end of January 1770 with the assistance of Barrie. (53) This is when they discovered Watt's error in his earlier quick survey which he realised would cause alterations to the line of the canal and some further expense. His new estimate for the construction came to £8,717.9s, compared with £8,653.10s. earlier. This covered £500 for banking and an aqueduct at Provan mill, eleven private road bridges at £30 each, seven tunnels for small brooks at £20 each and ten smaller ones at £4 each, as well as three weirs for controlling water levels at a total of £60. Watt pointed out the difficult conditions under which his earlier survey had been undertaken and also his own inexperience. He concluded his second report:-

> I must however beg leave to observe on my own part, that though I have seen and observed works of this kind, which have been done by others, yet I do not pretend to any practical knowledge of my own on this subject, and if more able or experienced artists should have reason to find any fault with what I have proposed, I hope the subscribers will excuse me, as I have laid before them the best judgement I have been able to form of this matter in the time, and have spared no pains to make myself master of it. It ought also to be observed; that although the idea of a level Canal is simple in itself, yet that it requires greater accuracy in leveling and planning than one with locks, as a small error may considerably increase the expence, or render it impracticable: whereas in the other case it can only multiply or diminish the number of locks. (54)

At least Watt was honest about his inexperience but he had 'the satisfaction of Mr. Barry's agreeing with me in the propriety of the above alterations, and in the practicability of the whole scheme'. (55) Barrie's support may have convinced the other subscribers to back Watt.

A petition for an Act was sent to Parliament at the same time as the one for deepening the Clyde. Watt declined to go to London to help secure it because he was busy with trials of his separate condenser and had also promised to survey the route of another canal up the Vale of Strathmore. (56) The Royal Assent was given on 12 April to 'An Act for making and maintaining a navigable Cut or Canal and Waggon Way, from the Collieries in the parish of Old and New Monkland, to the city of Glasgow'. (57)

It appears to have been tacitly agreed that Watt would supervise the construction of the Monkland Canal before his formal appointment because in March he was investigating water supplies around the Bishop Loch and Gartsheary areas. (58) Mackell must have been looking at these same places to supply the Forth and Clyde Canal so that clearly agreements had to be reached over this. (59) Since Watt himself acted as both consulting and resident engineer, he had to find contractors to carry out the work and he needed a supervisor. Golborne told the Lord Provost, 'I can't at present help you to a foreman for the Monkland canal but I will make inquiry, & in the mean Time, I wish Mr. Watt would write to me & let me know what kind of a man he wants & what wages he will give him'. (60)

He began to stake out the line of the canal at the end of May and work started under his supervision on 27 June 1770. (61) In addition, he had to make agreements over the purchase of the land, apportion work to contractors, bargain with them to settle a price, measure their work, check their accounts and make payments with the assistance of only one clerk. So his position was no sinecure but it was only for three or four days of the week, leaving him free to undertake other work. (62) He began this employment without any agreement concerning his salary and it was not until August that this was fixed at £200 per annum (63) but he did not receive payment for the first half year until the following February. (64) Similarly, further instalments were late in being paid owing to lack of money. He should have been due £500 until the end of 1772. Somehow £20 was lost by Mr. Harley which Watt took on himself. He received £50 on 27 September 1775 which brought the total to £410, leaving a balance of £90. He agreed then not to make any further demands on the Company until its affairs should improve. When the Canal was being extended with a series of locks to take it down to Glasgow in 1782, the new proprietors enquired whether Watt wanted to make any further claims for the remainder of his salary. (65) He replied,

> The bad success of the Monkland Canal has all along been a source of great uneasiness to me even though I am convinced that want of success proceeded from causes which I could not remedy and which I hope are not imputed to me. If I had been rich enough to have been able to have attended without salary as I did for six months in 1773, I would have readily laboured to have brought it to a better issue; but that was out of my power and the will must now be taken for the deed. (66)

Watt pointed out that he was still owed £90 but 'considering the circumstances I think the utmost that I ought to receive from the company should amount to £50'. (67)

Watt had to learn quickly in his new job. The structures for the canal he planned as and when they were required. For ordinary bridges under roads, he developed a standard design, possibly based on that for one at Cotes bridge which was his first. (68). These had a span of 16 ft., with uprights at either side 9 ft. high and an arch above rising 4 ft. (69) He found that the bridge at Airdrie 'will be considerably askew to the canal'. (70) and spent some time drawing this out to achieve the correct thrust lines. (71) Then he had to draw plans for an aqueduct over the Gartsheary burn, 16 ft. wide, 72 ft. long and 6 ft. high in the side walls which was 'a subject of great speculation to the country folks'. (72) Finally there was a drawbridge 'of new construction' with equilibrium weights at Netherhouse, which, when completed, moved 'very easy'. (73)

In order to construct his canal, Watt experimented with various ingenious devices. Cutting started at the eastern end and, when one section was finished, water was let in so that it could be used to transport materials. Sometimes the end of one section had to be excavated to join it to the next when both were full of water. He experimented with various types of ploughs in June 1771, partly on behalf of Golborne for dredging the Clyde. (74) Some resembled square bucket excavators. Later in the year, he tried loosening the clay along the line of the Monkland Canal with a more conventional form of plough drawn by two men working a capstan. (75) Heavy rainfall slowed progress so, to move loads over soggy ground, he tried an experiment with a barrel instead of a wheel on a waggon. (76) Finally there is mention of a 'model of a new bonny dredging machine' which he employed J. Dickinson to make. (77)

Smeaton visited the canal at the end of July 1770 and recommended that its depth could be increased to four feet simply by raising the banking. (78) In general he approved of the work that had been done so far. But Watt soon ran into difficulties. The cut at Muttonhole over five hundred yards long caused great trouble through 'running sands' and, in spite of having over a hundred men engaged on it, took the greater part of a year to complete. (79) He described this part later as 'covered with turf and have now a beautiful effect'. (80) In Drumpellier Moss, 'the most troublesome of all our works', again great difficulties were encountered because the bed of the canal, being soft peat, sometimes rose three feet overnight. (81) In June 1771, a reservoir broke out and was leaking. (82) On another section, the clay sides slipped, 'bank and all'. (83) A tunnel failed which 'luckily did no great damage except affronting me', (84) followed by another in July 1773 when Peggy had to write to her husband while he was surveying at Campbeltown telling him that Johnson's bank and tunnel had burst. (85) Solving all these technical problems gave Watt good experience in civil engineering.

If the physical construction of the canal caused Watt few major problems, it was otherwise with his labour force. The length of the canal was divided into sections of 110 yds. which were put out to tender. There were many bidders (86) with contracts going to the lowest offer. One of the first was let to James Johnston who was soon in difficulty cutting through stiff clay in heavy rain. However he later undertook other contracts on this canal and again at Port Glasgow for Watt. Soon there were 150 men at work but Watt had only one overseer besides the contractors or 'under-takers, who are mere tyros, and require constant watching'. (87) Skilled men were scarce so contractors poached each other's men as well as some from the Forth and Clyde Canal with the result that regulations had to be introduced to prevent this. (88) At first verbal agreements were made but Watt quickly found that he had to provide written instructions for each section or contract. He introduced General Regulations as well as further conditions for specific sections where necessary. (89) These anticipated his printed *Directions for Erecting and Working the Newly-Invented Steam Engines* which he wrote in 1779 to guide his engine erectors. (90)

In spite of all the precautions he took on the Monkland Canal, there were many disputes and he found that he was very quickly out of his depth in supervising and controlling this large workforce. He plaintively wrote to Small in September 1770,

> Nothing is more contrary to my disposition than bustling and bargaining with mankind: - yet that is the life I now constantly lead. Use and exertion render it rather more tolerable than it was at first, but it is still disagreeable. I am also in a constant fear that my want of experience may betray me into some scrape, or that I shall be imposed upon by

> the workmen, both which I take all the care my nature allows of to prevent. I have been tolerably lucky yet'. (91)

Two years later, we find that matters had not improved. Having to continue with work he disliked, could not have helped his physical state and could have been one cause of his headaches throughout this period. He again wrote in anguish to Small, who possibly was looking out for a job for him in England.

> I decline only being the manager, and not being engineer. I wrote you before how grievous that first part of the business was to me, and it daily becomes more so. Everything has been turned over upon me, and the necessary clerks grudged to me; I am also indolent, and fearfully terrified to make bargains, and hate to settle accounts. Why therefore shall I continue a slave to a hateful employment, while I can otherwise, by surveys and consultations, make nearly as much money with half the labour, and, I really think, with double the credit? for a man is always disgraced by taking upon him an employment he is unfit for. I have no quality proper for this employment but honesty, which reproaches me for keeping it so long.
>
> Remember in recommending me to business, that what I can promise to perform is, to make an accurate survey and a faithful report of anything in the engineer way; to direct the course of canals; to lay out the ground, and to measure the cube yards cut, or to be cut; to assist in bargaining for the price of work, to direct how it ought to be executed, and to give my opinion of the execution to the managers from time to time. But I can upon no account have anything to do with workmen, cash, or workmen's accounts, nor would I choose to be so bound up to one object that I could not occasionally serve such friends as might employ me for smaller matters. Remember also I have no great experience and am not enterprising, seldom choosing to attempt things that are both great and new; I am not a man of regularity in business, and have bad health. Take care not to give anybody a better opinion of me than I deserve; it will hurt me in the end. (92)

A few days later, there followed a similar letter.

> I am extremely indolent, cannot force workmen to do their duty, have been cheated by undertakers and clerks, and am unlucky enough to know it. The work is done slovenly, our workmen are bad, and I am not sufficiently strict... What provokes me most is, that I am sensible that most people could in the same time have done much more and better work, possibly with as little trouble to themselves. I would rather face a loaded cannon than settle an account or make a bargain. In short I find myself out of my sphere when I have anything to do with mankind. (93)

Even allowing for Watt being in a fit of despondency, we can see that he had resolved to undertake surveys, do the planning and write reports, but not to be responsible for direct supervision of labour and the physical execution of contracts. (94) We will see him following this resolution during both the design of Hamilton bridge and improvements to the harbour at Port Glasgow and we will examine a similar shift in the way he organised the erection of atmospheric engines in Scotland. (See p359) This was also his firm attitude when he moved south to commence the partnership with Boulton which he organised as essentially consulting rather than manufacturing engineers for their steam engines. His experience while constructing the Monkland Canal coloured his immediate career as a civil engineer and afterwards his more famous one as supplier of more efficient steam engines.

As the Monkland Canal progressed towards Glasgow, so sections were filled with water. On 26 November 1771, a boatload of coals was taken to Langloan and others followed. (95) Watt was able to show his friends his handiwork for, in June 1772, Clerk and Hutton, presumably George Clerk of the Forfeited Estates and James Hutton the geologist, approved 'in general of our work excepting the bridges which they think poorly done'. (96) Watt did not record Black's comments when he visited that July (97) but wrote to Small in November, 'Our canal has not stopped, but is likely to do so, from our having expended the subscription of £10,000 upon seven miles of the navigation, and having about two miles yet to make'. (98) He was proud that they had dug a canal four feet deep instead of the planned three and shortly after this he sent Small full details of what he had achieved. (99) Small had already told Watt that he was amazed at the low construction costs of the Monkland Canal. Watt had spent £450 on what would have cost £2000 or more in the south. (100) He had completed over seven miles for about £10,000, (101) which Small thought would have cost them £4,500 a mile. (102) Watt must have realised that work would soon stop through lack of cash even though the full length had not been completed. He made a survey on 7 May 1773 when he 'measured the whole canal which found to be 7 miles & 15 chains long'. (103) A boat carrying coals for Glasgow travelled along a large part on 10 May. (104) The last entry in Watt's Journals is for 17 July 1773; 'At canal inspected works, dam out at Mr. Dougal's, gate hung, Drawbridge counterpoised'. At that point, Watt's direct connection with the Monkland canal ceased. (105)

Watt thought that it would be 'of immediate and profitable use because even from that termination we can afford to undersell others' but, at first, the canal was not a success and traffic was so light that it failed to pay its way. (106) Gibson, Glasgow's historian, wrote in 1777, 'The Monkland Canal, which was undertaken with a view to reduce the price of coals, has not answered the end of which it was intended'. (107) The canal languished until the summer of 1781 when new proprietors purchased the shares. One of them, William Craig, wrote to Watt at Birmingham,

> I do not know if you have information enough of what is passing in this Part of the World to know that your own child the Monkland Canall has lately got a new sett of Godfathers. The fact is that it has been sold by the original proprietors... What the [new] proprietors would sincerely wish is that it could be consistent with your other business to pay a visit to your friends here & then they could have your advice at leisure... We have no person here as I know on whom we can relye, and the proprietors from their confidence in your integrity & ability are willing to deferr coming to any resolution for 3, 4 or even 6 months if they could have you here in that term. (108)

But Watt, other than trying to obtain payment for the arrears of his salary (See p216), was too busy with his steam engines to become involved again. Once a flight of locks had been completed from the end of his canal and the bottom joined to the Forth and Clyde Canal around the eastern side of Glasgow, the Monkland Canal became highly profitable. (109)

NOTES: THE MONKLAND CANAL

1 B. & W. Col., MI, 2/80, J. Watt, 'A Scheme for Making a Navigable Canal from the City of Glasgow to the Monkland Coalierys, December 1769'. Another copy is available at NLS, R234.b.1(5).

2 Muirhead, (1854), Vol I, p. 83, Letter 70, J. Watt to W. Small, 12 Dec. 1769; Boulton 348.14.

3 Hamilton, (1932), p. 45.

4 JWP, C7/7.24, fragment of a letter, Glasgow, Feb. 1771.

5 Muirhead, (1854), Vol I, p. 84, Letter 72, J. Watt to W. Small, 3 Jan. 1770; Boulton 348.15.

6 Ibid, Vol II, p. 2, Letter 83, J. Watt to W. Small, 9 Sept. 1770; Boulton 348.20.

7 B. & W. Col., MI, 2/62, J. Watt to 'Sir', 31 March 1770.

8 Ibid, MI, 2/42, 'Expenses on Surveys', 28 Oct. 1769.

9 Ibid, MI, 2/54, Note Book, J. Watt to J. Smeaton, no date.

10 Muirhead, (1854), Vol II, p. 2, Letter 83, J. Watt to W. Small, 9 Sept. 1770; Boulton 348.20.

11 B. & W. Col., MI, 2/80, Monkland Report, p. 7.

12 SRA, C1.1.34, Glasgow Council Act Book, Jan. 1768 - March 1772, p. 246, 2 Jan. 1770.

13 Muirhead, (1854), Vol I, p. 83, Letter 70, J. Watt to W. Small, 12 Dec. 1769; Boulton 348.14.

14 B. & W. Col., MI, 2/80, Monkland Report, p. 6.

15 Muirhead, (1854), Vol I, p. 83, Letter 70, J. Watt to W. Small, 12 Dec. 1769; Boulton 348.14.

16 B. & W. Col., MI, 2/80, Monkland Report, p. 6.

17 Ibid, p. 8.

18 See B. & W. Col., MI, 2/54, J. Watt to J. Smeaton, no date.

19 Duckham, (1970), p. 210.

20 Lewis, (1970), p. 133.

21 See for example J. Watt's being at Yoker in MI, 3/14, Clyde surveys.

22 Duckham, (1970), p. 211 and Lewis, (1970), p. 134.

23 Duckham, (1970), p. 215.

24 JWP, 4/32.38, J. Watt to Dr. J. Roebuck, 23 Aug. 1765. '...nished a model of your waggon... way which shall send pr first... are done the fore wheels & the hind... manners that you might examine... I also made a pair... wheels on one axle as common in a... that you may make a fair tryal'.

25 Adamson, (1988), p. 5 and Lewis, (1970), p. 255.

26 B. & W. Col., MI, 2/3, 'Cash Memorandum', 3 June 1766.

27 Ibid, MI, 2/1, 'Cash Memorandum', 10 Aug. - 1 Oct., 'suppose 1761' and Lewis, (1970), pp. 126, 252 & 266.

28 Desaguliers, (1763), Plates 21 & 22.

29 Leupold, (1724), plate 22 and Lewis, (1970), p. 53.

30 Belidor, (1737 - 1753). For Watt's purchase of Belidor, see B. & W. Col., MII, 8/7.2, 5 Jan 1767, 'Bt of Robert Urie, 4 vols Belidore, £6.0.0, carriage 8d.'.

31 He consulted Belidor on 7 Feb. 1773, B. & W. Col., MI, 1/19, 'Journal'.

32 B. & W. Col., MI, 2/80, Monkland Report, p. 9.

33 Ibid, p. 10.

34 Lewis, (1970), p. 302.
35 B. & W. Col., MI, 2/60, J. Smeaton to ?, 31 Jan. 1770.
36 Smeaton, (1812), Vol II, p. 63, 'The Forth and Clyde Navigation'.
37 B. & W. Col., MI, 2/54, J. Watt to J. Smeaton, no date.
38 Ibid, MI, 1/18, 'Journal', 13 Dec. 1771.
39 Ibid, 2, 14 & 15 May 1772.
40 Ibid, 2 & 8 June 1772.
41 Boulton 348.25, J. Watt to W. Small, 24 Nov. 1772. Watt said he had six boats including his special one.
42 B. & W. Col., MI, 2/42, 'Expenses on Surveys', 27 - 29 Feb. 1770.
43 Ibid, MI, 1/17, 'Journal', 28 Sept. 1771.
44 Ibid, MI, 1/18, 'Journal', 17 & 31 Jan. and 1 - 5 Feb. 1772.
45 Ibid, 3 March 1772.
46 JWP, 4/19.26, J. Hunter jun. to J. Watt, 20 Feb. 1772.
47 B. & W. Col., MI. 2/54, Note Book, 'Newcastle Prices of Waggon Way'.
48 Ibid, 2/68, 'Queries for Mr. Golborne', no date.
49 Ibid, MI, 2/54, Note Book, 'Estimate of Waggon Way from Jermiston to Broomylaw'.
50 Thomson, (1945), p. 2.
51 SRA, C1.1.34, Glasgow Council Act Book, Jan. 1768 - March 1772, p. 246, 18 Jan. 1770.
52 Ibid, p. 258, 2 March 1770.
53 B. & W. Col., MI, 2/42, 'Expenses on Surveys', 27 Jan. 1770.
54 JWP, 3/71, 'J. Watt's second report on the Monkland Canal', 2 Feb. 1770.
55 Ibid.
56 Muirhead, (1854), Vol I, p. 96, Letter 79, J. Watt to W. Small, 10 March 1770; Boulton 348.18.
57 Priestley, (1969), p. 449, 10 George III, cap. 105, Royal Assent 12 April 1770
58 B. & W. Col., MI, 2/42, 'Expenses on Surveys', 21 - 23 March 1770.
59 Muirhead, (1854), Vol I, p. 100, Letter 80, J. Watt to W. Small, 16 March 1770; Boulton 348.19.
60 JWP, 4/50.4, J. Golborne to Lord Provost, 19 May 1770.
61 B. & W. Col., MI, 1/15, 'Journal', 27 June 1770.
62 Muirhead, (1854), Vol II, p. 3, Letter 83, J. Watt to W. Small, 9 Sept. 1770; Boulton 348.20.
63 B. & W. Col., MI, 1/15, 'Journal', 1 Aug. 1770.
64 Scottish Historical Review, Vol XXIX, No. 108, Oct. 1950, G. Thomson, "James Watt and the Monkland Canal", p. 126.
65 JWP, JW/22.32, W. Craig to J. Watt, 9 May 1782.
66 Ibid, Letter Book 1, J. Watt to W. Craig, 2 June 1782.
67 Ibid, J. Watt to the Proprietors of the Monkland Canal, 3 June 1782.
68 B. & W. Col., MI, 1/16, 'Journal', 12 Dec. 1770.
69 Boulton 348.25, J. Watt to W. Small, 24 Nov. 1772.
70 B. & W. Col., MI, 1/17, 'Journal', 9 May 1771.
71 Ibid, 23 May 1771.
72 Ibid, 22 June 1771 and Boulton 348.25, J. Watt to W. Small, 24 Nov. 1772.
73 Ibid, MI, 1/19, 'Journal', 17 July, 16 Aug. and 27 Nov. 1771.
74 Ibid, MI, 1/17, 'Journal', 8, 26 & 29 June 1771 and MI, 1/18, 11 July 1771.
75 Ibid, MI, 1/17, 'Journal', 2 & 15 Oct. 1771.

76 Ibid, 8 Nov. 1771.
77 Ibid, MI, 1/20, 'Journal', 1 May 1773.
78 Ibid, MI, 1/16, 'Journal', 27 - 31 July 1770.
79 Ibid, 18 Jan. 1771 and Muirhead, (1854), Vol II, p. 21, Letter 93, J. Watt to W. Small, 24 Dec. 1771; Boulton 348.23.
80 Boulton 348.25, J. Watt to W. Small, 24 Nov. 1772.
81 Ibid.
82 B. & W. Col., MI, 1/17, 'Journal', 12 June 1771.
83 Boulton 348.25, J. Watt to W. Small, 24 Nov. 1772.
84 Ibid.
85 B. & W. Col., MI, 1/20, 'Journal', 3 July 1773.
86 Ibid, MI, 2/54, J. Watt to Mr. Golborne, no date.
87 Muirhead, (1854), Vol II, p. 3, Letter 83, J. Watt to W. Small, 9 Sept. 1770; Boulton 348.20.
88 Thomson, (1950), p. 125.
89 I could not find these at Birmingham, see SRA, TD 729/36, Birmingham Public Libraries, 1 June 1942, to G. Thomson, 'List of Archives held about the Monkland Canal'.
90 Dickinson & Jenkins, (1927), p. 375.
91 Muirhead, (1854), Vol II, p. 3, Letter 83, J. Watt to W. Small, 9 Sept. 1770; Boulton 348.20.
92 Ibid, Vol II, pp. 30 - 31, Letter 100, J. Watt to W. Small, 7 Nov. 1772; Boulton 348.24.
93 Ibid, Vol II, p. 34, Letter 102, J. Watt to W. Small, 24 Nov. 1772; Boulton 348.25.
94 Smeaton, (1812), Vol II, p. 122, for his summary of duties of the officials responsible for the Forth & Clyde Canal.
95 Thomson, (1945), p. 16.
96 B. & W. Col., MI, 1/18, 'Journal', 9 June 1772.
97 Ibid, 9 July 1772.
98 Muirhead, (1854), Vol II, p. 30, Letter 100, J. Watt to W. Small, 7 Nov. 1772; Boulton 348.24.
99 Boulton 348.25, J. Watt to W. Small, 24 Nov. 1772.
100 Muirhead, (1854), Vol II, p. 5, Letter 84, W. Small to J. Watt, 17 Sept. 1770; Boulton 340.14.
101 Ibid, Vol II, p. 34, Letter 102, J. Watt to W. Small, 24 Nov. 1772; Boulton 348.25.
102 Ibid, Vol II, p. 35, Letter 103, W. Small to J. Watt, 3 Dec. 1772; Boulton 340.25.
103 B. & W. Col., MI, 1/20, 'Journal', 7 May 1773.
104 Thomson, (1945), p. 16.
105 B. & W. Col., MI, 1/20, 'Journal', 17 July 1773.
106 Thomson, (1950), p. 131.
107 Ibid.
108 JWP, JW/22.43, W. Craig to J. Watt, 23 Aug. 1781.
109 Lindsay, (1968), pp. 52 - 65.

More Abortive Canal Surveys

Most of Watt's surveys for canals never resulted in their construction. Because he was employed for only part of the time supervising the building of the Monkland Canal, he could undertake other work. We do not know how profitable this was but it brought him to the attention of many of the prominent people in Scotland and would have helped to advance his career as civil engineer. These surveys also reveal his methods.

Baron Mure's or Caldwell Canal

After returning home from working on his steam engine at Kinneil towards the end of March 1770, Watt called on Baron Mure at Caldwell about a canal to serve his colliery at the side of Loch Libuch. (1) The Baron, who was a member of the Board of the Forfeited Estates, (2) wanted coal to be taken to his lime kilns and also the town of Dunlop, a little to the north of Kilmarnock. As it was intended to carry only a small quantity of coals to a country market, Watt proposed that the canal should be quite small with flat-bottomed boats 30 ft. long, 6 ft. wide and 18 ins. draught which would carry 4 ½ tons. Two of these could be hauled by one pony and the canal could be brought into the level of the mine. The cost of a canal two miles long with passing places, towing paths and other appropriate works was £524.18s.8d. which included £30 for two boats. (3) Such a scheme shows that, at this period, canals were the latest fashion but no doubt a few years later some form of railway would have been suggested instead.

The Strathmore Canal

About ten miles to the north of Perth, the River Tay is joined by the Isla which has flowed down a broad valley, the Vale of Strathmore, from Forfar about twenty miles to the north-east. The foothills of the Highlands form the northern side of this valley while the Sidlaw Hills cut off access to the south and Dundee. The valley itself was fertile and good for agriculture but had no supplies of fuel. A canal was mooted to help prevent depopulation by reducing the cost of coal, which it was expected would form the bulk of the traffic. This coal, together with lime and timber, would be carried up the canal. It was unusual to have most of the traffic in the upwards direction, (4) but some corn would be taken down as the area produced a surplus. (5)

On Smeaton's recommendation, George Young, a merchant in Coupar Angus who sometimes acted for the Board of the Forfeited Estates, (6) had employed at his own expense William Keir to carry out a quick survey to assess the practicality of a canal. He wrote to Lord Findlater with a copy of Keir's report in September 1769 suggesting that a small canal could be built for £20,000. Findlater forwarded these letters to the Board asking that the Board should request Smeaton to undertake a full survey at their expense. (7) This Board of Commissioners, which had been set up under the Annexation Acts of 1747 and 1752 to administer the estates of dispossessed Jacobites, acted as a clearing house for ideas and was able to sponsor at least preliminary investigations. (8) It was disbanded in 1782.

James, sixth Earl of Findlater and third Earl of Seafield, held many important posts in the government of Scotland. Among others, he became one of the Lords of Police in 1765 so would have known about Watt's survey of the Lomond Passage. He was an enthusiastic agriculturalist and transformed the face of the lands he owned. He also promoted all kinds of industry in the north of Scotland and was regarded as 'a man who in his day and generation proved himself one of the most substantial benefactors of his species'. (9) Unfortunately he died when only fifty-six on 3 November 1770 before Watt had submitted his report on this canal, thus losing one of its most influential supporters.

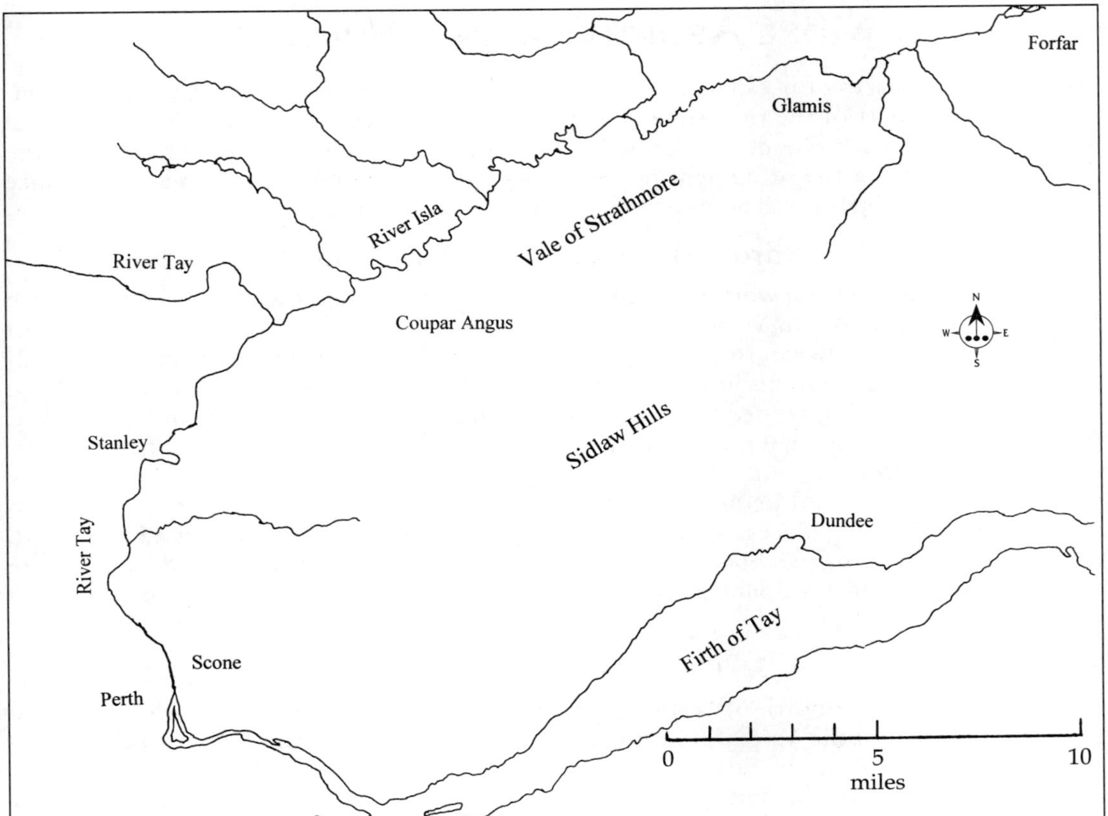

The Vale of Strathmore.

Having failed to secure either Smeaton, Mackell or Brindley to carry out this further survey, Young was authorised by the Board to employ Watt but no terms or rates of pay were discussed. (10) At the beginning of March 1770, Watt told Small that he had promised to execute it that month or early in April. (11) At this time, he was desperately testing his steam engine with the separate condenser at Kinneil and appeared to be very close to success. When he arrived at Perth on 8 April 1770 to start the survey, he wrote to Roebuck with instructions about alterations to the engine. (12) More instructions followed a week later written after he had viewed the area for the canal and realised that the survey would be more difficult than he had anticipated. (13)

Watt confirmed that the agriculture was more advanced than in other parts of Scotland and there was a great spirit of improvement but no fuel, which had to be brought over the hills from Dundee. Between Perth and Coupar Angus, there was a great deal of rough ground but from there to Forfar it was much easier. A letter to Roebuck shows that he was hoping to be able to construct one of his level canals to go from the high ground at the east of Perth to Forfar without a lock, a distance of thirty-six miles. (14) It was clearly the survey of this line for the canal which caused Watt so much trouble and took much longer than expected. He later told Small,

> I had to examine and survey a country of 36 miles in length, and to hunt about for a course for a canal through country where Nature had almost done her utmost to prevent it; - indifferent health, and weather viciously cold and stormy, were the attendants on my survey. The winds from the snowy Grampians, and snow even in the valleys a foot thick on the 10th of May, convinced me of the utility of what I was about; - for nothing can be more dismal than such weather in a country which nature and art have deprived of fuel'. (15)

From the start of this survey until he reached Birmingham in May 1774, Watt kept a daily Journal from which we can follow his movements. His survey work introduced him to a much wider range of people and industries than he would have met through his Glasgow instrument business and shop. This gave him invaluable experience and knowledge particularly of those industries in which he installed his steam engines later. For example, on the day after he arrived in Perth, he joined Sandeman at Luncarty to look at the whole of his bleaching enterprise with waterwheel, beetling mill, calender as well as the fields themselves. (16) On 11 May, during a quick survey for a canal between Perth and Crieff, he visited Ruthven paper mill and made sketches of some of the machinery. (17) When Watt wanted boats for the Monkland Canal, he asked a Mr. Turnbull, probably Sandeman's manager, about the person who built theirs at Perth. (18) It is possible that this person constructed Watt's double hulled boat. Watt would also have inspected other mills which he came across during his surveys, thus adding to his knowledge.

The Crieff Canal Survey

The request to survey a canal to Crieff probably came from people in the locality who, taking the opportunity of having Watt in their area, asked him to undertake this after he had finished the one up Strathmore. It may have been done with the support of the Board of the Forfeited Estates as later he said he was writing a report for them. (19) He carried out his survey from 11 to 15 May, starting at Perth, following the leat bringing water to the town's mills from the north round to Ruthven, before turning west up the valley of the Pow Water below Methven. (20) Once, when it drizzled all day, he was accompanied by Sir William Murray but he did have a day of sunshine and met other influential people in the area. Watt's surviving notes contain no heights but the summit at Methven Moss would have been about 130 ft. and the canal would have continued at this height on the northern side of the valley into Crieff. The distance from Crieff to the summit was 9 miles 42 chains, from there to the Tay 6 miles 16 chains and then into Perth 1 mile 58 chains, making a distance of 17 miles 36 chains. (21) No costings have survived. Nothing further happened at that time but interest in the scheme revived in 1806 when a Mr. R. Fraser wrote to Watt on behalf of the Earl of Breadalbane asking if Watt could help with any details of his scheme because a committee wished to ask for money from the residue of the balances of the Board of the Forfeited Estates. (22) Once again this scheme failed.

The Strathmore Canal Survey and Report

Watt took with him to Perth as assistants Morison and 'servants'. (23) Even for a locked canal following the course of the rivers, Watt found that the River Tay itself was too dangerous to make navigable through floods and ice as well as the strength of its current. Therefore a canal would have to be dug on the east side of it, locking out of it at Scone a little upstream from Perth with further locks up to Isla Bridge. The line of this part followed the bottom of an escarpment and would require considerable embanking. The Tay makes a large horseshoe bend round the sandstone bluff on which the village of Stanley has been built with a mill below. The river flows through a constriction in the valley between this bluff and a very steep section of the escarpment on the far side. The whole of this hillside appears unstable with trees slipping down. Once through here, the course of the River Isla would be followed with further locks up past Glamis Castle to reach the Earl of Strathmore's new navigable cut into the Loch of Forfar at a height of 160 ft. above the Tay at Scone. Watt said this would require sixteen locks of 10 ft. fall each but his final estimate was for twelve locks, 64 ft. long, 10 ft. wide and 10 ft. fall at £600 each. The total estimate for a locked canal between Scone and Glamis bridge, a distance of fourteen miles, was £41,537.13s.7d. (24)

After the success of his concept of the level canal and connecting railway at Monkland, Watt investigated repeating this again at Strathmore. A rough plan shows that he must have surveyed two other routes, one following a middle line which he did not pursue further because it would have passed along the side of the steep escarpment and the other his level canal. (25) All three routes become superimposed on top of each other at the escarpment opposite Stanley. It must have been the height of this spur and the difficult nature of the soil at this point which Watt tried to avoid by taking his highest route, his level canal, along the top. Various disadvantages followed from this. The canal would have been 183 ft. above low water in the Tay at Perth. Between Perth and the point opposite Stanley, many deep hollows had to be crossed with embankments and the canal still had to be cut along the side of the hills necessitating further embanking. Past Stanley, the canal had to bend back towards the south in a large sweep along that side of the valley which meant that it was too remote to serve the towns and villages on the northern side. Finally at Forfar, it would be higher than the loch so could not draw water from it. This route from the Rock of Kinnoul opposite Perth to Coupar Angus only, a distance of nineteen miles, would have cost £27,214.4s. (26) Watt gave no costs for the section from Coupar Angus past Glamis to Forfar.

The terminus of the level canal at the Rock of Kinnoul would be linked to the Tay and Perth by an inclined railway costing £500. Here we see another of Watt's ingenious proposals for which we have no proof whether the idea originated with him. He envisaged a water-balance incline with a system of hydraulic brakes on the waggons containing the water to control the speed of descent. Watt wrote in his journal that, from the place where the canal would terminate,

> ...there is here a gradual slope to the river in a space of 17 chains with an easy opportunity of making a double sliding waggon way to the river where by sending down a box of [water] a waggon of 2 or more tuns may be brought up I propose to have on the axle of the [water] waggon a crank to work a pump in the [water] this pump is to have only a small opening at bottom so that the wheels of the waggon may turn no faster than the [water] getting in and out at that hole so that by no carelessness of the people the waggon can run amain & do mischief this waggon way can be easily carried over the high road on a bridge & a conven't harbour may be made in the inside of the inch a little below the Kirk of Kinoul where the waggon way terminates. (27)

In his official report, Watt wrote that 'a double waggon way can be constructed, upon which the goods can be conveyed from the Tay to the canal, by the force of the superfluous water of the latter applyed to proper machines'. (28)

Watt realised that his scheme for a level canal was impracticable in this instance. Its circuitous route increased the expense so that the difference in costs between it and the other routes would be 'inconsiderable'. (29) After allowing for the time to pass the locks, he found that the journey on the locked canal would take only an hour and a half longer than on the level canal and that this disadvantage was more than outweighed by problems of transhipping cargos and working the railway incline. Barges on the locked canal could sail out onto the Tay certainly to Perth and even to Dundee where they could lie alongside and tranship their cargoes only once. With his railway, goods would have to be transhipped twice, from the canal barges into the waggons and from those into the sea-going ships. 'It is obvious that this additional loading and unloading is of itself much more than an equivalent for the longer time of passage by the locked course'. (30) He reluctantly abandoned his concept of the level canal and wrote, 'These things taken all together, make the Locked Canal to appear so much superior to the other in point of *utility*, that I have no hesitation in saying that it ought to be preferred, even although the first cost was very great'. (31)

Although the account of Watt's report has been included here immediately after his survey, he failed to submit even a preliminary draught for over a year. On his return to Glasgow, he was soon engulfed in the Monkland Canal. He received a reminder about the Board wanting his report in November, (32) but pleaded his work on the Monkland as an excuse and also 'I have wished by a little experience to become a better judge of the expence etc before I gave in my Report which I shall with all diligence complete'. (33) We can see from his Journals that he began this report in January, spending about six days on it then but he did not resume until 10 March 1771. (34) He continued working on it through the rest of that month and into April but the first copy was not finished until 3 June and acknowledged by the Board on the 6th. (35) This time he pleaded ill health as a cause of the delay. (36) A fair copy was sent on 28 September, the first having been delayed by Mr. Menzies of the Board reading it. (37) This report was submitted to the Board on 25 November 1771 and 'ordered to Lye on the table'. (38) That was the end of the matter as far as any canal was concerned.

Watt's accounts for his survey, which amounted to £211, were not submitted to the Board until 2 March 1772. (39) The Secretary to the Board, George Clerk, was recommended to examine them and report to the next meeting when he said,

> I am happy to find that Mr Watt discovers a great share of Genius in this Performance and that he is particularly fortunate in arranging his thoughts - but am sorry to observe that he is not so good at stating his accompt, and therefore proposes that it should be sent back in order to be reconsidered - However as the Money has been long resting I humbly move that Mr Watt should be immediately paid £100 in part & desired to give in to the Board a new state of his accts. (40)

It was agreed that Watt should receive part payment of £100. (41) Later, Roebuck told Watt that he ought to have charged his 'pay neat & no expenses'(42) and must have tried to smooth things over with Clerk when he pointed out,

> Perhaps his very scrupulous honesty may make his operations rather more tedious to himself and expensive to his employers than is thought absolutely necessary. Though in expensive and difficult undertakings a just and accurate plan and estimate which may cost something considerable will be thought judicious oeconomy in the execution. (43)

There is a paper showing Watt's expenses on this survey but, as it has no date, we do not know whether it is his first or revised estimate. He noted Horse Hire, 44 days at £3.3s. but gave no length of time for his own charges of £80 for the survey and £30 for writing the report and directing the map. Morison would receive £40 for 44 days assisting and £18.18s. for 36 days drawing the plan. There were sundry other expenses but it is not clear how the whole account was added up. (44) Neither is it clear whether Watt was ever paid any further sum by the Board.

The Bo'ness Canal

Bo'ness, with its custom-house, was the major port at the western end of the Firth of Forth. However this position of importance was threatened first by the Carron Company shipping its wares direct from Carron and then by the development of Grangemouth with a harbour for the Forth and Clyde Canal. To accommodate these fears, the Act of Parliament authorising the construction of the Forth and Clyde Canal in 1768 also made provision for a branch from the mouth of the Grange Burn to Bo'ness, which was to be built by a separate company. (45) It is not clear what happened during the following three years because it was not until the end of August 1771 that Watt viewed the ground with the committee (46) and carried out a full survey between 18 and 24 November with John Stewart and Osborne. (47) He started writing his report just before Christmas and spent a large part of January 1772 drawing plans, working out estimates and composing his report. In all he spent twenty four days on this project at a cost in one account of £32.2s. (48)

This canal was intended to link the Forth and Clyde Canal above the first lock with Bo'ness harbour. Most of the region that it had to cross was prime agricultural land which would be expensive to purchase. The canal would not lie much above high tide level which presented problems with crossing the Grange Burn and more particularly the Water of Avon. The aqueduct over the Grange Burn was to be 72 ft. long, 16 ft. high, and the arches to have a circumference of 18 ft. In one scheme, Watt proposed damming the Avon to cross it on the level but feared this might cause flooding of adjacent land. His alternative was a wooden aqueduct.

> I would propose the Aquaeduct to be a very strong wooden trough, Twelve feet wide, and seven feet deep, supported upon rows of strong posts which should be set upon piles driven in the Bottom of the river. It should also be braced by diagonal *Shores* fastened to piles and to the bottom and top of the *Trough*... I would propose the clear water way for the passage of the River or length of the openings of the Bridge, to be one hundred & twenty feet and the height to be eleven or twelve. From what observations I have been able to make, this will be abundantly sufficient for the passage of the waters, in any state which they ever are in.
>
> I have preferred this wooden fabrick, to one of stone, principaly [sic] on acount of the expence, which will be much less particularly, in such precarious foundations as the Kerse lands afford. (49)

He was confident such a structure would survive the combination of a high tide, a river flood and ice. This would be the main civil engineering feature on the canal.

Below Kinneil and from there to Bo'ness harbour, the foreshore was restricted so that some cutting through rock would be necessary as well as walling off from the sea. The canal itself would be quite narrow, sixteen feet wide at the bottom, and five feet deep. This was less than the Forth and Clyde Canal so this one must have been designed for barges and not boats that could sail in the Firth. Boats could not pass out of it into the Firth because, to save expense, there was at first to be no lock from the canal into Bo'ness harbour. Therefore Watt recommended, 'the loading the Boats by Carts or wheel barrows for sometime; until the extent of the trade passing upon the Canal being fully known, it may be found whether or not it will be advantageous to make one [a lock]'. (50) Once again Watt gave no costs for transhipping. He proposed two routes, the longer one by Jynkabout, just over four miles at a cost of £3,210.4s.6d. and a shorter one by Bearcroft about three and a half miles at £2,932.9s.9$^{1}/_{2}$d. (51)

Presumably Watt submitted his report soon after it was completed at the end of January 1772 but the scheme must have been halted by the collapse of the Ayr Bank that June. He had not been paid. He had difficulty in getting his claims settled for this and other projects which shows the insecurity of this type of surveying work. In March 1773, Roebuck junior forwarded Watt's letter asking for payment of £30 but reported that the funds of the canal committee were exhausted although some subscribers still owed money. (52) In May over a year later, Roebuck junior thought Watt would have to go to court to obtain his money. (53) This was the month when Watt set out for Birmingham and entrusted his Scottish affairs to Hamilton who tried to obtain payment at least in June and July 1774 and again in March and May 1775, wishing to avoid recourse to legal action. (54) Persistence was finally rewarded for in October 1775 Watt received £25 (55) but the records are silent about the outstanding £5.

Some work, but not under Watt, was started in 1783. The money soon ran out. Another Act was obtained but again the funds quickly became exhausted. A new survey was made by Robert Whitworth in 1789 who estimated that it would cost over £17,000 to complete the canal and the whole scheme was abandoned in 1796. (56)

The Renfrew Canal

In March 1772, John Maxwell junior told Watt, 'The Magistrates of Renfrew and Alexander Spiers Esq. of Eldersley have resolved to make a canal near Renfrew into the Clyde and they want that you should see the ground through which this Canal is to pass and give them your opinion as to the Depth, Levell, etc.' (57) Maxwell hoped that Watt could make his survey the following Saturday which he did on 5 April and made out his report on the 14th. This must have been a very small scheme because he must have sent it off the same day noting 'whole expense £134', presumably referring to the cost of the canal. (58) This scheme also was never executed.

NOTES: MORE ABORTIVE CANAL SURVEYS

1 B. & W. Col., MI, 2/42, 'Expenses on Surveys', 24 - 27 March 1770. See also JWP, 4/19.12, Baron Mure to J. Watt, 19 March 1770.

2 SRO, Forfeited Estates, E 721/9, Minute Book General, Nov. 1765 - Aug. 1766, 18 Nov. 1765.

3 JWP, 3/71.2, 'Cauldwall Report', 29 March 1770.

4 *Transport History*, Vol I, 1968, S.G.E. Lythe, 'James Watt and the Strathmore Canal Project', pp. 67 - 70.

5 Watt, (1771), p. 3.

6 Smith, (1982), p. 210.

7 SRO, Forfeited Estates, E 727/37/1, (1) Ld. Findlater to H. Barclay, 15 Sept. 1767, (2) G. Young to Ld. Findlater, 4 Sept. 1767, (3) W. Keir to G. Young, 2 Sept. 1767.

8 Lythe, (1968), p. 67.

9 Paul, (1905), Vol II, pp. 39 - 40.

10 SRO, Forfeited Estates, E 727/37/3, Letters Relating to the Strathmore and Crinan Canals, no date.

11 Muirhead, (1854), Vol I, p. 95, Letter 79, J. Watt to W. Small, 10 March 1770; Boulton 348.18.

12 Ibid, Vol I, p. 101, Letter 81, J. Watt to Dr. J. Roebuck, 8 April 1770; JWP, 4/32.8.

13 Ibid, Vol I, p. 102, Letter 82, J. Watt to Dr. J. Roebuck, 15 April 1770; JWP, 4/32.14.

14 Ibid.

15 Ibid, Vol II, pp. 1 - 2, Letter 83, J. Watt to W. Small, 9 Sept. 1770; Boulton 348.20.

16 B. & W. Col., MI, 1/14, 'Journal', 9 April 1770 and see Ibid, MI, 1/15, 'Journal', 16 May 1770, when Watt went for a trip in Sandeman's experimental sailing boat.

17 Ibid, 11 May 1770. See also *The Quarterly*, British Association of Paper Historians, No. 14, March 1995, R.L. Hills and J. Reece, 'British Paper Mills: Ruthven and Methven', pp. 18 - 20.

18 JWP, 4/80.3, J. Watt to Mr. Turnbull, 23 Feb. 1771.

19 Ibid, 4/62.7, no name but must have been J. Watt, 28 Sept. 1771, 'I have a short report to make concerning some levels I took for a Canal to Crieff by Mr Menzies order'. There is no trace of this report.

20 B. & W. Col., MI, 1/15, 'Journal', 11 - 15 May 1770.

21 Ibid, MI, 2/73, 'Note of Distances of the Crieff Canal'.

22 JWP, 4/85.15, R. Fraser to J. Watt, 8 May 1806.

23 SRO, Forfeited estates, E 727/37/6(3), J. Watt's Accounts.

24 J. Watt, 'Strathmore Report', p. 52.

25 B. & W. Col., MI, 4/46, 'Map of Strathmore Canal'.

26 J. Watt, 'Strathmore Report', p. 49.

27 B. & W. Col., MI, 1/14, 'Journal', 22 April 1770.

28 J. Watt, 'Strathmore Report', p. 29.
29 Ibid, p. 56.
30 Ibid, p. 59.
31 Ibid, p. 60.
32 JWP, 4/62.1, J. Morison, Clerk, to J. Watt, 29 Nov. 1770.
33 Ibid, 4/82, J. Watt to A. Menzies, 3 Dec. 1770.
34 B, & W, Col., MI, 1/16, 'Journal', 10 March 1771.
35 J. Watt, 'Strathmore Report' and JWP, 4/62.5, J. Morison, Clerk, to J. Watt, 6 June 1771.
36 JWP, 4/33.3, J. Morison, Surveyor, to J. Watt, 6 April 1771.
37 JWP, 4/62.7, no name, but must be J. Watt, 28 Sept. 1771.
38 SRO, Forfeited Estates, E 721/11, Minute Books General, Feb. 1768 - March 1782, p. 73, 25 Nov. 1771.
39 Ibid, p. 84, 2 March 1772.
40 Ibid, E 727/37.6(1), G. Clerk to J. Morison, Clerk, 27 Feb. 1772.
41 Ibid, E 721.11, Minute Books General, Feb. 1768 - March 1782, p. 88, 9 March 1772.
42 B. & W. Col., MI, 1/18, 'Journal', 13 March 1772.
43 JWP, 4/32.73, Dr. J. Roebuck to J. Watt, 22 March [1772], copy to G. Clarke Esq.
44 JWP, 4/62.19, 'Expenses upon the Strathmore Survey'.
45 Lindsay, (1968), p. 190.
46 B. & W. Col., MI, 1/17, 'Journal', 30 Aug. 1771.
47 Ibid, MI, 2/42, 'Expenses of Surveys'.
48 Ibid and also JWP, C7/7.16 & 27 for sketch drawings of Bo'ness Canal.
49 B. & W. Col., MI, 3/59, J. Watt, 'A Report to the Committee of Proprietors of the Borrowstowness Canal, concerning the expence of making that Navigation'.
50 Ibid.
51 Ibid.
52 JWP, 4/32.55, J. Roebuck jun. to J. Watt, 31 March 1773.
53 Ibid, 4/32.41, J. Roebuck jun. to J. Watt, 9 May 1774.
54 Ibid, 4/20.1, 17 June 1774; 4/20.2, 27 July 1774; 4/20.13, 19 March 1775 and 4/20.16, 22 May 1775, G. Hamilton to J. Watt.
55 B. & W. Col., MI, 2/6, 'Cash Memorandum', 10 Oct. 1775.
56 Lindsay, (1968), p. 190.
57 JWP, 4/19.28, J. Maxwell jun. to J. Watt, 26 March 1772.
58 B. & W. Col., MI, 1/18, 'Journal', 5 & 14 April 1772 and MI, 3/75, J. Watt's 'Report to the Magistrates of Renfrew', no date.

Canals Across Argyll

The peninsula formed by Knapdale and Kintyre, which juts out towards the coast of Ireland and forces the Firth of Clyde to turn south, forms a barrier, cutting off direct communication from Glasgow towards the West and the Atlantic Ocean. People in the Western Isles and Highlands trading with the Clyde also had to make the voyage round its southern end, the Mull of Kintyre, which has always been a dangerous place for sailing ships because it is a point where many tides meet. To sail round the Mull of Kintyre could mean facing a contrary wind on part of the journey. Ships might be held up for several days when the winds were against them while there were few harbours for shelter, particularly on the west coast of the peninsula. But the hills forming the backbone of the peninsula gave way to low passes or isthmuses at three places. The southern one was at Campbeltown. Further north, West Loch Tarbert almost cut through to the village of Tarbert itself and East Loch Tarbert. The most northerly one stretched from Crinan in the west to Lochgilphead in the east. These last two were considered as possible routes for canals to link the Firth of Clyde with the Atlantic.

Early Proposals

The earliest proposal for a canal across one or other of these isthmuses in Argyll which has been discovered so far is contained in a letter written by Cathcart from his home at Schaw Park to Dougald McTavish of Dunardry in 1764. It is quoted in full here because it shows why there was interest in making a canal in this region.

> I have seen your Letter to Mr Freebairn and a sketch he had taken (without instruments) of the ground lying betwixt the Lochs Crenan and Gilp, at your Desire: I perfectly well remember the Conversation at Inverara, to which you allude, and am much obliged to you for the particular Information you have sent me, in consequence of which I have directed the Bearer Willm Morris to take an ?Octnal? Survey of those grounds with the true levels, and also of the grounds betwixt the East and West Tarbert. The advantages to the Highlands from either Communication are obvious, but if a passage could be gained sufficient for Herring Busses, the additional advantage that important branch of the Fisheries would receive ought to be a further Inducement, and from the notion I have of the ground from Mr Freebairn's sketch and the plentyfull supply of water in the highest part, I should imagine that in your Neighbourhood, if it is not practicable to do it by cutting it may be certainly done by locks.
>
> I recommend him (Wm Morris) to you for your assistance during his survey. (1)

The importance of the fishing industry as well as providing a communication with the Highlands should be noted. The Crinan route seems to have been preferred. However, if Morris did carry out a more detailed survey, his results were unknown to Watt.

There must have been continuing interest in such a scheme because Watt senior wrote to his son in Glasgow during January 1767. 'Last weeke their was a farder declaration by Shypmasters & others that Terbet was ye safest & quickest passage for vessels to or from Clyde the expected to be pitched on & Lord Cathcart is inquasative about ye menner of macking ye cutt & securing the passage.' (2) During the summer of 1767, Watt senior sent Cathcart some comments about these routes (3) and wanted to go to Tarbert in July 1768 but wrote to Peggy that he had delayed the trip through bad weather until the following week, hoping also that Watt would be ready by that time. Presumably his father thought that Watt would be able to assist with any surveying needed. Charles MacDowall, a friend of Watt's,

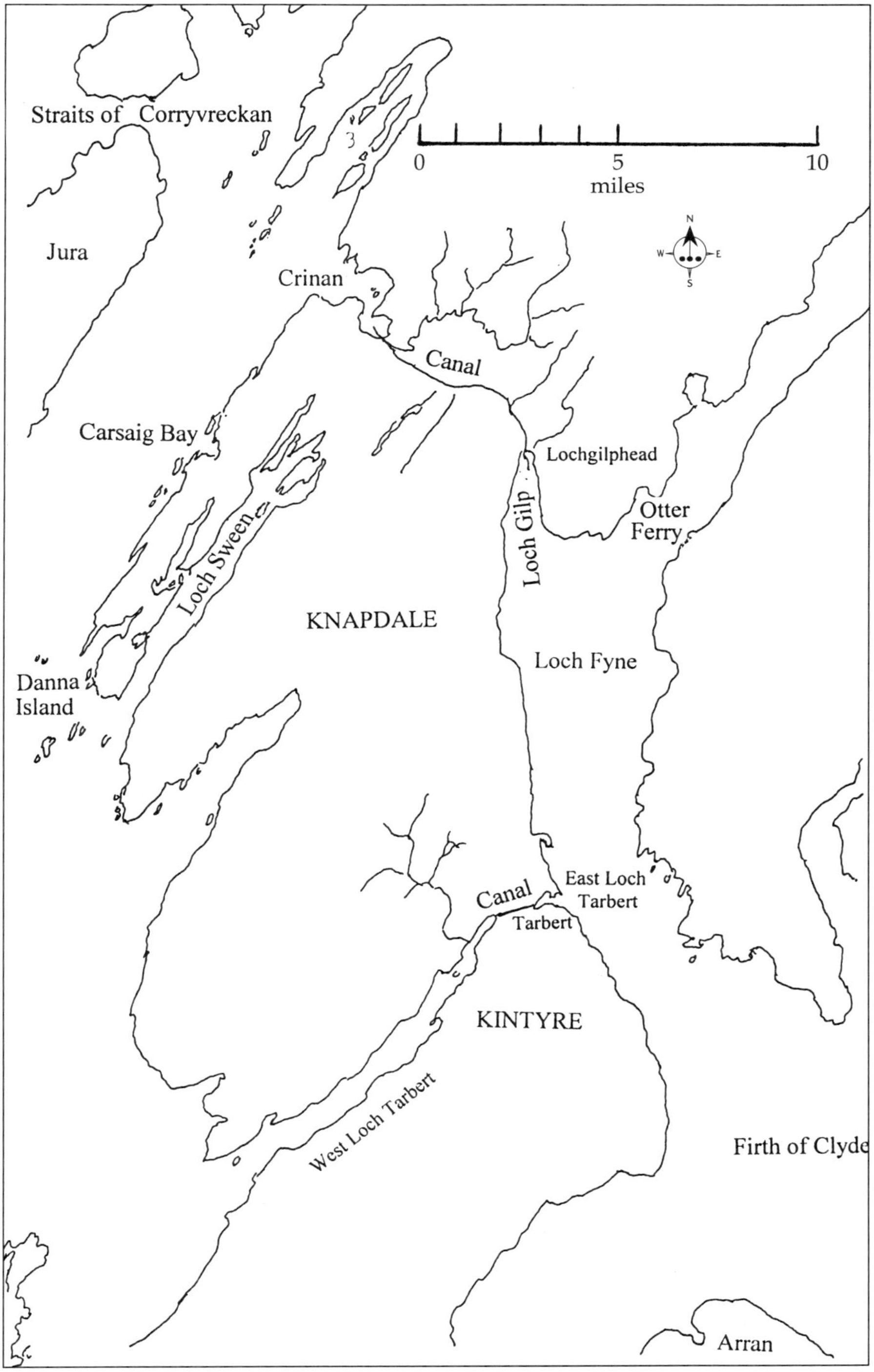

Knapdale and the northern part of the Kintyre peninsula.

warned him in 1771 that there was opposition to his carrying out a survey owing to the support his father had given to the Tarbert scheme.

> I wish your father had not gone with you, I have difficulty to convince the well wishers of Loch Gilp that he will not be able to mislead you about Loch Tarbert, for my own part, I know you too well to doubt of the accuracy and faithfullness of your Report, and therefore go on and prosper. (4)

Watt Invited to Carry Out a Survey

There is no mention of any further schemes until the early part of 1771 when the City of Glasgow became involved. Charles MacDowall wrote to Watt at the end of February that Glasgow Council was considering writing to the Commissioners of the Board of the Forfeited Estates about making a canal from Loch Fyne to the 'West Sea' and recommended him, if he thought he could complete a survey before the beginning of July, to write to the Baillie at Glasgow so that he would be mentioned. Charles MacDowall warned that 'Some of the Commissioners think you very dilatory'. (5) This advice was probably given because Watt had so far failed to submit his report on the Strathmore Canal. Watt must have told Charles MacDowall that he was too busy with the Monkland Canal to be able to undertake this new survey. (6)

In the meantime, the Glasgow Magistrates had sent their memorial to the Board in Edinburgh setting forth the advantages of a canal from Loch Fyne and pointing out that there were two potential passages. They asked that a proper person be appointed to carry out a survey in order that the practicality and costs could be determined. (7) The memorial was read to the Board on 4 March 1771. (8) No preference for either passage was stated. It was pointed out that a canal of seven feet depth could accommodate half of the herring busses and that it would open up a shorter communication between the Clyde and the West Coast as far north as Cape Wrath which would help to promote trade in timber, bark, kelp, grain, fish, 'or whatever these Countrys could spare'. A report, dating from 1768 by Mr. Menzies, the Board's General Inspector, was also read which emphasised the utility of such a scheme. (9) A copy of a map of Tarbert by Menzies in the archives at Birmingham has a grid of squares drawn on it which may have been used by Watt as a basis for his surveys. (10) The Board agreed to pay the expenses of the new survey. Their secretary was to send this information to the Glasgow Council and at the same time write to Watt saying that he would be employed to execute it. (11) Charles MacDowall communicated this good news to Watt and thought that it was greatly in his favour. (12)

However it was not until 27 May that the Board's secretary, Harry Barclay, sent Watt the official letter of authorization. (13) A couple of days later, Watt was informed by the Lord Provost of Glasgow that he had been appointed. (14) A committee meeting of the Monkland Canal was not held until 19 June when Watt was given leave of absence. (15) On the same day, he sent this news to the Board of the Forfeited Estates and said he required a little time to put his affairs at the Monkland Canal in order. He also requested information about the size of boats which would use the canal and any other particulars. (16) Accordingly instructions were drawn up by J. Campbell. (17) Watt was to prepare estimates for canals of either seven or ten feet depth through Crinan and Tarbert and, in addition, he was to investigate a through-cut at Tarbert. He was to report on the different heights and times of tides on both the east and west coasts of the peninsula as well as the state and convenience of harbours on the west coast. The Board wanted to know whether coasting vessels coming from the north used the sound between Islay and Jura or between Scarba and the mainland. (18)

The depth of the proposed canals does not seem to bear any relation to the ships which were meant to pass through them. Seven feet was presumably based on the depth of locks which were being built at that time for the Forth and Clyde Canal. (19) In his report, Watt thought that the herring busses would be the vessels most commonly using the canal. Even the smallest of these at twenty tons burthen drew eight feet of water when loaded and five when in ballast. One of thirty tons drew up to ten feet when loaded, six and a half feet in ballast and was about thirty five feet long at the deck. One of forty tons was forty feet long; when loaded it drew eleven feet and in ballast seven and a half feet. Hence a canal seven foot deep would seem to have been of little use for most of these ships, certainly not half of them as has been suggested above. Watt estimated that there were 300 busses 'belonging to the Firth, great part of which belong to the Port of Campbellton' (20) but he gave no indication of their sizes.

Preparations for the Survey

In the meantime, Watt was making his preparations for the survey. His chief assistant was once again Morison, who would be ready to leave early in July. Morison intended to take his assistant and Watt three more; John and Robert Johnston, at 2s.6d. per diem and the apprentice, Osborne. (21) Morison arrived in Glasgow on 5 July and assisted with the preparations for what was to be a thorough survey. (22) We have noted earlier the list of equipment they took. (23. See p188) How this, and the surveying instruments, were transported across the rough tracks at Crinan and Tarbert is not clear but there is an entry 'To corn miller long line for cartage home wit. rods' and 'To rope bought', possibly for tying equipment onto horses. (24) Most of that equipment would be used to ascertain the nature of the ground through which the canals might pass. In addition, Charles MacDowall had told 'Mr. Campbell of Duntroon' who lived on Crinan Loch about the intended survey and wrote to Watt, 'You will find he [Campbell] will show you all manner of civility, and will assist you as far as he can, by finding a proper boat to carry you on your expeditions & proper people to give you information & show you what you wish to see'. (25)

The Survey of the Crinan Isthmus

The expedition set sail from Greenock on 11 July in a hired boat as the packet was not ready. After beating against a gusting south-westerly wind and rain, they put in to Holy Loch at about 2 o'clock and, at six, made for Dunoon as the weather had moderated a little. The next day, as there was no sign of a fair wind, they hired horses to reach the Otter Ferry across Loch Fyne and arrived at Lochgilphead in the night. The cost of the ferry was 4s. and the horse hire 19s.6d. (26)

On Saturday, they were taken by Mr. Campbell of Askomelle in Campbeltown along the line of the present road and canal from Lochgilphead over to Crinan. They went through the almost level mossy stretch to Cairnbaan and then up a sudden rise to Dunimuck mill dam near Dunardry with steep rocks on either side. The valley broadened at the summit where there appeared to be a considerable depth of soil. Then it was contracted again at the far end by a ridge of rocks. They descended to a 'most extensive mossy plain', the present Moine Mhor, and about a mile further on came to the sea loch, the 'Cock of Crinan', the inner bay of which dried out at low water. The outer Crinan Loch was said to be deep water except for some rocky islands which remained visible at high water. Watt noticed a couple of burns falling into the summit at Dunimuck which were said to be fed by lochs in the mountains. He also noticed the Water of Add which joined the Dunardry stream at the inner Crinan Loch. They must have returned to Lochgilphead where Watt spent the Sunday, 14 July, writing up his journal and, in the evening after the weather had cleared up, walked along the shore. He

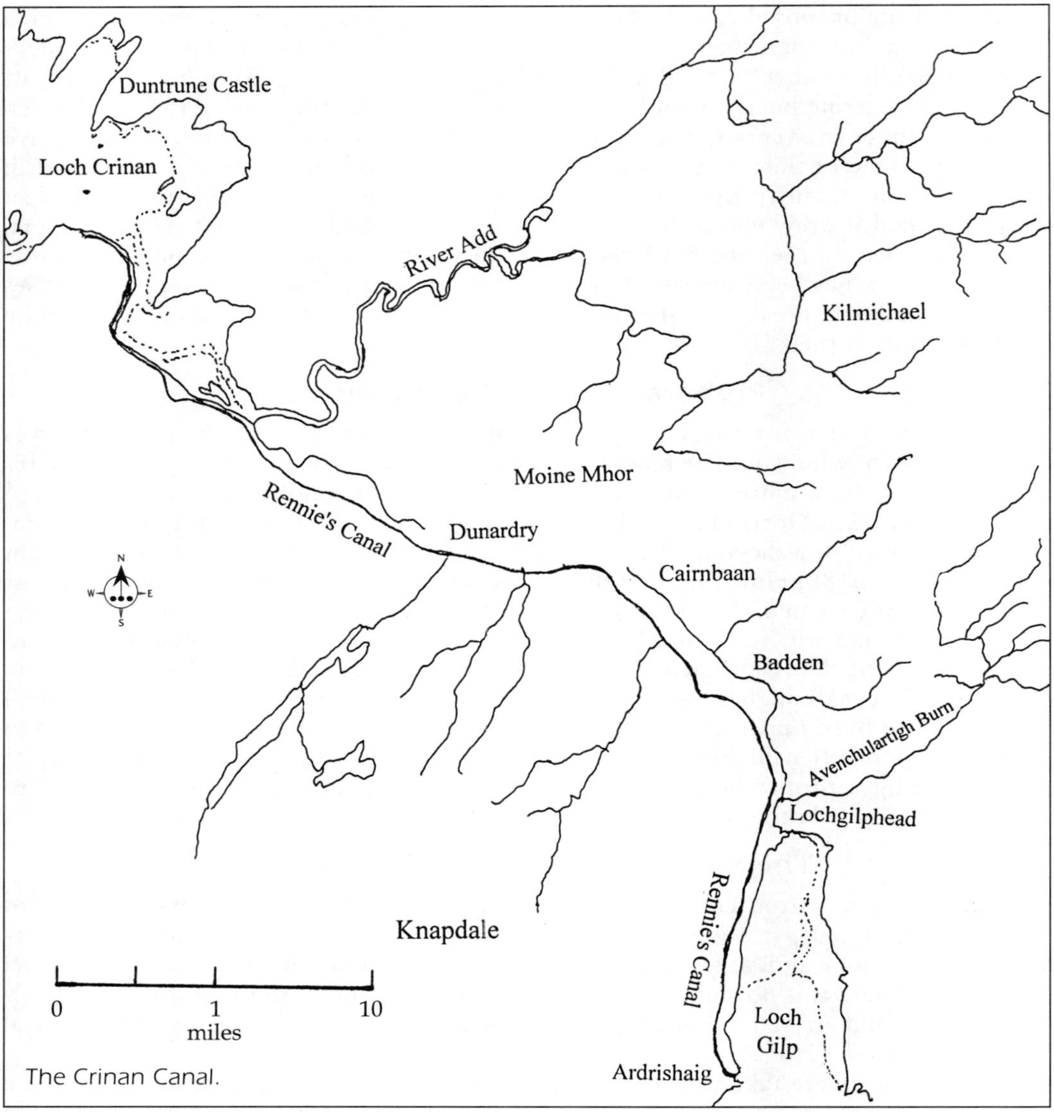

The Crinan Canal.

took back some of the mussels he found on the rocks and boiled them but found they had a bad taste. He also saw a place on the east side of Loch Gilp where a dry harbour might be made.

On the Monday when it rained heavily in the evening, the northern end for the canal was surveyed and Watt dined with McTavish of Dunardry and others. He also put his men to investigate the depth of the soil at the summit where, after digging down five feet, they were drowned out by water and had to use the boring rods. On this, and some other occasions, more men were hired to help, including quarrymen to examine the rock. (27) It was probably on this day that Watt sent Morison to investigate the size of the lochs supplying the streams at the summit. He himself discovered the glen leading northwards from Cairnbaan towards Kilmichael where the present main road to Oban goes. This looked a more promising route. Not only did the summit seem lower, but a canal might be able to follow the large Water of Add into Crinan Loch. 'I resolved to give this a more narrow Examination'.

Wednesday was spent at the Lochgilphead end when the fall from the summit to high water was ascertained to be 70 ft. He proposed to make a separate channel to the sea for the Avenchulartigh burn which flows down behind Lochgilphead because it carried a great deal of gravel in flood. The Water of Gilp from Cairnbaan was much clearer. In the evening, Watt used his perspective machine to make sketches of Loch Gilp. On Thursday, Morison measured Loch Gilp when there was more rain. Watt intended that his canal would follow the course of the rivers out to the sea at both ends so his route would have been much shorter than the one Rennie built later. No detailed drawings have survived but a pencil sketch has a line curving past Lochgilphead and probably down Loch Gilp itself.

Although it rained on Friday morning, the glen towards Kilmichael was surveyed and found to be eleven feet lower than at Dunimuck mill dam. However, while it would have been easy to carry a canal round by this track, the problem of a large enough supply of water for lockage proved to be insuperable. The levels on the other route were checked and the party stayed that night at Duntrune Castle, 'a fine old castle situated upon a jutting rock with a fine sandy bay betwixt it and Crinan'. (28) The next day was spent studying the tides and making drawings of the outer bay of Crinan.

The Survey at Tarbert

On the Sunday, they took the Mull packet for 'Tarbit'. Watt was required to report on the state of harbours in the region and seems first to have sailed north to look at the Gulf of Corryvreckan. Watt fully appreciated the difficulties of navigating it, having seen the breakers from Crinan. That evening, 'we got into an eddy near a little island where the vessel turned round three or 4 times before we could get her out of it'. This was near 'Barneshallock', described as a good harbour with an island in the mouth of it, probably the present Carsaig Bay near Tayvallich. They stayed the night either here or further south at 'Lough na Keil' with Mr. Campbell Dunah when there was more heavy rain. (29) On the Monday, after looking at Loch na Chastel and Loch Sween, they arrived at Tarbert at 4 o'clock and spent the evening inspecting the ground for the canal. (See p238)

On Tuesday, 23 July, they surveyed right across the isthmus. The rain that night continued into the next day when East Loch Tarbert was surveyed. For this, boats were hired. Most of West Loch Tarbert was surveyed from the water owing to the difficulty of access by land. Also boats were necessary for measuring the depth of the lochs. (30) Thursday was spent in more drizzly rain in the hills on the south side of the route searching for places where reservoirs might be built to supply the canal. One place was found near Tarbert itself but the burns were only small and it was doubtful if they would suffice. Rain continued on the Friday when the survey of the West Loch had to be abandoned after only a mile had been covered. Watt was interested by a 'lecky' or double tide in the Loch. On Saturday, the survey was carried on with Watt taking the north side and Morison the south, starting from the entrance at the west. They did not return to Tarbert until 11 o'clock at night! (31) Sunday was a fine day when there was no surveying. On Monday, another fine day, Watt joined up his surveys but found a very shallow place at the Eillan da Ghallagain where the deepest point was only 13 ft. On the Tuesday, while Morison checked the depths in West Loch Tarbert, Watt went searching on the northern side for a good water supply. He found this in the 'Avennagillian' burn. One of the lochs supplying it, Loch Huron, was said to be about half a mile long and as broad and could be dammed up to provide storage. Watt found two places

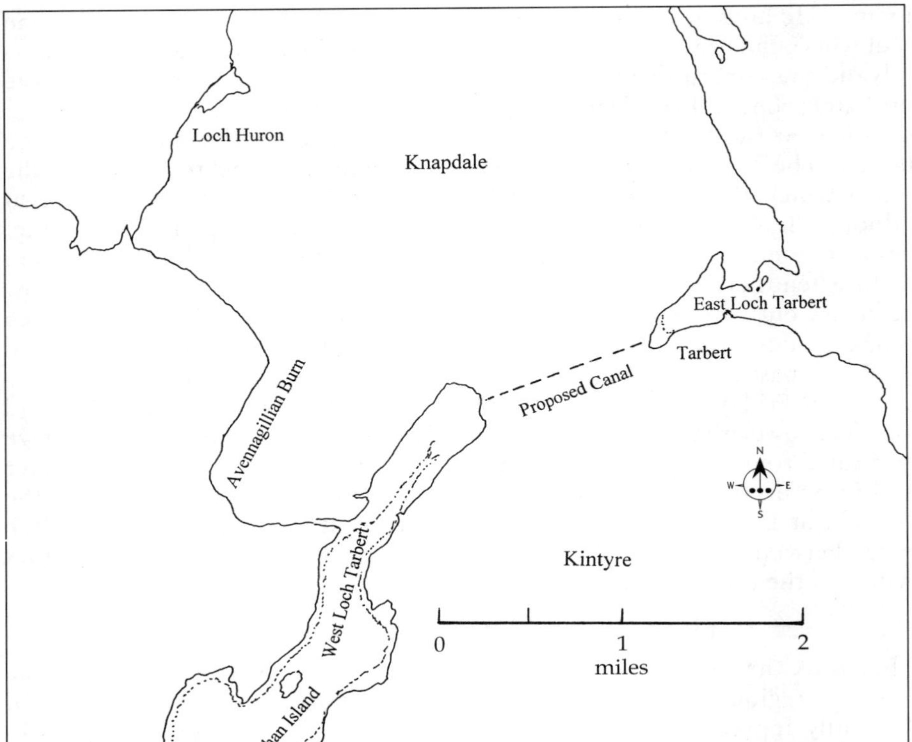

Proposed canal at Tarbert.

where this water could be tapped and carried along the sides of the hills in channels to the summit level of the canal above Tarbert. On the 31 July, they set sail in the Lochgilphead packet for Greenock, where they landed at 3 o'clock in the morning of 1 August. (32)

Writing the Report

From Greenock, Watt returned home to Glasgow and started preparing his drawings for the Tarbert part of his report while Morison did those parts which he had surveyed. (33) Charles MacDowall wrote to enquire about their results and whether a canal seven feet deep might be possible. He thought a canal at Tarbert for Atlantic ships would have to be twenty feet deep to take boats drawing eighteen feet. (34) Watt had sent him some particulars but Charles MacDowall was 'afraid from the heights the expence will run so high as to bar both'. He wondered whether thirty or forty feet might be cut out of the summit on the Crinan route. (35) Presumably more pressing business called because, after seven whole or part days on this report in August, Watt seems to have done nothing until he prepared an Abstract in the middle of November. There was more activity in February with a further nine days work and another day on 1 March. He may have submitted the abstract to the Board of the Forfeited Estates because, in the middle of April 1772, Campbell sent some clarification about what the report should contain. Campbell pointed out that it must be impartial and that 'ye persons interested in the Cutt at Crinan, who are abundantly keen should not have reason to complain of any bias in the instructions in favour of Tarbert'. The Board was particularly interested to know whether it was shorter to come by Tarbert or by Crinan into the Clyde and also the costs of a through-cut at Tarbert 'tho' it is a scheme that I am persuaded never will take place'. (36)

Watt continued with his report during July and August. Part of the delay was because Morison had mislaid his field pocket book and was unable to send the figures of soundings in East and West Tarbert Lochs until 22 August. (37) However, it was not until the latter end of November and early December that Watt spent a lot more time on the report. He completed it on 21 December 1772, having been occupied with it on 42 different days or parts of days. We can now see why Watt was accused of being dilatory but the time this report took is not really surprising because the full version covers 25 typed pages of modern A4 in single line spacing so it is a very long document, even without all the plans and sections.

The Canal across the Crinan Isthmus

In his report, (38) Watt began by describing the country through which the Crinan route would have to pass. The head of Loch Gilp was shallow with no natural harbour and little shelter. A channel leading to the start of the canal would have to be cut through the sands and secured by two jetties of rubble stones, one on either side. The Water of Gilp would be used to help keep this channel clear. At the shore line near Lochgilphead, a tidal lock would be constructed to allow access to the canal. The route would lie through the mossy plain with a second lock about half a mile from the first. The position of the third lock was to be at 'Bad-dana Cranavan', probably the modern Badden, where the canal would pass on the west side of the mount. This lock might be dug out of rock, as would the canal itself along part of this section. It would be level then to Cairnbaan where three locks would take it up to the summit. Cutting through solid ground here would be expensive but part could be formed by banking.

The height of the summit would be reduced by cutting sixteen feet down through the gravel; if much more were attempted, rock might be hit. This would make the summit level 54 ft. above neap tides at Lochgilphead. The canal would descend by the Dunardry burn where the locks would follow each other closely. It would be carried by extra cutting and banking through the moss where there would be a final lock into the sea. The channel of the Add would have to be deepened. The Bay of Crinan formed a good harbour except in winds from the north and west when vessels might take shelter at nearby Craignish. The few rocks in Crinan Bay were exposed at high tide. The bar in the outer bay here would have to be removed by dredging and a jetty built.

The dimensions of the locks were to be the same as those being built at that time on the Forth and Clyde Canal, 20 ft. 6 ins. wide, 72 ft. long, but with 8 ft. fall. The bottom of the canal would be 24 ft. wide so there would be places where boats could pass. Watt envisaged there would be a sufficient supply of water for lockage from the burns falling into the summit, especially if the lochs in the mountains were turned into reservoirs. He based his estimates for water supply on twelve vessels passing each day. The reservoirs would hold sufficient for 57 days, and he assumed that there would be plenty of rain to refill them.

If a canal were to be constructed with a depth of ten feet, then its summit level would have to be three feet higher than that for the seven foot canal because Watt did not think the bottom could be lowered there any further. The two tidal locks would have to be increased in their fall to compensate. Part of the line up to Cairnbaan might have to be altered to avoid a depth of ten feet cut through the moss. The fall in each lock would not be increased, only the depth, so water required for lockage would remain almost the same.

Watt drew up detailed estimates. Those for the seven foot canal show the most expensive items as the 14 locks at £1,000 each. There were only two drawbridges, one at Lochgilphead and another at the summit, each costing £400. The sea piers at each end were £100. Cutting the main line of the canal for four and one eighth miles would cost £5,898.15s. The total cost for the shallower canal added up to £34,879.0s.4d. while that for the ten foot canal came to £48,405.5s.3d. These estimates allowed for higher wages which would be necessary to entice men to work in so remote an area. 'The weather is more uncertain, and the men thereby being thrown idle, must have higher prices for what they do'. The waterlogged soil might also present difficulty, 'the boggy part is of the worst species of Earth for standing even with the most experienced Artist'. Watt thought the locks could be built from the local stone although it might be necessary to import some from the Island of Arran.

Watt tried to determine the financial viability. He calculated the interest which would have to be paid on the capital needed to build each canal and a toll at 1s.6d. per ton 'of the vessels burthen to be paid loaded or not; for the expence of navigating a vessel being in either case equal'. To pay interest at 7 ½ per cent for the seven foot canal would require the passage of 1,800 vessels of 20 tons but at 5 per cent only 1,200 vessels. The ten foot canal would need 1,622 vessels of 30 tons to pay 7 ½ per cent or at 5 per cent 1,088 vessels. Watt thought

> The advantages of the ten feet canal are evident - many of the present Busses could pass it, besides the ordinary sloops and Brigantines, that trade to the Isles and some of those that go to foreign parts. Upon the whole there can be no doubt that the number of Vessels passing upon the ten feet canal would greatly surpass the number passing upon the seven feet canal but whether it would exceed it in the necessary proportion, is what I cannot take upon me to determine. (39)

The Canal across the Tarbert Isthmus

Watt found that the town of Tarbert consisted of about one hundred houses, 'tolerably well built though mostly upon the ground floor'. Fishing and dealing in herrings were the main occupations. He thought the harbour 'one of the finest natural Harbours I have ever seen but am of opinion it might be much improved'. His ideas for the canal were drawn out on plans which have not survived. (See p244) The isthmus was about a mile wide. The land at Tarbert rose suddenly, starting close to the high-water line. The canal here would have to be cut out of rock but at the summit he found a plain consisting of a boggy moss 'so very wet and soft that it was with great difficulty we could pass through it'. Underneath this bog was more rock. The height here rose about forty-five feet above neap tides in the East Loch. On the west side, the boggy part narrowed and was interrupted by jutting rocks which would leave room for a crooked passage of a canal. The ground became gravelly as it descended towards the West Loch. Watt described the rock as 'hard undulated, silvery grey *schistus* or slate with thin veins of *Quartz*', which rotted in the air. He found it very difficult to quarry and thought the only way of excavating it would be by using gunpowder 'which would have triffling effect upon it'. In addition, the strata of the rocks ran edgeways to the bed of the canal.

Watt made a thorough investigation of West Loch Tarbert because this was reputed to be full of rocks. He found it to be ten and three quarter miles long and about three quarters broad. Generally there were five fathoms of water but, at 'Gallaghan Island', the channel narrowed to only quarter of a mile and at one point there was only nine feet depth at low water. Where the Avennagillian Water fell into the loch, there was a shallow region produced by sand being brought down by the stream. The head of the Loch ebbed dry and the bottom was deep mud 'through which even a person on foot cannot pass but with the greatest difficulty'. The hills on the south side were not as high as the generality of those in the Firth of Clyde so Watt thought there would not be much danger of 'Flanns' or sudden gales of winds descending from them which might endanger shipping. He found the tides to be extremely irregular but that there was a six hour difference between their times in the two lochs.

A seven foot canal would be cut through the sand in the East Loch where the channel would have to be defended with stones. The first lock would be situated at the end of this channel and it, as well as the three locks needed to climb to the summit, would have to be dug out of the rock. The summit would be lowered by ten feet. On the west side, there would again be four locks, also cut out of rock, so saving maintenance costs later as well as the expense of bringing in stone from somewhere like Arran for their construction. On this side, a channel would have to be made through the mud for more than quarter of a mile to reach deep water. For a ten foot canal, the dimensions would be little different. Watt discussed the problem of finding a water supply for the locks and noted that the rugged nature of the land would make building water courses through it very difficult. The only solution was to divert the Avennagillian Water along the sides of the hills, either in wooden troughs or tunnels or open leats as the situation demanded which Watt thought practicable though not easy. He thought that the supply from this river would be ample.

If any deeper canal were to be attempted, there would be the greater expense of cutting through the rock and the necessity to dredge an artificial channel through the bank at Gallaghan Island. In addition, the shallow by the Avennagillian Water might have to be dredged and that stream diverted. Watt thought that there would be little point in deepening the canal beyond ten feet unless it could be extended to sixteen or seventeen feet in which case it would admit American traders. At the summit, he thought the canal could be widened through the moss to serve as a reservoir should many vessels enter it at the same time at high water. A jetty would be needed at the western end both to protect the channel and to assist vessels turning into the cut.

For the through-cut, Watt gave two estimates, one based on a cut of forty feet wide and sixteen feet deep and the other thirty-three feet wide and twelve feet deep. Drainage of the land on either side might be a problem, as might somewhere to place the material excavated owing to the narrowness of the gap between the hills. Part of the stone might be used to form retaining walls. Watt could not easily reckon the cost of the excavation because his trials were relatively superficial for such a deep cutting.

Another problem was the difference in the times of high tides in the two lochs. This would necessitate an extra set of gates in each of the sea locks both to retain water in the canal and to prevent tides running through it. Some people had thought that enough water might flow through a cut at Tarbert to raise the tides in the Firth of Clyde itself, but Watt pointed out how impracticable was this idea as the canal would be so much smaller than the Firth. One problem at Tarbert would be that, having entered the canal on one tide, it would be necessary to wait six hours for the tide to be right to leave at the other end.

Watt again presented detailed estimates for all these canals. That for the seven foot canal with its eight locks and one draw bridge amounted to £17,988.10s.6d. Cutting the summit was £1,850.13s. The channel leading to deep water in the West Loch was £600 but to divert the Avennagillian Water would be £1,500, showing how difficult he thought this task. The costs of the locks at £1,000 each was the same as he had allowed at Crinan. For the ten foot canal, Watt gave a total figure of £23,884.7s. The through-cut of sixteen feet depth would cost £97,001 for cutting 388,004 cubic yards. Deepening the bank at Gallaghan Island would be £5,333.6s.8d. The six pairs of flood gates 33 ft. wide were £400 per pair or £2,400. Allowing for purchase of ground and contingencies, the total came to £120,789.9s.5d. A twelve foot deep through cut came to a mere £73,849.2s.9d. Tolls would be one shilling a ton. To pay interest at 7 ½ per cent would require the passage of 27,000 tons annually or

1,350 vessels of 20 tons each on the seven foot canal or only 900 vessels to pay 5 per cent. To pay 7 $^1/_2$ per cent on the ten foot canal, 1,200 of thirty tons each would be necessary but 800 vessels of the same capacity to pay 5 per cent. He gave no figures for the through-cut.

Comparison of the Two Passages

Watt finished his report by comparing the advantages which might accrue by the construction of either of these canals and wrote, 'No part of this task has been so disagreeable to me, as this'. He complained that he had neither experience of navigating in the area nor did he have an accurate map. He worked out the distances to various points on the West Coast going from the Clyde either by Tarbert or by Crinan. He estimated how long it would take to sail assuming that the winds were fair for he could not calculate time lost by adverse winds. He found that some vessels sailing from Greenock to Loch Don took eight days to cover 294 miles or 36 miles each day. In another instance, he found that boats went at the rate of 2 $^1/_3$ miles per hour. To go to the north with most winds, the Crinan passage was the most favoured because it avoided tacking round the Mull. Tacking would also be involved to a lesser extent through Tarbert.

Watt sought to determine the time that would be saved by a ship sailing from Greenock by either of the two passages. He allowed for the length of time it would take for a boat to pass through the locks of the both canals, and, in addition, the delay to a ship caused by the tides as well. He prepared a table showing the differences hours and miles.

Table of Savings of Time & Miles by the two Passages.

	hours	*miles*	*hours*	*miles*
From Greenock to Crinan by the Moil			84	127
From Greenock to Crinan by Tarbert	48	72		
Time lost at Tarbert passage	6		54	72
Saving by Tarbert Passage			30	55
From Greenock to Crinan by the Moil			84	127
From Greenock to Crinan by the Crinan Passage	32 $^1/_2$	52		
Time lost at the Crinan Passage	9 $^1/_2$		42	52
Saving by Crinan Passage			42	75

He estimated that, compared with going round by the Mull, a ship would take thirty hours less by the Tarbert passage and, by the Crinan, forty-two hours, a difference in favour of Crinan of twelve hours. He also calculated that the financial savings would be for a 30 ton vessel £1.5s. in the Tarbert passage and £1.15s. by the Crinan.

Watt drew up his conclusions.

> The advantages of the Crinan passage are First, the being more direct and a shorter passage to the North Highlands and Isles.
> Secondly, being more safe and certain than the Tarbert one, from its entering Loch Fyne sooner, and from its being accessible with the winds blowing from more points of the compass. Thirdly the having abundance of water naturally at the very summit.
> Its disadvantages are, First: the great length of the artificial navigation and the number of the Locks, Secondly, that the Harbours at both ends are rather more exposed than could be wished. Thirdly that the deep cut at the summit must be made in a very uncer-

tain & mixed soil. Fourthly, that several of the Locks must be founded in boggy, and other bad foundations. Fifthly, that for a considerable district, it must be the common drain of a hilly country and will require many precautions to prevent its being hurt thereby.

The advantages of Tarbert are, First That it may be a passage not only to the North but also to the islands of Isla and Jurah with the external coast of Knapdale, the west side of Kintyre. Secondly, the having very safe natural harbours at both ends. Thirdly the shortness of the artificial navigation. Fourthly, the possibility of cutting most, if not all of the Locks out of the rock, and the probability thereby making them at a smaller expence or making them more capacious at the same expence. the Sea Locks in particular may be thus much more easily constructed to a great face than where the Artist is perplexed with mud, sand, or similar materials. Fifthly, that there is no deep cutting in very soft materials, unless at the Summit, where they may be entirely removed. Sixthly, that a great part of the canal being rock, it will require few repairs.

Its disadvantages are, First, That it is not a direct passage to the North Highlands, and that the direction of the west loch being towards the south west vessels may be detained in it when the wind is otherwise fair for them. For the same cause, with other winds, vessels may not have it in their power to enter the west Loch and may be obliged to go into some other harbour when with the same wind they could have easily made their passage at Crinan. Secondly That the West Loch is a long narrow passage, in which vessels cannot conveniently tack; and that it is generally shallow near the shores and in its bays. Thirdly, the want of a sufficient quantity of water at the summit; and the being obliged to bring it from a considerable distance through rough Grounds. Fourthly, the great quantity of cutting in rock the expence of which is rather uncertain. (40)

Watt concluded with a table to show the differences between the two passages.

A Table Containing some of the Most Material Facts concerning both Passages.

	TIME OF PASSAGE	COMPUTED SAVINGS		LENGTH OF CANAL		NUMBER OF LOCKS	PERPENDICULAR HEIGHT		EXPENCE
		TIME	DISTANCE	WITHIN LOCKS	IN THE SEA		NATURAL	REDUCED	
		hours	miles	miles	yards		Feet	Feet	£ s. d.
Seven Feet Canal									
at Crinan	13	42	75	4 3/4	2000	14	70	54	34,879. -.4
at Tarbert	6	30	55	1	764	8	45	35	17,988.10.6
Ten Feet Canal									
at Crinan	13	42	75	4 3/4	2800	14	70	57	48,405. 5.3
at Tarbert	6	30	55	1	764	8	45	35	23,884. 7.-

It will be noticed that Watt made no mention of the through-cut at Tarbert in his conclusions. He must have realised that cost ruled it out. Also it will be noticed how the difficulties of navigating West Loch Tarbert ruled out that passage even for a smaller canal in spite of the poor harbour facilities at Lochgilphead and Crinan. However Watt did not make any positive recommendation but Clerk realised that the Crinan route would be preferable. Watt was also asked to inspect the harbours on the west coast of Kintyre and in that region. He included comments on them in a section of his report which may have been one reason why its preparation took so long.

Report Submitted to the Board of the Forfeited Estates

The Board of the Forfeited Estates had taken no action on Watt's earlier report for a canal up the Vale of Strathmore and also there had been a dispute over his payment. Possibly this time Warr wanted to make certain that there would be no further problems because he sent first the maps and sections followed a little later by the report itself to Black in Edinburgh for comment. He told Black that he had written a great deal of it by candlelight when he was suffering from headaches and his hand was shaking. Furthermore, he had spilt some ink on one of the Crinan plans and done his best to cover this up. (41) Black's first reactions to the plans and sections were favourable but, on further inspection once the report arrived, he raised some criticisms. 'I perceive that the Sections cannot be conveniently applyed to the maps or compared with them so as to read both at the same time - the east end being opposite the right hand in the one and the left hand in the other which might have been avoided'. (42) Black found parts of the report needed to be corrected for language but commented, 'Your Friends here have now read your Report and are extreamly well pleased with it'. (43)

The report was read at the meeting of the board on 1 February 1773. (44) Clerk wished that Watt would send in his account as soon as possible (45) and it was submitted on 22 February. Because his earlier bill for charges on the Strathmore survey had been disputed, Watt kept very full notes of the Crinan and Tarbert expenses. The following is a copy of the account he sent to Edinburgh.

The Honorable Board of Annexed Estates, Dr. to James Watt

To myself & assistant making survey of a Canal proposed between Loch Gilp & Crinan and between the two LochTarberts, 26 days		£81.18.0
To drawing up report of that survey		26. 5.0
To drawing up plans & sections of both places		26. 5.0
To our polemen & their charges	£6. 6.4	
To Quariers boring & digging the ground & their charges	6.16.2	
To incident expenses	6. 0.3	
To Freights & Horse Hires	9. 4.6	28. 7.3
	£	162.15.3

Money paid out by James Watt on that survey		
Mr Morison 28 days 21/-	£29. 8	
do. 22 days 10/6	11.11	
	40.19	40.19.0
Own personal charges		12.17.5
The above account of expenses		28. 7.3
His own time on Survey		50. 8.0
Money lost acct. of		2. 2.0
		134.13.8 (46)

Presumably Morison charged a guinea a day when he was in the field and half this amount when working at home or in Glasgow. He claimed that he had spent thirty days from the time he set out until he returned to Glasgow and a further thirty-seven days employed at Glasgow working on the plans. In addition, there was £45.12s.11d. paid out for travelling expenses. (47) Watt disputed these figures but the matter settled in February 1773 with a reduced number of days at £48.17s.7d. (48)

Watt's account was examined by Clerk and thought to be 'reasonable'. (49) It was passed for payment by the Board on 1 March (50) and Watt received his £162.15s.3d. on the ninth. (51) Clerk made the following comments on Watt's report:-

> I... have the pleasure to observe that these Surveys appear to be extremely accurate, and that his [Watt's] manner of drawing & shading his Plans communicates a very strong idea of the nature of the Country and forms of the Hills thro' the Intersticies of which the Canals are proposed to be conducted. His sections of these Tracts are very usefully conceived and neatly executed.
> The method of conducting the works upon the different canals appear to be properly adapted to the nature of the soil & situation of the place.
> His estimates are arranged in good order, and from what I have learn'd from the Execution of other works of the same kind, his values or prices are as near the Truth as Calculations of Cutting thro' a variety of Stones or soils (whose nature cannot be well known from small Trials on the surface) will admit.
> Tho' not properly the Business of an engineer Mr Watt has gone at great Length in stating facts to prove which of the two tracts is most adviseable.
> He does not positively say so, but I imagine his Sentiments are in favour of that from Loch Gilp to Loch Crinan, and indeed this tract tho' estimated much higher than that between the two Loch Tarberts appears to me preferable on several accounts.
> The making a clean deep Cutt between the two Loch Tarberts I imagine ought to be entirely rejected, as it would, tho' very expensive, answer no good purpose, and on considering all Circumstances, if money could be found, I would advice a Canal of Seven feet deep rather than one of Ten from Loch Gilp to Loch Crinan.
> The reading Mr Watt's papers & considering his Plans & Sections has taken up a good deal of time but I esteem myself fully repaid for this loss by the Entertainment & Instruction I have received in reading & Examining so masterly a Performance which does great honour to the Author and his Country. (52)

Watt is likely have been pleased with the favourable reception this report received from everybody.

The Crinan Route is Chosen

There was no further progress until Rennie was asked to draw up a scheme for the Crinan Canal in 1793, and turned to Watt for advice. Watt lent Rennie his report (53) and Rennie found that his survey and Watt's agreed very nearly but Rennie would have made more reservoirs in the hills and also used the Water of Add through a three mile long aqueduct. (54) That February, a bill was already before Parliament and money had been subscribed. Rennie's estimate for a fifteen foot canal was £107,600. (55) Part of this extra cost was caused by extending the line at both ends to reach deep water and avoid the shallow estuaries. Rennie reported that work was proceeding at Crinan in November 1794 (56) but progress was slow through lack of finance and difficulty in obtaining labour. (57) When Watt junior was on holiday at Inveraray in July 1801, he wrote to his father,

> We have been to Crinan. The Canal is so far compleated that two vessels have already sailed through it, but the banks are not yet in perfect order and there is some little work to be done about the locks. I walked the whole length of it to the Westward from Lochgilp Inn. (58)

There is no record of Watt's feelings about the completion of one of his early projects.

Notes: Canals Across Argyll

1 ARO, McTavish Records, Ld. Cathcart to Dougald McTavish, 12 Oct. 1764.
2 JWP, 4/7.29, J. Watt sen. to J. Watt, 21 Jan. 1767.
3 B. & W. Col., MI, 3/43, Ld. Cathcart to J. Watt, 24 Aug. 1767, 'I beg you will thank your father for me for his dispatch relating to Tarbert & Creinan Passages and tell him these things will be examined'.
4 JWP, JW/24.3, Charles MacDowall to J. Watt, 7 July 1771.
5 Ibid, JW/24.1, Charles MacDowall to J. Watt, 27 Feb. 1771.
6 Ibid, JW/24.2, Charles MacDowall to J. Watt, 13 March 1771.
7 SRA, C1.1.34, Glasgow Council Act Book, Jan. 1768 - March 1772, p. 385, 1 March 1771.
8 SRO, Forfeited Estates, E 728/32/1, 'From the Magistrates of Glasgow concern ing the Crinan Canal', 4 March 1771.
9 Smith, (1982), p. 210.
10 JWP, 4/77.6, 'Mr. McKenzie's Draught of Tarbert', no date.
11 SRO, Forfeited Estates, E 721/11, Minute Book General, Feb. 1768 - March 1782, p. 63, 4 March 1771.
12 JWP, JW/24.2, Charles MacDowall to J. Watt, 13 March 1771.
13 Ibid, 4/62.4, H. Barclay to J. Watt, 27 May 1771.
14 B. & W. Col., MI, 1/17, 'Journal', 29 May 1771.
15 Ibid, 19 June 1771.
16 SRO, Forfeited Estates, E 727/37/5, J. Watt to H. Barclay, 19 June 1771.
17 Ibid, E 721/11, Minute Book General, Feb. 1768 - March 1782, p. 68, 24 June 1771. See also JWP, 4/62.6, H. Barclay to J. Watt, 25 June 1771 and 4/62.14, J. Campbell to J. Watt, 16 April 1772.
18 SRO, Forfeited Estates, E 726/3.23,'Letters & Instructions to Factors, etc., 1776 - 1778', Instructions for Mr. James Watt, Engineer, 25 June 1771.
19 Lindsay, (1968), p. 19.
20 B. & W. Col., MI, 2/82, J. Watt, 'A Report to the Honourable his Majesties Commissioners for managing the annexed estates in Scotland concerning the Isthmuses of Tarbert and Crinan', Glasgow, 21 Dec. 1772.
21 JWP, 4/33.5, J. Morison to J. Watt, 25 June 1771 and B. & W. Col., MI, 1/17, 'Journal', 10 July 1771.
22 JWP, 4/62.20, 'General Account of Charges upon Tarbert & Crinan, July & August, 1771 - 2'.
23 Ibid, see also 4/82, A. Harley to J. Watt, 9 July 1771 and 6/28.11, J. Watt Dr. to John Wilson, 9 July 1771.
24 Ibid, 6/28.14, 'Johnston's Acct. upon Tarbert, 1771'.
25 Ibid, JW/24.3, Charles MacDowall to J. Watt, 7 July 1771.
26 Ibid, 'General Account of Charges upon Tarbert & Crinan'.
27 Ibid.
28 B. & W. Col., MI, 1/17, 'Journal', 19 July 1771.
29 Ibid, 21 July 1771. See JWP, 4/62.21, 'Time at Tarbert & Crinan Survey', 21 July 1771.

30 See JWP, 4/62.20 'General Account of Charges' and 4/62.21 'Time at Tarbert & Crinan Surveys' for details.
31 B. & W. Col., MI, 1/17, 'Journal', 27 July 1771.
32 JWP, 4/62.21, 'Time at Tarbert & Crinan Surveys', 1 Aug. 1771.
33 Ibid.
34 Ibid, JW/24.4, Charles MacDowall to J. Watt, 1 August 1771.
35 Ibid, JW/24.5, Charles MacDowall to J. Watt, 1 Sept. 1771.
36 Ibid, 4/62.14, J. Campbell to J. Watt, 16 April 1772.
37 Ibid, 4/82, J. Morison, Surveyor, to J. Watt, 22 Aug. 1772.
38 One version of J. Watt's Report on Crinan is filed at B. & W. Col., MI 2/82 and another at JWP, 3/22.1.
39 Ibid.
40 Ibid.
41 Robinson & McKie, (1970), p. 34, Letter 28, J. Watt to J. Black, 15 Dec. 1772.
42 Ibid, p. 36, Letter 30, J. Black to J. Watt, 26 Dec. 1772.
43 Ibid, p. 37, Letter 31, J. Black to J. Watt, 31 Dec. 1772.
44 SRO, Forfeited Estates, E 730/21/3, 'Papers concerning the construction of canals...', G. Clerk's Report was written on 30 Jan. 1773 after reading Watt's.
45 Ibid, E 721/11, Minute Book General, Feb. 1768 - March 1782, p. 105, 1 Feb. 1773.
46 JWP, 4/62.23, J. Watt's Account, no date.
47 Ibid, 4/33.11, J. Morison, Surveyor, to J. Watt, 9 Feb. 1772.
48 Ibid, 4/33.12, J. Watt to J. Morison, Surveyor, 18 Feb. 1773. See also 4/33.17, 'J. Watt Dr. to James Morison Surveyor for sundries upon Tarbert'.
49 SRO, Forfeited Estates, E 732/20/1, 'Account of James Watt', 25 Feb. 1773.
50 Ibid, E 721/11, Minute Book General, Feb. 1786 - March 1782, p. 109, 1 March 1773.
51 JWP, 4/62.16, J. Morison, Clerk, to J. Watt, 9 March 1773.
52 SRO, Forfeited Estates, E 730/31.3, G. Clerk's Report, 30 Jan. 1773.
53 JWP, 6/23.36, J. Rennie to J. Watt, 7 March 1793.
54 Ibid, 6/23.40, J. Rennie to J. Watt, 23 Aug. 1793.
55 Ibid, 6/23.41, J. Rennie to J. Watt, 23 Feb. 1793.
56 Ibid, 6/32.1, J. Rennie to J. Watt, 26 Nov. 1794.
57 Lindsay, (1968), pp. 117 - 121.
58 JWP, 6/29.3, J. Watt jun. to J. Watt, 23 July 1801.

Watt at Campbeltown

The Background

Watt visited Campbeltown in the summer of 1773 at the instigation of his friend, Charles MacDowall, to survey a canal when he also seized the opportunity to undertake extra commissions for others. This visit should be seen against the background of the prevailing interest in the improvement and development of the agricultural and industrial resources of Scotland. (1) It may be regarded as a microcosm of his surveying career and shows how he had developed as a consulting engineer. Our interest in his surveys here lies in their variety, covering so many aspects of civil engineering. There was improvement of land through better drainage of the Craigs Water and Loch Sanish. Then there was improvement of water supplies to drive mills at Campbeltown and Glen Saddell as well as the mine pumps at Drumlemble. Finally, there were improvements for transport with the canal between Drumlemble and Campbeltown as well as harbours at Campbeltown and Carradale. He laid out his reports first to explain the problems, then to outline the remedy and finally to give instructions about carrying out the recommendations. These instructions show the civil engineering techniques which he employed and which presumably were those which other people were also using at that time. His reports reveal how he was evolving the techniques of a consulting engineer, finally giving the instructions which would be carried out by others with only nominal, if any, supervision from himself. We may note as well the delays in his journeys at sea caused by the weather, which made travelling so hazardous before the days of the steam ship.

Even at the beginning of the eighteenth century, Campbeltown was the seaport and centre of a rich agricultural district. (2) The preparation of malt for brewing ale was a skill practised for centuries by the farmers of Kintyre and, in 1770, Orr, Ballantyne & Co. built a brewery near Campbeltown Mill. In 1743, twenty-one malsters signed a petition relating to the supply of barley from neighbouring farms and the existence of this industry, with as many as forty malt barns in the town, was an important factor in the growth of whisky distilling here. (3) Whisky was distilled in Kintyre during the seventeenth century and there is a reference in 1706 to 'Distillers in the Burgh' (Campbeltown). (4) By 1713, there was a need for three inspectors and in 1792 there were twenty-two licensed distillers. This was to meet the demand both locally because the Scots preferred whisky to their home-brewed beer and also in England where around 1776 whisky became popular and where large quantities were sent. (5) During the years 1782 - 3 and 1795 - 7, the Commissioners of Supply for Argyll banned the manufacture of whisky due to failure of the harvests. In 1785, a harsh Licence Duty was imposed and it became less profitable to make whisky on a small scale.

Presumably, to begin with, most of the malting and indeed distilling, was carried out with peat fuel which abounded in the region known as the Laggan between Campbeltown and Machrihanish. The great majority of the inhabitants burnt peat. The freight for a pony load of peat from the Moss to the town was two pence. (6) But there was also coal near Machrihanish for, as early as 1498, it was being mined to supply the castles maintained in Kintyre by King James IV. A survey was made in the summer of 1745 when it was decided that coal could be won, provided that some mechanical means of draining water from the mines by a steam engine or a windmill was employed. Mines owned by the Duke of Argyll were in existence by 1750 at 'Drumlemon', a small village just east of Machrihanish now called Drumlemble. (7) One of the seams was six feet thick. A report of 1752 refers to diverting neighbouring springs and rivulets in order to provide power for driving the pumps. (8) From at least 1763, some coal was being sent by horse and cart to Campbeltown for use by the many distilleries. (9) It was thought that the harbour there had greater potential for

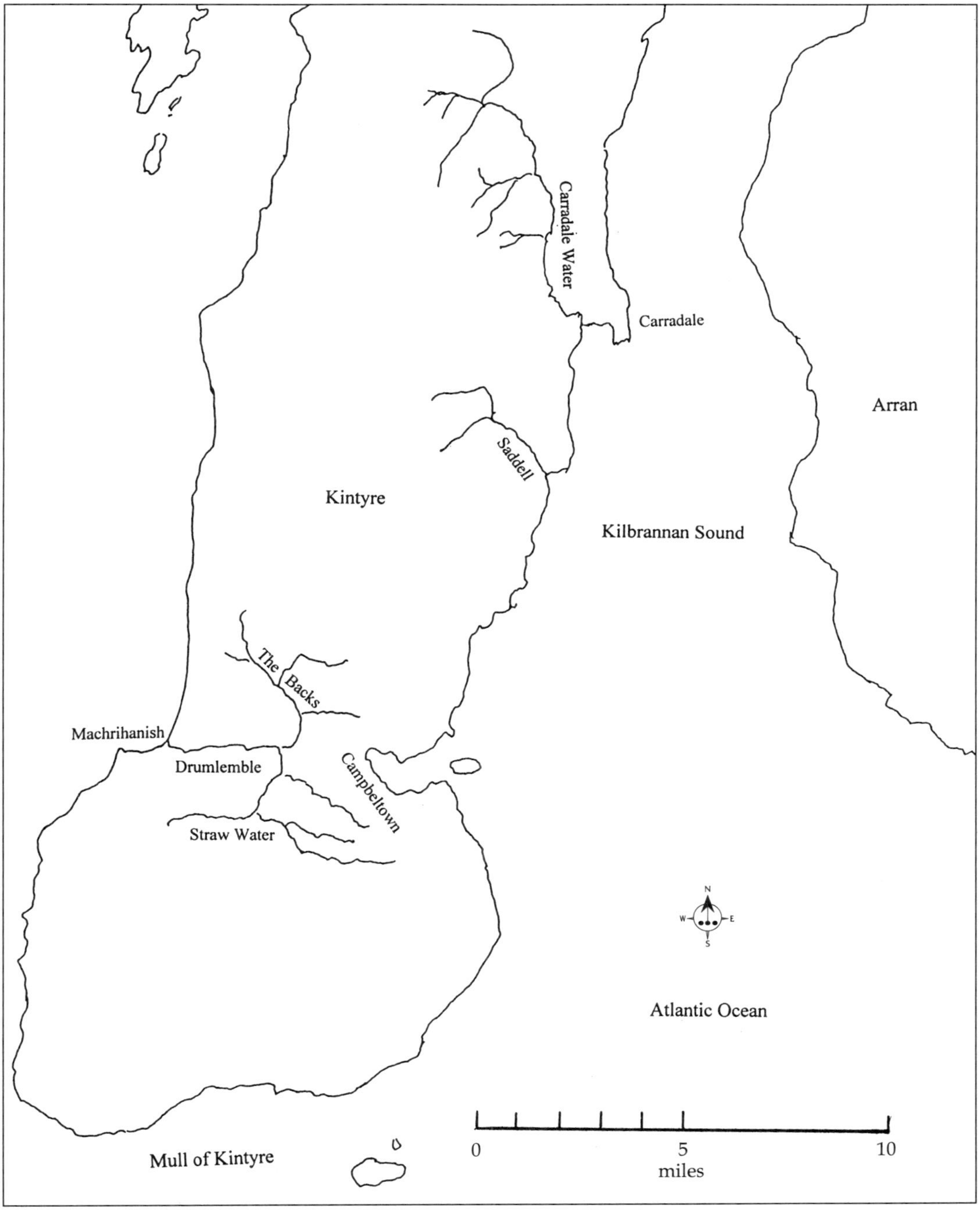

The southern part of the Kintyre peninsula.

exporting coal than those at Saltcoats, Irvine or Ayr. However the Drumlemble coal was poor quality and, in 1807, it was not to be sent to Inveraray because 'We get Clyde Coal as cheap and 2 carts of them are better than 3 of Kintyre". (10)

Charles MacDowall

In 1771, the Duke of Argyll leased the Campbeltown coal and salt works to Charles MacDowall for twenty-seven years and agreed not to take any rent or royalty before Whitsuntide 1773 in order to encourage development and expand operations. (11) Charles MacDowall of Crichan was admitted advocate in 1734 and later became Sheriff of Renfrewshire. (12) That he was both influential in the community and knowledgeable about coal mining is clear because he and Peter Colville were approached by the Carron Company in 1767 to act as arbitrators in a dispute about mining practices. (13) His correspondence shows that he knew Watt well and it is possible that there was a connection through Robison who, around 1765, was tutor to two MacDowall boys. (14)

In 1770, Charles McDowall was investigating how to develop the coal mines near Machrihanish because he wrote from Campbeltown to the Carron Company in August about having wrought iron cranks forged which they were unable to do although they had had a forge since 1762. (15) He was installing a waterwheel to pump out his mines, and cranks would have been a necessary part of such a machine. He urged Watt in February 1771 to write to Glasgow Council expressing his willingness to carry out a survey for canals at Crinan and Tarbert before the following July. He hoped that Watt, having finished at the top end of the Kintyre peninsula, would come south and survey the line of a canal to link his mines at Drumlemble to Campbeltown. Watt completed the Crinan and Tarbert surveys, leaving the latter place on 31 July 1771, and returned to Glasgow. He and Charles MacDowall missed each other because the latter was delayed by winds for eight days at Ayr and wrote from Campbeltown on 1 August in an express letter which failed to reach Watt. (16)

Charles MacDowall wrote again on 1 September, for, having missed Watt in Glasgow as well, he wanted to know whether Watt had to return to Tarbert to finish his surveying and 'Whether or not I can expect you to make my survey. This I am anxious to have done whether you have anything to do at Tarbert or not'. (17) Watt must have answered that he had completed his surveys there and had no time to go to Campbeltown. Charles MacDowall replied that, while he was vexed that Watt could not go, he was still determined to have him and no one else. (18) Either Charles MacDowall was convinced of Watt's competence, or possibly he wished to help a friend.

Watt and Charles MacDowall dined together in Glasgow on 22 March 1772 when we can assume that the subject of this survey was raised. Over a year later, Charles MacDowall was still urging Watt to go to Campbeltown and, in April 1773, even told him that the packet boat from Campbeltown would wait one night at Greenock so that Watt could catch it as it returned. (19) By this time, Watt's supervision of the construction of the Monkland Canal was drawing to a close, but it was not until 8 June 1773 that he set sail from Greenock at night in the 'vinegar boat' but was becalmed off Pladda and so did not finally arrive at Campbeltown until the tenth, a Thursday. (20) After he had dined, he made a preliminary view of the line for the canal.

Watt's First Surveys at Campbeltown

Watt was occupied with surveying the canal for the next couple of days and took Sunday off but continued with the canal again on the Monday. Tuesday was spent looking at the Craigs Water, a stream starting in the hills to the north of Campbeltown but flowing in a broad zig-zag to reach the sea by Machrihanish, which caused flooding in the meadows west of the town. It was proposed to use this stream to help power Campbeltown mill. Then he returned to Charles MacDowall's problems and looked to see if the water supply for working the mine pumps could be improved and again inspected the line of the canal. The beginning of the week 21 June saw him planning how to drain Loch Sanish which lay in the low ground between Campbeltown and Machrihanish. On Tuesday, the report about the water supply for Campbeltown mill was sent to Inveraray and a rough estimate made of the canal. The rest of that week was spent checking levels at Loch Sanish and details of the canal including a possible alternative, more southerly route. On the Saturday, he noted that, 'This night Mr Macdowal set his sliding work agoing which went very smoothly, only the swords being too slender bobbed up and down - recommended them to be made about 9 inches broad by 4 inch thick'. (21) This refers to the pumping engine for the mines. Its erection had been delayed through bad weather. An undated map in the Argyll Record Office shows the waterwheel placed just west of West Drumlemble farm on the north side of the Machrihanish road. The leat supplying the water came from the hills to the south and was carried over the road on some sort of wooden structure. (22) The pumps in the mine would have been driven by a series of flat rods, through cranks, to change the rotary movement of the waterwheel into reciprocating motion for the pumps. (Plate 10)

The following week, Watt went up to Colonel Donald Campbell's house at Saddell and surveyed improvements to the river there. He then rode on to Carradale to advise about building a jetty for making the harbour safer. He returned to Campbeltown and took the packet for Greenock on 1 July. He was again becalmed, this time between 'Cantyre and Arran' where he saw some whales. After another calm day, he finally reached Greenock at 9 o'clock on the morning of 3 July. This time, he did not delay finishing his reports and wrote those he had not completed at Campbeltown during August.

Examination of Watt's surveys around Campbeltown will start by reviewing his reports for Carradale and Saddell, then look at those around Campbeltown and conclude with Charles MacDowall's canal.

Carradale River

There was, and still is though now seldom used, a small harbour on the river Carradale at Waterfoot in Carradale Bay and not at the present village. The Bay of Carradale lies in a horseshoe of hills which protect it from all winds except southerlies. The river flows out on the western side of the bay where a small hill gives some protection to the harbour there from the southerly winds. Watt thought that this harbour, with its narrow mouth, would be very safe for small vessels. However storms, particularly south-easterlies, blew sand into the river mouth, formed a bar and blocked it which not only prevented ships from entering the harbour but also caused the river to flood lands upstream.

Watt looked for an inexpensive solution and sent instructions to Dugald Campbell at Carradale which included diagrams to show how the work should be carried out. (23) He recommended constructing what he called a 'jetty' which was to be built 'from common quarry stones laid together in a heap as solid as may be, and the largest & best of them placed regularly upon the outside so as their least dimensions or their ends shall be outmost'. The cheapest way he suggested was to build a mound of stones three feet high with a base of twelve feet out into the sea which would be left all winter when the sea would pile sand up against it. Smaller jetties at right angles would stop the sea washing the sand away. When the sand had built up level with the top, a second jetty could be built on top, against which the sand would accumulate again. He wrote,

> It may be objected that these jettys are only founded upon sand, but I have observed that

> a heavy sea which would undermine & beat down a strong upright wall, will beat over a heap of stones and drive sand up against them, and thereby render them more firm. At any rate if a break should happen in those jettys it is easy repaired, the workmanship being a mere trifle'.

Watt's method is similar in principle to modern concrete 'stabits' for building harbour protection works which break up the force of the waves. His jetty would also help direct the river to create a better channel. Compared with a cost of £250 for a single large jetty, Watt's small ones would cost £174. The plan for this jetty shows that it was to be situated on the east side of the river mouth and stretch eight hundred feet southwards into the sea with a beacon at its end. (24) It is doubtful if this jetty was ever built because other contemporary maps show no signs of it and neither does the first edition of the Ordnance Survey. Certainly there are no signs of any stones there today. (25, see p256 and Plate 8)

Glen Saddell

The valley bottom at the seaward end of Glen Saddell is about half a mile wide and fairly flat. For improving the river here, Watt proposed simple remedies which would allow the river to straighten itself naturally. In his report, Watt said he would 'confine myself to practical directions touching their execution and Estimates of their expence'. He staked out the lines of the cuts to be made to straighten the river and sent his instructions with diagrams about how the work should be done so that it could be carried out without any further direction from himself. The cuts were to be nine feet wide at the bottom and three feet deep. The earth taken out was to be used to make embankments, behind which there could be drains to prevent the rain that fell on the land flooding it. The banks were to be faced with turf and seeded. The cuts were to be started at their lower ends, leaving a solid piece of earth at the highest part that could be easily removed when everything else was finished. To block the old channels of the river, the higher ends were to be filled with stone dams but the lower ends left open for the first winter so that the old channel might silt up. To direct the flow of the river at bends, he proposed building small jetties of stone out into the stream to enable the river to scour out its own channel. Watt's estimate for four cuts, dams, drains, etc., was just under £60. While he was there, he also suggested improvements to the road and to the mill which had an overshot wheel twelve feet diameter. He concluded his report for Col Campbell of Saddell, 'I have now discussed those things of which I have taken notes when with you If I have omitted or too lightly touched upon any part of the subject I shall be glad to hear from you & give you an answer - either upon these or any other subject I am conversant in'. (26)

Today, at the top of the valley, the river meanders across the valley floor while at the lower end it follows a straight course, suggesting that it has been straightened at some time. The earliest map is the Ordnance Survey of 1866 where the river has the same course as today so we cannot tell whether this is the result of Watt's recommendations or someone else's later. The cornmill stood by the east side of the road on the south side of the valley and was probably powered by a small stream running straight off the hill which would have been appropriate for the overshot wheel that Watt was proposing. (27)

Campbeltown Harbour

At Campbeltown, the construction of the 'Old Quay' was begun in 1722 but was not finished until about 1765. The adjacent 'New Quay' was started in 1754. (28) Watt was asked to report about building a 'breast', or sea wall against which ships could lie, between the two quays. His idea was to construct a wall, slightly inclined on the seaward side, higher than these two quays because they were sometimes overflowed by the sea. The wall would be faced at intervals with timber rubbing strips. It was to be built some distance from the existing houses on the waterfront so that a new space would be created by filling up the intervening area with dredgings from the harbour or rubbish from the town. Thirty feet would be devoted to the new quay and thirty feet for a row of new houses. The expense of building the breast could be defrayed by renting out the ground for these houses or other buildings and yards. Two estimates were given, one of £459 and the other £481, depending upon the type of stone used. (29) His proposals were never carried out.

Craigs Meadows and Campbeltown Town Mill

To understand Watt's other reports, it is necessary to look at the geography around Campbeltown. The southern end of the mountain range down the Kintyre peninsula is broken by a gap running across between Campbeltown and Machrihanish. Campbeltown itself lies at the foot of a small range of grassy hills with three passes through them. The lowest of these is the most northerly, where the present main road from the north enters the town. Here a small stream ran down from hills in the north, past Drumore House and was used to power the Town Mill. This burn was liable to heavy spates and, on several occasions during the eighteenth century, swept away the bridges over it. (30) Watt was asked to advise about better ways of controlling the water and providing more power at the mill.

Schemes for the Town Mill could be linked to improving the drainage of Craigs Meadows and also to the canal from Drumlemble because the canal would enter the town through this same pass. The hills north of Campbeltown were drained through the Craigs Meadows by the Backs Water. This burn pursued a very erratic course, starting with small streams flowing westwards out of the hills which joined to form the Backs Water as it flowed eastwards. The Backs Water then turned south and westwards before making its way out to the sea at Machrihanish through Loch Sanish. This burn flooded large areas of otherwise good agricultural land. In this report, Watt was concerned with the meadow area to the north-west of Campbeltown, running alongside the present road. His solution was to straighten the Backs Water from Craigs Farm and build banks along the new cut, with appropriate side drains and sluices. The cut would not exceed nine feet wide at the bottom and would be four feet deep. To receive the full benefit of improving the upper section of this burn, it was necessary that the channel should be widened down to Loch Sanish. Full instructions were sent to the Duke of Argyll about the angle of the slopes for the new channels, how to build the banks and so on. This scheme, or a similar one, must have been carried out because the Craigs Canal or channel was measured by George Langlands a little later. (31)

From his surveys, Watt realised that the Backs Water in the Craigs Meadows was at a high enough level to be diverted through the pass to the Campbeltown Town Mill. A leat could be constructed with sluices to control floods and water for the mill. There would be also a settling pond on the Craigs Meadows to prevent the channel silting up. The bank of the mill dam could be raised to provide much greater storage. Once again Watt gave dimensions of the works and how to construct them but it does not appear that this proposal was ever carried out in its entirety. (32) At some time, a new leat was dug through the pass to the mill dam but it seems to have tapped only the easterly Auchalochy Burn and not the rest of the Backs Water.

Loch Sanish

Watt had a good eye for assessing the country in which he was working and his description of the area to the west of Campbeltown is quoted here because it is still accurate today - that is except for the modern military airbase.

> Machrihanish Bay upon the West coast of Cantyre is bounded upon the south and north by mountainous grounds but to the east lyes an extensive plain surrounded by an amphitheatre of hills. near the bay the plain is composed of hills of sand blown up from the sea - behind these lay a lake called Lough *Sanish* now drained and turned into meadow land - the *Backs* runs through this meadow or lough the bottom of which is much lower than Mr Macdowals water engine which is situated not far from the south side of it - A part of the plain is moss [peat] with a sandy bottom, near the engine but further N it is lower has a clayey bottom and is partly arable meadow land - that part of it is remarkably level & subject to be overflowed by the Backs & *Straw* [now known as Strath] waters which runs through it On the south side of the mossy part of the plain lyes a higher but level gravely plain mossy & moorish on the surface through which the Campbellton road to Machrihanish passes - still further south lye lower grounds of a good soil and arable behind these the country is hilly - The Straw water rises in these hills - in dry weather it is inconsiderable - but having a very crooked and narrow channel through the lower part of the plain, whenever a flood happens it lays the whole under water The Backs water enters the N.E. part of the plain by a valley with a flat bottom between the farms of *Moy* and *Backs* It joins the Straw water near to the houses of *Backs* and both pursue their course through Lough Sanish to the west sea'. (33)

The Straw Water rises in the hills to the south of Machrihanish Bay and flows north, under the road to Campbeltown before joining the Backs and then both turn westwards to the sea. The sand hills to the west kept out the sea but also prevented the waters of the Backs and Straw burns from flowing out easily so that a large area of fresh water peat formed between the sand-hills and the range of hills behind Campbeltown. It would be interesting to know the height above sea level of this peat area in Watt's day which has probably sunk since due to better drainage. On the southern side of this lay Loch Sanish. Charles MacDowall's coal mines were situated on the carboniferous outcrops of the hills south of Loch Sanish, a mile east of Machrihanish where the later row of miners' cottages at Drumlemble still survives. The road from Machrihanish to Campbeltown passes along the side of these hills before following a low ridge down to Chiskean or Straw Water and climbing again through the hills behind Campbeltown by the southernmost pass. Watt surveyed the hills behind Drumlemble in which Charles MacDowall had built a reservoir with a twelve foot high dam to supply his water-engine and found that several more small streams could be diverted into it. A map indicates that the wheel was fed by a leat from the west, possibly connecting with the stream flowing today past the ruined chapel at Kilkivan. (34)

At the present time, the sand-hills by the sea stand about fifteen feet high and protect the valley behind but also block the exit of the river. In 1773, Loch Sanish had already been drained but, as the level of the peat when dried out would fall through bacteriological action, the land level would sink. Watt reported to Col. Charles Campbell of Barbreck that there was sufficient fall in the river between Loch Sanish and the sea to secure better drainage for this area. His recommendations were to straighten the channel of the Backs between Loch Sanish and the sea and also to widen that channel, removing from its bed the water grass which grew there in abundance forming somewhere for sand to lodge, thus blocking the channel. He also proposed embanking the new channels and constructing proper sluices in these banks to prevent the river floods entering the meadows. To help make these drains, Watt said that 'A bevel or slope level should be used for carrying the slopes regularly'. This was a right-angled triangle made from wooden bars with the hypotenuse showing the slope which was determined by a plumb-line hanging down the vertical side. Such a device was typical of Watt's lively mind. A back drain was to be cut at the edge of the high ground on the south to take away water coming from the hills before it could enter the Loch Sanish area. This drain would discharge its waters into the tail race from Charles MacDowall's engine. All this and other ancillary works would cost over £200. Once again it is impossible to tell how many of Watt's proposals were carried out, but it is evident that the course of the river through the sand-banks has never been straightened. (35)

The Campbeltown and Machrihanish Canal

Charles MacDowall had invited Watt to Campbeltown for the purpose of surveying a canal to link his collieries at Drumlemble with their main market in Campbeltown. Horses and carts were being used along the existing roads but these became impassable in winter. (36) Some coal was shipped from the exposed beach at Machrihanish until 1774 when the Customs Surveyor seized an open boat carrying two tons of coal because it did not have a customs docket. (37) It is significant that a canal was preferred to a wooden waggon way and in his survey we see the last of Watt's attempts to recommend a level canal complete with an incline at the Campbeltown end. His estimates included the cost of 242 yds. of double waggon way by which he proposed to join one branch at the eastern end of the canal to sea level at Dalintober pier. (38)

Watt never mentioned how he intended to fill his canal with water. The Backs Water lay to the north and its waters through the Craigs Meadows might have been used to supply the canal but presumably Watt thought that a level canal without locks would not need major reservoirs. This may have been the reason for a waggon way down to the sea at Dalintober to avoid locks. There were two main obstacles, the Straw Water and the range of hills separating Campbeltown from Machrihanish Bay. To avoid having to build an aqueduct over the Straw Water, Watt looked at an alternative route further south around the basin of this burn but found that not only would this add about a mile and a quarter to the length but also that four branches of the Straw Water would have to be crossed, one of which was nearly as big as the whole river. (39)

The route that Watt chose was to follow the most northerly of the passes through the grassy hills into Campbeltown. There it could continue along the edge of the hills eastwards to

> ... Dallantober... there is a pierhead here distant from the Dam [of the Town Mill] 35 chains The canal can be carried within 11 chains of this pierhead, partly along a projected street & partly along a steep brae where it may end 34 feet above high water of neaps.
>
> If that is preferred the canal may be carried to the south of the Dam along the brae face towards Campbellton to above the new Church - or it may be locked down to the lough'. (40)

We do not know whether Charles MacDowall envisaged a connection with the harbour which would enable his coal to be exported more easily. However the small size of the canal meant that no sea-going vessels could use it and hence the suggestion for a waggon way as a link. The Dalintober branch was never constructed and the canal, when built, passed on the west side of the mill dam, following the contour of the hills in a southerly direction with the Town Mill on its east to a terminus near the present gas works. The above quotation from Watt's Journal is the only clue we have about the height above sea level of the canal, 34 ft. above neap tides. The Ordnance Survey Map shows a milestone on the road at the present bridge over the Straw Water as 37 ft. above sea level which confirms the height of the canal.

Continued on page 258...

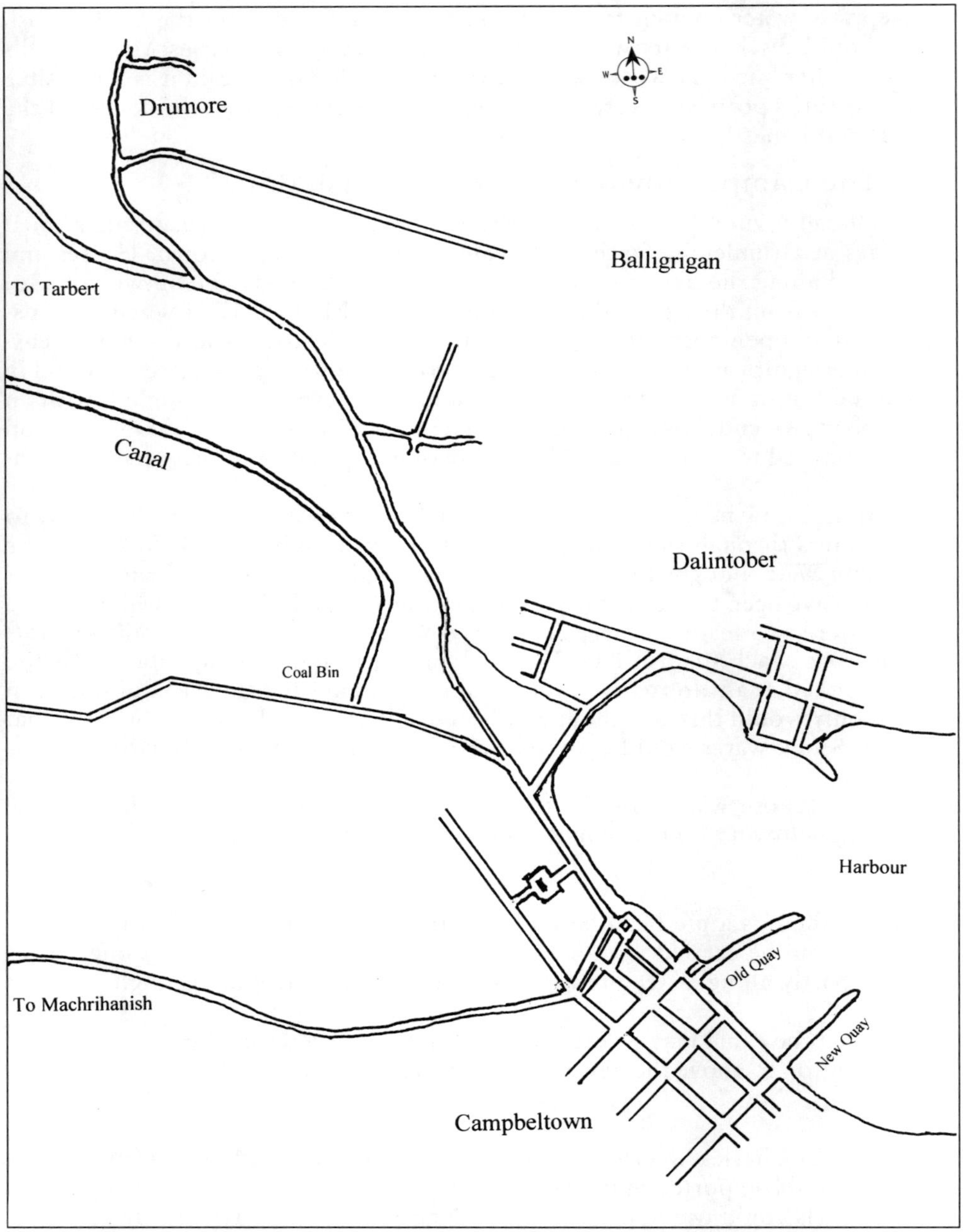

A Plan of Campbelltown harbour with its quays and end of the canal, 1792

Opposite page: Western End of the Campbeltown Canal

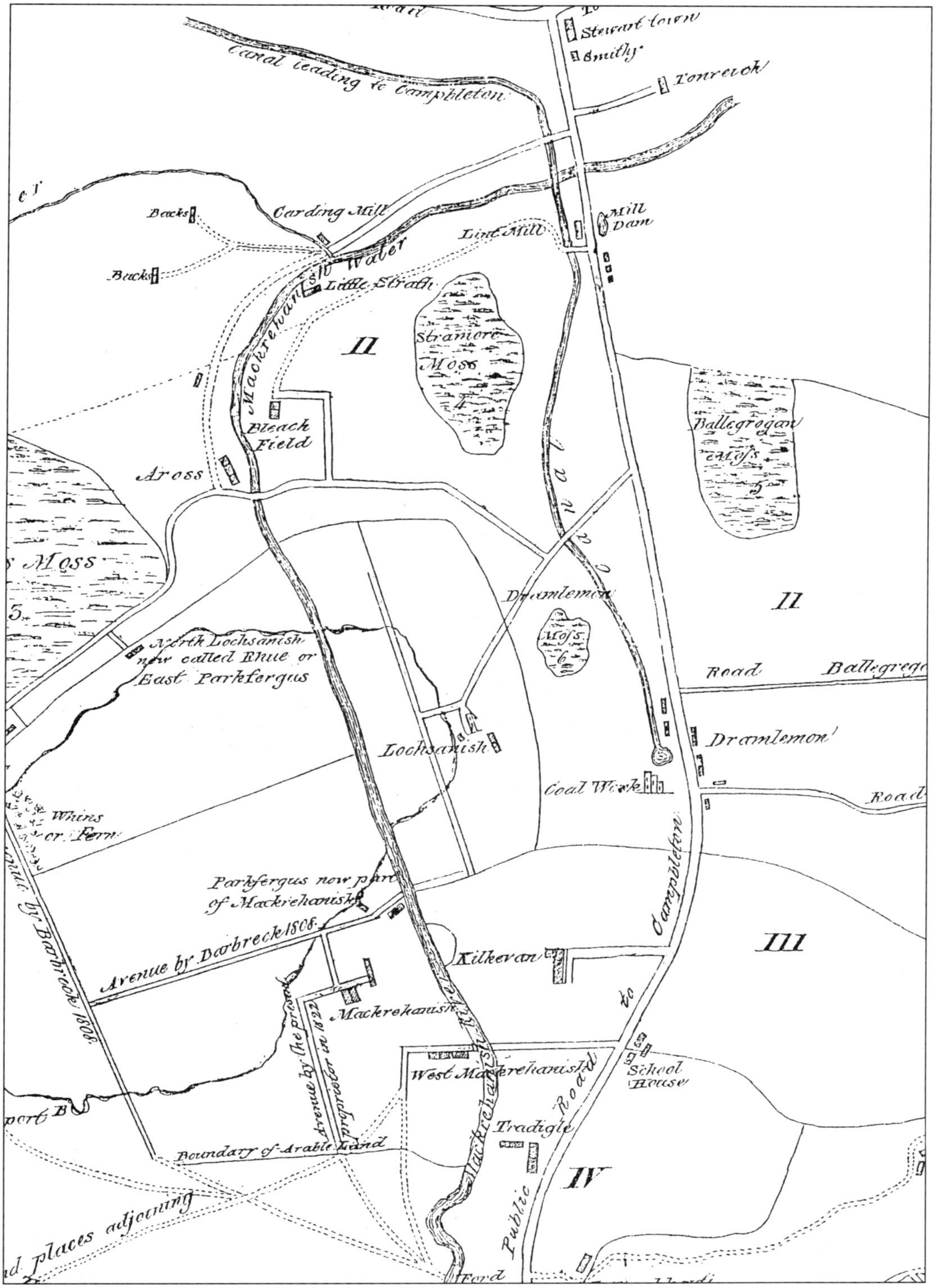
Stewart town
Smithy
Tonreoch
Canal leading to Campbleton
Carding Mill
Backs
Backs
Mill Dam
Lint Mill
Mackrehanish Water
Little Strath
Stramore Moss
4
II
Bleach Field
Aross
Ballegrogan Moss
5
Moss
5
Dramlemon
Moss
6
Canal
II
North Lochsanish now called Blue or East Parkfergus
Road
Lochsanish
Dramlemon
Coal Work
Road
Whins or Fern
Parkfergus now part of Mackrehanish
Avenue by Barbreck 1808
Kilkevan
to Campbleton
III
Mackrehanish
Mackrehanish River
School House
West Mackrehanish
Avenue by the present proprietor in 1822
Tradigle
Public Road
Boundary of Arable Land
IV
places adjoining
Ford

From the north end of Campbeltown, the canal was to follow the pass in a westerly direction but how much cutting would be necessary here is not clear. Then it would contour round the hills past the Backs and Moy farms turning to a southerly direction to meet met the Campbeltown to Machrihanish road at the Straw Water where there was a mill. A more or less westerly course around the foothills would take it to the mines at Drumlemble. Therefore the whole length of 4 $^3/_8$ miles could be built on the level without any locks. At Drumlemble, the canal terminated beside the mineshaft which today lies under the farm buildings at West Drumlemble. Charles MacDowall's waterwheel was further to the west. In 1797, an advertisement was circulated for sinking a new pit nearer to the village, (41) where a steam pumping engine broke its beam in the same year. (42)

The canal was to be broad enough to take only a single boat and so passing places equal to a quarter of a mile in total length were to be built. The depth was 4 ft. 6 ins. and width at the bottom 9 ft. (43) In March 1798, there were three flat-bottomed barges on the canal valued at £69 and the waterwheel at £169. (44) The boats were 6 ft. overall width at the bottom and 8 ft. at the gunwale but of unknown length. (45) Watt allowed extra costs for cutting through difficult ground such as steeply sloping parts around Moy but there was no allowance for puddling to keep water in the canal. This may have been because it was cut around the peat lands which would originally have been level with it. Shrinkage of these through better drainage would have left the canal high and dry. One of the reasons for its closure in 1856 was the poor state of its earthworks which caused it to be a nuisance to the tenants of adjoining grounds. (46) Three bridges, presumably for roads passing over the canal, were estimated at £30 each while two foot bridges cost only £10 together. The estimate for the whole canal was £1,562.16s, but, on the reverse of this document, there are some other figures, such as an alternative pricing for the aqueduct at the Straw Water which might cost either £104.13s. or £159.5s.8d. (47)

The aqueduct over the Straw Water was the main engineering feature and was probably one of only two wooden canal aqueducts to have been constructed in the British Isles. Watt probably based his design on those he had suggested near Stirling or for the Bo'ness Canal. (See p202 and 228) The canal was to pass north, or downstream, of the road bridge over the Straw Water but the water level of the canal was so low that the river bed had to be dug out underneath the aqueduct to allow passage of floods. The bottom of this excavation, if not a very firm gravel, was to be floored with beams covered with old ships' planks or secured with large stones. Charles MacDowall questioned whether the large beams forming the foundation of the flooring were really necessary, particularly if the understratum were good, 'as all I want is to serve my time of it I wish to be at no expense but what is absolutely necessary'. This excavation meant that the foundations of the road bridge had to be protected with two arches of rough stones braced against the aqueduct pillars.

The trough, or tunnel as Watt called it, was supported on three piers or pillars built of large stones, laid with the best mortar that Charles MacDowall could find. The piers were to be 9 ft. high above the ground with good foundations. For the trough, five beams 12 ins. square, 36 ft. long were required. For the bottom of the trough, Watt preferred 2 ins. or 2 $^1/_2$ ins. planks of oak to deal, and treenailing instead of 'spilning' or nailing. At either side of the flooring were to be laid beams which would support the sides, with oak uprights mortised into them. Those parts where the boats might rub against the sides should be sheathed with oak. There was to be one gangway, presumably for the towing path, and stop gates at either side of the aqueduct to prevent the water running out of the canal in the event of an accident. An unusual provision was one or two hatches in the bottom to empty the water out occasionally and also to let water in to prevent the trough floating should any flood rise higher than the level of the canal. The trough was to be caulked with oakum and covered

Reconstruction of Watt's proposed wooden aqueduct carrying the Campbeltown canal over the Straw Water. (also see page 256)

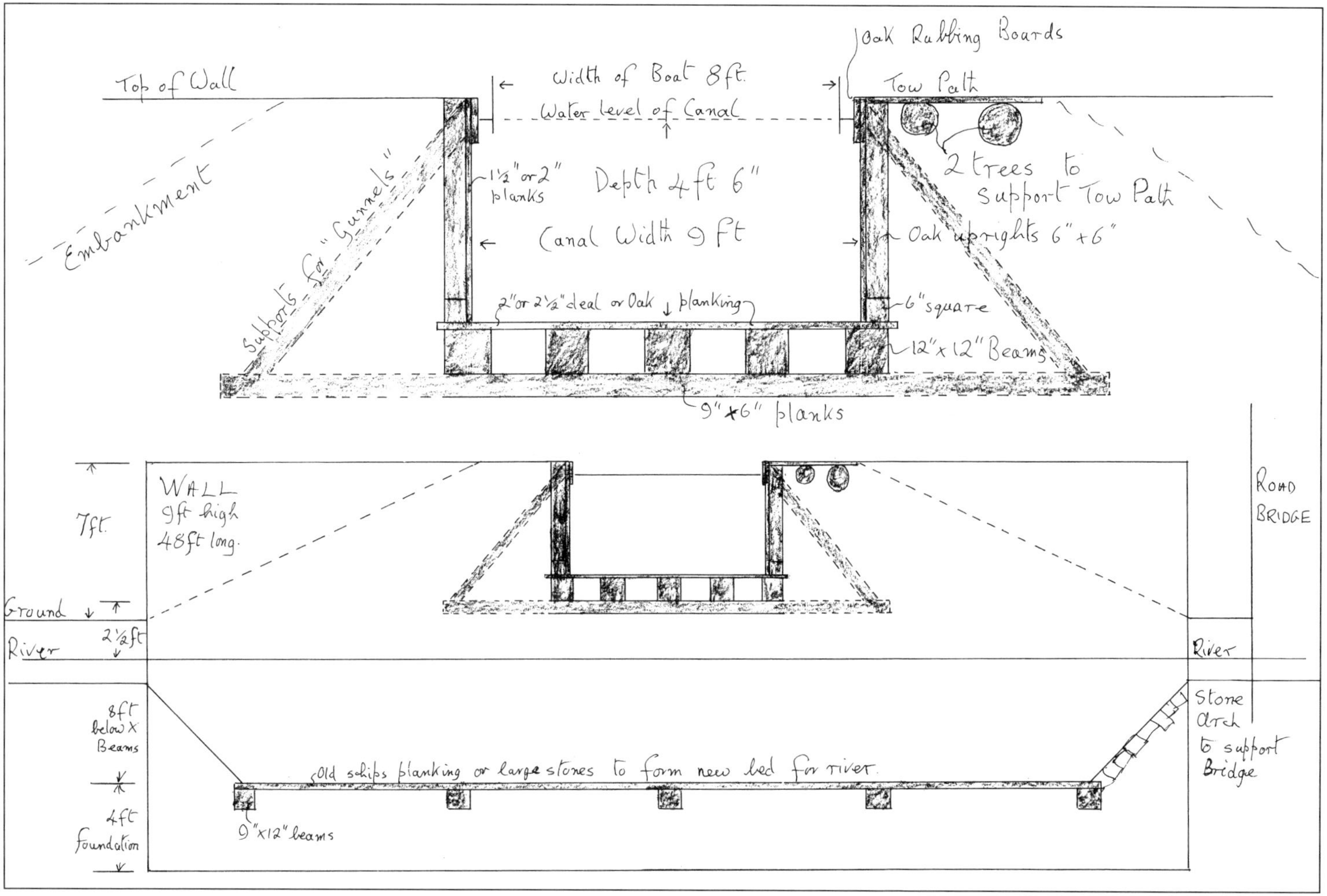

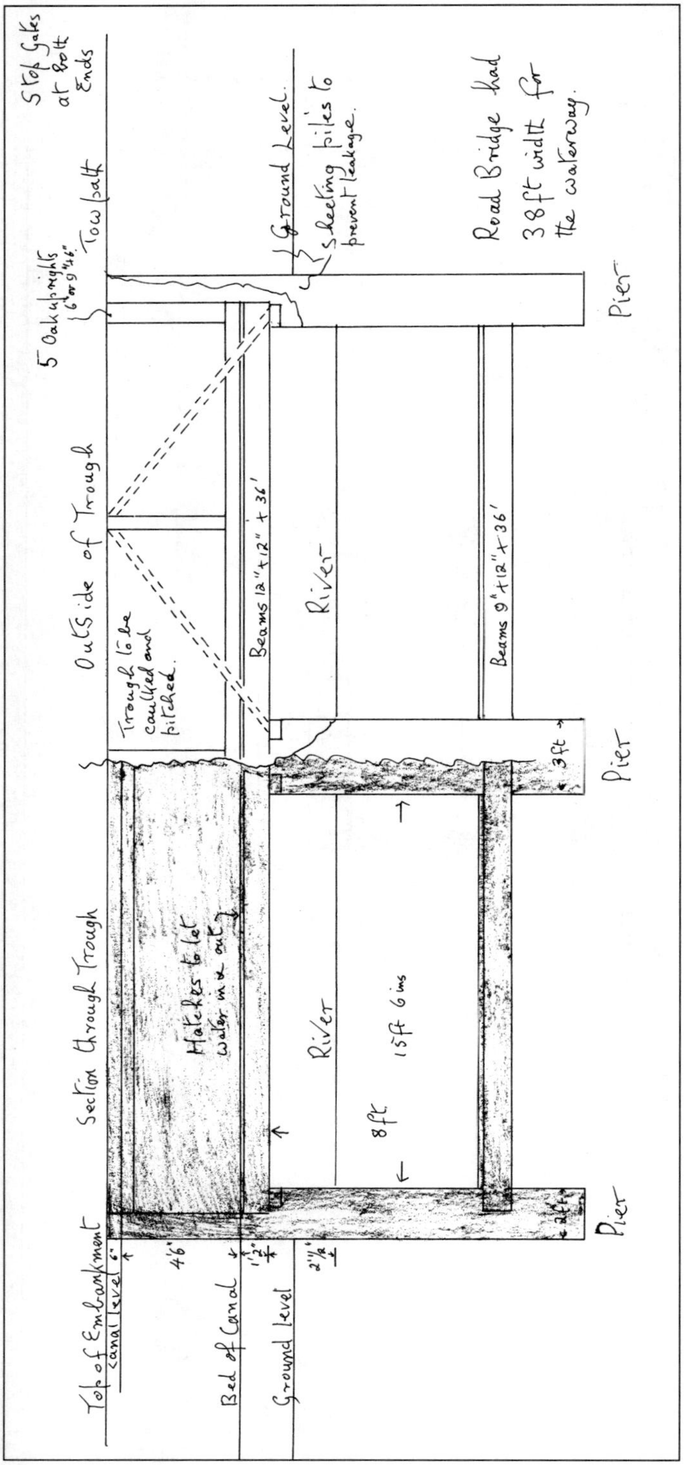
Top of Embankment
canal level
Bed of Canal
Ground level
Section through Trough
Hatches to let water in & out
River
15ft 6ins
8ft
Pier
3ft
Outside of Trough
Trough to be caulked and pitched.
Beams 12" + 12" + 36'
Beams 9" + 12" + 36'
River
Pier
5 Oak uprights
Towpath
Stop Gates at both Ends
Ground Level.
Sheeting piles to prevent leakage.
Road Bridge had 38ft width for the waterway.
Pier

with tar or pitch like a boat, while the ends where it joined the embankment had to be well puddled. It is not known when the wooden trough was replaced by one built with stone arches. One abutment of this later aqueduct still survives.

Charles MacDowall's need for the canal must have become urgent when shipment of coal from Machrihanish was stopped by the Customs in 1774. There was some correspondence in 1776, when Watt sent a new estimate, but then nothing until a letter from Charles MacDowall in August 1782 asking for further details about the construction of the aqueduct and Watt's reply early in the following January. (48) Charles MacDowall wrote,

> The last letter I had the pleasure to receive from you was dated 30th Oct, 1776. In it you sent a new estimate of my Canal, with some directions about the aqueduct bridge & promised me a compleat drawing of it, but in the meantime sent me the above to stay my stomach. Had I been sharp sett I would have put you in mind of this before now, but want of leisure, & some other causes blunted my stomach, so as not to begin the canal till spring last year, & perhaps it would have been finished by this time had I not met with interruptions from a quarter which will surprise you. I mean Col. Charles Campbell, with whom I am just now at law, altho I have a power under his own hand to carry the canal thro his farms, on paying damages. In the meantime the canal is going on and with your leisure you will oblige me with the above drawing. It will require a full explanation, for I have no engineer but my own Overseer. Hitherto it has gone on pretty well & without any undertaker, excepting about 50 falls in each bargain, & I believe I shall be within your estimate, which indeed you gave me reason to expect. (49)

This letter tells us when construction of the canal was started, in the spring of 1781. That it could be built with Charles MacDowall's own overseer and without any resident engineer must reflect credit on Watt's instructions. There is the additional point, that it was being built within Watt's estimates, confirming the quality of his methods. The completion of the canal has been given as 1794 (50) but there is a letter written by William Ker in Knochrioch dated 21 July 1791 saying 'The canal is to be finished this season'. (51) There is no way of telling whether this delay was due to the lawsuit with Campbell mentioned by Charles MacDowall. In addition, the Campbeltown town council took out an interdict to prevent him building his canal next to the mill dam and various tenants in Drumlemble also sent claims for damages to land and property caused by the coal mining. (52) Three of the Duke of Argyll's tenants at Drumlemble complained of subsidence. (53) Failure of the harvests in 1783 may have delayed construction. Then Charles MacDowall's death on 11 September 1791 could have postponed the opening further. (54)

Yet coal was produced from these mines and the output for the years 1788 to 1791 was 31,418, 40,987 and 34,397 cart loads respectively (a load was $^1/_3$ ton). (55) After the canal had opened, coal was being used in Campbeltown by the better class of people by the end of the century. About fourteen tons, the equivalent of forty cart loads, were imported daily at a cost of seven shillings a ton. (56) In 1855, the Drumlemble pit was reduced to a care and maintenance basis. The canal had become little used and was abandoned in the following year. (57) In 1861 - 3, the Drumlemble pits were finally abandoned but a new pit opened at Kilkivan before 1866. (58) The colliery changed hands in 1875 and the new owners constructed a narrow gauge railway to Machrihanish in 1877. New pits were opened in 1881 about half a mile further west of the old. The last colliery, at Trodigal, Machrihanish, was opened in 1946 and closed in 1968. (59) Thus ended a small saga in the history of both coal mining and coal transport for these pits were unusual in sending out the coal by horse and cart, canal, railway and finally motor lorries.

NOTES: WATT AT CAMPBELTOWN

1 A longer version of this section was published in *The Kintyre Antiquarian and Natural History Society Magazine*, No 41, Spring 1997, pp. 3 - 8; No 42, Autumn 1997, pp. 3 - 8 and No 43, Spring 1998, pp. 26 - 30.
2 Mactaggart, (1923), p. 40.
3 Newton, (1991), p. 3.
4 Ibid, p. 6.
5 Bremner, (1869), p. 445.
6 Mactaggart, (1923), p. 34.
7 DAP, Bundle 192.
8 Ibid, Bundle 746.
9 MacMillan, (1970), p. 21 ff. This contains the fullest account of the Machrihanish coal mines. See also Farr, (1969), p. 7.
10 DAP, Bundle 744.
11 Duckham, (1970), p. 160. Duckham states that Charles MacDowall leased the mines from the Duke of Hamilton but this seems unlikely.
12 Butt, (1967), p. 90.
13 SRO, Carron Co., GD58/6/1/6, Letter Book 1766 - 1767, p. 173, to P. Colville, 23 Jan. 1767 and p. 200, to Charles MacDowall, 16 Feb. 1767.
14 JWP, Letter Book 4, p. 102, 'J. Watt's Recollections of his Friend Dr. J. Robison', April 1805.
15 SRO, Carron Co., GD58/6/1/11, Letter Book 1770 - 1771, to Charles MacDowall, 22 Aug. 1770 and Shaw, (1984), p. 432.
16 JWP, JW/24.3, Charles MacDowall to J. Watt, 1 Aug. 1771.
17 Ibid, JW/24.5, Charles MacDowall to J. Watt, 1 Sept. 1771.
18 Ibid, JW/24.6, Charles MacDowall to J. Watt, 24 Oct. 1771.
19 Ibid, JW/24.7, Charles MacDowall to J. Watt, 18 & 27 April 1773.
20 B. & W. Col., MI, 1/20, 'Journal', 8 - 10 June 1773.
21 Ibid, 26 June 1773.
22 ARO, DR4/9/112, 'Plan of Campbellton Canal', no date.
23 B. & W. Col., MI, 3/73, J. Watt, 'Carradel River', 20 Aug. 1773. The quotations in this section are taken from this report.
24 SRO, RHP 37947, 'A Plan of Caradell Bay, by Robt. Pate, July 1773'.
25 ARO, DR1/1203, 'Plan of the farm of Glen Carradell in the property of Dugald Campbell, Esq., 1775'.
26 B. & W. Col., MI, 3/71, J. Watt, 'Report on Glen Saddel', 21 July 1773.
27 ARO, Ordnance Survey Map, 1st. edn., 1866 and B. & W. Col., MI, 3/71, Glen Saddel Report.
28 Mactaggart, (1923), pp. 43 - 44.
29 B. & W. Col., MI, 3/74, J. Watt, 'Report concerning a new Breast in Campbellton Harbour', no date.
30 Mactaggart, (1923), p. 13.
31 B. & W. Col., MI, 3/70, J. Watt, 'Report concerning the Drainage of the Craigs Meadows, Campbellton', 20 June 1773, and DAP, Bundle 534.

32 Ibid, MI, 3/68, J. Watt, 'Drainage of the Craigs Meadows and of bringing the water to Campbellton Mills', no date.
33 Ibid, MI, 1/20, 'Journal', 11 June 1773.
34 Ibid, 16 June 1773 and ARO, DR4/9/112, 'Plan of Campbeltown Canal', no date.
35 B. & W. Col., MI, 3/66, J. Watt, 'A Report to Coll. Charles Campbell of Barbreck concerning the more perfect Drainage of his lands of Lough Sanish', no date, and ARO, DR4/9/128, 'Plan of Lough Sanish', c. 1805.
36 MacMillan, (1970), p. 22.
37 Ibid.
38 JWP, JW/24.8, 'Estimates for Campbeltown Canal', no date.
39 B. & W. Col., MI, 1/20, 'Journal, 26 June 1773.
40 Ibid, 11 June 1773.
41 DAP, Bundle 806.
42 MacMillan, (1970), p. 23.
43 JWP, JW/24.8, 'Estimates for Campbeltown Canal'.
44 MacMillan, (1970), p. 24 and DAP, Bundle 806.
45 JWP, JW/22.30, Charles MacDowall to J. Watt, 12 Aug. 1782.
46 MacMillan, (1970), p. 25.
47 JWP, JW/24.8, 'Estimates for Campbeltown Canal'.
48 Ibid, JW/22.30, Charles MacDowall to J. Watt, 12 Aug. 1782 and Letter Book 1, J. Watt to Charles MacDowall, 3 Jan. 1783.
49 Ibid, JW/22.30, Charles MacDowall to J. Watt, 12 Aug. 1782.
50 Lindsay, (1968), p. 211 and Newton, (1991), p. 23.
51 The Royal Commission on the Ancient and Historical Monuments of Scotland, *Argyll, Vol I, Kintyre*, Edinburgh, 1971, p. 201.
52 MacMillan, (1970), p. 23.
53 DAP, Bundle 805.
54 MacMillan, (1970), p. 23.
55 Ibid.
56 Mactaggart, (1923), p. 34 and DAP, Bundles 254 and 743 for figures of coal production in the early nineteenth century.
57 MacMillan, (1970), p. 25; Lindsay, (1968), p. 211 and Newton, (1991), p. 11.
58 ARO, Ordnance Survey Map, 1st. edn., 1866.
59 Butt, (1967), p. 212.

Watt's Last Canal Surveys for Lord Cathcart

The Caledonian Canal

The Great Glen, that geological fault stretching south-westwards from Inverness to almost Fort William, had long been seen as a possible route for an artificial navigation to allow shipping to avoid the dangerous journey round the north of Scotland. The route seemed quite practicable because it would pass through the natural lochs of Ness, Oich and Lochy which would comprise more than half of the distance. Yet even in 1725, Burt, in his letters from the Highlands, realised that there were formidable natural obstacles especially with the winds.

> The Spot the projectors [of the canal] say, is a Level between the two Seas, pointed out, as it were, by the hand of Nature, and they pretend the Space of Land to be cut through is practicable. But it would be an incredible Expense to cut 14 navigable Miles in so rocky a Country, and there is yet a stronger Objection, which is, that the whole Opening lies in so direct a Line, and the Mountains that bound it are so high, the Wind is confined in its Passage, as it were, in the Nozzle of a pair of Bellows; so that, let it blow from what Quarter it will without the Opening, it never varies much from East or West within. This would render Navigation so precarious that hardly anybody would venture on it, not to mention the violent flurries of Wind that rush upon the Lake by Squalls from the Spaces between the Hills, and also the rocky Shores, want of Harbour and Anchorage; and, perhaps there might appear other unforeseen Inconveniencies and Dangers if it were possible the work could be completed. (1)

Burt was proved correct for, after the formal opening in October 1822, the canal proved to be little used owing to the winds being contrary for long periods, (2) but this did not deter projectors in the second half of the eighteenth century. Morison had been employed by the Court of Police to make a preliminary survey in 1764 while Watt had been approached to make a fuller one in 1767 but had not carried it out. (3)

Small told Watt in May 1773 that there had been talk in England of making a passage for ships between Inverness and the Western Sea which could be easily done because the ground between Lochs Ness and Oich was level. He suggested that Watt might carry out a survey (4) to which Watt replied,

> In relation to your Inverness navigation, I know something about it. It would cost much money and time to make such a survey as I could set my hand to, and I am afraid the estimate would frighten you. The ground between Loch Oich and Loch Lochy is 50 feet high, and probably hard ground; the whole height, deducting the said 50, is 110 feet, the country hills and rocky, few inhabitants, and labour very dear. I would like well to survey it, if paid for it; but I like better to come and visit you. (5)

Small thought the object of great importance and that Watt's survey would be the first and could be warmly commended to

> ... Lord Dartmouth, who is head of the Council of Trade, to Lord Sandwich, the First Lord of the Admiralty, and to Lord North... Advantages might arise from your having only proposed such a scheme to those great men, and having been shown to be most fit for executing it. (6)

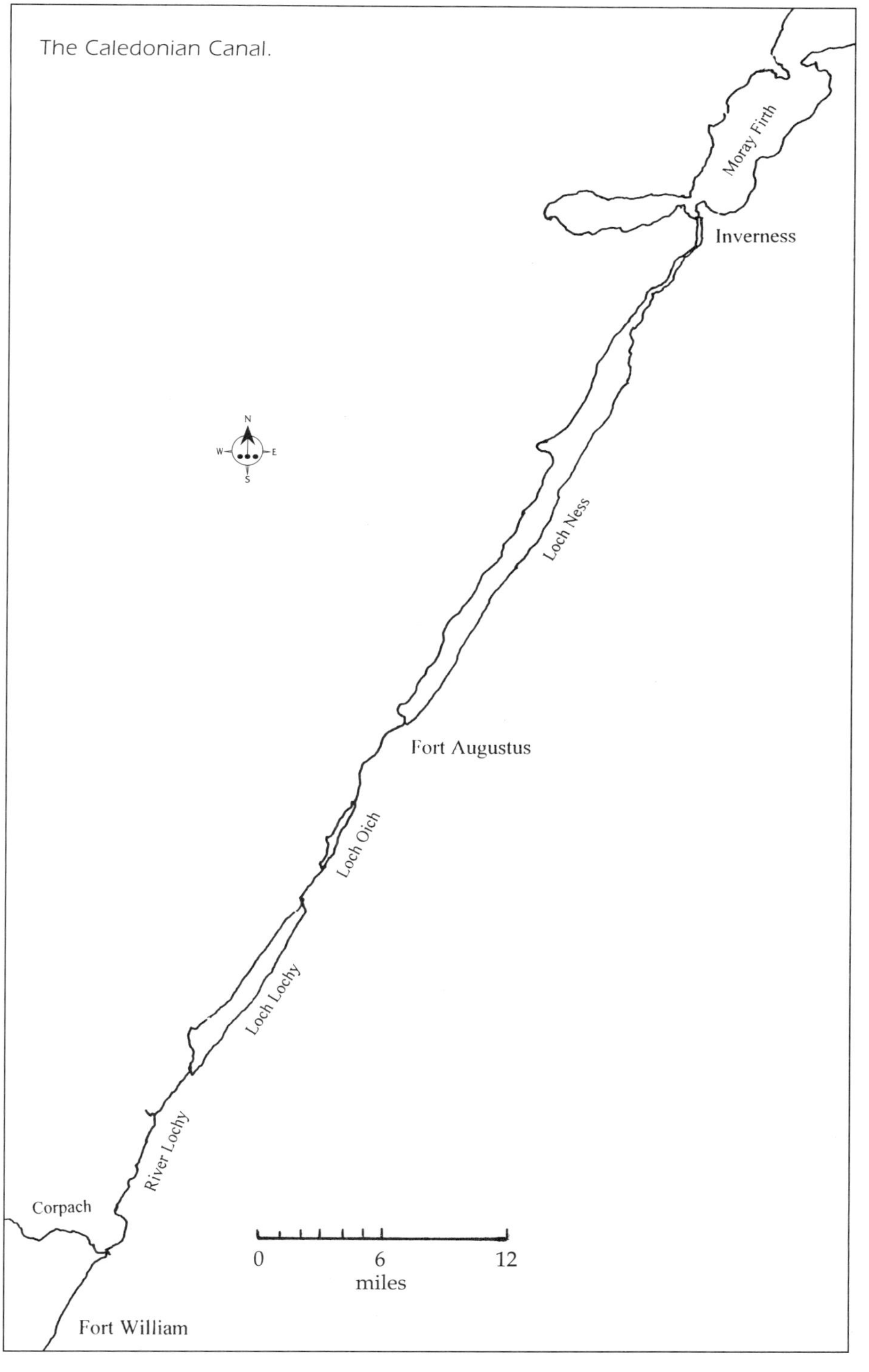
The Caledonian Canal.
Moray Firth
Inverness
N
W
E
S
Loch Ness
Fort Augustus
Loch Oich
Loch Lochy
River Lochy
Corpach
0
6
12
miles
Fort William

While Small was kindly trying to promote Watt's career, it is probably lucky that Watt never became involved in the construction of this navigation. The more experienced Thomas Telford had to overcome many very severe difficulties when he built the Caledonian Canal in the opening years of the nineteenth century.

It is possible that Watt had some knowledge that a survey was being planned but it was not until the middle of August when he told Small, 'I am appointed by the Court of Police to make a survey of the canal from Inverness to Fort William, and set out the week after next. I accepted of this merely in consequence of your desire, otherwise I should have delayed it until next season'. (7) This was the first survey which Cathcart asked Watt to undertake for him after his return from Russia in 1771 (8) possibly due, first, to the failure of the Ayr Bank in June 1772 and, second, to Watt being occupied with the construction of the Monkland Canal until at least May 1773. Watt must have been approached about this survey before August because he noted in his Journal for the seventh, 'My Lord [Cathcart] renewed order for North Country survey'. (9) A couple of weeks later, Morison, who would assist again, told Watt that Cathcart was about to send instructions. (10) At the end of the month, Watt was making his preparations and checking his instruments before setting out for Schaw Park on 31 August where he received Cathcart's orders. (11) He left home very anxiously because Peggy had not been well through her latest pregnancy.

Watt was to examine the ground between Fort William and Inverness, including the Lochs Lochy, Oich and Ness together with their lengths and levels. He was to investigate the practicality and expense of making 'a communication for vessalls drawing nine feet of water', or, if possible, for larger vessels. He was to report to the Board of Police and could charge his expenses on the scales allowed by the Commissioners of the Forfeited Estates, as well as taking another surveyor, 'in which capacity the Board recommend you to employ Mr James Morison'. (12)

They left Schaw Park on 1 September and, after passing through Perth and Brechin, arrived at Stonehaven where Morison's horse broke its back. They had had two rainy days but Watt's horse had survived very well. (13) After passing through Banff, they had to spend a day at Fochabers waiting for the Spey to subside since it had been swollen by the rain. (14) They stayed at Elgin and reached Inverness at three o'clock on the afternoon of the ninth. Watt wrote again to his Peggy that he had found no letters waiting for him and he would be very glad to hear from her. 'Every day of our survey will be so far on our road home being now at the furthest end of it'. (15)

After adjusting their instruments, they started surveying the following day, checking on harbours and working south along Loch Ness. Morison carried on alone on the sixteenth because Watt was ill but they arrived at Fort Augustus on the nineteenth where they established a base for levelling around Loch Oich. On the twenty-fourth, Watt received a letter from Peggy saying that she was recovering (See p127) but on the twenty-sixth, a Sunday, he received another letter '... from Mr Muirhead advising that Mrs Watt was dangerously ill & her life despaired of & desiring me to come home with all speed. I immediately set out, with a very sad heart & came to Fort William that night having a heavy rain all day'. (16) When he reached Dumbarton a couple of days later, he was met by Hamilton with the melancholy news of Peggy's death. He later described this journey to Small, 'I had a miserable journey home, through the wildest country I ever saw, and the worst conducted roads: an incessant rain kept me for three days as wet as water could make me. I could hardly preserve my journal-book'. (17)

Morison was left to complete the survey alone. At the beginning of October, he reported from Fort William that it had rained every day since Watt had left which had delayed his observations. Through the bad weather, he could not say when he would return to Glasgow, but he thought that the canal would have to follow the north side of the valley which he would explain. (18) In fact, Morison came to Glasgow on 25 October and from then until 2 November Watt spent most of the time at the Caledonian Canal report. But then on the third he had to go to Schaw Park to begin a fresh survey for the Rivers Forth and Devon. This work and its report took precedence over the Caledonian Canal which was only resumed in December. Other commitments caused further delays so the report was not finished until 30 April and Watt learnt on 19 May of Cathcart's approval. (19) Watt had spent two days preparing for the Caledonian Canal survey, twenty-seven days on the survey itself and a further twenty-six days on plans and the report. He charged £2 each day in the field and one guinea while working at home. With other expenses for ferries and assistants, his account came to £100.13s.1d. (20) Morison's private expenses totalled £30.14s.6d. but this could not have included his daily payment. (21)

In his report, Watt pointed out how very unfavourable the weather had been so that he was 'prevented from entering into the minutia of the subject with that accuracy I intended'. (22) At the northern end, he planned that the canal should finish in the harbour of Inverness itself, just north of a ford where the present railway bridge crosses the river Ness. One quay had been built here in 1675 which had been extended in 1738 and allowed vessels of 150 tons to berth. (23) The lochs appeared to be deep enough for navigation except for a few places where a channel would have to be cut. Loch Oich would act as a reservoir to supply water for lockage. At the southern end, the canal would descend on a more gentle route than the present flight of locks called Neptune's Staircase. The canal linking the lochs would be 30 ft. wide at the bottom and 70 ft. at the top. The locks would be 90 ft. long and 25 ft. wide to be capable of taking vessels drawing 10 ft. In contrary winds, vessels would be hauled by horses. The length of the cuts would be 22 1/2 miles out of a total of 59 miles for the full distance. To surmount and descend the rise of 100 ft., 32 locks would be needed. Watt's estimate for the whole project was £163,263.19s.3d. (24) The high cost must have deterred any thoughts of its construction in a country recovering from an economic depression.

In October 1784, the Lords of the Treasury tried to revive the scheme and instructed one of the Boards of Trustees in Edinburgh to have a survey carried out. To save expense it was suggested that Watt's report and plans might provide the necessary information but they could not be found in the Police Office. (25) A meeting was held in Edinburgh and Watt's name was put forward as engineer. (26) While Watt was pleased to have been asked, he was unwilling to have his report printed as it had been prepared such a long time ago. (27) If he had any thoughts about taking up a surveying career again, he realised that he was too heavily involved with his steam engine business for this was the period when he was developing his rotative engine. He would have to consult his partner Boulton and wrote that '... setting aside the damage which the distractions of my attention might do the partnership, that my own share of the loss in the engine business might exceed my gains by the canal direction'. (28) In a further letter, he recognised that he had been absent too long from civil engineering.

> I should for some time at least be much inferior in that line of business to many others which might be found and cut a sorry figure in coming after a man of Mr Smeatons abilities and experience in that branch and it is now too late in life to study matters foreign to the common course of my ideas the powers of my memory and the ardour to obtain knowledge are with me greatly diminished. (29)

Whatever the temptations might have been, this episode terminated any ambitions Watt might have harboured about resuming his civil engineering career.

However he was consulted again by Telford when construction of the Caledonian Canal became a reality in 1802. Telford had searched in Edinburgh for Watt's report and, not surprisingly, was unable to find a copy, (30) although later Rennie discovered one in the 'Reports of the Committee of the House of Commons on the British Fishery'. (31) Even Watt could not lay his hands on his own copy of his rough draft but was able to send Telford some figures for distances and heights. (32) Telford replied,

> I found the whole of your statements particularly correct except in the fall from Loch Oich to Lochness where we differ a few feet, but the states of the water in the Lochs make some difference - and the only bad weather I experienced was during this part of the Survey, so that I am not positive that I am right'. (33)

Considering the bad weather which Watt experienced, this shows the accuracy of his surveys. Watt's next letter to Telford stressed the dangers of the winds, both those blowing up and down the length of the canal and also the sudden cross winds. He recognised the problem of ships trying to battle against the great strength of adverse winds and suggested boats might be hauled by horse capstans placed every 200 yds., or even windmills but the wind 'is but an unruly power, small steam engines would answer infallibly & might be at greater distances'. (34) The days of the steam ship based on Watt's own engines were still in the future.

Surveys Around Stirling in 1773

For most of October 1773, Watt tried to console himself after the loss of Peggy with experiments on telescopes and barometers and, after Morison's return towards the end of the month, drawing part of Loch Ness. (35) But, just before Morison came back, Cathcart wrote to Watt on behalf of the Commissioners of Police to encourage him after his bereavement. He suggested that Watt and Morison should survey 'the Tract from the lint miln at Menstry on ye Devon to the upper Forth & give your opinion respecting a canal to join those rivers and respecting the navigation of the upper Forth'. (36) The tidal Forth below Stirling was to be included as well. Watt arrived at Schaw Park on 3 November and left on the next day for the lower part of the Forth for what would prove to be his last survey of any navigation. Except for three Sundays spent at Schaw Park, Watt was surveying the Forth for the whole of November and, after settling accounts with Morison on 5 December, sent an abstract of his report to Cathcart on 7 December. (37) Watt commented to Small,

> I have been about a month upon a survey of some improvements of the Upper Forth. I had many cold fingers and feet, and have much boiled my brains suiting a report to Lord Cathcart's genius, which I do assure you is no easy matter. He is however, a most honest and friendly man, and much my friend. (38)

Watt should have taken note of Small's warning, 'I am glad you are employed by Lord C., although he may give you trouble'. (39)

The report was printed by 17 December and corrected the following day. (40) Golborne had also been asked to carry out a survey in November and sent a four page report on the Forth below Stirling, commenting that 'it has almost every defect possible', (41) but also saying that he needed better weather to carry out a proper survey. It seems that Cathcart must have been principally interested in improving the transport of coal from his collieries. Watt was asked whether there should be a lock for shipping at Craigforth or whether a navigable canal should be built to some port lower down the river. Cathcart also wanted a report about making the 'Goodie Water' navigable from the Forth to Mentieth which Morison had surveyed in April 1769. (42) To make the Gudie navigable would mean passing the mills of Thornhill and Cardross and straightening the course of the river.

For the Forth and Devon, Watt seems to have recommended his earlier scheme of a canal from Menstrie to Craigforth to avoid navigating the difficult lower reaches of the Forth. This time, there were to be locks instead of an aqueduct over the Allan. Otherwise, he followed his earlier reports of 1767. Above Sauchie, the canal would have to be raised 20 ft. by three locks and a separate channel to reach the collieries higher up-stream around Coalsnaughton which would take it at least to Mallack Foot. (43) The canals would be 16 ft. wide at the bottom and 3 ft. depth. The locks were to be 13 $^1/_2$ ft. wide, 72 ft. in length so that the boats on the Forth which were 56 to 60 ft. long and 13 ft. wide could be accommodated. The cost for 'Rendering Navigable the Rivers Forth & Devon' was £13,040.13s.6d. (44)

This scheme seems to have been favourably received by Cathcart at first, who wrote to Watt on 7 December, 'I had yesterday ye pleasure of reading your summary & conclusion to my friends at ?Tillibody? & have ye satisfaction to assure you that they commend it much so you may rest satisfied it will be well received by others'. (45) Cathcart made some comments about the Devon above Sauchie which are illegible so it is evident that he was referring to Watt's report on the Forth and Devon and not the 'Northern' or Caledonian Canal. But soon after this, Cathcart must have received an alternative report, possibly from Morison, who recommended a saving of £4,147 by terminating the canal from the Devon at Queenshaugh opposite Stirling and transhipping at Craigforth which would avoid the expensive locks. A shorter alternative was to end the canal a little lower down the Forth at Manor Pow which would avoid the Thrask. This was certainly Morison's suggestion. (46)

Watt stuck to his original proposals but one wonders about his reaction to Morison submitting a rival scheme. Also harsh comments were to follow from Cathcart. Once again he must have been referring to Watt's Forth and Devon report and not the 'Northern' canal when he wrote on 9 January 1774, 'I have lately read two reports of Mr. Smeaton's; I think his manner very concise clear & satisfactory & a good model for such compositions: your report is most incorrectly printed which I am persuaded you will attend to in future publications'. (47) Again there must be speculation about the effect of this letter on Watt from his main patron. Part of Cathcart's anxiety may have derived from the proximity of a meeting in Edinburgh to be held on 12 January about obtaining a bill in Parliament for the Forth and Devon scheme. (48) Watt was in Edinburgh from 8 to 22 January, possibly with Morison, when he had a further report printed in which he modified his suggestions. (49) He returned to Edinburgh again from 27 January to 4 February about the Forth & Devon. On 31 January, there was a meeting when £4,500 was subscribed. (50)

Cathcart wrote to Watt on 14 February that there were problems with the Forth and Devon bill and on 9 March that there was lack of support for it. (51) Cathcart also commented,

> I am not surprised at the bad account you give me of poor Morison's map. It is very desirable to employ people of one's own country but I am much afraid it must be allowed to be better in the lights of certainty despatch & [word illegible] to bring in artists from a distance. I am sorry your Headaches still continue and have so long retarded your Report of the Northern Canal. (52)

Did Watt take this as a warning? Many of his own reports had been long delayed so his reputation for being dilatory may have become widespread. He may have felt that Cathcart would turn to others for surveys in the future.

This was the period when Small was pressing Watt to go to Birmingham where his steam engine could be built at Boulton's Soho Manufactory. Watt's surveying career was not proving very profitable because he had not been paid for much of his work, including that for the Forth and Devon. In the middle of May 1774, he packed up his things for Birmingham, left Glasgow on the 16th and arrived in Edinburgh a couple of days later. Here he met Cathcart who 'was satisfied with my Account of Survey of Inverness Canal'. (53) There had been no meeting about the Devon Canal since Watt was last in that city so he left Edinburgh without being paid. Hamilton tried to obtain payment in December 1775 with no success (54) and arguments with Morison over parts of the accounts continued into October 1776 at least. (55) Cathcart died on 14 August 1776 after a long illness (56) so the chief protagonist for making the rivers Forth and Devon navigable was no longer there to give his support. In any case, by that time, the wooden waggon way was beginning to be seen as a cheaper alternative for collieries to transport their coal to the staithes for shipment by sea and neither of these rivers was ever made navigable.

NOTES: WATT'S LAST CANAL SURVEYS FOR LORD CATHCART

1 *Transactions of the Gaelic Society of Inverness*, Vol XIII, 1886 - 1887, A Ross, 'Notes on the Formation of the Caledonian Canal and its Effects on the Highlands', p. 314, quoting from *Burt's Letters from the North of Scotland,* (1876), Vol II, p. 321.
2 Lindsay, (1968), p. 156.
3 B. & W. Col., MI, 3/97, J. Watt, 'Report on the Northern Passage'.
4 Muirhead, (1854), Vol II, p. 47, Letter 112, W. Small to J. Watt, 1 May 1773; Boulton 340.31.
5 Ibid, Vol II, pp. 49 - 50, Letter 114, J. Watt to W. Small, 20 May 1773; Boulton 348.31.
6 Ibid, Vol II, p. 51, Letter 115, W. Small to J. Watt, 2 June 1773; Boulton 340.32.
7 Ibid, Vol II, p. 58, Letter 120, J. Watt to W. Small, 17 Aug. 1773; Boulton 348.36.
8 Paul, (1905), Vol II, p. 521.
9 B. & W. Col., MI, 1/20, 'Journal', 7 Aug. 1773.
10 JWP, 4/33.15, J. Morison to J. Watt, 21 Aug. 1773.
11 B. & W. Col., MI, 1/20, 'Journal', 30 & 31 Aug. 1773.
12 Ibid, MI, 3/96, 'Lord Cathcarts Instructions to Mr. James Watt, Engineer, 30 Aug. 1773'.
13 JWP, 4/11.5, J. Watt to Margaret Watt, 3 Sept. 1773.
14 B. & W, Col., MI, 1/20, 'Journal', 7 Sept. 1773.
15 Ibid, 4/4.4, J. Watt to Margaret Watt, 9 Sept. 1773.
16 B. & W. Col., MI, 1/20, 'Journal', 26 Sept. 1773.
17 Muirhead, (1854), Vol II, pp. 62 - 63, Letter 123, J. Watt to W. Small, no date; JWP, 4/59.1.
18 JWP, 4/33.16, J. Morison to J. Watt, 2 Oct. 1773.
19 B. & W. Col., MI, 1/21, 'Journal', 19 May 1774.
20 Ibid, MI, 2/42, 'Expenses on Surveys'.
21 JWP, 4/33.18, J. Morison's bills on Caledonian survey.
22 B. & W. Col., MI, 3/97, J. Watt's Report on 'Northern Passage'.
23 Pollitt, (1981), pp. 193 - 194.

24 B. & W. Col., MI, 3/97, J. Watt's Report on 'Northern Passage'.
25 JWP, 4/43.33, J. McGrigor to J. Watt, 22 Oct. 1784.
26 Ibid, 6/21.10, J. McGrigor to J. Watt, 8 Nov. 1782.
27 Ibid, Letter Book 1, p. 86, J. Watt to J. McGrigor, 30 Oct. 1784.
28 Ibid.
29 Ibid, p. 95, J. Watt to J. McGrigor, 21 Nov. 1784.
30 Ibid, 4/27.78, T. Telford to J. Watt, 14 April 1802.
31 Muirhead, (1854), Vol II, p. 287, Letter 345, J. Watt to Hon. C.F. Greville, 19 Nov. 1804.
32 JWP, Letter Book 3, p. 224, J. Watt to T. Telford, 26 April 1802.
33 Ibid, 4/27.77, T. Telford to J. Watt, 3 May 1802.
34 Ibid, Letter Book 3, p. 228, J. Watt to T. Telford, 7 May 1802.
35 Parts of this section were published in *Forth Naturalist and Historian*, Vol 21, 1999, R.L. Hills, 'James Watt's Surveys Around Stirling', pp. 71 - 84.
36 JWP, 4/10.19, Ld. Cathcart to J. Watt, 24 Oct. 1773.
37 B. & W. Col., MI, 1/20, 'Journal', 7 Dec. 1773.
38 Muirhead, (1854), Vol II, pp. 67 - 68, Letter 127, J. Watt to W. Small, 11 Dec. 1773; Boulton 348.38.
39 Ibid, Vol II, p. 72, Letter 129, W. Small to J. Watt, Dec. 1773; Boulton 340.38.
40 B. & W. Col., MI, 1/20, 'Journal', 17 & 18 Dec. 1773.
41 JWP, 3/69, 'Report on Rivers Forth and Devon', 9 Dec. 1773.
42 Ibid, 4/33.1 & 2, J. Morison to J. Watt, 23 & 29 April 1769.
43 Ibid, 4/82, J. Morison to J. Watt, 7 Dec. 1773.
44 B. & W. Col., MI, 2/81, J. Watt, 'An Account of the Scheme for Rendering Navigable the Rivers Forth & Devon with Estimates of the Expence', 13 Jan. 1774.
45 JWP, 4/82, Ld. Cathcart to J. Watt, 7 Dec. 1773.
46 Ibid, 4/10.10, Ld. Cathcart to J. Watt, 6 Jan. 1774.
47 Ibid, 4/10.9, Ld. Cathcart to J. Watt, 9 Jan. 1774.
48 Ibid, 4/10.18, J. Callander to J. Watt, 28 Dec. 1773.
49 B. & W. Col., MI, 1/20, 'Journal', 8 - 22 Jan. 1774. See JWP, 3/22 for J. Watt's modified report.
50 B. & W. Col., MI, 2/42, 'Expenses on Surveys' and MI, 1/20, 'Journal', 27 Jan. - 4 Feb. 1774.
51 JWP, 4/10.6 & 7, Ld. Cathcart to J. Watt, 14 Feb. & 9 March 1774.
52 Ibid, 4/10.6, Ld. Cathcart to J. Watt, 9 March 1774.
53 B. & W. Col., MI, 1/21, 'Journal', 18 May 1774.
54 JWP, 4/20.24, G. Hamilton to J. Watt, 11 Dec. 1775.
55 Ibid, 4/33.21, J. Morison to J. Watt, 27 Oct. 1776.
56 Paul, (1905), Vol II, p. 522.

Road and Rail Surveys

Hamilton Bridge

Watt's new method of practising as a civil engineer can be observed in the way he acted over the request to design a new bridge to cross the Clyde at Hamilton. The Duke of Hamilton at the time was a minor so it was probably his tutors or agent who asked Watt to make a preliminary survey. Roebuck had leased Kinneil House from the Duke of Hamilton so it is possible that this was the connection. Watt visited the site on 7 June 1770, a month before he started with his position as Resident Engineer on the Monkland Canal (1) and must have suspected that the soil for the foundations of the piers was inadequate because tests were carried out by some preliminary boring which were stopped until Watt gave his opinion. The Duke's tutors wanted Smeaton's advice but realised that would involve extra expense. (2)

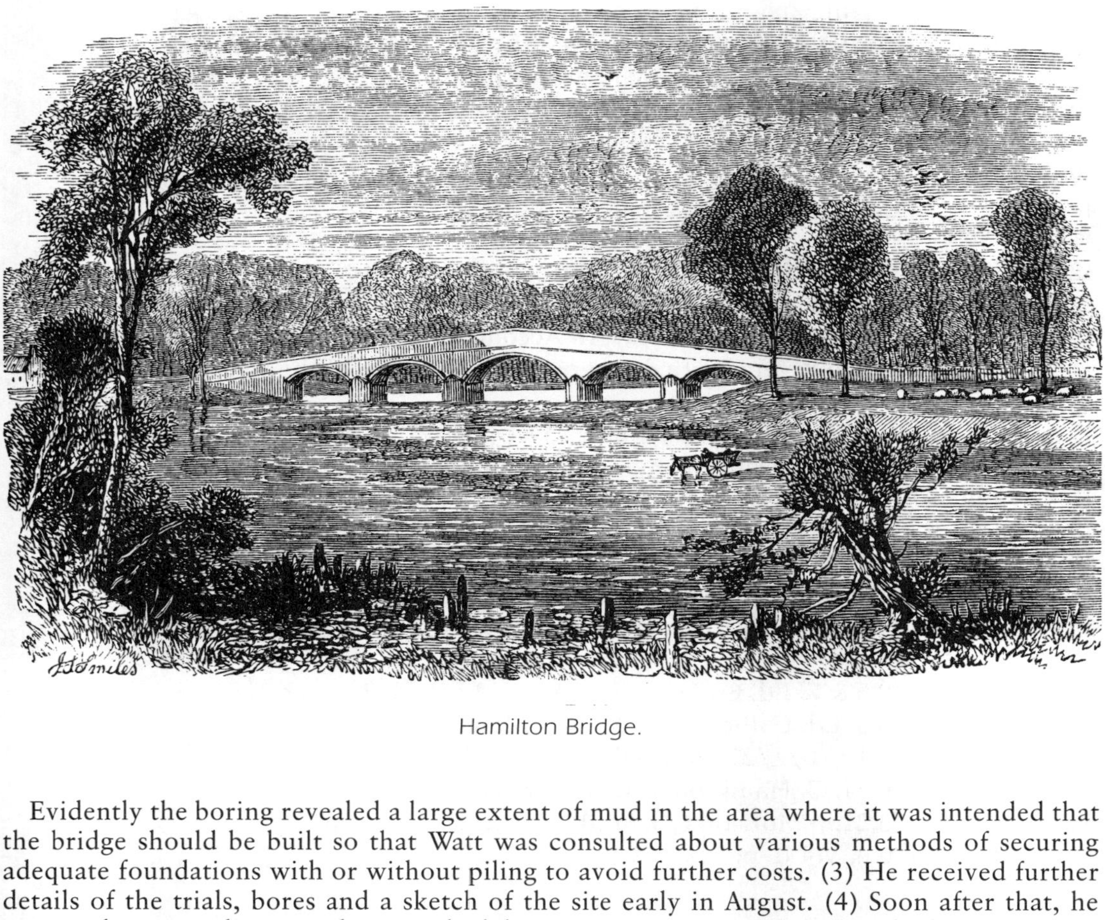

Hamilton Bridge.

Evidently the boring revealed a large extent of mud in the area where it was intended that the bridge should be built so that Watt was consulted about various methods of securing adequate foundations with or without piling to avoid further costs. (3) He received further details of the trials, bores and a sketch of the site early in August. (4) Soon after that, he prepared some preliminary designs which he sent to Smeaton. Smeaton replied on 8 September 1770 and thought that, from the particulars which Watt had sent, it did not appear to be possible to build a bridge 'upon solid principles' with the two or three thousand pounds available owing to the difficult nature of the site. He enclosed a copy of his method of constructing the piers at Coldstream with piles and also explained how he had built that bridge with one set of centring. He thought most of Watt's sizes were adequate but recommended that the bridge would have more agreeable proportions if the two side arches had spans of 46 instead of 45 ft. (5) In this way, the Duke of Hamilton's tutors obtained their consultation with Smeaton free!

Watt looked at a quarry for the supply of stone at the beginning of October, (6) but it was another month before he prepared new drawings (7) and yet another before he finished his estimates. (8) He was employed on this project for about forty-five days (9) and told Small, 'I have lately made a plan and estimate of a bridge over our river Clyde, eight miles above this; it is to be of five arches and 220 feet waterway, founded upon piles on a muddy bottom'. (10) Watt submitted a full break-down of his estimate in which the piling and foundations were £950.4s.2d. out of a total cost of £2,555.17s.8d. (11) William Purdy challenged this, claiming that a true estimate for Watt's design would be £2,991. He offered to build the bridge himself more cheaply. (12) This may have been one reason for delays because it was not until 7 July 1771 that Watt was advised that he would be asked to supervise its construction. (13) By this time, he had had a year's experience of labour problems on the Monkland Canal which could have been his reason for declining. Instead he recommended a mason from Perthshire and the reply shows the high esteem in which Watt was held even though this was his first venture into bridge design. 'Both Duke Hamiltons Tutors and the Magistrates & Councill have so high an Opinion of your Abilities and Integrity that your recommendation in favour of any person will have a very considerable weight with them'. (14)

Watt was asked to write to this mason who presumably accepted the task because Watt drew up 'Regulations for the Undertaker of Hamilton Bridge' in addition to plans and drawings. These are our earliest surviving set of instructions prepared by him and cover twenty sections. The materials for the different parts were specified, such as beech for sheeting piles, as well as heights of different parts of the foundations. The outside of the bridge was to be faced with good rough ashlar from Windmill Hill quarry. The methods of construction were outlined, with the two middle piers being started first, and no pier was to be commenced until there was sufficient stone assembled on site to build it up to the spring of the arch. There were to be overseers at both bridge and quarry, while the Commissioners were to appoint a surveyor who would always be present. The contractor would maintain the bridge for three years after it had been finished. Methods of payment were laid down as well as how to settle disputes. (15) This document may have been based on some form of standard contract available at the time but it shows Watt's attention to detail and attempt to give comprehensive instructions. He was not involved with the construction and had left Scotland before it was finished in 1780. (16)

The Dumbarton Road

Some of Watt's other commissions show how he broadened his experience and diversified into other spheres of civil engineering. In his report on the 'Road proposed from Dumbarton to Yoker bridge', he wrote,

> I was desired to assist in making an Estimate of the above road & wish[ed] I had it in my power to be of any service in that way the making of High roads is a thing I have had no experience off & I would be sorry to mislead by pretending to be more knowing than I am. (17)

He went to Dumbarton on 12 October 1770 and had finished his report within a couple of days. (18) He probably wished to gain experience in a new field of civil engineering by assisting a Mr. McCorkall. He commented in his Journal that 'I think they are laying out this road too narrow I advised them thirty foot at least'. (19) He gave his reason in his report.

> Where it can be done with any manner of conveniency I would recommend making the road thirty foot wide, a narrow road can never be long a good one as the carriage will go always in the same Rutts. In all cases where the Ground is not inclosed it will now cost little more than the present purchase of the additional breadth whereas at some future period when the Public Conveniency may render this breadth absolutely necessary the expence will be encreased by the value of the fences necessarily removed. (20)

Once again, Watt was not involved in the construction. A bill was laid before Parliament in February 1772 for a road between Glasgow and Dumbarton which must have included the stretch Watt surveyed. (21)

Gartsheary Road

Watt was soon surveying roads on his own account. In August 1771, he was consulting with 'Aichison and Hamilton' about a turnpike road. (22) This could have been the Gartsheary Road which he 'viewed' at the end of September and 'laid out' in the middle of November. (23) No further details have survived but the construction of the Monkland Canal may have necessitated an alteration to the line of this road.

Clyde and Argyll Streets, Glasgow

Watt was not concerned with roads again until he had finished on the Monkland Canal. In June 1773, Glasgow Council resolved that Alex Campbell, the town's causewayer, should finish Miller Street and then begin Clyde Street, completing that before starting on any other. (24) Clyde Street ran from the Water port to the Broomielaw. Accordingly, Watt was asked in July to inspect it and draw up specifications for paving it because, 'thro some Division of oppinions it has been delayed too long'. (25) He carried out this task immediately and had levelled it by the end of the month. (26) Towards the end of August, there followed a similar survey of Argyll Street. (27) There is no mention of his supervising the work while it was being carried out, only inspecting it after it had been finished in January (28) and submitting his measurements to the Town. (29) We see him here in the position of a consulting engineer and possibly quantity surveyor.

Cathcart Road

We know that Watt spent the day 'viewing line' for the Cathcart road on 9 March 1774 with a Mr Hill. (30) The report was written the following day but no details have survived. These and other small surveys must have brought in some cash but it has been impossible to determine what sort of income Watt gained from his various commissions.

Rutherglen Bridge

Watt's final involvement with roads as a civil engineer concerned a bridge at Rutherglen. He was approached in March 1774 by a committee which had already considered several designs. Watt was asked to examine these and judge the 'most proper' so that a contract could be drawn up. (31) These plans and proposals were also to be laid before Thomas Brown, mason in Renfrew, and James Barrie, land surveyor in Glasgow. Brown and Watt were to choose the best and lay down specifications. (32) Various meetings were held that April when it must have been decided that these early suggestions were unsatisfactory (33) because, on 19 April, we find Watt himself drawing up a scheme for Rutherglen Bridge with five arches (34) which he had finished by the end of the month. (35) He also produced 'Regulations for the Undertaker of Rutherglen Bridge', similar to those for Hamilton Bridge. In the obituary of his father, Watt junior noted this bridge as having been built. (36)

Railways

During 1770, Watt surveyed, designed and superintended the construction of a waterwheel to drain a coal pit at Clunie for a William Alexander of Edinburgh who also asked about the line of a waggon way to take the coal down the the Firth of Forth. This seems to have been the first time Watt was involved with this form of transport other than in connection with his canals. Clunie was situated on the River Ore about four miles north-west of Kirkcaldy, but between the two lay a ridge of hills. Transport of coal either to that town or to the coast had to follow a circuitous route along poor roads. Alexander wrote to Watt, 'With regard to a Wagon way it is what we wish but have some doubt whether it wou'd answer the Expence which will be high as it must pass through other peoples ground - but if you can give us an Estimate & plan I beg you will come prepared for that purpose'. (37)

Roebuck had pits in this area and he too asked Watt to survey a line for a waggon way, which Watt must have declined, probably because he was surveying Ayr harbour and drawing up reports and plans for improvements there. Instead he suggested Morison. (38) While there are no records of Morison's survey, we know that he completed it because he complained to Watt in February 1774 that he had not been paid. (39) By that time, Roebuck had been declared bankrupt. How Roebuck junior also became involved with collieries in this area is unknown. He wrote to Watt in August 1771 optimistically about his prospects; that he was selling all the coal he could raise; was paying his rent and hoped to sink another pit.

> When this additional Pit is got properly agoing I expect to be quite easy and independent of the Waggon Way proposed by my father - with respect to that Plan the carriage of Coals to Kircaldie by carts is at *most* 1/6d pr cart of 9 cwt equal to the price of the coals upon the Hill & as the Cartroad is badly constructed & Hilly & my principal sale towards Kirkcaldie by land or sea, a waggon way would (if practicable & within a reasonable outlay) bring the success of this undertaking to a certainty, but Mr Morrison will inform you of it, who I think is a sensible judicious person... I am afraid my father is [hole] sanguine about the Waggon Way & when you write to him if you find it a feasible plan do not encourage him too much you know as well as I what I mean & how to manage matters quietly concerning it. (40)

We hear no more about this scheme and presumably all thoughts about it were abandoned with the crash of the Ayr Bank in the following year which precipitated further bankruptcy proceedings against Roebuck.

Two years later, Watt was again involved with a scheme for a waggon way and this time he drew up estimates comparing one with a canal. He noted in his Journal on 23 July 1773, 'Canal to cost £4609 and waggon way £1400' (41) but then appears to have obtained further figures which were used to confirm his first. A couple of days later, he recorded that he 'wrote to Mr Colevile wanting to know the length of his waggon way with the expence of carriage & upholding'. (42) This must have been for his mine at Torryburn because Colville replied with the information, 'Waggonway... it costs £31 pr mile to maintain it and £3.10 for each waggon a waggon with wooden wheels costs £7.7.0'. (43) Further information was obtained from a William Wilson;

> I was at Wights wood waggon way yeasterday and I was told by Hart who made it that a

> waggon way can be laid at 2d pr yard when the wood is prepared, and at 5d pr yeard when the wood is to cut square & the road kept up at £70 pr Annum their is about a mile & a half of it. (44)

With these and other figures, Watt 'Drew comparison for Ld. C between wagn way & canal'. (45)

These notes and correspondence seem to refer to estimates for either a canal or a waggon way for a Mr. Ross Macky from Hurlet to Paisley. Hurlet lies a couple of miles south-east of Paisley, on the other side of a hill. The estimate for a canal 5,060 yds. long, 12 ft. wide at the bottom was £4,629.4s.8d. A waggon way 5,680 yds. long including 3 stone bridges, 27 'tunnels' (presumably what today we would call culverts), 9 cwt. for 50 yds. of iron on the waggon way and 240 yds. 'bysets' for empty waggons, would cost only £1,373.8s. (46) Each waggon would carry four carts or two tons and be drawn by one horse with a man. Watt worked out costings for either two or three trips a day to examine whether the scheme would be practicable. With three trips, he estimated that the volume of coal moved would be 9,750 tons per annum. He also took into account that the railway could be maintained at 6 pence per waggon trip, based on figures from Carron. What he called the 'perishable part of the waggon way' would last for five years. He seems to have based his costs for the rails of fir, with a 10 ins. diameter tree making four rails. Presumably the iron was needed to plate sections of rail where heavy wear was expected. He anticipated that there would be four waggons in use which would take 100 'hutches' weekly or 3,250 waggon loads per annum. Interest on these figures and the allowance for repairs show that economic viability was as important then as it is today. This was the final involvement of Watt with waggon ways while he was in Scotland.

Canal versus Rail

Watt based his designs and estimates on information supplied from both the Newcastle and Cumberland coalfields as well as other people in Scotland who had experience of building and operating waggon ways. Sinclair, in his *Statistical Account*, noted that coals were taken from the Clackmannan pit to the harbour at Alloa in waggons of a ton and a half capacity by one horse. (47) The gauge of this line was 3 ft. 3 ins. (48) and its size seems to correspond with those proposed by Watt. It was constructed of wood and carried around 7,000 tons per annum. Watt planned to use entirely wood for both rails and waggons even though Ralph Allen had fitted iron wheels and axles to his waggons at Bath. Wood was carefully selected, for example, with cheaper fir and fewer sleepers on the 'by-ways' or return lines where empty waggons ran instead of stronger, more costly, oak for the main line. Watt thought that the 'perishable' part of the Hurlet - Paisley waggon way, probably the rails, would last only five years.

These estimates and costs explain why Watt and his contemporaries preferred canals but we cannot draw a true comparison because we have no details about maintenance costs for canals. The maximum a horse could haul on a wooden waggon way seems to have been two tons. On the Monkland Canal, Watt designed boats hauled by one horse to carry twenty tons and, on broader canals, one horse could draw up to fifty tons. Therefore the carrying capacity on canals was much greater than on these early waggon ways. Watt based his costings for the Monkland Canal on first 20,000 tons and later on 30,000 or even 35,000 tons of coal being carried annually. (49) It would be surprising if a wooden waggon way could have handled that tonnage because on those at both Hurlet and Alloa the figure was under 10,000 tons. But, after Watt had ceased to be a civil engineer, the advantage began to swing towards waggon ways as more and more iron was introduced. The Alloa waggon way was relaid with iron plates in 1785 (50) and haulage capacity increased. In 1814, one horse could take down 8 waggons loaded with 8 tons of coal. (51)

In the instances examined so far, the coal was carried directly from the pit to its destination without transhipment. On the Duke of Bridgewater's Canal, the coal was loaded in special containers for ease of transhipment but Watt seems to have shied away from following this example through the cost of the containers. In other instances where goods had to be transhipped en route, canals proved to be unprofitable. There was a contemporary system in Scotland at Stevenston where a canal should have linked collieries near the sea with Saltcoats harbour, a couple of miles away. The canal was dug between 1770 and 1772 but, owing to difficult terrain, terminated 600 yds. short of the harbour. The coals had to be transferred to the ship by cart and it was claimed that it would have been cheaper to have dispensed with the canal and used carts for the whole distance. (52)

The problems at Stevenston and the fact that the Monkland Canal itself remained little used in its original state with maintenance costs exceeding income, (53) must lead us to doubt the viability of Watt's schemes for level canals and connecting railways. In England, the Peak Forest Canal only became a flourishing concern after it had raised sufficient funds to replace the railway at Marple with locks. On the other hand, the Lancaster Canal, which never succeeded in replacing the railway over the Ribble at Preston with an aqueduct, did pay small dividends. It was only after locks had been built in the early 1790s connecting the Monkland to the Forth and Clyde Canal that it became one of the most profitable canals in the country with its coal trade. This suggests that, although the initial cost of cutting one of Watt's level canals may have looked attractive, the expenses and problems of transhipment would have made them uneconomical in spite of the ingenious devices he thought of employing on his railways. But we do not have enough details of his methods of transhipment to form a true judgement.

Notes: Road and Rail Surveys

1 B. & W. Col., MI, 1/15, 'Journal', 7 June 1770.
2 JWP, 3/43, J. Naismith to J. Watt, 7 July 1770.
3 Ibid, 21 July 1770.
4 Ibid, 8 Aug. 1770.
5 B. & W. Col., MI, 3/60, J. Smeaton to J. Watt, 8 Sept. 1770. See JWP, C7/7.25 for a drawing of the centring of Perth bridge.
6 B. & W. Col., MI, 1/15, 'Journal', 5 Oct. 1770.
7 Ibid, MI, 1/16, 'Journal', 2 Nov. 1770.
8 Ibid, 21 & 22 Nov. & 8 - 10 Dec. 1770.
9 Ibid, MI, 2/42, 'Expenses on Surveys'.
10 Muirhead, (1854), Vol II, p. 15, Letter 88, J. Watt to W. Small, 21 Dec. 1770; Boulton 348.22.
11 B. & W. Col., MI, 3/61, 'An Estimate of the Expense of Building a Bridge over the River Clyde near Hamilton'.
12 Ibid, MI, 3/62, W. Purdie to Baillie Naismith, 20 June 1771. See also JWP, 4/19.14, J. Naismith to J. Watt, 28 March 1771.
13 JWP, 4/19.15, W. Mather to J. Watt, 7 July 1771.
14 Ibid, 4/19.16, W. Mather to J. Watt, 9 Aug. 1771.

15 Ibid, 3/43, J. Watt, 'Regulations for the Undertaker of Hamilton Bridge', no date.
16 Smiles, (1878), p. 107.
17 JWP, 3/39.2, 'Report of James Watt concerning the road proposed from Dumbarton to the Yocker bridge', 13 Oct. 1770.
18 B. & W. Col., MI, 2/42, 'Expenses on Surveys', 12 - 14 Oct. 1770.
19 Ibid, MI, 1/15, 'Journal', 12 Oct. 1770.
20 JWP, 3/39.2, 'Report... on road... from Dumbarton to the Yocker bridge'.
21 SRA, C1.1.34, Glasgow Council Act Book, Jan. 1768 - March 1772, p. 563, 25 Feb. 1772 and p. 566, 4 March 1772.
22 B. & W. Col., MI, 1/17, 'Journal', 6 Aug. 1771.
23 Ibid, 24 Sept. and 15 Nov. 1771.
24 SRA, C1.1.35, Glasgow Council Minute Book, March 1772 - Sept. 1777, p. 174, 22 June 1773.
25 JWP, 4/50.30, R. Finlay to J. Watt, 20 July 1773.
26 B. & W. Col., MI, 1/20, 'Journal', 26 & 27 July 1773 and MI, 2/42, 'Expenses on Surveys'.
27 Ibid, MI, 1/20, 'Journal', 25 Aug. 1773.
28 Ibid, 6 Jan. 1774.
29 JWP, 4/82, 'Mensuration of Clyde & Argyle Streets', 25 Jan. 1774.
30 B. & W. Col., MI, 2/42, 'Expenses on Surveys', 9 March 1774.
31 JWP, 3/57, W. Weir & J. Fleming to J. Watt, 18 & 19 March 1774.
32 Ibid.
33 B. & W. Col., MI, 1/21, 'Journal', 16 April 1774.
34 Ibid, MI, 4/26 - 29, Drawings of Rutherglen bridge.
35 Ibid, MI, 1/21, 'Journal', 19, 27 & 28 April 1774.
36 JWP, 3/26, J. Watt jun., 'Memorial of James Watt', p. 8.
37 Ibid, 6/38.11, W. Alexander to J. Watt, 11 April 1771.
38 Ibid, 4/32.89, J. Roebuck to J. Watt, 3 Aug. 1771. See also B. & W. Col., MI, 1/17, 'Journal', 4 Aug. 1771.
39 JWP, 4/33.19, J. Morison to J. Watt, 21 Feb. 1774.
40 Ibid, 4/32.88, J. Roebuck jun. to J. Watt, 9 Aug. 1771.
41 B. & W. Col., MI, 1/20, 'Journal', 23 July 1773.
42 Ibid, 25 July 1773.
43 Ibid, 1 Aug. 1773.
44 JWP, 3/39.1, W. Wilson to J. Watt, 28 July 1773.
45 B. & W. Col., MI, 1/20, 'Journal', 7 Aug. 1773.
46 JWP, 3/39.3, 'Estimate of a Waggonway from Hurlitt to Pasley'.
47 Sinclair, (1978), p. 725. See also Graham, (1814), p. 401.
48 Lewis, (1970), p. 267.
49 SRA, TD 727/38, J. Watt, 'A Scheme for making a Navigable Canal from the City of Glasgow to the Monkland Coalieries'.
50 Duckham, (1970), p. 212.
51 Graham, (1814), p. 401.
52 Duckham, (1970), pp. 220 - 222.
53 Lindsay, (1968), p. 36.

HARBOURS AND WATER SUPPLY

Port Glasgow Harbour

After the Monkland Canal, Watt's most important civil engineering undertaking was the improvements to Port Glasgow harbour. Presumably the magistrates of Glasgow had been satisfied with his survey of the Clyde carried out during the autumn of 1769 (See p205) because we find him in December writing 'Proposal for Cleansing the harbour of Port Glasgow and Improving the dry Dock'. (1) As had been feared, the harbour was indeed silting up, and manually digging out the mud and carting it away or taking it away by boat had proved to be both expensive and ineffective. Watt proposed adding an extra pair of gates to the dry dock to retain the water at high tide. At low tide, a sluice gate could be opened to allow the penned up water to scour away the mud. (2) No action followed, probably because the magistrates were concentrating on deepening the river Clyde during the next couple of years.

Building a dry dock at Port Glasgow had been suggested in April 1758 and, in September, the magistrates resolved that a ship carpenter, John Webb, should inspect the dry docks at Liverpool and then travel into Wales to purchase suitable oak timber for their dock which was under construction. (3) When finished, it was described as 'in a great measure useless', because it could not be drained properly with over two feet of water left in the bottom. Accordingly, Smeaton designed a pump which was to be built by 'Mr McKell' who was also to carry out any other necessary repairs. (4) Smeaton was paid eleven guineas. (5) Over the following years, the harbour deteriorated and the dock timbers rotted so that the magistrates resolved on 7 May 1771 to 'visit and inspect the harbour... [and] to report on the state and condition thereof & what ought to be done'. (6) Accordingly a visit was arranged for 27 May 1771 when Watt 'at 12 o'clock sett out for Pt. Glasgow with the magistrates - viewed the harbour & dry dock the cross beams of the flooring of the latter are said to be quite gone but the planks are good'. (7)

Watt remained at Port Glasgow a further day, looking for a site for a quay as well as sources for water which could be used to scour out the harbour. (8) He wrote his report at the beginning of July before setting off on his Crinan and Tarbert surveys. (9) On his return, he met the magistrates 'at night' and returned to Port Glasgow again for a further inspection and 'bargain with some person about quarrying the stones for the mid quay and dock... the quarrys are bad'. (10) He was back in Glasgow in time for a meeting of the Council on 30 August at which it was resolved that the 'Harbour at Port Glasgow should be cleaned, improved and inlarged agreeable to a report of James Watt engineer laid before the meeting'. (11)

Watt opened his report by saying, 'The removing & keeping away of the mud from the harbour of Port Glasgow is not only much desired but absolutely necessary otherwise the harbour may be lost or at least rendered only fitt for small craft'. (12) He thought that it would be too expensive to remove the mud through dredging it with men and horses or boats, costing £824.6s.9d. He suggested flushing it away either by tidal water collected in a reservoir and discharged at low tide, or, if the dry dock were used, extra sluices would cost £161.4s.4d. and pay for men to divert this water about £300. Another scheme was to use an abandoned mill with a couple of dams above the town. Its water could be diverted through a tunnel at a cost of £505.5s.1d, exclusive of purchase of the site and the necessary labour. To extend shipping facilities, a new quay could be built in the middle of the harbour costing £ 852.10s. He wrote a separate report on the state of the dry dock where, although the stone work was in good condition, the wooden bottom had rotted so it would have to be replaced with new piles and flooring. (13)

Having outlined the work he thought necessary, he sent a further letter to the Lord Provost stating the conditions under which he would be willing to act. This shows how he was determined to become a consulting engineer, drawing up the specifications and then checking on the work when it was finished but not becoming responsible for managing the labour force.

> I would undertake to direct the repairs in the following manner I would make agreements with different contractors for the several branches of the repairs, & would apoint some proper person to oversee the execution masonry or any other part the faults of which could not be seen when finished. The other works I would review myself when finished & see that they were properly executed in terms of their respective agreements. The town of Glasgow are to be bound to advance the money to carry on the works as required & I to deliver sufficient vouchers for these disbursements for which service I trust the good Town will allow me a reasonable gratification. I wish I could have an answer soon because if this work is to be done I would go down to Port Glasgow next spring tides in order to examine more narrowly the bottom of the Dock & to prepare matters for carrying on these repairs. I would at the same time have an opportunity of trying more accurately the depth of mud in different parts of the harbour'. (14)

This letter reveals both Watt's later methods of working and also his thoroughness of approach. His conditions were accepted and work started during that autumn. Letters from William Laird and Andrew Drysdaill have survived showing progress as well as claims for money which Watt passed for payment. (15) Drysdaill was the carpenter since he obtained and laid the timber for the flooring. The triumphal reopening was held on 29 November in the presence of the magistrates as well as Watt. (16) The cost had been £680.7s.10d. (17)

Construction of the mid-quay was delayed, probably through lack of money, and a petition for an Act of Parliament was submitted in January 1772 for raising the harbour rates and dues. (18) Once this was passed, the magistrates could begin in May to search for contractors to carry out Watt's designs. (19) Watt had drawn up specifications that April in readiness (20) and a thirteen page contract was prepared between the city and two masons in Port Glasgow, William Anderson and Robert Adam, for the stonework at £1,146. (21) It contains meticulous details about the size of this quay, how it should be built, the foundations, sizes of stones which must be without flaws, and so on. It was to be finished by 11 November 1773. The magistrates would nominate a 'skilled person for superintending and inspecting the building and erecting of the pier'. Watt was to inspect the work and any questions or disputes would be referred 'to the Descreet Arbitral and final sentence and determination of James Watt Engineer at Glasgow sole arbiter Mutually Elected and chosen by the aforesaid partys and failing of him by decease or non acceptance to John Brown Mason Baillie of Dumbarton'. (22) Not only did this confirm Watt in his role as consulting engineer but it shows the esteem in which he was held.

A second contract was drawn up with James Johnston who had been one of Watt's contractors on the Monkland Canal. (See p217) As the walls of the quay were finished, he was to fill up the interior with mud and stones taken out of the harbour. He was paid weekly by the yard of solid fill. His work was also to be 'declared to be well and properly finished and completed by James Watt' before any payments were made. (23) All of this took a considerable amount of Watt's time. In May 1772, he was given a part payment of £20 having spent over thirty days or part days for the City on Port Glasgow alone. Then in May 1774, he received a further £21 for another twenty-four days or part days (24) but this may not represent his total expense. (25) On 10 December 1773, Watt went with the magistrates to Port Glasgow 'to examine the new quay' when presumably it had been finished. (26)

The Broomielaw Crane

Watt senior had designed and built a crane for unloading ships at Greenock which could be the reason why Watt was asked to construct another on the Broomielaw. As no drawings have survived of either, it is impossible to determine if they were similar but that at the Broomielaw was for 'weighing boats'. (27) The crane must have been designed to lower cast iron weights into boats and hence determine their displacement. The first knowledge we have of this crane is when Watt wrote to the Carron Company on 30 May 1770 about weights which he wanted casting with a wrought iron handle at the top. Carron asked if they should be straight sided or barrel shaped, the latter being more expensive. (28) Watt needed twenty-five tons, in pieces weighing about 2 cwt. 2 lbs. each, 33 ins. long, about 5 $^{1}/_{2}$ ins. square. (29) Carron finally agreed to supply these as ballast iron of the toughest quality cast with a hole for lifting cast at £7 per ton. (30) He was also sent dimensions of the crane wheels used on 'the canal', presumably the Forth and Clyde. This enabled Watt to finalise his drawings and estimates which he took to the magistrates on 13 August 1770. (31)

Watt was supervising the construction during September after Carron had sent wheels and pinions. (32) However some of these were found to be the wrong pattern so others were substituted. (33) Then Watt complained about some pieces of the ballast weights being too light, which again Carron agreed to replace. (34) The cost of the ballast was £175.9s.9d which was to be paid through Hamilton who was their Glasgow agent for a while. (35) The crane was finished before January 1771 when Watt went to look at it and 'proposed covering it with a roof'. (36) He charged the City of Glasgow for his time spread over three days, writing letters and directing the work. (37) As we hear no more about it, it presumably worked successfully.

Ayr Harbour

In March 1771, Robert Montgomerie, whom Watt had advised about an atmospheric steam engine at his coal mine, (See p359) told him that he (Watt) and Smeaton would be asked to give estimates for improving Ayr harbour. (38, see p282) Golborne also submitted a report so that this was a significant request for Watt because here he was being asked to compete with two of the leading civil engineers of his day. From a position of obscurity, he had risen to equality with them. On 13 May, a Dr. John Campbell called on Watt in Glasgow inviting him to Ayr because the town had resolved to consult engineers about deepening their river. (39) Watt spent a couple of days there in the middle of June and took notes. (40) He worked on his report and drawings during parts of October and November, not finishing until February 1772, having spent nine days on them. (41)

Smeaton's printed report fills four pages, (42) while Watt's consists of twenty-four hand-written sheets. (43) Smeaton and Watt agreed about the problem and its solution. Gravel brought down by the river formed a bar across the mouth of the harbour. Also the existing training walls on either side were too low and allowed winds to blow sand over them, filling up the harbour. River floods failed to clear out this sand and gravel because the walls were so low that even at mid-tide water overflowed them and spent its force uselessly. Both agreed that these retaining walls needed to be raised. Watt pointed out that the best position and direction for these walls was difficult to determine on a single visit but thought that, once it had been deepened and the bar removed by dredging, the river would by itself maintain five or six feet depth over the bar and more in the harbour. In the harbour, there was a lack of berths for loading ships so Watt suggested extending the quays and building a wet dock.

For the harbour works, Watt's estimate came to £7,246 and the wet dock £5,430.19s.2d, with a total including drainage for the wet dock of £13,210.19s.2d. (44) Golborne's estimate for extending and building piers was £12,720. While Golborne was inspecting the

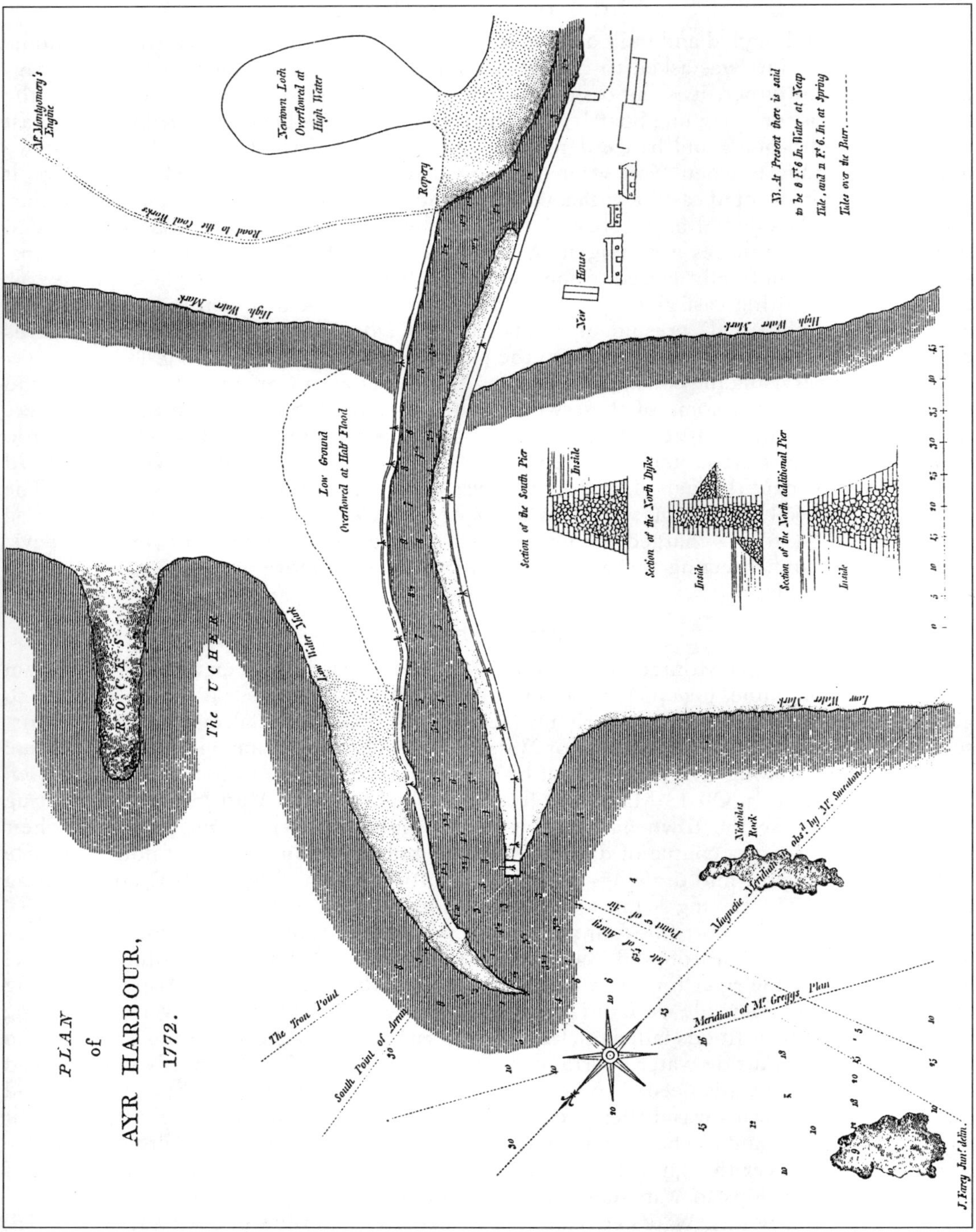

Smeaton's plan of Ayr Harbour.

harbour, he saw a ship wrecked at the mouth, dramatically emphasizing the need for improvements! (45) Smeaton recommended raising the existing dykes or retaining walls and lengthening the northern one at a cost of £2,776.0s.4d. (46) Golborne and Watt were invited to go to Ayr at the beginning of June 1772 to a meeting of the subscribers. (47) We should not be surprised that Watt, the most expensive of the three, was turned down and sounded very disappointed that all his careful preparations had been to no avail; 'The committee gave me 40 guineas for my trouble in their business & we returned to Glasgow'. (48)

The principles on which Watt based his estimates and plans for this type of work are shown in a letter to some unknown person, possibly Montgomerie, who was considering building a quay near Ayr. Watt wrote,

> I think we agreed on the situation of your wharf the principal maxims to be observed in building Quays are in the first place securing the foundation if naturally bad - secondly building the wall with Large & Long stones not only in its front or facing but throughout its whole thickness. it should be carried up perfectly perpendicular in the Backside & the front should batter back an eight of its whole height that it should be that much thinner at top than at Bottom behind this wall it should be filled up first with Quarry shivers or Gravell the grosser the better & behind that the natural sand the use of the Gravell being to prevent the sand from washing through. the height of this breast should be at least that of the Deck of such vessels as you intend to load & that at high water you will proportion the length of the Wharf to the number of vessels you intend to load at once for the quay at Ayr being so near you have no need to make places for them to lye at except during the time of loading The front of the Wharf should be secured from the vessel side by what is called sliders that is trees of about six or 8 inches square with one end fixed in the ground close to the foot of the wall, & the other end bolted or spiked to a tree that is laid along the fore edge of the Quay & bound to the stones of it by cramps of Iron (I have seen loose ill shaped stones secured in this manner stand a very heavy sea) the distance from one of these sliders to the other may be from 4 to 6 feet or if your stones are large & well shaped the sliders may be so far as 8 feet asunder. (49)

This was the type of harbour wall Watt planned for most of his schemes.

As far as Watt was concerned, that was not quite the end of the Ayr harbour affair. In spite of his bankruptcies, Roebuck continued to help Watt advance his surveying career with advice and through what influence he still had. In March 1772, he tried to give Watt some advice.

> I have reflected much on your undertaking the improvement of the Harbour of Air. Accept of it on such terms annually as they are willing to give provided it is only one hundred a year or more But then do not engage to employ yourself any number of days in the year more than you yourself shall judge absolutely necessary to give the principle directions in which case you may like Brindlay have four or five undertakings at the same time. (50)

In fact, Watt must have already realised this when he accepted the post of resident engineer on the Monkland Canal. Because this was not full-time, he had been able to undertake other commissions such as this one for Ayr harbour and so develop further his engineering skills and practice into the many different spheres which we have seen already. Then in July 1772, Roebuck asked Watt to send copies of his reports on the 'Cambleton Canal' and 'the River and Harbour of Ayr' to a Mr Russel in Edinburgh as soon as possible. (51) The reason for this remains a mystery.

Greenock Harbour and Water Supply

It was not until November 1772 that Watt began to undertake any civil engineering work for his home town of Greenock. First he was asked about ways of improving the harbour and, after a preliminary survey on the eighteenth, returned to finish it a month later. (52) His father supplied further details after Watt had returned to Glasgow. Then Watt was approached about laying out a water supply for the town. (53) But he finished his plans and report for the harbour first by working on them into February. A new pier was needed and he consulted Belidor on its construction. (54) Some dredging work was carried out in the spring of 1773. (55) Further estimates for a new breast were submitted in December 1773. A breach occurred in the West pier in the following February and Watt advised about the repair. (56) He charged £16 for fourteen days work and recorded that he was paid in full on 8 April 1774. (57)

In the meantime, he had been busy with his scheme for the water supply. A preliminary survey was made in December 1772 and plans drawn up followed by discussions with landowners. (58) One was Sir Michael Stewart on whose land the reservoir was to be built. (59) Another was Cathcart about a cistern. (60) A bill was submitted to Parliament and Watt carried out further levelling for the pipes in June 1773. Watt met the town's magistrates at the end of July when supplies of pipes were ordered. (61) Laying out the sites of the dams followed that August. (62) Watt did not design the cistern which had a circular vault until the end of March 1774. He spent twenty-two days on this project and charged £25 which he also received on 8 April 1774. (63) This was his only venture into supplying water for a town during his civil engineering career but there would have been many similarities with supplying water for canals and the skills and knowledge could have been transferred from one to the other.

Waterwheels

Waterwheels were the chief source of industrial power during Watt's career as civil engineer. We have seen earlier how he was involved with waterpowered mills at the Delftfield pottery, (See p168) the bleach fields of Sandeman at Luncarty and McGrigor near Glasgow, (See p134) as well as cornmills at Glen Saddell and Campbeltown. (See p253) Following his customary practice, he read the literature on the subject and was aware of Smeaton's experimental work on improvements to wheel design. (64) During his visits to Carron, Watt could have inspected Smeaton's improvements to the wheels carried out after 1769 (65) and was sent details of one in June 1771. (66) Smeaton installed a cast iron axle there but existing records are silent about Watt following his example. In order to give advice about the suitability of any site, Watt would have needed to know the theories and so relate the available head and volume of water to the power required.

Dalnotter Iron Company was founded in 1769 on the site of a corn mill at Duntocher (67) where there was an 18 ft. diameter wheel. Early in 1771, Watt was consulted to see if there were sufficient water to drive a tilt hammer and two more grinding stones but noted,

> She has not sufficient force to roll a piece of iron at the proper heat & it is proposed to increase the depth... As the company propose erecting a tilt hammer here I think it would be quite wrong to do it upon the same fall & propose they should purchase the corn mill at the road side & remove it further up where they could have a fall for it independent of what it now has[.] by erecting a good dam it would be better supplied with water. (68)

He based this advice on the calculations he made on the output from an overshot wheel, 'the water in fall having one fourth part of the effect'. (69) This was lower than the potential suggested by Smeaton. From Watt's comment, it sounds as if he were already familiar with rolling wrought iron, and his calculating the power needed for a tilt hammer may have been a help later at Birmingham when he erected a steam engine to drive one. This Company acquired two more mills on their river in 1773 in order to enlarge their works, suggesting that Watt's assessment of lack of power at the first was correct. (70)

Clunie Coal Mine

As he was returning from the Strathmore survey, Watt diverted through Clunie at the request of William Alexander in order to advise about draining a coal mine there with a waterwheel operating the pumps. Alexander had just purchased the mine and wanted to deepen it, thus needing more powerful pumps. The execution of this scheme hit many problems and it must raise doubts about Watt's method of working as a civil engineer where he often did not closely supervise the work himself. On the Monkland Canal and at Port Glasgow, supervisors were appointed who kept an eye on the contractors, but in this case, after Watt had written out his report and prepared his drawings, the work was carried out by Alexander's own millwright with no supervision by Watt.

A water pumping engine had been erected at Clunie Bridge by Stephen Row in 1738. (71) While the exact site of this mine is unknown, the valley here is quite broad and flat so that the main river could have had little fall. Alexander had written to Watt in the middle of April about his need for a water engine and 'as in matters we do not understand, we chuse to be directed by the best advice we can obtain, I use the freedom of applying to you & must beg if you agree to go, that it be immediately, as the work must be set about instantly'. (72) It is surprising to find Watt being held in such high regard so early in his new career. Alexander wanted a machine that could cope with the water in the mine at all times, even in case of a drought when the river might lack water. In this instance, Watt might allow for it to be driven by horses or some other way. 'When you are master of ye subject we must request a plan for the whole work, so as it may be executed by an inferior hand with perhaps another visit from you on any unforseen emergency'. (73) Watt inspected the site on May 17 with Alexander's factor, John Berry from Leslie, and sent directions for the dam and leat before the end of the month. (74) Although construction of the dam had started, Watt suggested capping it by laying stones over the top in a form of an arch. He also suggested that they should begin to construct the leat at the dam end, ensuring that on corners it was double the width. The level could be checked if each section were filled with water as it was finished. (75) Drawings for the engine were prepared in July and August, and finished at the beginning of September. (76) The lift was in two sections. The iron barrels for the pumps and pipes were ordered by Watt from the Carron Company soon after. (77)

It is at this point that Watt received a warning about Alexander's millwright, Methven.

> Methven has the reputation of a good tradesman but I think him a little conceited - I have given him the most peremptory orders to execute blindly what you direct - It will however occur to you that a fellow of this kind is in more danger of going wrong than a man that is more diffident. I therefore wish much you cou'd afford a day when the work is somewhat further advanced to examine the way he goes on, & measure his dimensions & examine the quality & materials as he has had everything of his own chusing. (78)

A problem arose almost immediately over the position of the two cranks which Methven would have put opposite or at 180 degrees to each other. (79) Watt recommended Smeaton's method of having the cranks at right angles. (80) Towards the end of October, Methven had erected the wheel 'exactly according to your directions' but found that the 'water strikes a foot lower than the plan'. (81) It is not clear whose levels were wrong. However construction continued so that, by early April, the engine had been pumping for a month but was not really satisfactory. By this time, Methven had been replaced and Alexander wanted to know whether mistakes had been committed. First of all there were not enough pipes to reach to the bottom of the pit. Then the new millwright 'informs me to my great surprise that the powers of the Engine are by no means equal to what Methven gave out, or what was expected - That in a days work she did not clear above one foot in the pit'. (82) Alexander wanted to meet Watt as he did not understand these machines so that Watt started out for Clunie on 28 April 1771. He arrived there the next day and commented, 'Methven had entirely neglected my directions'. (83) Watt proposed various improvements including altering the shape of the buckets of the wheel. (84) He made further calculations by comparing the radius of the wheel to a lever in order to find the weight of the water acting through the crank against the water in the pumps and concluded that the wheel ought to have sufficient power. (85) As we hear no more about it, presumably it performed satisfactorily after it had been altered.

Culross Tide Mill

Culross was a very ancient coal mining area, stretching back into the fifteenth century. By 1592, George Bruce had pioneered many innovations in mining techniques and such was the prosperity generated by the mines there that James VI accorded Culross the status of a Royal Burgh. (86) As the more accessible seams became exhausted, so miners had to dig deeper with the consequent problems of water removal. Lord Cochrane sought Watt's advice about building a tide mill on the foreshore which is flat but muddy, stretching a long way out into the Firth of Forth. Watt crossed over from the Falkirk side on 31 August 1771 and found that, at spring tides, there was a rise and fall of 9 ft. but at neap tides only 5 ft. 6 ins. (87)

Watt sent off Cochrane's report on 21 September (88) and pointed out that, in his opinion, it would not be a success.

> I am afraid it is short of what you had supposed but it is the result of serious considerations founded upon experiments that have been accurately made by Mr Smeaton, which made the maximum effect of an undershot wheel above double of what was assigned by theory - It gives me pain to differ in opinion from your Lordship, but it would give me much more if by flattering your hopes I should mislead you in a matter of so much moment - I have not considered the expense... I may however observe that so large a water engine will be very expensive & that the stonework of such large sluices ought to be executed in the most substantial manner with the best jointed ashlar laid in water cement. (89)

The plan was to build a reservoir in the tidal area. Watt estimated that the area would need to be twelve acres with a wheel eighteen feet diameter and ten feet broad. He pointed out that the force of the water would diminish as the reservoir emptied and that the wheel would thus lose power, even with the water acting in the most advantageous manner. The pumps would be a constant load, lifting the same volume of water against a given head all the time. But, even if parts of the machinery had no friction, there would still be insufficient power. Cochrane also wanted the embankment containing the reservoir to be built so it would become a quay against which shipping could moor but Watt pointed out,

> I do not pretend to much knowledge in maritime affairs but must beg leave to give it as my opinion that when the wind blows fresh from the other shore (particularly from the mouth of Carron) that no vessel will dare to lye at it. I say this from some observations I made upon the force of the sea during a six weeks stay I made at Torryburn one winter, in the whole of which time there were very few days in which a vessel could have lain in safety on the outside of the pier. (90)

Cochrane did not proceed with a tide mill and could not have been annoyed at Watt's rejection of this scheme because later he ordered Watt to build a Newcomen steam engine for him.

Notes: Harbours and Water Supply

1 B. & W. Col., MI, 2/54, Note Book, Dec. 1769.
2 See JWP, 4/50.1, A. McKeakany to J. Watt, 28 Dec. 1769 with dimensions of Port Glasgow harbour and dock.
3 SRA, C1.1.32, Glasgow Council Act Book, Sept. 1755 - Jan. 1763, p. 254, 27 April 1758 and p. 277, 29 Sept. 1758. See also JWP, C7/7.11 for Glasgow port regulations and C7/7.12 for dry dock regulations.
4 SRA, C1.1.33, Glasgow Council Minute Book, Jan. 1763 - Jan. 1768, p. 495, 2 July 1767. See also Smeaton, (1812), Vol I, p. 323, 'DESCRIPTION of a Pump for the dock at port *Glasgow*' and plate 20 for a drawing. The mean depth to be pumped out was 2 ft. 5 ins.
5 SRA. C1.1.34, Glasgow Council Act Book, Jan. 1768 - March 1772, p. 316, 12 Sept. 1770.
6 Ibid, p. 400, 7 May 1771.
7 B. & W. Col., MI, 1/17, 'Journal', 27 May 1771.
8 Ibid, 28 May 1771.
9 Ibid, 7 & 8 July 1771.
10 Ibid, 22 - 24 & 29 Aug. 1771.
11 SRA, C1.1.34, Glasgow Council Act Book, Jan. 1768 - March 1772, p. 446, 30 Aug. 1771.
12 B. & W. Col., MI, 3/57a, J. Watt, 'Report concerning the Harbour of Port Glasgow', 9 Aug. 1771 and MI, 2/54, Note Book.
13 Ibid, MI, 3/58b, 'Report concerning the Dry Dock and Harbour of Port Glasgow'.
14 JWP, 4/50.7, J. Watt to Ld. Provost, 17 Aug. 1771.
15 Ibid, 4/50.8 - 20, W. Laird and A. Drysdaill to J. Watt, 12 Sept. - 27 Nov. 1771.
16 Ibid, 4/50.21, Ld. Provost to J. Watt, 28 Nov. 1771.
17 B. & W. Col., MI, 3/58a, 'Expenses of Repairs of Pt. Glasgow Dock', 1771.
18 SRA, C1.1.34, Glasgow Council Act Book, Jan. 1768 - March 1772, p. 526, 15 Jan. 1772.
19 SRA, C1.1.35, Glasgow Council Minute Book, March 1772 - Sept. 1777, p. 20, 12 May and p. 28, 28 May 1772.

20 JWP, 4/50.26, Dean of Guild to J. Watt, 25 April 1772.
21 SRA, B10/15/7640, 'Contract betwixt the City of Glasgow and William Anderson and Robert Adam', reg. 3 Aug. 1773.
22 Ibid.
23 SRA, B10/15/7590, 'Contract betwixt the City of Glasgow and James Johnston', reg. 5 April 1773.
24 B. & W. Col., MI, 2/42, 'Expenses on Surveys'.
25 See JWP, 6/28.54, The Town of Glasgow Dr. to James Watt, 1770 - 1772.
26 B. & W. Col., MI, 1/20, 'Journal', 10 Dec. 1773.
27 Ibid, MI, 1/15, 'Journal', 30 May 1770.
28 SRO, Carron Co., GD 58/6/1/10, Letter Book 1769 - 1770, p. 350, to J. Watt, 31 May 1770.
29 JWP, 4/51.2, Carron Co. to J. Watt, 23 July 1770.
30 SRO, Carron Co., GD 6/1/10, Letter Book 1769 - 1770, p. 523, to J. Watt, 23 July and p. 525, 25 July 1770.
31 B. & W. Col., MI, 1/15, 'Journal', 13 Aug. 1770.
32 JWP, 4/51.5, Carron Co. to J. Watt, 1 Sept. 1770.
33 Ibid, 4/51.7, Carron Co. to J. Watt, 18 Sept. 1770.
34 Ibid, 4/51.9, Carron Co. to J. Watt, 19 Sept. 1770 and SRO, GD 58/6/1/11, Letter Book 1769 - 1770, p. 147, to J. Watt, 19 Sept. 1770.
35 SRO, Carron Co., GD 58/6/1/11, Letter Book 1769 - 1770, p. 11, to J. Watt, no date.
36 B. & W. Col., MI, 1/16, 'Journal', 7 Jan. 1771.
37 JWP, 6/28.54, 'The Town of Glasgow Dr. to J. Watt'.
38 Ibid, 4/19/10, R. Montgomerie to J. Watt, 26 March 1771.
39 B. & W. Col., MI, 1/17, 'Journal', 13 May 1771.
40 Ibid, 16 - 17 June 1771.
41 Ibid, MI, 2/42, 'Expenses on Surveys'.
42 Smeaton, (1812), Vol III, p. 151, 'The REPORT of John Smeaton, upon the Harbour of Ayr'.
43 B. & W. Col., MI 3/54 and JWP, 3/39.5, J. Watt, 'Harbour of Ayr Report', Dec. 1771.
44 Ibid. Watt gave two slightly differing sets of figures.
45 B. & W. Col., MI, 3/55a, J. Golborne, 'Report on Ayr Harbour', 1 Jan.1772.
46 Ibid, MI, 3/55b, Smeaton's Report on Ayr Harbour.
47 JWP, 4/19.29, Dr. J. Campbell to J. Watt, 30 May 1772.
48 B. & W. Col., MI, 1/18, 'Journal', 5 June 1772.
49 JWP, 4/82, 'Dear Sir, 5 Jan. 1771'.
50 Ibid, 4/32.75, J. Roebuck to J. Watt, 18 March 1772.
51 Ibid, 4/32.64, J. Roebuck to J. Watt, 18 July 1772.
52 B. & W. Col., MI, 1/19, 'Journal', 18 Nov. & 19 Dec. 1772.
53 JWP, 4/40.8, T. Cochrane to J. Watt, 26 Dec. 1772.

54 B. & W. Col., MI, 1/19, 'Journal', 31 Dec. 1772 and 1, 18, 19 Jan., 1, 7 & 10 Feb. 1773.
55 Ibid, 2, 3, 12, 13, 15 & 17 April 1773.
56 Ibid, MI, 1/20, 'Journal', 9 - 12 Feb. 1774.
57 Ibid, MI, 2/42, 'Expenses on Surveys'.
58 JWP 4/40.6, J. Gammell to J. Watt, 16 Jan. 1773.
59 Ibid, 4/40.3, J. Gammell to J. Watt, 28 Jan. 1773.
60 Ibid, 4/40.7, J. Campbell to J. Watt, 8 Jan. 1773.
61 B. & W. Col., MI, 1/20, 'Journal', 29 July 1773.
62 Ibid, MI, 2/42, 'Expenses on Surveys'.
63 Ibid.
64 See Smeaton, (1794), the lectures were given on 3 & 10 May 1759.
65 Smeaton, (1812), Vol I, p. 367, 'The REPORT of John Smeaton, Engineer, concerning the quantity and disposition of the water at *Carron*', 7 July 1769.
66 B. & W. Col., MI, 1/17, 'Journal', 16 June 1771.
67 Shaw, (1984), p. 424.
68 B. & W. Col., MI, 1/16, 'Journal', 30 March 1771.
69 JWP, 4/82, Report on Kilpatrick Burn, Jan. & April 1771.
70 Shaw, (1984), p. 424.
71 Ibid, p. 381.
72 JWP, 6/38.1, W. Alexander to J. Watt, 14 April 1770.
73 Ibid.
74 B. & W. Col., MI, 1/15, 'Journal', 17 & 20 May 1770 and JWP, 6/38.2, W. Alexander to J. Watt, 24 May 1770.
75 B. & W. Col., MI, 2.54, Note Book, J. Watt to W. Alexander, 21 May 1770.
76 Ibid, MI, 1/15, 'Journal', 19 July, 8, 9 Aug. & 2 Sept. 1770
77 JWP, 6/38.7, J. Watt's Order to the Carron Co., 8 Sept. 1770.
78 Ibid, 6/38.6, W. Alexander to J. Watt, 5 Sept. 1770.
79 Ibid, 6/38.9, W. Alexander to J. Watt, 10 Sept. 1770.
80 B. & W. Col., MI, 1/15, 'Journal', 17 Sept. 1770.
81 JWP, 6/38.8, W. Alexander to J. Watt, 25 Oct. 1770.
82 Ibid, 6/38.10, W. Alexander to J. Watt, 8 April 1771.
83 B. & W. Col., MI, 1/17, 'Journal', 29 May 1770.
84 Ibid, 29 April & 11 May 1771.
85 B. & W. Col., MI, 2/65 & 66, Notes on the Cluny engine.
86 Sked, (1995), p. 3.
87 B. & W. Col., MI, 1/17, 'Journal', 31 Aug. 1771.
88 Ibid, 21 Sept. 1771.
89 Ibid, MI, 3/67, 'Lord Cochrane's Water Engine'.
90 Ibid.

WATT AS CIVIL ENGINEER

Watt's career as a civil engineer was remarkable for his quick rise to a prominent position in Scotland and he must have felt that he had a promising future. His practice soon covered a wide variety of different types of civil engineering, principally canals but also harbours, water power, waggon ways and much more in addition. His own surviving archives are sparse before 1770, so that, from the start of his new career in 1766 to 1770, we have records only of his reports for various waterway surveys and some steam engines. Then, from 1770 to 1774, his Journals show that, while he spent the greater part of his time supervising the construction of the Monkland Canal, other canal surveys were the most important commissions which he received.

It has been possible to supplement Watt's own reports on these navigations with information from other sources, such as the archives of Glasgow City Council and the Board of the Forfeited Estates, which sometimes show other people's assessment of Watt's work. These archives reveal that Watt was soon accepted as a competent civil engineer and was asked to submit reports on schemes alongside other prominent engineers like Smeaton and Golborne. Indeed, he seems to have replaced Smeaton as adviser to the Glasgow City Council about improvements at Port Glasgow. Both the Board of the Forfeited Estates and the Commissioners of Police commissioned major surveys by Watt alone.

But Watt quickly recognised that he needed practical experience in executing the recommendations he had proposed in his reports. This may be why he tried so hard to produce satisfactory reports on the Clyde surveys and suggested a partnership with Golborne to carry out the improvements on that river. This may also be why he accepted the task of constructing the Monkland Canal. Perhaps it was mere coincidence, but, once Watt had been appointed resident engineer on the Monkland Canal, he was inundated with further commissions. It looks as if that were the turning point in his civil engineer career.

There were other financial inducements for accepting the position on the Monkland Canal, which can be seen in Watt's letter to Small in September 1770.

> I had just got to Kinneil to finish my experiments on the engine, when, on a day's warning, I was called to Glasgow to attend a survey of the river [Clyde]. Before that was finished, the canal [Monkland] I had projected last winter, for which an Act had been obtained, was wanted to be begun under my inspection. I had now a choice whether to go on with the experiments on the engine, the event of which was uncertain, or to embrace an honourable and perhaps profitable employment, attended with less risk of want of success :- to carry into execution a canal projected by myself with much trouble, or to leave it to some other person that might not have entered into my views, and might have had an interest to expose my errors; (for everybody commits them in those cases.) Many people here had conceived a much higher idea of my abilities than they merit;- they had resolved to encourage a man that lived among them rather than a stranger. If I refused this offer I had little reason to expect such a concurrence of favourable circumstances soon. Besides, I have a wife and children, and saw my self growing gray without having any settled way of providing for them. There were also other circumstances that moved me not less powerfully to accept the offer; which I did. (1)

We have seen that Watt had begun to run down his scientific instrument business from possibly 1763 when he launched out into his enlarged shop. The shop itself was closed in either 1769 or 1770. Certainly it was Watt's intention to give it up even before he obtained the position of resident engineer on the Monkland Canal, although this may have precipitated its final closure. His desire 'to embrace an honourable and perhaps profitable employment, attended with less risk of want of success' is understandable. We know that he received £100 for going to London for a couple of months in order to support the 'Ditch' canal bill through Parliament. In 1767, Smeaton raised his daily fees when away from home to an unprecedented five guineas. The financial rewards must have seemed very attractive to Watt who was asking only one guinea. Later at the beginning of 1772, Roebuck suggested that he ought to charge two guineas. (2) No wonder Watt must have looked on this as a way of reducing his indebtedness and providing for his family.

We must also note that some people in Scotland 'had resolved to encourage a man that lived among them rather than a stranger'. Cathcart may have used Watt to check on Smeaton's and Morison's surveys and Glasgow Council may have followed this example to check on Golborne. Watt acquired a reputation for thoroughness and honesty even though some people thought him dilatory. So, in his native country, he may have been in a favourable position to gain work, especially after 1768 when Mackell became resident engineer on the Forth and Clyde Canal, effectively removing from the scene a potential competitor, although at one time instructor and partner. There seems to have been no other Scotsman involved in civil engineering in Scotland at this period.

To have the chance of constructing the Monkland Canal, which he had laid out and which indeed embraced his own particular conception of building and operating economically with no locks but a connecting railway, must have seemed to Watt a golden opportunity and the chance of a lifetime. 'If I refused this offer I had little reason to expect such a concurrence of favourable circumstances soon'. If he were successful in showing that this combination provided an economical solution to transport problems in a poor hilly country like Scotland, then further openings might present themselves. While we do not know the other circumstances that moved Watt to accept this offer, we can understand why this seemed to have the possibility of firmly launching him into a new career.

But the experience as resident engineer on the Monkland Canal changed his attitude towards engineering. He discovered that he could not handle men. He realised that he must become a consulting engineer, carrying out the preliminary surveys and designs. He put the specifications he had drawn up out to tender and then inspected the finished work. While doubts remain about his methods of supervision, the importance of this change was that this was the way he evolved the organisation of the construction of the last Newcomen steam engines he built in Scotland and this was basically the way he built his own type of steam engine once he had entered into the famous partnership with Boulton at Birmingham.

After a couple of hectic years, we can see how his earlier vision of optimism had vanished when Watt wrote to Small at the end of 1773, telling him he was

> ... heart-sick of this country...
>
> The engineering business is not a vigorous plant here; we are in general very poorly paid. This last year my gains do not exceed 200 £., though some people have paid me very genteelly. There are also many disagreeable circumstances I cannot write; in short I must, as far as I see, change my abode. There are two things which occur to me, either to try England, or endeavour to get some lucrative place abroad; but I doubt my interest for the latter. (3)

All he thought he was fit for was to be a 'surveying engineer' and he wondered whether Small knew of any openings in England. The economic crisis after 1772 and the subsequent slow recovery followed by the American War of Independence which devastated Glasgow's important tobacco trade meant that major capital civil engineering projects were few and far between in Scotland. In fact Watt does not mention being asked to undertake any more surveys in 1774. The engineering business would take a long time to become a vigorous plant. In the 1790s, it was necessary for the government to advance a loan to enable Rennie's construction of the Crinan Canal to be completed, while, after 1800, it was still essential for the government to provide the funding for the Caledonian Canal.

If Watt realised that Scotland would be unable to sustain the career of a civil engineer, we can understand why he turned his eyes to the more prosperous south. But would he have been accepted? Small thought that, had Watt been resident in Birmingham, he might have been consulted over difficulties on a canal, but also warned, 'we cannot persuade the people to send into Scotland for you, or for anything else'. (4) Had Watt remained in Scotland, it seems unlikely that he would have found much work south of the border.

But, after the death of Peggy and possibly through Cathcart's criticisms as well as the lack of prospects for his career in Scotland, Watt seems to have seriously entertained establishing himself as a civil engineer in England. Small encouraged him, 'Should you come hither soon, pray consider aqueducts in the meantime, and the conveyance of water in pipes, and the expenses of both... Also consider the price of lockage very minutely'. (5) In fact, when he knew that he would be going to Birmingham in the spring of 1774, Watt intended 'to bring copies of several of my reports and plans with me to show you and friends', (6) possibly suggesting that he might carry on in this field.

Yet, if he had started to practise in England, he would have lost his comparatively sheltered position in Glasgow and been exposed to the full competition of English engineers. While Brindley had died in September 1772, (7) there were still people around like Golborne, John Grundy and above all, Smeaton. While Watt had a good eye for the lie of the land, it is very difficult to judge if he would have reached the top of this profession. The works he completed in Scotland are too few to give an indication of how he might have developed. In 1774, the Monkland Canal languished, the Campbeltown Canal was not started and Hamilton Bridge was not finished. On the other hand, improvements to both Greenock and Port Glasgow harbours and the Greenock water supply seem to have been successful. To use a Scottish legal term, the case for Watt's civil engineering career remains unproven. However, all was not lost for at least there still remained his steam engine. On taking up his post on the Monkland Canal, he had resolved 'at the same time not to drop the engine, but to prosecute it the first time I could spare'. (8) That time had come at last.

NOTES: CIVIL ENGINEER

1 Muirhead, (1854), Vol II, pp. 2 - 3, Letter 83, J. Watt to W. Small, 9 Sept. 1770; Boulton 348.20.
2 Skempton, (1981), p. 26 and JWP, 4/32.81, J. Roebuck to J. Watt, 28 Feb. 1772.
3 Muirhead, (1854), Vol II, p. 69, Letter 127, J. Watt to W. Small, 11 Dec. 1773; Boulton 348.38.
4 Ibid, Vol II, p. 71, Letter 129, W. Small to J. Watt, Dec. 1773; Boulton 340.38.
5 Ibid.
6 Ibid, Vol II, p. 78, Letter 134, J. Watt to W. Small, 29 April 1774; Boulton 348.42.
7 Smiles, *Early Engineering*, (1874), p. 288.
8 Muirhead, (1854), Vol II, p. 3, Letter 83, J. Watt to W. Small, 9 Sept. 1770; Boulton 348.20.

5
Engineer of Steam Engines

The First Steam Engines and Watt's Introduction to Them

The First Steam Engines

When Watt first turned his attention to steam engines around 1758, two types had been invented. The earlier was that of Thomas Savery who took out a patent in 1698 for 'Raising water by the impellent force of fire'. (1) An illustration published in his treatise, *The Miners Friend*, (2, see p296) shows a boiler in which steam could be raised for supplying a pair of receivers which were operated alternately. When one receiver was full of water, its steam valve would be opened manually so that the steam, under pressure, forced the water out of that receiver, through a non-return valve and up a tall pipe to the required height. When this receiver was empty, the steam valve was closed and cold water poured over the outside to create a vacuum inside. Atmospheric pressure forced more water from a reservoir below the engine up a lower pipe and through another non-return valve to refill the receiver. While the first receiver was filling with water, steam was let into the other to commence the cycle in that. For pumping water out of mines, this engine had two disadvantages. One was because the boiler, and hence its fire, had to be situated down the mine and the other was the high steam pressure necessary to overcome the lift for which it was difficult to construct boilers at that time. Desaguliers improved Savery's engine by injecting the condensing water directly into the receiver, copying Thomas Newcomen's engine. (3) The Savery engine, with direct injection and having virtually no upper standpipe, was used to a certain extent in the latter part of the eighteenth century as a returning engine to recirculate water for waterwheels in competition with Watt's improved engines. (4) The Savery engine could be constructed cheaply and had few moving parts which needed machining and fitting, so its capital costs were low. While the early ones had to be operated manually, the last were made to work by the waterwheel.

The type of engine which Watt would improve was that invented by Newcomen. The first definite date for a successful Newcomen engine is 1712, given on the engraving by Thomas Barney of one draining a coal mine near Dudley Castle. (5, see p297) Pumps down the mine shaft were operated by rods to raise the water. These were connected to one end of a beam or lever pivoted in the middle and were heavier than the steam piston at the other end. The pivots supporting the beam rested on the main wall of the building which housed a steam boiler situated underneath the steam cylinder. This cylinder had an open top so that the pressure of the atmosphere could push against the steam piston. The cycle started with the piston at the top of the cylinder. A valve was opened to allow steam from the boiler to enter the cylinder, warm it and blow out the air through a snifting valve. The steam valve was closed and the injection valve opened to allow cold water to spray into the cylinder and condense the steam, cooling both cylinder and piston. On top of the piston was a layer of water to act as a seal and prevent air entering past the cylinder walls and the edge of the piston. As the steam and the cylinder itself were cooled, a partial vacuum was formed inside until the weight of the atmosphere could force down the steam piston, raising the pumps with the water out of the mine. Towards the end of the stroke, the injection valve was closed and the steam valve opened which destroyed the vacuum and stopped the piston descending. The incoming steam blew any residual air out through the snifting valve and the condensate down an eduction pipe away through another valve. As the steam reheated the cylinder destroying any residual partial vacuum, the heavier weight of the mine pumps allowed them to descend, raising the steam piston ready for the next stroke. Newcomen's great achievement was to create the first really practical source of mechanical power in the world that did

WORKING PARTS OF SAVERY'S ENGINE

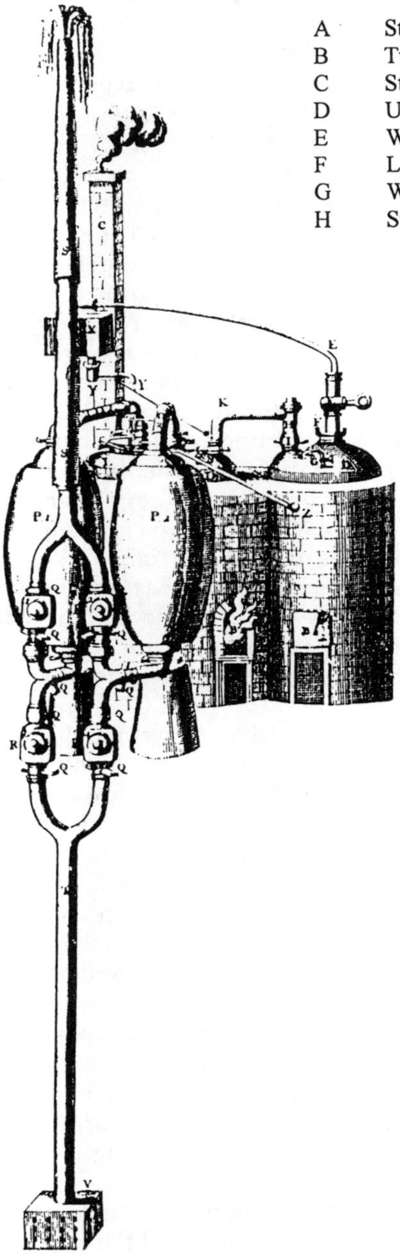

A Standpipe to take water out of the mine
B Twin Receivers
C Steam Pipes from Boiler (not shown)
D Upper Non-return Water Valves
E Water Pipes into bottom of Receivers
F Lower Non-return Water Valves
G Water Pipe from bottom of mine
H Strainer

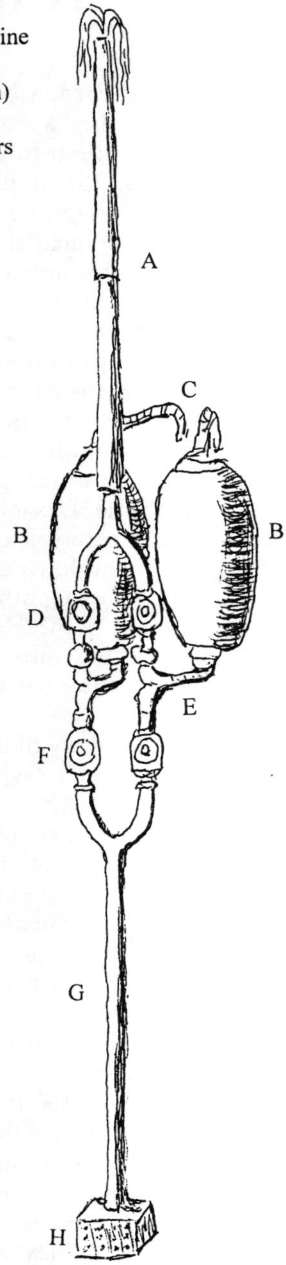

Left: Savery's illlustration of his engine, published in The Miners Friend.
Right: Working parts of Savery's engine.

not rely on variable forces such as muscles, wind or water. It was a true heat engine but, like the Savery engine, it consumed a vast amount of fuel.

In spite of its large fuel consumption, Newcomen's invention had been introduced quickly into Scotland for one was erected by Peter Walker and John Potter for Stevenston Colliery, Saltcoats, Ayrshire, in 1719. The York Buildings Co. was responsible for another around that same year for the Tranent Colliery, East Lothian. By the time that Savery's patent, which covered Newcomen's engine, expired in 1733, eight Newcomen engines have been traced in Scotland. (6) The coal industry was no exception to the general transformation of industrial activity in Scotland after the middle of the eighteenth century. (7) The development of better harbours such as Ayr together with canals such as the Forth and Clyde and the Monkland as well as the introduction of wooden waggon ways, all helped to lower the cost of transport, stimulating the demand for coal. (8) Earlier methods of pumping by men, animals, water or wind power increasingly proved inadequate. Examples were Watt's attempt to improve the Clunie engine and the tide mill for Cochrane at Culross. (9) The first Newcomen engine which we definitely know to have been erected in the neighbourhood of Glasgow was one at Shettleston during 1763 - 64. A very conservative estimate for the year 1769 gives fourteen or fifteen engines installed north of the border. (10)

Barney's engraving of Thomas Newcomen's first engine built near Dudley Castle, Staffordshire, in 1712.

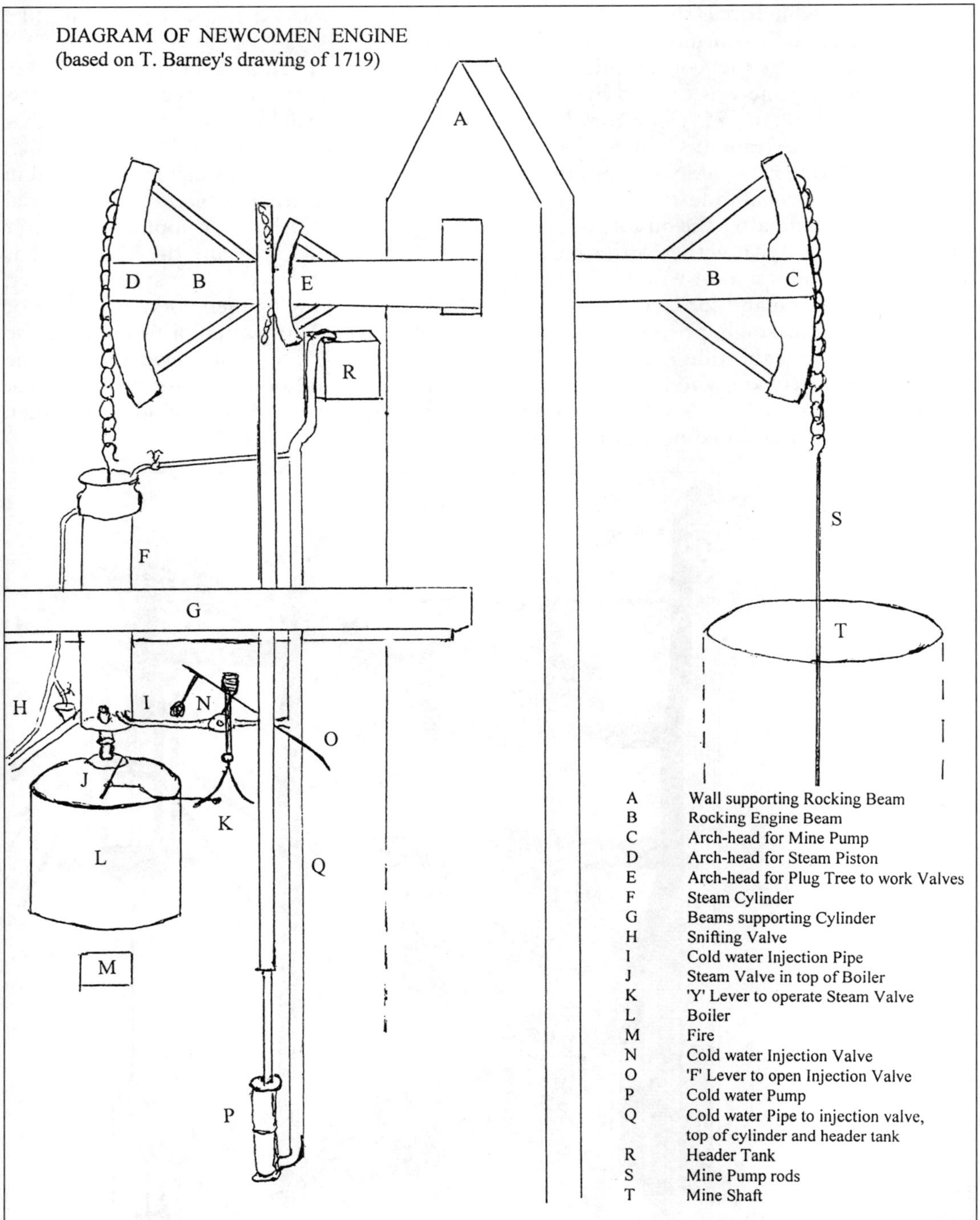

Diagram of Thomas Newcomen's engine.

Watt's First Contacts with Steam Engines

We do not know when Watt may have seen his first Newcomen engine pumping water from a mine. Possibly it was when he travelled south with Marr while he was going to London in 1755. If so, it presented little attraction, because, fifty years later, he wrote,

> My attention was first directed, in the year 1759, to the subject of steam engines, by the late Dr. Robison, then a student in the University of Glasgow, and nearly of my own age. He at that time threw out an idea of applying the power of the steam-engine to the moving of wheel-carriages, and to other purposes; but the scheme was not matured, and was soon abandoned on his going abroad. (11)

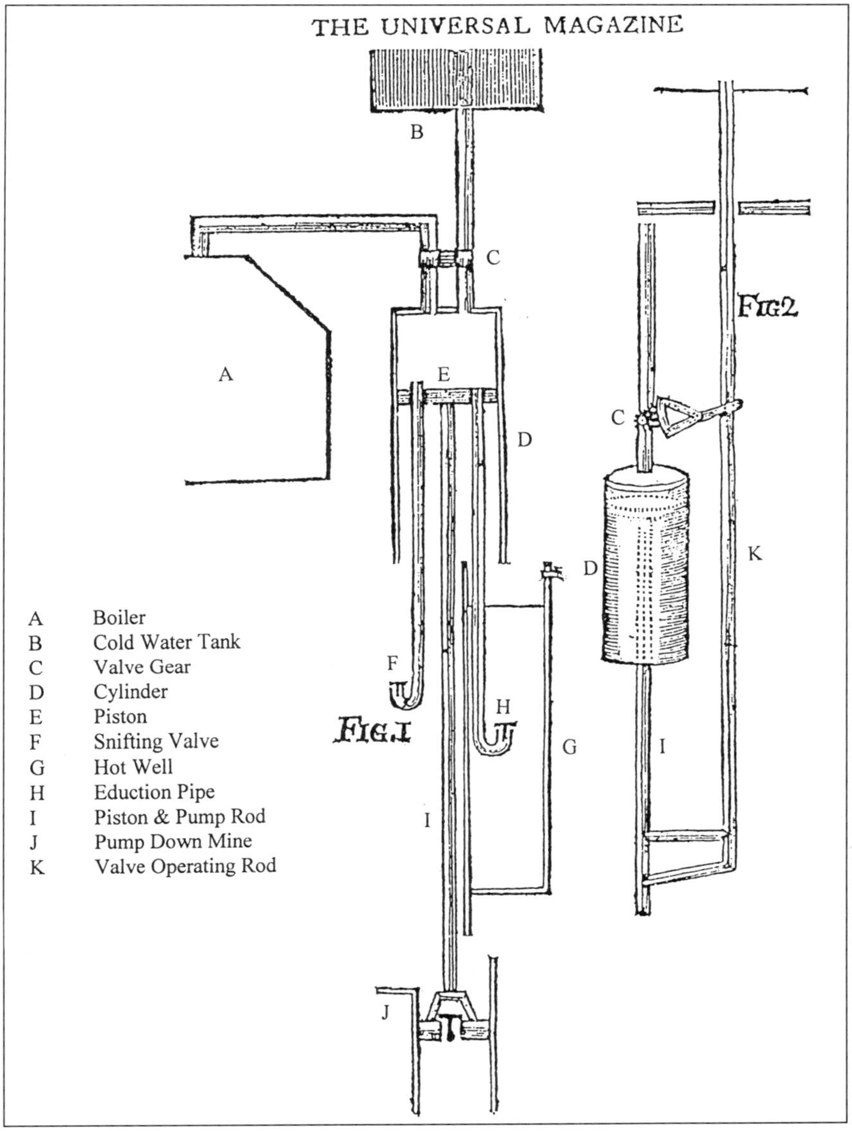

John Robison's inverted Newcomen engine based on his illustration in The Universal Magazine, 1757.

Some of Watt's recollections do not tally with an earlier undated account which he wrote possibly in 1765 which is also inaccurate.

> About 6 or 8 years ago My Ingenious friend Mr. John Robison having ['contrived' deleted] conceived that a fire engine might be made without a Lever - by Inverting the Cylinder & placing it above the mouth of the pit proposed me to make a model of it which was set about but never Compleated he going abroad & I ['being' deleted] having at that time ['ignorant' deleted] little knowledge of the machine. (12)

Robison had submitted a design for a new type of steam engine to *The Universal Magazine*, based on Newcomen's principles but with an inverted cylinder to be placed over the pit shaft in order to work the pumps directly without any lever beam thus reducing the expense. He said that 'it has succeeded extremely well in a model which I have by me'. His drawing and descriptive text are dated 4 August 1757 and were published in the issue for November 1757. (13) Watt did not establish his workshop in Glasgow until that same August as Robison wrote to *The Universal Magazine*. There is nothing in Watt's existing business records to show that he built anything for or sold anything to Robison as early as this and also it is most unlikely that he would have had enough experience to build Robison's model. Therefore Watt's comments that he attempted to make Robison's model be wrong. Who made it remains a mystery, perhaps Mackell who made some models for the Duke of Argyll or perhaps Cumming who, in 1796, said he also made models for the same Duke. (14)

We will briefly examine Robison's engine because it shows great ingenuity and most likely influenced Watt's early models and engines in ways which have not been recognised heretofore. In fact Robison probably gave Watt practical advice about experiments on Newcomen engine models. Robison wrote,

> Those who are acquainted with Newcomen's fire-engine, will easily see that one great increase of the charge and expence, in erecting them, arises from the height and solidity they are obliged to give the building, on account of the cylinder's being placed above the boiler, and the injecting cistern a considerable height above the cylinder; and that the proper strength of the building is greatly increased (especially on the side next the pit) by its being loaded with more than double the weight of water to be raised, viz. by the water itself; by a preponderating column of air upon the piston in the steam cylinder; by the resistance arising from the friction in the pumps; and by the weight of the great beam with the plug frame; all which press upon the centre of motion of the great beam, and consequently upon the supporting wall. I think less of the additional expence arising from the difficulty of getting a piece of wood proper for the great beam. (15)

While Robison's layout looks as if it should have worked, it might not have had the finesse of Newcomen's but it must have shown that other designs would function. He inverted his cylinder and placed it with the open end downwards. Watt copied this on some of his early models and probably for his first attempt at Kinneil in 1766. Robison had to alter Newcomen's design of valve gear and move the boiler to one side. Watt's boilers were never placed under the cylinders even on the more conventional Newcomen engines he erected in Scotland nor on his own later engines. In this way, smaller engine houses were needed. Robison avoided the great cost of the main working beam. Watt experimented on his Newcomen engines in Scotland with a wheel instead of a beam, again a cheaper construction.

We have no record of Robison's design ever being erected at a mine but its success as a model may have encouraged him to see if it could be tried in some other form and this was when he turned to Watt with the idea of driving wheeled carriages. When Watt was writing after his friend's death in 1805, he said before Robison left the University in 1758,

> During this period he turned my attention to the steam engine, a machine of which I was then very ignorant, & suggested that it might be applied to giving motion to wheel carriages, & that for that purpose it would be most convenient to place the cylinder with its open end downwards, to avoid the necessity of using a working beam... In consequence I began a model with two cylinders of tin plate to act alternately by means of rack motions upon two pinions attached to the axis of the wheels of the carriage but the model being slightly and inaccurately made did not answer our expectation, new difficulties presented themselves. (16)

At least it seems that Watt was introduced to steam engines by Robison through making some model with inverted cylinders which was not a success and which was abandoned when Robison left Glasgow.

If this interpretation is correct, then it suggests that Robison had forgotten that it was he who had a much superior grasp of the steam engine at first. But, while Robison was away at sea, Watt continued experimenting, augmenting his knowledge of many different subjects, and returned to steam again in 1761 or 1762 when he carried out some experiments with a Papin's digester, a form of boiler capable of sustaining high steam pressures. He

> ... formed a species of steam-engine by fixing upon it a syringe, one-third of an inch diameter, with a solid piston, and furnished also with a cock to admit the steam from the digester, or shut it off at pleasure, as well as to open a communication from the inside of the syringe to the open air, by which the steam contained in the syringe might escape. When the communication between the digester and the syringe was opened, the steam entered the syringe, and by its action upon the piston raised a considerable weight (15 lbs.) with which it was loaded. When this was raised as high as was thought proper, the communication with the digester was shut, and that with the atmosphere opened; the steam then made its escape, and the weight descended. The operations were repeated, and, though in this experiment the cock was turned by hand, it was easy to see how it could be done by the machine itself, and to make it work with perfect regularity. But I soon relinquished the idea of constructing an engine upon its

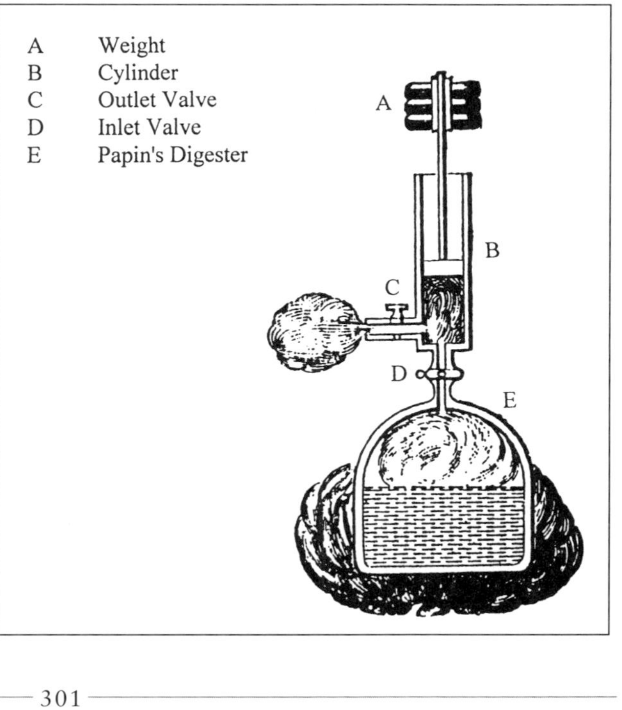

Smiles's drawing of Watt's high pressure engine.

> principle, from being sensible it would be liable to some of the objections against Savery's engine, viz., the danger of bursting the boiler, and the difficulty of making the joints tight, and also that a great part of the power of the steam would be lost, because no vacuum was formed to assist in the descent of the piston. (17)

Watt included this 'engine' in his patents of 1769 and 1784. To raise a weight of 15 lbs. on a piston of one third of an inch diameter would require a pressure of around 170 p.s.i. so it is probably lucky that Watt gave up these experiments. But one feature is worth noting, which is that he worked the valve or valves by hand. He certainly did this on some of his later models and these high pressure experiments may have shown him that he could experiment with quite simple models without the complexity and sophistication of Newcomen's valve gear. There the matter might have been left because,

> The attention necessary to the avocations of business prevented me from prosecuting the subject further; but in the winter of 1763 - 4, having occasion to repair a model of Newcomen's engine belonging to the Natural Philosophy class of the University of Glasgow, my mind was again directed to it. (18)

Notes: The First Steam Engines and Watt's Introduction to Them

1 Patent 356, 25 July 1698.
2 Savery, (1702).
3 Desaguliers, (1763), Vol II, p. 467.
4 Hills, (1989), p. 16 ff.
5 Hills, (1989), p. 21 and Rolt & Allen, (1977), p. 44 ff.
6 Rolt & Allen, (1977), p. 147.
7 Hamilton, (1932), p. 1.
8 Duckham, (1970), pp. 82 - 3.
9 Ibid, p. 23 ff. for a discussion on the industrial expansion of Scotland.
10 Hills, (1994), pp. 200 - 4 for windmills at Scottish coal mines.
11 Muirhead, (1858), p. 74 from J. Watt's 'A Plain Story'. See also Robinson & McKie, (1970), p. 416, Letter 259, 'Copy of Mr. Watt's Remarks on Mr. Robison's Edition of Dr. Black's Lectures', 4 Jan. 1809.
12 Robinson & McKie, (1970), p. 434, Watt's Note Book, JWP, W/14, p. 7 r.
13 *The Universal Magazine of Knowledge & Pleasure*, Vol 21, Nov. 1757, Letter from J. Robison written 4 Aug. 1757, pp. 229 - 31.
14 JWP, 3/29.2, Court of Common Pleas, Boulton and Watt v. Hornblower and Maberly, 1796, p. 47, A. Cumming's evidence.
15 *Universal Magazine*, Vol 21, Nov. 1757, p. 229.
16 JWP, Letter Book 4, p. 107, 'J. Watts Recollections of his Friend, Dr. J. Robison', April 1805. See Smiles, (1878), p. 68.
17 Muirhead, (1858), pp. 74 - 5.
18 Ibid, p. 75 and Robinson & McKie, p. 418, Letter 259, 'Copy of Mr. Watt's Remarks on Mr. Robison's Edition of Dr. Black's Lectures', 4 Jan. 1809.

Contemporary Theories of Heat

When Watt started to study the steam engine seriously, he must have been aware that fire or heat played an essential role. But it is very difficult to discover what knowledge he had at that time because, first, he never wrote any treatise about either the steam engine or what today we would call thermodynamics so that we have to try and piece together his thoughts from fragments in his correspondence and, second, the study of natural philosophy was advancing so rapidly during his lifetime that what he may have known and understood at one period may have been changed at another. It would be a long time before the true relationships between, and the true theories of, fire, heat, combustion and modern thermodynamics would be established. All that can be attempted is to reconstruct some ideas current in the 1760s which Watt may have known and to see how they might have influenced his understanding of the steam engine.

Aristotle's Elements and Boerhaave

At the beginning of the eighteenth century, both medical practitioners and chemists were interested in the properties of fire and heat. The medical world realised that life in animals seemed to depend on heat. Chemists recognised that heat played a crucial role in many chemical reactions. Boerhaave was both doctor and chemist. We have seen how he retained Aristotle's four elements as the basis of his chemical theories. Fire was of course one of those elements. He thought that fire consisted of subtle, immutable bodies that were capable of insinuating themselves into the pores of other bodies and that it was 'the great changer of all things in the universe while itself remaining unchanged'. (1) He considered, 'Now fire is two-fold; the one pure, and *elementary*; which alone we here call *fire*: the other *vulgar*, raised and kindled from the former; which is what agitates and affects ignited, combustible, and moveable bodies'. (2) His elementary or metaphysical 'fire' was uniformly distributed in space, no matter what substances occupied it. But his vulgar 'fire' was more important for chemists and others because it was more closely allied to what we consider as heat.

> By *fire* we mean any body that rarifies the air in a thermometer: this fire is the universal instrument of all nature, as well as of the art of chemistry. By this it is, that all bodies are moved, and it is by this that they discover their powers: If this were away, all bodies would immediately become solid and immoveable... it is absolutely impossible to exclude the fire out of any body whatever. (3)

Here we can see that Boerhaave realised that fire was essential for life itself. Chemical reactions depended upon fire (4) as did motion (5), yet fire itself could not be created *de novo* by any artificial means. (6) He considered fire as 'corporeal', made up of an infinite number of very small bodies which were immutable but smooth. Hence it could penetrate other bodies easily (7) and it moved with the greatest velocity of all bodies. (8) He considered that there was a fixed amount of fire in the world and that it could not 'either be created, or destroy'd, or produced'. (9)

The Phlogiston Theory

Boerhaave influenced not only Cullen and Black but also a pupil of his, Georg Stahl. Stahl further developed this corpuscular or material basis of fire into the phlogiston theory which rose to popularity during the 1740s. Therefore a brief outline of this theory is included here. Stahl based his ideas on the theories of J.T. Becher which he saw as a means of explaining combustion, but his phlogiston theory covered much more. Stahl followed Boerhaave in believing that what he called phlogiston was some sort of material substance. It was 'the matter and principle of fire but not fire itself'. (10) It was found in all combustible bodies like soot, oils, fats and resins, in animal and vegetable bodies as well as in metals. It escaped from burning bodies with a rapid whirling motion. The burnt product could be restored by supplying more phlogiston from any other material which contained it. It was the cause of metallic properties as well as of some colours and odours. Alcohol was a compound of water and phlogiston, from which, on burning, the phlogiston escaped and the water remained. Stahl inverted the true theory of combustion and calcination when dealing with metals. He failed to recognise that, what he claimed was adding phlogiston, was really removing oxygen and vice versa. He paid little attention to current atomic theory as well as what was known about gases. Yet he managed to link together a larger number of facts into a coherent corpus but unfortunately his doctrines were false. However this was the predominent theory during the middle years of the eighteenth century. It was the one which Watt followed and which indeed he continued to hold at least up to 1800 for Humphrey Davy told Davies Gilbert, 'I had a great deal of chemical conversation with them [Keir and Watt]. Both he and Mr. Watt are still phlogistians'. (11)

The Centrality of Fire in Black's Lectures

The phlogiston theory also influenced Black through Cullen. However, in his lectures, Black used the term heat which we can see was central to his teaching. He defined chemistry as

> the science or study of those effects and qualities of matter which are discovered by mixing bodies variously together, or by applying them to one another with a view to mixture, and by exposing them to different degrees of heat... The chemist studies the effects produced by heat and by mixture... with a view to the improvement of arts, and the knowledge of nature. (12)

In his opening lecture as recorded by Robison, Black delivered a panegyric on heat.

> Whatever that is which we call heat, and in whatever manner produced, it is certainly the chief material principle of activity in nature. Upon its gentle action, as confined within the bounds prescribed to it in this part of the universe, depend the production and preservation of all animal and vegetable life. Take away heat, to a certain degree, and they must all perish: a total stop would be put to all the operations of nature. But beside this extensive influence, by which it supports action and life in this great system of beings, the manner in which the various particular kinds of matter are affected by the different degrees of heat, is a subject of inquiry which we shall find most fertile of surprising and useful discoveries. (13)

But, in dealing with heat, Black was looking for general laws which would explain changes in the states of matter and was not trying to suggest a new theory of heat. He applied general laws of heat, flow, change of state and so on to particular cases dealing with specific substances and in this he had much more affinity with nineteenth century physicists who developed the study of thermodynamics than with his contemporary fellow chemists. (14) Black drew this section of his lectures on heat to a close with a further four pages on the subject! (15)

It would appear that Black carried out few, if any, experiments on heat after he accepted the professorship at Edinburgh University in 1766, (16) so that it is probable that his lectures on heat published by Robison would be the same as those he gave while still at Glasgow.

Also these views of Black are mostly the same as those which can be found in the earliest surviving notes taken down by one of his students, Thomas Cochrane, in 1767. Therefore his opinions stated at Edinburgh would be those he discussed with Watt when Watt sought explanations about the problems he faced with his steam engine. It seems doubtful whether Watt showed much interest in the philosophical notions about fire and heat before his involvement with the steam engine even though he had made Black's design of furnace. (See p79)

Black's Concept of Heat

Black himself never seems to have defined what he considered heat was. Possibly this was because he wanted to give a balanced view to his students in his lectures and therefore he never declared whether he was in favour of the material or the kinetic theories of heat. But he was more inclined to the first and was not convinced by the arguments of Lord Verulam, Boyle and others that heat was a form of motion or rapid tremor of particles. (17) He also pointed out that heat might be considered as a chemical substance and that the French and many other chemists thought that heat was a certain 'Vibration of an Elastic Fluid called Oether & supposed to be the cause of Electricity, Magnetism & Gravitation'. (18) Here we see another form of the phlogiston theory. While he supposed that this opinion agreed more with the 'Phenomena of Nature... it is a Cause involved in obscurity, therefore shall not enter into it further'. (19) Black thought that, 'Neither of these suppositions were fully and accurately considered by their authors, or applied to explain the whole of the facts and phenomena relating to heat. They did not, therefore supply us with a proper *theory* or *explication* of the nature of heat'. (20)

If indeed heat were some form of material substance, then it ought to behave like other chemicals. Black supposed that it entered into a chemical combination with another body and hence could cause a change of state. (21) For example, he thought that, during boiling, the heat did not escape along with the vapour but *in* it. (22) He saw heat as a positive, active agent, (23) the most communicable quality which one body could give to another. Hot bodies would communicate their heat to colder ones in their vicinity while density and compactness would be no obstacle in its progress. (24) He refuted the notion that the colder body could be the active mass and that a hotter body received cold from the cooler one. (25) This would be vital information for Watt when dealing with his steam engine because he could start at the right point, by considering how to use the heat in the best possible way.

The Communication of Heat

That fire was essential for working a steam engine was obvious because the steam was generated by burning fuel. When Savery patented his engine in 1698, he described it as 'raising water by the impellent force of fire', (26) while Newcomen's engine was often called a fire engine. Watt quickly realised the importance of communicating as much heat as possible from the fire to the water in the boiler to raise steam and his design of boiler may have been influenced by Black's three propositions.

> I am persuaded that when different kinds of matter are equally exposed to the same heating cause, the matter of heat enters into them all at first with nearly equal celerity...
> The second proposition... is that the celerity with which heat is communicated from hotter bodies to colder ones is, when other conditions are equal, nearly proportional to the difference of their temperature, or to the intensity of the heat in the hot body above that of the cold.
> The last of the three propositions... The celerity with which heat is communicated from

> hotter bodies to colder ones, when all other conditions are equal, is proportionable to the extent of contact, and closeness of communication, between the bodies. (27)

Black saw the advantages of large-scale furnaces, foundries and distilleries to give the best economy. (28) Watt would see the advantages of having a brightly burning fire placed as close to the boiler as possible. Black also recognised the importance of surface area for conducting heat. According to Thomas Cochrane, Black pointed out that a greater surface allowed heat to escape more quickly. (29) This would have helped Watt understand some of the problems with the great condensation of steam in the cylinder of the University model Newcomen engine.

The Thermometer and Its Use

The thermometer was the crucial piece of equipment which enabled theoretical concepts of heat to be put to the test experimentally. The mercury thermometer was introduced and thermometric standards fixed about 1715 by Daniel Gabriel Fahrenheit who adopted fixed points for graduating the scales on his thermometers. He soon discovered that the freezing point of water was a very settled and constant degree of heat and so appropriate for one fixed point while that of boiling water for a second varied with the pressure of the atmosphere and so was not quite as reliable. (30) His thermometers were described in a communication to the Royal Society in 1724. (31) Prior to his invention, thermometers could show the relative heat or cold of things but no comparison could be made between the readings of one thermometer with another. (32) Hence scientists could not share accurate temperature measurements with each other. Irvine commented in 1805,

> The invention of the thermometer superseded all the former vague and uncertain methods of estimating the degrees or qualities of caloric. It showed that all kinds of heat were the same, and differed from each other only in intensity, an opinion very dissimilar to that entertained formerly, when the grandeur or baseness of the source was supposed to influence and alter the nature of the caloric itself. Experiments made in one country could be repeated and extended in another, when the thermometer was employed; and a degree of accuracy was introduced into the philosophy of heat, of which the subject had not been imagined to be susceptible...
>
> About the middle of the last century, by the help of the instrument,... some observations were made and some consequences deduced, which have produced unforeseen and immense changes in the whole science of chemistry. (33)

The question that Black asked was, what did the graduations on the thermometer represent? Did equal increments of the expansion of the fluid in the thermometer represent equal increments of heat? In 1760, Black repeated experiments similar to those published by Dr. Brook Taylor in the *Philosophical Transactions* in 1723, although at the time he was unaware that he had been preceded. He took known volumes of hot and cold water. He measured the temperature of the hot water and then the cold. Then he mixed the water together and again measured the temperature. He found that the temperature of the mixture was the arithmetical mean, confirming that the mercury was 'expanded by different heats, in proportion to the differences of these heats'. (34) By this, Black was able to show that variations in the degrees on the scale of a thermometer measured quantities of heat imparted to a definite quantity of matter and in this way he laid the foundation for the quantitative science of heat. But his argument was circular; the thermometer would have been calibrated by mixing known proportions of boiling and freezing water. (35) Black introduced the foundation of the quantitative science of heat through measuring quantities of heat by the number of degrees of temperature imparted to a definite quantity of matter. In this way, he showed clearly the difference between heat and temperature, or between the quantity and intensity of heat.

Mixing known weights of water at different temperatures, this 'most simple and satisfactory experiment', as Black called it, (36) would be the key to other discoveries about heat. When Watt tried to mix known quantities of steam and water, he found that the usual law broke down, so he asked the reason why. Black then explained his theory of latent heat which will be examined later. Experiments based on similar principles could determine the heat capacities of different substances. A known weight of say a metal at a known temperature could be immersed in a known weight of water at a different temperature and the change in temperature of the water recorded. Comparisons with different substances could then be made.

Fahrenheit had tumbled upon this phenomenon when he mixed equal weights of water and mercury, each at a different temperature. (37) He found that mercury, despite its greater density, exerted less effect on the temperature of the mixture than did water. This puzzled Black, and he looked into Martine's experiments on the same problem as well. He soon realised that the result showed that mercury had a smaller capacity for the fluid of heat than had an equal weight of water. He concluded that, in general, the capacity of bodies to retain heat did not vary either with their bulk or density, but in a different fashion for each substance. He thought about this in 1760 but could assign no general principle or reason. (38) He recognised that 'to investigate the capacity of different bodies for heat, we can learn it only by experiments', (39) but he went no further into this phenomenon at that time. Watt was to use these methods in his own investigations on the principles of the steam engine and would carry out experiments into the heat capacities of various materials. Without this knowledge, he could not have found solutions to some of his problems.

The Change of State in Substances caused by Heat

It had long been recognised that a number of different solid substances could be rendered fluid by the addition of heat. With the addition of even more heat, they could be changed into a vapour. (40) However the intensity of the heat necessary to produce this effect differed between various materials. Then it was necessary that the temperature be maintained if the substance were to remain in its changed state. Black pointed this out in the case of steam.

> If the heat be again diminished, and the vaporous form is destroyed, the vapour loses its elasticity, and again collapses into that form from which it was produced. You must all have observed this in the case of water, the steam of which is condensed into water again, whenever it strikes any mass of cold matter, such as a lump of iron. (41)

Black could not have heard of Watt's childhood experiments with a kettle but might have said that here was the principle behind atmospheric steam engines.

The thermometer revealed that this change of state always occurred at the same temperature for the same substance under the same conditions of atmospheric pressure. Fahrenheit based his whole method of graduating thermometers upon this principle. He went on to find that water was capable of a greater or lesser degree of heat at its boiling point depending upon the greater or lesser weight of the pressure of the atmosphere as recorded by a barometer. (42) Boerhaave thought that the degree or point of boiling was

> ... perfectly stable, and permanent; nor is there any increasing thereof by any art. As soon as water has attain'd the degree of boiling, provided it be pure, it is not capable of

> conceiving any further heat; tho' an hundred times the fire were applied; as was first discovered by the help of the thermometer by Mons. Amontons. (43)

This was as far as other people too like Martine (44) and Cullen progressed. (45) They failed to ask where all that heat from the fire went without heating the water or steam further. Black provided the answer.

Latent Heat

Black's interest in heat had been aroused while he was still a student at Glasgow through the lectures and speculations of Cullen. On his return to Glasgow, Cullen communicated to him in 1755 the results of an experiment in which ether, boiling in an evacuated receiver, froze the water in a vessel surrounding it, although the room temperature was above freezing. Cullen's experiments on evaporative cooling and the phenomenon of freezing turned Black's thoughts to the question of what was happening to the temperature of substances when their state was changed. He pondered whether, 'Is it not owing to this that all bodies, in becoming fluid, have occasion for more heat than in their solid state?' (46) At that period, it was generally assumed that, when a solid was warmed to its melting point, only a small additional increment of heat was sufficient to melt the entire mass. Equally, when say water was cooled to freezing point, it was thought to solidify at once when a small increment of heat was taken away. On considering this, Black was particularly struck by the length of time ice or snow took to melt even after the atmospheric temperature had risen well above freezing point. Robison later wrote how Black realised that, if it were true that only a small increment of heat were necessary, then a fine sunny winter day ought to clear the hills of snow immediately or a frosty night 'suddenly cover the ponds with a thick cake of ice'. (47) Black became convinced that such changes in state required the transfer of considerable amounts of heat.

Black was convinced of the reality of the latent heat of fusion as early, perhaps, as 1756, but this was not based on any of his own experiments. He presented his theory in his Glasgow lectures in 1757 - 58, or possibly the following year. (48) It may have been about this time that he outlined a simple experiment to confirm the latent heat of vaporization and to measure the latent heat of steam but he carried this no further at that time. (49) It was not until the summer of 1761 that he conceived how to measure the latent heat of the fusion of water 'by comparing the time of raising a pound of water one degree in its temperature, with the time required for melting a pound of ice, both being supposed to receive the heat equally fast'. (50) As there was no ice house then in Glasgow, he had to wait until that December before he could carry out this experiment. He found that it required as much heat as would have raised the water more than 140° to liquify the same amount of ice. He then tried adding a quantity of boiling water to an equal quantity of water frozen into ice and found that the resulting temperature was not the mean but much less. 'Much of the heat had disappeared in a moment; as much indeed as should have heated it to 144° more, which is nearly the same as had been lost in the former experiment'. (51) He termed this 'lost' heat *latent* because it could not be measured by the thermometer. Irvine made the comment,

> Dr. Black thus established the truth of a general and most important law of nature, that all bodies require for their fusion and evaporation an immense, and formerly unsuspected quantity of caloric. To the knowledge of these facts, it cannot be doubted, that we owe much of the modern advancement in chemical theory. (52)

It was not until nearly a year later, in October 1762, that Black began to explore the latent heat of steam. He poured measured quantities of water into flat-bottomed tin vessels and heated them on a cast-iron stove. He measured the time required to heat the water from 50°F. to the boiling point and then the time necessary for the water to boil away. He calculated from the average of three experiments that the heat absorbed turning the water into steam would have raised the same amount of water to 810°F, if that were possible. (53) He said later that, from these experiments, 'The latent heat of steam seemed to me to be so completely established, that I was little solicitous of more experiments for my own conviction, or for making the doctrine clearly comprehensible to others'. (54) There Black let the matter rest until Watt came to him, having discovered that he could not make his figures add up when he used steam instead of water in his mixtures. This caused Black to reinvestigate the subject and start more experiments with Irvine on 9 October 1764 using an improved still. (55) We shall see that Watt used some of Black's techniques in his experiments.

Water Boiling in a Vacuum

Cullen had been strongly influenced by Boerhaave and so expounding problems of heat was an important aspect of his lectures at Glasgow. He was particularly interested in the variations in temperature resulting from different chemical combinations. By 1754, he had put a Mr. Dobson, one of his students, to work studying the effects when various substances were added to spirit of wine. Dobson observed that a thermometer registered a sudden drop in temperature when removed from the spirit of wine. Cullen decided to study this phenomenon in a vacuum. His single paper on heat, 'Of the Cold Produced by Evaporating Fluids, and Some Other Means of Producing Cold', (56) dealt with the phenomenon of cooling by evaporation, particularly under reduced pressures. He pointed out the fact, first noted in the seventeenth century, that water boiled at lower temperatures as the pressure was reduced.

It is not clear what figures Cullen produced to show the relationship between temperature and pressure at which water boiled in circumstances of reduced atmospheric pressure since none are contained in this paper. Writing many years later in his notes for Brewster's edition of Robison's *System of Mechanical Philosophy*, Watt himself said that one cause of the waste of steam in his engine 'seemed to be the boiling of water in vacuo at low heats, a discovery lately made by Dr. Cullen and some other philosophers, (below 100°, as I was then informed)'. (57) Cullen's experiments on the boiling point of ether in a vacuum and some conversations with Black and Robison encouraged Watt to undertake experiments. (58) He soon found that water would boil in a vacuum at a lower temperature than this which proved to be one of the keys to his understanding what was happening to the steam inside the cylinders of Newcomen engines.

Watt's Experiments

Watt's experiments rising from his trials with the University model Newcomen engine would, first, establish a more accurate determination of the volume of steam derived from water; second, establish the value of the latent heat in steam and stimulate study of latent heat in other substances; and third, cause Black to undertake more serious study of specific heats of substances. Yet in those experiments to discover why this model engine used so much steam, Watt did not need to understand contemporary concepts about heat such as the phlogiston theory and so he made no contributions in this sphere. While he never began to unravel the basic laws of thermodynamics nor the relationship between heat and energy, his practical experiments elucidated some basic properties of physics and stimulated further research. The detailed study of steam, which he initiated to gain an insight into how steam engines worked, led to a deeper understanding of the nature of gases and the effects of heat on them, which would be best explained by the kinetic and not material theories of heat. The effective impulse to the study of thermodynamics came from Watt's study of the steam engine and, in this way, he inspired the vast extension in the understanding of physics, an achievement which must rank as important as that of Newton. (59) While Black's discoveries of latent and specific heats were fundamental as the foundations of this new knowledge, their significance was confirmed by Watt who used them to develop the source of power which boosted the dramatic expansion of the Industrial Revolution.

Notes: Contemporary Theories of Heat

1 Brock, (1992), p. 77.
2 Boerhaave, (1727), p. 225.
3 Ibid, p. 219.
4 Ibid, p. 51.
5 Ibid, p. 220.
6 Ibid, p. 233.
7 Ibid, p. 234.
8 Ibid, p. 257.
9 Ibid.
10 Partington, (1957), p. 87.
11 Cited by Clow, (1952), p. 496, H. Davy to D. Gilbert, 22 Feb. 1799.
12 Robison, (1803), Vol I, pp. 11 - 12.
13 Ibid, Vol I, p. 12.
14 Donovan, (1975), p. 221.
15 Robison, (1803), Vol I, p. 243 ff. See also Cochrane, (1966), p. 99.
16 Simpson, (1962), Anderson, p. 9 rt.
17 Robison, (1803), Vol I, p. 31 and Cochrane, (1966), p. 4.
18 Cochrane, (1966), p. 5 and Simpson, (1962), Guerlac, p. 13 rt.
19 Robison, (1803), Vol I, p. 33 and Cochrane, (1966), p. 5.
20 Robison, (1803), Vol I, p. 3.
21 Donovan, (1975), p. 226.
22 Ibid, p. 230.
23 Ibid, p. 227.
24 Robison, (1803), Vol I, p. 23.
25 Ibid, pp. 26 - 27.
26 Patent 356, 25 July 1698.
27 Robison, (1803), Vol I, pp. 86 - 89.
28 Ibid, Vol I, p. 90.
29 Ibid, p. 90 and Cochrane, (1966), p. 8.
30 Martine, (1780), pp. 11& 16, Robison, (1803), Vol 1, pp. 53 - 54 and Middleton, (1966), p. 68.
31 Singer, (1941), p. 298.
32 Martine, (1780), p. 7.
33 Irvine, (1805), pp. 42 - 44.
34 Robison, (1803), Vol I, pp. 56 - 57 and Middleton, (1966), p. 50.

35 I am indebted to the late Professor D.S.L. Cardwell for this observation.
36 Robison, (1803), Vol I, p. 57.
37 Prof. Cardwell noted; 'Specific heats are always measured and given in terms of equal weights (for the obvious reason that volumes change with temperature, weights do not). The specific heat of mercury is 0.035 while its specific gravity is 13.8. Its volumetric specific heat is therefore 0.4623, just below half that of water'.
38 Robison, (1803), Vol I, p. 79.
39 Ibid, Vol I, p. 82.
40 Ibid, Vol I, p. 105.
41 Ibid, Vol I, p. 147.
42 Martine, (1780), p. 11.
43 Boerhaave, (1727), p. 272.
44 Martine, (1780), p. 174.
45 Donovan, (1975), p. 137.
46 Simpson, (1962), Guerlac, p. 15 rt.
47 Robison, (1803), Vol I, p. xxxvi.
48 Muirhead, (1854), Vol II, p. 119, Letter 179, J. Black to J. Watt, 15 March 1780.
49 Simpson, (1962), Guerlac, p. 16 rt.
50 Robison, (1803), Vol I, p. xxxvi.
51 Simpson, (1962), Guerlac, p. 17 rt. See Robinson & McKie, (1970), p. 76, Letter 59, J. Watt to J. Magellan, 1 March 1780 and Irvine, (1803), pp. 46 - 47.
52 Irvine, (1805), p. 47.
53 Simpson, (1962), Guerlac, pp. 17 - 18.
54 Robison, (1803), Vol I, p. 171.
55 Ibid.
56 Cardwell, (1971), p. 34.
57 Muirhead, (1854), Vol I, p. lxx.
58 Robinson & McKie, (1970), p. 418, Letter 259, 'Copy of Mr. Watt's Remarks on Mr. Robison's Edition of Dr. Black's Lectures', 4 Jan. 1809. See also Robinson & McKie, (1970), pp. 377 - 378, Letter 221, J. Robison to J. Watt, 25 Feb. 1800, in which Robison described experiments he conducted while teaching Anderson's class.
59 Crowther, (1962), p. 98.

THE UNIVERSITY OF GLASGOW MODEL NEWCOMEN ENGINE

Description of the Model

Anderson had been authorized on 25 June 1760 by Glasgow University to 'lay out a sum not exceeding two pounds sterling to recover the steam engine from Mr. Sisson instrument maker at London'. This University model had been sent there because it had 'been in such a situation that it did not answer the end for which it was made'. (1) However it was not until the winter of 1763 - 64 that the engine was passed over to Watt to see if he could make it work properly. Watt said it was made by Sisson and that he 'mett with considerable difficultys in the execution owing to the very bad construction of some of its parts'. (2) It was only on 10 June 1766 that he was paid £5.11s. by the University for 'altering the steam-engine with copper pipes and cisterns'. (3) This engine is displayed today in the Hunterian Museum of Glasgow University. (Plate 11)

Robison claimed, 'This model being an exact copy of a real engine, the motion of the piston behoved to be the same, and the strokes to be much more frequent'. (4) Although it is impossible to know now what parts have been altered by Watt or subsequently, we can see that it has been well constructed with very accurate lathe turning and screw cutting. There are some very delicate and beautiful parts, which should not surprise us as the Sissons, father and son, were and still are held in high regard as scientific instrument makers. (5) But it also has features which suggest it is not a model of an actual engine. The main difference is that the cylinder, which has a diameter of 2 ins. and stroke about 3 $^{1}/_{2}$ ins., is a little shorter in proportion to ones depicted in contemporary pictures. Then the cylinder is not supported by any cross beams but by the dome of the boiler. Although there are one or two schematic drawings showing unsupported cylinders, (6) all other engravings show cross beams. It is very doubtful if a boiler on a full size engine could have been strong enough to withstand the strains of holding down the cylinder.

Watt said that the boiler was nine inches in diameter. As it is shrouded in a casing, it has not been possible to determine the actual diameter which must be nearer six inches. The casing encloses the fire grate underneath and there should be a chimney to take away the smoke. The grate itself is about the same diameter as the boiler but the firebars are missing so the size cannot be determined accurately. The ashpan is also encased and was probably joined to the upper casing but today is separate. He said that 'the fire was violently urged with bellows' (7) and through the ashpan door is a hole into which the nozzle of the bellows might have been inserted. On top of the boiler are five fittings. (Plate 12) One is a conical funnel with a cock through which the boiler could be filled since the boiler cannot be topped up from the overflow of the water seal on top of the cylinder as on full-size engines. Two are try cocks to determine the water level. The fourth is a tube or small cylinder $^{3}/_{4}$ ins. diameter by 1 $^{1}/_{2}$ ins. high mounted on a pipe leading directly into the boiler. A conjectural suggestion is that this might have been some pressure or level indicator. The final fitting may have been a safety valve. If this assessment is correct, it would be possible to obtain some idea of both steam pressures and consumption.

Steam is admitted to the cylinder through an horizontal oscillating plate valve in the same way as on full size engines. The inlet hole seems large in proportion to the diameter of the cylinder and a short pipe conveys the steam well above the level of the water injection hole and eduction pipe. The snifting valve, mounted on the side of the cylinder, is a beautifully made circular flat valve on a thin stalk. It would certainly allow steam and air to escape easily. The steam piston is a solid disc screwed on the end of the rod. The seal is a leather cup or 'garland' secured in place by a junk ring which is screwed on to the piston. Around the top of the cylinder is a large cup with an outlet pipe taking the overflow down to the eduction well. With the piston in its highest position, there would be a large amount of water on top of it, giving a positive seal which was topped up from the upper water pump cistern. The piston is suspended from the rocking beam or lever by a linked chain around an archhead. The chain is nicely made, rather like a clock fusee chain.

So far the parts that have been described have the feeling of being original but there are suspicions about the valve operating mechanism. The plug tree is suspended from the main beam by a thin rod passing through it with an adjusting screw on the top. This can only pull up and not push down, which must be necessary to operate the valve gear. The wooden plug tree is a strange shape and so possibly a replacement. To return it to its lowest position, there is a heavy lead weight, 2 ins. by 1 $^1/_8$ ins. by 1 $^7/_8$ ins., which looks suspiciously like an addition. Perhaps Watt added this to overcome one of the faults, that the original valve gear did not operate properly. The steam valve gear has screwed pegs to adjust the timing so that the plug tree can move the primary levers without actuating the valve itself until the end of the stroke when a heavily weighted lever overbalances and throws the valve over, like that on Newcomen's original engine. The water injection valve probably opened in a similar way but closed slowly. Some of these parts seem to be later modifications. The condensate is carried away by the eduction pipe leading down to a tank, the eduction well, beside the ashpan where there is another flat circular valve similar to the snifting valve.

The cistern and pipe supplying the water to the injection valve are definitely later additions which must be attributed to Watt. While the pipe leading from the bottom of the cistern is a fine piece of tapering copper work, it is crudely fixed to the framing. The cistern is almost certainly one mentioned in an invoice sent to Watt by James Fleming on 29 February 1764,

	s.d.
To a copper funel & Tube	3. 6
To a Glue pot	3. 0
To a tin cistern & ?...?	4. 0
To a copper pipe	0.10
To a copper pan	2. 0
To mending a boyler	0. 2
To a tin cover parced with holls	0. 6
	14. 0 (8)

This cistern stands high above the engine and has a cover pierced with holes. It has to be filled by hand, which would facilitate easier recording of how much water was used. At the bottom end of the copper pipe, the injection cock seems very small, as are the holes in the bottom of the cylinder itself to admit the injection water.

At the other end of the main beam is a second archhead from which the water pump piston is suspended by a chain similar to that for the steam piston. The diameter of this pump is also two inches with the same stroke as the steam cylinder. Here again the piston is sealed by another leather cup or packing and a screwed junk ring. In the middle of the piston is a conical plug valve with central locating stalk. Its diameter and lift are sufficiently large to allow easy passage to the water. There is a similar valve at the bottom of the pump casing so the water passages would not have caused much obstruction. The casing is in two parts, the bottom of the lower one pierced with holes to allow entry of water. The total height of this pump casing is 29 $^3/_4$ ins. so that the water would have been raised about 26 ins. in a single lift. The water flows out of the upper end into a lead lined cistern above the level of the top of the cylinder from where some water can flow into the top of the cylinder to act as the seal. An overflow pipe allows the rest of the water to return to the bottom tank. The weight of the pump bucket, rod and chain, by themselves, would be insufficient to return the beam to its 'outdoor' position with the pump in its lowest position. This may have been another of the problems which Watt discovered because, normally concealed by the pump casing, there is now a large lead weight fixed on the pump rod. The reason for this will be discussed later.

The Glasgow University model has been compared with a larger, near contemporary one in the George III Collection at the Science Museum, London. Differences suggest that they have not been made by the same person. On the London engine, there are two sets of mine pumps giving a double lift of about 36 ins. The greater weight of these might make this engine run in a way more like large engines. While the steam cylinder has the same bore of 2 ins., it is much taller and the stroke is nearly twice as long at about 7 ½ ins. The boiler is again encased to contain the heat from the fire. This casing is much larger than the one on the University model, suggesting that the boiler is larger as well. The fire grate is certainly very much bigger, being about 12 ins. by 7 ins. So this engine seems to have much greater steam-raising capacity. However, the pipes to the snifting and eduction valves are much smaller as are the valves themselves, with the result that this engine might not work so freely. The steam inlet valve has the weight on an arm to overbalance it and open it sharply, but the water injection valve is quite different and probably would not have functioned as well as Newcomen's. The steam cylinder has provision for being mounted on cross beams but there are none in position today. A replica has been made of this model which has worked satisfactorily.

Watt's Approach to the University Model

Watt said that 'I set about repairing it [the University model] as a mere mechanician' (9) and, once he had it running regularly, started to be puzzled by the high consumption of steam and water. Robison, who by this time had returned to Glasgow, commented about Watt,

> Whenever any puzzle came in the way of any of us, we went to Mr. Watt. He needed only to be prompted; everything became to him the beginning of a new and serious study; and we knew that he would not quit it till he had either discovered its insignificance, or had made something of it. No matter in what line, - languages, antiquity, natural history, - nay, poetry, criticism, and works of taste; as to anything in the line of engineering, whether civil or military, he was at home, and a ready instructor... Everything became to him a subject of new and serious study, - everything became science in his hands. (10)

The autumn of 1763 was when Watt was launching out into his new venture of the enlarged shop in the Trongate, which clearly engaged a great deal of his time. During this period, he was also courting Peggy since they were married in the following summer. So he would probably have been grateful for Robison's assistance with the Newcomen model. Robison says,

> Our acquaintance at this time became very intimate, and I believe neither of us engaged far in any train of thought without the other sharing in it. I had the advantage of a more regular education: this frequently enabled me to direct or confirm Mr. Watt's speculations, and put into a systematic form the random suggestions of his inquisitive and inventive mind. This kind of friendly commerce knit us more together, and each of us knew the whole extent of the other's reading and knowledge. (11)

Robison helped by searching out literature on steam engines because Watt,

> was confined to his business; I was more at large, and going about the College. I ransacked the libraries for every book that he wanted; and every quotation that he met with made him impatient till he got at the original. I saw every book that he got by any other channel besides the public libraries so I may safely say that I knew the whole extent of his reading. (12)

Here again we see the importance to Watt of studying the literature but this was probably not either as extensive or as helpful as Robison implied.

Robison said that Watt 'learned the German language in order to peruse Leupold's "Theatrum Machinarum": - so did I, to know what he was about'. (13) Leupold was a gun founder in Vienna who published his book in 1725. (14) It has a drawing of a twin cylinder single acting high pressure engine working twin pumps with a rotary valve which would have been similar in principle to Watt's engine with the Papin digester and preceded it. (15) But, beyond drawings and descriptions of the Newcomen engine, Watt would probably not have learnt anything about the theory. Another book to which he may have had access, S. Switzer, would have given him no help at all. (16) Neither would he have gained much more from Belidor's *Architecture Hydraulique* which he did consult. (17) Belidor had measured up very accurately the Newcomen engine at Fresnes and his book contains a long description of the various parts with their sizes. He also explained how to start it and the method of operation. But he gave no theory of heat or steam or explanation of how the engine might work. (18) Belidor mentioned Desaguliers and so may have led Watt astray. While Watt would have found more help about contemporary theories of heat, evaporation, vapour and the atmosphere in Desaguliers's *Course of Experimental Philosophy*, (19) he would not have found very much to assist him with his struggles to fathom out why the University model engine would not work. Desaguliers also gave a description of the working parts of an engine and pointed out the problem of air being entrained with the steam and so partially vitiating the vacuum. However, after experimenting, Watt thought that the figure was wrong which Desaguliers gave for the volume of steam derived from water and it was this which set him off on his path of scientific investigation.

Watt Starts Experimenting

Watt said that, after repairing the model,

> It was set to work. I was surprised to find that its boiler could not supply it with steam, though apparently quite large enough, (the cylinder of the model being two inches in diameter, and six inches stroke, and the boiler about nine inches diameter). By blowing the fire it was made to take a few strokes, but required an enormous quantity of injection water, though it was very lightly loaded by the column of water in the pump. It soon occurred that this was caused by the little cylinder exposing a greater surface to condense the steam, than the cylinders of larger engines did in proportion to their respective contents. It was found that by shortening the column of water in the pump, the boiler could supply the cylinder with steam, and that the engine would work regularly with a moderate quantity of injection. It now appeared that the cylinder of the model, being of brass, would conduct heat much better than the cast-iron cylinders of larger engines. (20)

If the walls of the cylinder caused quicker cooling, then the engine ought to have needed less injection water rather than more, as Watt states here. Robison's account also confirms that the boiler was able to supply only enough steam for the engine to take a few strokes, but, after it was made to boil more violently and give a plentiful supply of steam, the engine stopped altogether! That somebody has had a good fire beneath this engine at some time is confirmed by the wooden side of the bottom water tank being charred next to the ashpan. What cannot be done on these engines is to increase the boiler pressure because then the steam may blow the piston out of the cylinder.

Watt quickly devised ways of determining the performance of this engine. By taking the surface area of the piston, he could calculate the weight which should be lifted when there was a perfect vacuum in the cylinder. Also if the surface area and the weight lifted were known, then the degree of vacuum in the cylinder could be ascertained. For the University model, he calculated that the actual column of water in the pump would weigh about three pounds, giving a load equal to 1 p.s.i which was much less than an equivalent load in larger engines. (21) The surface area of a two inch diameter piston is just over three square inches so that, if a perfect vacuum could be formed in the cylinder, the pressure of air against the steam piston would be around forty-five pounds. No wonder he realised that the pump was lightly loaded! But then we find him shortening the column of water in the pump to reduce the load with the result that the engine ran regularly. It has been pointed out recently that the present casing of the pump is made from two parts and that this shows where Watt restored it to its original condition after shortening it before returning the model to the University. (22) The pump could not be dismantled for a closer inspection but the join appears to be part of the original construction. If the pump were shortened, the upper water cistern would have to be lowered so it could no longer supply sealing water for the steam piston and a longer chain would have to be added to the pump piston. Shortening the column of water would make little difference to the load on the engine which could have been done much more easily by removing part of the large lead weight on the pump piston. Later Watt said he 'tried the Model under various loads, or higher and lower columns of Water in the Pump'. (23) His comment about various loads suggests he may not have shortened the pump barrel but rather altered the lead weight.

Therefore we have come to the conclusion that Watt must have forgotten how he succeeded in making the engine run regularly and we must consider the effect of this large lead weight on the pump rod. (Plate 14) Although the existing pistons in both steam cylinder and water pump would counterbalance each other, some weight must have been added to the pump end to raise the plugtree with its lead weight, operate the valve gear, force the pump piston down through the water and overcome friction generally. But a lump of lead 1 $^{3}/_{8}$ ins. diameter and 6 ins. long weighing over 7 lbs. seems excessive. Why is it there and who may have put it there? On model Newcomen engines, it is difficult to simulate the operation of the mine pumps. In a proper mine, the weight of the spears or pump rods will be sufficient to bring the engine 'out of the house'. They may be even too heavy and need counterbalancing. But they would weigh a great deal less than the column of water which had to be raised. In the instance of Newcomen's first engine, this was 51 yds. Bucket or lift pumps would raise nearly all the water on their up stroke, the power stroke of the steam cylinder. In the case of small model engines and the one-third size version of the Dudley Castle engine built by the Mechanical Engineering Department, UMIST, weights had to be added to the pump rods above the pumps to force the water pumps down and also to simulate the load. A sort of compromise was reached whereby the weights gave sufficient load to prevent these engines raising their pumps too quickly but this meant that the weights returned the engines too fast on the down stroke. Therefore these engines worked more quickly than their full scale counterparts. This also may have happened when Watt tried the University model so it ran out of steam.

On the University model, the pump casing extends too close to the beam for any weight to be added above the level of the pump so the existing lead weight disappears into the pump barrel as the engine goes out of the house. The effect of this is to turn the pump into a form of plunger pump with most of the water being lifted on the down and not the up stroke. A similar weight was tried on one of the small model engines in Manchester. Water was certainly raised on the down stroke and, in addition, the engine seemed to run more smoothly and slowly. So what effect might the heavy lead weight have had on the University model? A slower descent would have given more time for steam to enter the cylinder and heat it up to destroy the vacuum. The ample steam port would have allowed plenty of steam to pass through. There would also have been more time for the boiler to generate steam.

But the corollary of this must have been that more steam might be needed to reheat the cylinder because, to raise this heavy weight, a greater degree of vacuum would be necessary. The power of the engine must be linked to the change in temperature between the steam and the degree of vacuum. Then, to generate this greater vacuum on the down stroke of the steam piston, more injection water might be necessary. But the injection valve and openings of the jet into the cylinder seem small. It could have been that Watt was relying more on surface cooling of the outside of the cylinder to produce the vacuum. When experimenting with the model engines in Manchester, it was often felt that much of the cooling occurred in this way. While this would reduce the amount of injection water, the problem would remain, and did in Manchester with the one third size engine, of providing enough steam.

The lead weight slowing down the descent of the pump piston and a smaller injection with slower cooling on the other stroke may have reduced the speed of the University engine sufficiently to have enabled the boiler to produce enough steam to keep it running. The lead weight would be most effective at the beginning of its down stroke when the greater part of it would be exposed and hence have the greatest weight to overcome inertia, friction, etc. As it descends and is immersed, it will lose effective weight through the bouyancy of the water so its effect will be least at the end of the stroke when the momentum needs to be reduced to bring the engine to a halt. The lead weight, or at least the present large one, is most likely to have been added by Watt and shows either a profound understanding of its properties as well as of the Newcomen engine or perhaps it was just a lucky chance.

The construction of the lead weight is much cruder than other parts of the engine and, if it were added or enlarged by Watt, then it would have increased the load on the engine. This would have meant that, in order to achieve the greater range of pressure difference in the steam cylinder, there had to be a greater range of temperature to correspond. Now Watt later used the law of proportions of mixtures of different temperatures in his experiments. When one unit of boiling water at 212°F is added to another unit of water at room temperature, say 60°F, the temperature of the mixture is 136°F. But, when the next unit of water at room temperature is added, the temperature of this mixture drops to only 111°F. In other words each additional unit yields diminishing returns. The same reasoning must apply to the steam at the upper end of the scale so that it requires a lot of steam to bring the temperature back up to boiling point because the steam itself would be only at or just above boiling point. On the other hand, it would be easier to create a good vacuum because a near perfect vacuum can be attained at about 90°F whereas the temperature of the water is only 60°F, a significant difference. We know that Newcomen achieved quite a high vacuum in his first engine with a loading of 9.4 p.s.i. on the piston. Soon it was discovered to be more economical to load the engine less heavily at about 7 or 8 p.s.i. and make more strokes per minute. In 1770, Smeaton found that the optimum working conditions on his model engine were about 17 strokes per minute with 8 p.s.i. below atmospheric pressure. (24)

The dilemma of the Newcomen engine is that, at the beginning of the ascent of the steam piston, there must be sufficient incoming steam at sufficient pressure and temperature both to reheat the cylinder bottom, the lower part of the cylinder and the piston *and* to blow any residual air out through the snifting valve against atmospheric pressure - otherwise the engine becomes windlogged and soon stops. Therefore it is essential to have an adequate supply of steam. Then as the piston rises, the amount of reheating lessens and, on light loads, the piston could ascend with a partial vacuum and heavy weight on the pump side. Watt's boiler might have been temporarily drained of steam as the stroke began and then, as the piston descended on the power stroke, recovered sufficiently for the next stroke. If the load were increased, the supply of injection water had to be increased proportionately to achieve greater cooling. Watt summed up the problems in a draft specification for his steam engine patent.

> I have found Common steam engines are Imperfect 1st because the Cylinder having been Cooled by the Injection in the preceding stroke Condenses a Considerable Quantity of Steam besides what is necessary to fill it.
> 2dly because the Vacuum is Imperfect without the Cylinder and all the water in it be cooled down to 90° of Fahrenheits Thermometer. If this was done the Vacuum could be made perfect but the Condensation of Steam the next stroke would much Over ballance the power gained, therefore experience has taught Men that an engine works most advantageously when loaded about half the pressure of the atmosphere. (25)

So Watt concluded that the Newcomen engine was far from perfect.

Watt's Concept of a Perfect Engine

Watt said that, in 1758, neither he nor Robison 'had any idea of the true principles of this Machine', the steam engine, but that, in 1763, he 'set about the matter more seriously & discovered the principles upon which my improvements on the Steam Engine are founded'. (26) At some point he conceived his idea of the perfect engine. This would use only one cylinder full of steam each stroke and the steam would be condensed to a perfect vacuum. The incoming steam would have been at atmospheric pressure. Although we can now see that this would be impossible to achieve, the importance of this concept was that it gave him a standard of performance against which he could judge his experiments and also that he had developed a concept based on science which explained how the engine must work. This convinced first Black, then Roebuck, followed by members of the Lunar Society such as Darwin, Small and finally Boulton to support Watt in his lengthy trials to develop his separate condenser. Watt frequently referred to his perfect engine throughout these years of experimentation, and even later, which shows how important he thought this concept.

But from where did Watt derive this idea of perfectability? It seems to have been a common concept during the eighteenth century in many different spheres such as some aspects of religion or architecture with Greek temples. Here we will discuss two links with mechanical engineering. The first is the design of waterwheels. In 1704, the Frenchman, Antoine Parent, had shown that, at best, an undershot waterwheel could generate only 4/27 of the total available power, or 'effort', of the stream. This implies that there must be some maximum, which in this case was the ability to raise back to the same height the same volume of water that had driven the wheel. This suggested that there could be a perfect waterwheel which could utilise all the effort of the stream. The debate was continued in France by others and then in England by Smeaton. Among those books which Robison ransacked from libraries was without doubt Smeaton's paper on *Experimental Enquiry concerning the Natural Powers of Wind and Water to turn Mills and Other Machines depending on a Circular Motion*, read before the Royal Society in May 1759. (27) By using models, Smeaton showed that Parent's figure was wrong and that an undershot wheel could develop one third of the effort of the stream but that an overshot wheel could raise this to two thirds. (28) We know that Watt was definitely aware later of Smeaton's work. (See p286) For Watt, the importance of Smeaton's papers in 1763 would have lain in showing that the performance of a machine might be improved and that this could be judged against a conceptual optimum.

Another example of an almost perfect machine at this time was John Harrison's chronometer. For navigating ships across the oceans, it was essential to know their longitude. Indeed, the Royal Observatory at Greenwich had been founded 'to find out the so-much desired Longitude at Sea, for perfecting the art of Navigation' by astronomical methods. (29) Harrison attempted the same solution with a series of mechanical clocks. The trial of his model number 4 was carried out in 1761 and 1762. It was put on board H.M.S. *Deptford* for the Atlantic crossing to Port Royal, Jamaica, which took nearly three months. On arrival, the representative of the Board of Longitude, Watt's friend Robison, set up his instruments and established local noon. (30) Robison and Harrison then synchronized their watches to fix the longitude of Port Royal by the difference between them. Harrison's chronometer had lost only five seconds after 81 days at sea. (31) Here was near perfection in a mechanical time piece. Soon after this, Robison returned to Glasgow to take up again his friendship with Watt and assist with the University model steam engine. Surely he would have described to Watt Harrison's wonderful near perfect machine just as he did the flora of Jamaica. (32)

The Quest for Improvement to Steam Engines in the Middle of the Eighteenth Century

Savery's engine, particularly in its original form, was little used owing to its heavy consumption of fuel. (33) While the Newcomen engine required less, its consumption was also so great an expense that it too was employed mainly where coal was available cheaply or at mines for the valuable ores of tin, lead or copper. Many people looked at these engines and tried to enhance their performance at the same time as Watt, so he was not alone in seeking improvements. The inventions of these rivals shows that Watt's experiments should be seen against a general search for improvement at that period and it will be instructive to see how he succeeded where others, with whom he came into direct contact, advanced the design of steam engines but failed to make the crucial advance with the separate condenser.

James Brindley, 1716 - 1772

One of the earliest of these improvers was Brindley. In 1756, he undertook to erect a steam engine near Newcastle-under-Lyme upon a new plan. It had a boiler shell made of brick and stone instead of iron plates with a cast iron stove in the middle and internal iron flues. (34) This cut fuel consumption by a half, partly through the reduction of the loss of heat from the outer surface of the boiler. The cylinder was made of wood, but Brindley quickly found that this did not work so he reverted to a metal one which was lagged with a wooden casing containing ashes. Presumably he was trying to lessen loss of heat caused by the external cooling of the cylinder walls. (35) While he succeeded in reducing steam consumption, he probably increased the amount of injection water needed.

William Blakey

The patent which William Blakey obtained in 1766 (36) for an improved form of Savery engine is difficult to understand because it seems to be referring to three different engines. (37) In one of them, he was trying to reduce the condensation of steam by having oil floating on top of the cold water to be raised in the operating receiver. In another, he confined the water to be raised in a lower vessel and carried out the condensation with a jet of water in an upper one. However the steam still had to pass into the lower vessel to force the water out. Then he tried to separate the steam from the water by a layer of air but of course the steam became mixed with that air. He possibly could have been moving towards the concept of condensing in a separate vessel but it is doubtful whether he would have gained any great savings with his layout. It may be significant that Blakey raised the chief opposition to the extension of Watt's patent during the Parliamentary debates in 1775 to prolong the famous 1769 patent. (38) Between 1772 and 1776, Blakey improved his engine and experimented with a new type of boiler which may have been the cause of an explosion when his engine was tried in a Cornish mine. (39) He was still extolling the virtues of his engine before the St. Petersburg Academy as late as 1781! (40)

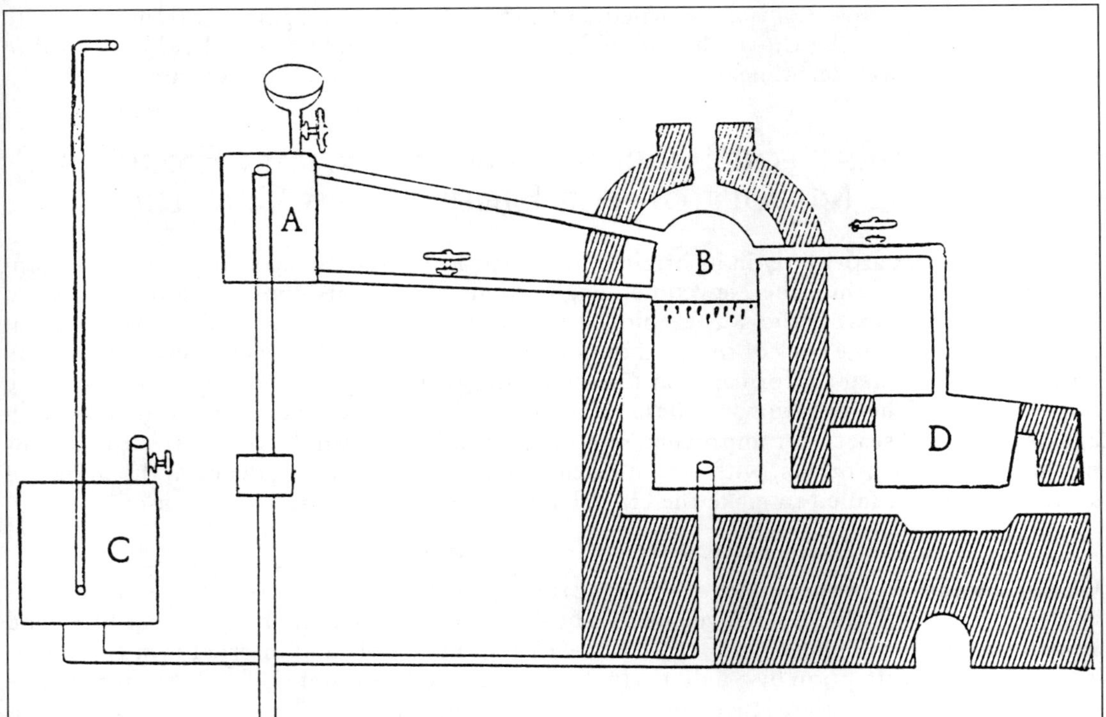

In Blakey's first engine, steam from the boiler 'D' was admitted to the main receiver 'B'. Water was drawn up into the receiver 'A' and finally forced out of receiver 'C' up the standpipe.

Humphrey Gainsborough, ? - 1776

Humphrey Gainsborough also thought of a form of separate condenser but almost definitely sometime after Watt. Unfortunately we know very little about his engine. He regretted that he had not opposed the extension of Watt's patent in Parliament. (41) He had submitted an application for a patent on 6 February 1775 which Watt contested. (42) Watt claimed it was an infringment (43), but Gainsborough wrote to him in May 1775 saying that, 'Those who know both inventions have assured me that mine is totally different from yours'. (44) Thirty years later, Jabez Hornblower said that,

> It occurred to Mr. Gainsborough... that it would be a great improvement to condense the steam in a vessel distinct from the cylinder, where the vacuum was formed... by placing a small vessel by the side of the cylinder, which was to receive just so much steam from the boiler, as would discharge the air and condensing-water, as was the practice from the cylinder itself in the Newcomenian method; that is by the snifting valve and sinking pipe. (45)

When he wrote to Boulton and Watt, both Gainsborough and his wife were very ill and, although he met Boulton at their London solicitors in 1776, the affair was certainly drawn to a close by Gainsborough's death on 23 August that year. (46) The examples of both Blakey and Gainsborough indicate that people were searching for methods of improving the condensation of the steam and were examining ideas for condensing away from the working cylinder.

Alexander Cumming, c. 1732 - 1814

Scotland was swept by this same vogue for improvement and a good example is Cumming. The date of Cumming's first interest in steam engines remains unknown but it may have been a little before 1760. In his evidence at the trial of Watt's patent in 1796 he stated,

> That having a natural inclination to mechanical pursuits & having the use of an extensive Library & every Accommodation for Experiments & Information under the Patronage of Archibald the Duke of Argyle he Dep[onen]t. devoted the whole of his life from 16 to 28 in acquiring an intimate knowledge of the most useful & curious Engines & Machines then known in Great Britain & to render his Knowledge the more perfect a Number of Artificers were employed at the said Duke's Expence & under Depts. Direction in constructing Models of such machines from the printed or other Descriptions & with such improvements on each as occurred to him Dept. & that among these were working Models of Steam Engines of different construction. (47)

Through all his experiments on models, Cumming became familiar with steam or fire engines but admitted that he 'knew Newcomen's Engine and had made improvements upon it but so far inferior to Mr Watts that I consider my own improvements as nothing'. (48) Here was a person with ample financial, academic and technical backing who failed to make the vital discovery and it is interesting to compare his and Watt's positions. Watt was, at first, self-financed and most likely made his early models himself.

John Smeaton

The best known example of a person making improvements to the Newcomen engine at this period was Smeaton but this was after Watt had invented the separate condenser and had

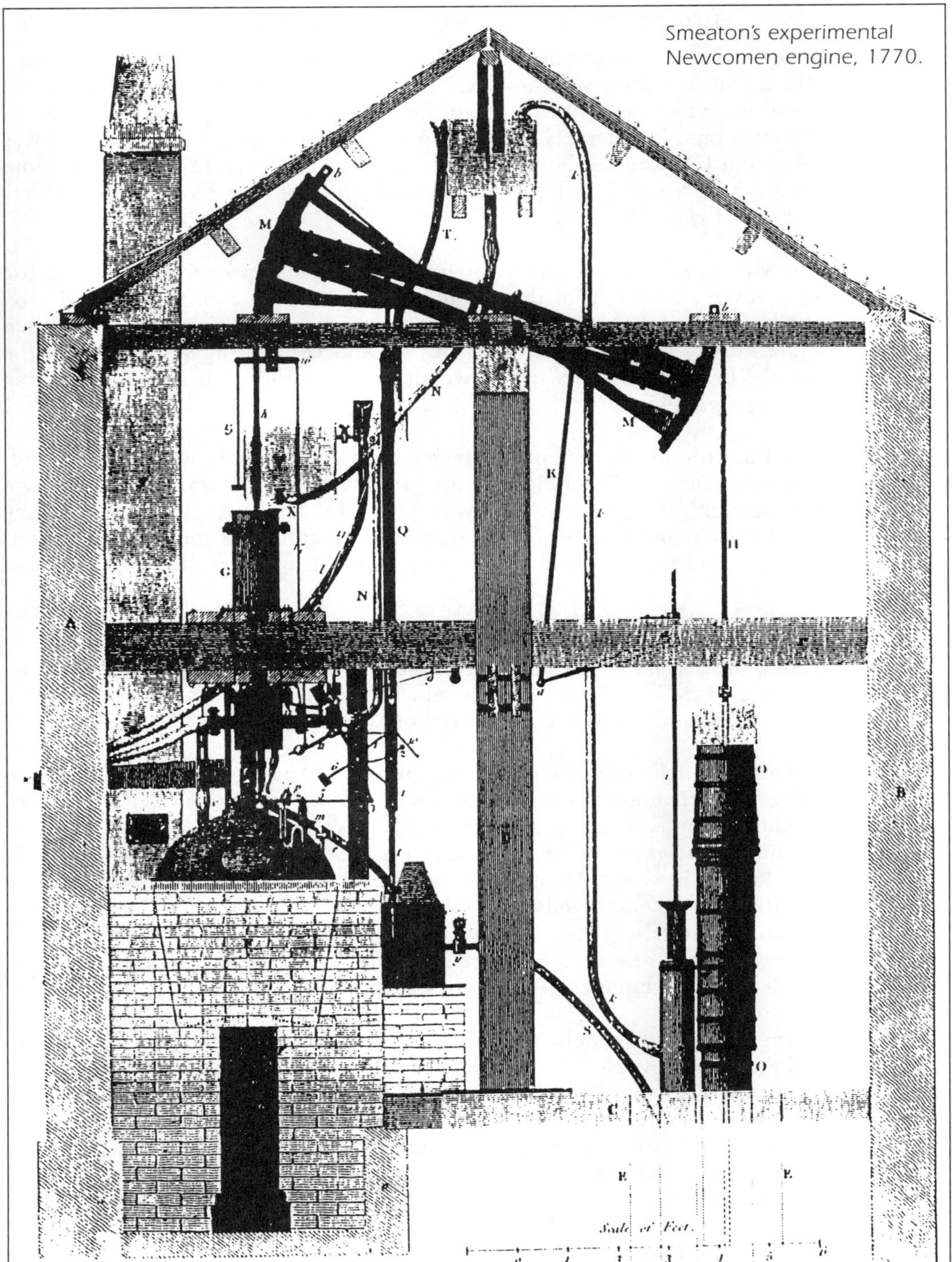

Smeaton's experimental Newcomen engine, 1770.

stopped his experiments at Kinneil. Smeaton had installed an engine for the New River Company which supplied London with water. When set to work early in 1769, it did not perform as well as he had hoped. He said, 'By good luck, the Engine performed the Work it was expected to do, as to the raising of Water, but the Coals by no means answered calculation'. (49) Therefore he constructed an experimental engine with a 10 ins. cylinder at his home at Austhorpe, near Leeds. Parts were supplied by the Carron Company from Scotland and it was erected during the summer of 1770. (50) Smeaton hoped to set it to work in less than a week after he had written to Watt on 8 September. (51)

Watt reported to Small about a meeting during the summer of 1770.

> I had... a visit of Mr. Smeaton on our canal, and that he behaved to me in a most gentlemanly, friendly manner. He is trying experiments on fire-engines... I told him as much about my engine as is contained in the specification, or thereabouts; principally with a view to check the ardour of his invention... He told me of an experiment he had made on the New River engine which seems to say that common engines can be made to do more than I imagined. (52)

The canny Scot gave nothing away and must have realised that he would have to make his own engine as perfect as possible. A little later, Watt informed Small that 'Smeaton started some objections about the air that would enter with the steam, which I answered; he then said a great deal in favour of the contrivance, and desired me to push it'. (53) Writing some years later, Smeaton said that, although he had thought of condensing in a separate vessel, he did not see any way of clearing the condensate and air out of that vessel other than by using steam, which would of course have heated up the condenser. He thought,

> For if he [Watt] snifts *that*, he will lose as much steam in that vessel, as he will save in the cylinder. And it was not until the summer following, viz of 1771 (in which interim I had made some fundamental experiments) that I learnt from Mr. Watt himself, that his method of clearing the condenser of air and water, was by an air pump. (54)

Watt told Smeaton that, through other problems with his engine and being wholly occupied in another career, he had laid aside his engine. As a result of this, Smeaton said he was 'determined to prosecute my original intentions of finding out the true *Rationale* and making the best of Newcommen's construction'. (55) This he proceeded to do, so Small, urging Watt on, warned in 1773, 'Smeaton has only trifled hitherto; but he may perhaps discover something'. (56)

As a quid pro quo for Smeaton's advice on Hamilton Bridge in 1770, (See p272) Watt helped Smeaton by sending him details of tests on various engines in Scotland to which Smeaton replied,

> I am obliged to you for the proportions you have given me of different Engines. I have not yet had time to compare their effects, with that of the New River. All the proportions there were taken from my own observations except the measure of coals, but this was done by a person appointed by the Company for weeks together who I suppose would ?...? give in more coals expended than really were than less. I find in looking over my notes that I made a small mistake in the quantity I gave you...
>
> The performance of the Wemyss Engine seems extraordinary, but it is usual to rate the effects such as can be produced for a few minutes by pushing the fire to the utmost; and the Coals by the ordinary consumption when the engine does not make so many strokes

in a Minute by 2, & is perhaps a foot short of her full compass. My experimental Fire Engine I expect will be started in less than a week & probably will furnish me with some facts I am not aware of. (57)

Relations between Smeaton and Watt always remained cordial.

From the performance of the New River Engine, Smeaton suspected that there was a great waste of steam when it first entered the cylinder. He hoped to take advantage of condensation through the cylinder walls to reduce the volume of injection water by having a long narrow cylinder. These were some of the problems he hoped to elucidate with his experimental engine. (58) He made over 130 tests, altering one particular factor and again observing the performance. In this way he learnt how to optimise the piston loading, the timing of the valves, the form and size of the injection nozzle and so on. He also made some experiments with a Florence flask as Watt had done to determine the correct volume of steam produced from water at boiling point as well as the correct volume of steam condensed by the cold surface of the cylinder. From this Smeaton deduced that large diameter cylinders would be more economical through having less surface area in proportion to volume. He found that the amount of steam needed to reheat the cylinder was at least the equivalent to the volume of the cylinder itself. (59) He also studied coal consumption both in relation to the boiler and to the different layouts of his engine. He had to reduce the loading on the piston from 10 $^1/_4$ lbs. to 8 $^1/_4$ lbs., and so may have been investigating pressures and not temperatures. Watt on the other hand, was looking at temperatures and so was on the right track for thermodynamic principles.

Smeaton's boring machine.

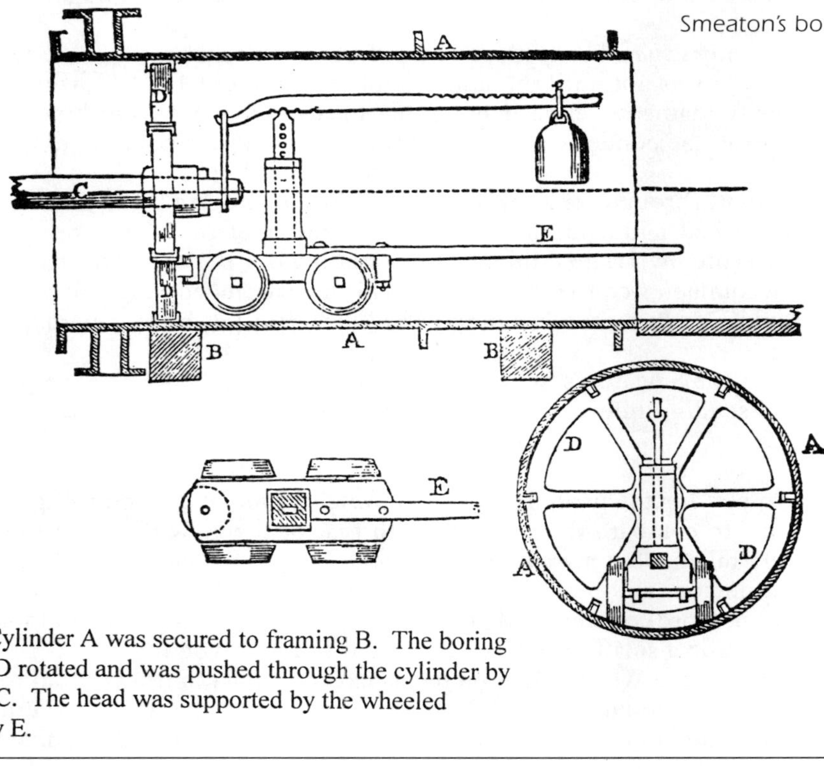

The Cylinder A was secured to framing B. The boring head D rotated and was pushed through the cylinder by shaft C. The head was supported by the wheeled trolley E.

Smeaton drew up a table of his results which showed the best proportions for the different parts of engines of different sizes and power. (60) Based on these, he designed a pumping engine for Long Benton colliery, Northumberland, in 1772. Tests on the old engine here with a 52 ins. cylinder showed it developed 30 h.p. and a duty of 4.6 million. This duty figure was calculated from the number of pounds weight of water raised one foot high with a bushel of coal. The new engine, which had the same cylinder diameter, was set to work in February 1774 (61) and tests that May gave 40 h.p. with an impressive duty of 9.1 millions. This represented an increase of 25 per cent on the most efficient engine previously built and an improvement of 40 per cent on the typical good performance of the average Newcomen engine at that time. (62)

Although he never built Newcomen engines in Scotland during Watt's time there, Smeaton did act as technical adviser to the Carron Company from at least June 1768 when he saw their fire engine. (63) His first surviving report in February 1769 concerned improving the blowing apparatus for No. 4 furnace (64) and in the following summer he proposed alterations to the waterwheels to make them more efficient and reduce water consumption. (65) It was not until October 1770 that he submitted his proposals for a new double boring mill for steam engine cylinders and guns. (66) This was after he had received the cylinder for his experimental engine as well as after Watt had finished his trials at Kinneil. Smeaton's cylinder boring machine had a cutting head that was supported only at one end by the drive shaft. Therefore the head lacked guidance as it was being pushed down the bore and, while achieving a reasonably circular bore, would not have given a straight one. However it seems to have satisfied Carron in 1774 when they told Smeaton,

> With regard to the cylinder she is bored quite thro from Top to Bottom, the 66 inch cylinder I mean. she is perfectly true and smooth, but we shall put her on the mill and scower her a little with emery...
> We shall proceed with the Cast Iron Boiler agreeable to your letter... We shall certainly give the first Boiler we cast a trial. (67)

Earlier, in April 1773, the idea of cast iron boilers to Smeaton's designs had been put before the main governing Board of the Carron Company and was examined with great care because 'If the Boiler succeeds it may introduce an important new sale at Carron'. (68) Cast iron boilers were supplied for the Kronstadt engine but the material was not suitable for this application. Watt never used cast iron boilers, which may have been due to experience gained in Scotland.

Smeaton had probably improved the Newcomen engine as far as was practicable. Yet he failed to discover the vital step in their improvement. John Farey, writing in the early years of the nineteenth century, considered that the development of the steam engine would have been more rapid had it not been for Smeaton's improvement in waterwheel design and construction, 'whereby the performance of almost all the mills then existing, could be doubled with the same supplies of water'. (69) His improvements to the water supply and waterwheels at Carron enabled them to dispense with steam engines. (70) It has been claimed that Smeaton's paper to the Royal Society on wind and watermills based on his experiments made in 1752 to 1753, was 'one of the greatest contributions to engineering science made in the eighteenth century'. (71) Indeed, his tests on windmills were quoted in 1895 as being 'still the most reliable data extant'. (72)

Smiles claimed that Smeaton 'limited his professional employment, that he might be enabled to devote a certain portion of his time to self-improvement and scientific investigation'. (73) Smeaton described his methods as 'experiments made on working models which I look upon as the best means of obtaining the outlines in mechanical enquiries'. (74) He used what has been termed more recently as 'parameter variation' where one particular aspect could be varied and measured at a time. In the case of a waterwheel, these parameters might be the head of water, quantity of water acting on the wheel, the speed of the wheel, the load, the number of blades or buckets, and so on. He followed the same pattern with the experiments on his model Newcomen engine. In addition, he conducted systematic trials on engines in the Newcastle area. He was able to improve performance by proportioning the working parts better, but his experiments did not lead to any fundamental change in the way of operation or lead to any new theories.

Smeaton had realised that 'It is very necessary to distinguish the circumstances in which a model differs from a machine in large; otherwise a model is more apt to lead us from the truth than towards it'. (75) He went on to point out,

> The best structure of machines cannot be fully ascertained, but by making trials with them, when made of their proper size. It is for this reason that the models [watermills and windmills] referred to, and the greatest part of the following experiments were made in the years 1752 and 1753, yet I deferred offering them to the Society, until I had an opportunity of putting the deductions made therefrom in real practice, in a variety of cases. (76)

While these methods were successful with waterwheels and answered the question, 'Which is the more efficient type of wheel?' they ultimately limited the scope of Smeaton's work, for they were bound to exclude the possibility of examining, or even disclosing, some questions and issues of a precise and fundamental nature. (77) The Smeatonian method tends to the perfection of known inventions and not to the production of new ones. (78) John Farey commented, 'Mr. Smeaton must be mentioned as a most eminent example of sound judgement in mechanical combinations, without great invention or power of creating original ideas'. (78) Watt's scientific approach resulted in the crucial breakthrough.

Notes: The University of Glasgow Model Newcomen Engine

1 Muirhead, (1854), Vol I, p. xciv and Dickinson & Jenkins, (1927), p. 22.
2 Robinson & McKie, (1970), p. 434, J. Watt's Note Book, JWP, W/14, p. 7 r.
3 Muirhead, (1854), Vol I, p. xciv.
4 Muirhead, (1858), p. 65.
5 Goodison, (1992), p. 242.
6 See Rolt & Allen, (1977), p. 70 for an example of Leupold's drawing of the 1722 Konigsberg engine where the cross beams are not shown.
7 Robinson & McKie, (1970), p. 418, Letter 259, 'Copy of Mr. Watt's Remarks on Mr. Robison's Edition of Dr. Black's Lectures', 4 Jan. 1809.
8 B. & W. Col., MII, 5/2, Invoice from J. Fleming, 29 Feb. 1765.
9 Muirhead, (1858), p. 75.
10 Ibid, p. 63.
11 Ibid, p. 62.
12 Ibid, p. 72.
13 Ibid, p. 64.
14 Leupold, (1725).

15 Rolt & Allen, (1977), p. 23.
16 Switzer, (1729), Vol II, p. 325 ff.
17 Muirhead, (1858), p. 75.
18 Belidor, (1782), Vol II, pt. 1, pp. 308 - 329.
19 Desaguliers, (1763), Vol II, p. 312 ff.
20 Muirhead, (1858), p. 75. See also Robinson & McKie, (1970), p. 418, Letter 259, 'Copy of Mr. Watt's Remarks on Mr. Robison's Edition of Dr. Black's Lectures'. 4 Jan. 1809.
21 Robinson & McKie, (1970), ibid.
22 Law, (1969), p. 17.
23 Robinson & McKie, (1970), p. 419, Letter 259, 'Copy of Mr. Watt's Remarks on Mr. Robison's Edition of Dr. Black's Lectures', 4 Jan. 1809.
24 Hills, (1989), p. 29 and Rees, (1972), Vol V, p. 92.
25 Robinson & Musson, (1969), p. 50, and JWP, JW4/24.5, 'Draft Steam-engine Specification'.
26 JWP, Letter Book 4, p. 102, 'J. Watt's Recollections of his Friend, Dr. J. Robison', April 1805.
27 Smeaton, (1794).
28 Cardwell, (1971), p. 69.
29 Sobel, (1996), p. 31.
30 Muirhead, (1858), p. 62.
31 Sobel, (1996), p. 120.
32 Robinson & McKie, (1970), p. 411, Letter 257, 'J. Watt's Recollections of his Friend Dr. J. Robison', April 1805.
33 Rees, (1972), Vol V, p. 68.
34 Ibid, Vol V, p. 93.
35 Stuart, (1829), p. 225, note. See also Boucher, (1968), p. 40.
36 Patent 848, 9 Oct. 1766.
37 Hills, (1989), p. 37.
38 JWP, 4/69, 'Comparison of J. Watt's Steam Engines and answer to Blakey, 1775'.
39 *Transactions of the Newcomen Society*, Vol XVI, J.J. Bootsgezel, 'William Blakey - a rival to Newcomen', p. 102.
40 JWP, 4/48.11, J. Magellan to J. Watt, 28 Aug. 1781.
41 JWP, 4/78, H. Gainsborough to J. Watt, letters 16 May 1775 to 2 Aug. 1776.
42 Dickinson & Jenkins, (1927), pp. 301 - 303.
43 JWP, 4/78, J. Watt's notes.
44 JWP, 4/78, H. Gainsborough to J. Watt, 16 May 1775.
45 Dickinson & Jenkins, (1927), p. 302 from Gregory's *Mechanics*.
46 Ibid, p. 302. See also JWP, 4/20.30, G. Hamilton to J. Watt, 9 Oct. 1776.
47 JWP, 3/29.2, p. 47, Court of Common Pleas, Boulton and Watt v. Hornblower and Maberly, 1796.
48 JWP, 3/9, p. 126, Court of Common Pleas, Boulton v. Hornblower, 1796.
49 Skempton, (1981), p. 181. See also B. & W. Col., MII, 4/9.7, J. Smeaton to J. Watt, 5 Feb. 1778.
50 SRO, Carron Co., GD58/6/1/9, Letter Book, 1769 - 1770, pt. 1, p. 478, 16 March

1770, to J. Smeaton esq., Austhorpe, 'The above working barrell & cap is directed to the care of Mr. John ?...?, Hull with instructions to forward it to you on arrival of the vessel'.

51 B. & W. Col., MI, 3/60, J. Smeaton to J. Watt, 8 Sept. 1770. See also B. & W. Col., MII, 4/9.7b, J. Smeaton to J. Watt, 5 Feb. 1778.

52 Muirhead, (1854), Vol II, p. 9, Letter 30, J. Watt to W. Small, 30 Sept. 1770.

53 Ibid, Vol II, p. 14, Letter 87, J. Watt to W. Small, 20 Oct. 1770; Boulton 348.21.

54 B. & W. Col., MII, 4/9.7b, J. Smeaton to J. Watt, 5 Feb. 1778.

55 Ibid.

56 Muirhead, (1854), Vol II, p. 64, Letter 124, W. Small to J. Watt, 8 Oct. 1773; Boulton 340.35.

57 B. & W. Col., MI, 3/60, J. Smeaton to J. Watt, 8 Sept. 1770.

58 B. & W. Col., MII, 4/9.7b, J. Smeaton to J. Watt, 5 Feb. 1778.

59 Rees, (1972), Vol V, pp. 89 - 90.

60 Skempton, (1981), p. 182.

61 B. & W. Col., MII, 4/9.6, J. Smeaton to J. Watt, 5 Jan. 1778.

62 Skempton, (1981), p. 186.

63 SRO, Carron Co., GD58/6/1/7, Letter Book, 1767 - 1768, p. 33, to J. Smeaton, 2 July 1768.

64 Smeaton, (1812), Vol I, pp. 359 - 361, 'Concerning the Improvements of the Blast at the furnace at *Carron*, No. 4'.

65 Ibid, Vol I, pp. 364 - 375, 'Explanation for the Design of the Blast Machine for the Furnace No. 2 at *Carron*'.

66 Ibid, Vol I, pp. 376 - 379, 'Some Remarks concerning the design for a double boring mill for cylinders and guns, to be erected upon the tail-water of the *Carron* Works'.

67 SRO, Carron Co., GD58/6/1/13, Letter Book, 1773 - 1774, p. 332, to J. Smeaton, 2 Feb. 1774.

68 Ibid, GD58/1/1, Minutes of General Court, 1760 - 1774, p. 484, April 1773.

69 Farey, (1827), p. 296.

70 SRO, Carron Co., GD58/2/1/1, Minutes of General Court, 1760 - 1774, p. 486, from J. Gascoigne, April 1773.

71 Skempton, (1981), p. 15.

72 Hills, (1994), p. 91.

73 Smiles, (1874), *Harbours, Lighthouses, Bridges*, p. 171.

74 Smeaton, (1794), p. 1.

75 Ibid, p. 2.

76 Ibid, p. 1.

77 Skempton, (1981), p. 38.

78 Rousseau & Porter, (1980), p. 470.

79 Farey, (1827), p. 652.

Watt's Experiments on Newcomen Engine Models

We have left Watt and Robison puzzling over why the University model Newcomen engine should have consumed so much steam and condensing water. There are four main sources describing how Watt would turn his investigations into serious scientific studies which would inspire his invention of the separate condenser. One of his note books survives which has a date of 1765 at the top of a relevant page. (1) While the primitive spelling may indicate an early date, the way in which some entries are written suggests that they were transcribed into this book after the events had taken place. Robison gave his version of what happened at the Hornblower and Maberley trial in 1796, but that was thirty years later and so may not be always correct. At the time of this same trial, Watt compiled what Muirhead called 'A Plain Story', which concentrated more on the later experiments with his own model Newcomen engines. (2) Finally, during the winter of 1813 - 14, Watt wrote some notes for Robison's *System of Mechanical Philosophy* which were edited by Brewster but not published until 1822. (3)

Watt's First Model Newcomen Engine

In his evidence at the patent trial in 1796, Robison implied that he helped Watt from the beginning with experiments on the University model but, according to his note book, Watt started experimenting before Robison returned to Glasgow and in fact had managed to make it work reasonably. (4) Watt's 'Ledger of Personal Accounts' (5) shows that he began to sell Robison some hair powder and other similar items only from December 1763, so it is likely that this reflects the date of his friend's return. Watt had been surprised at the immense quantity of fuel consumed in proportion to the size of the cylinder which he 'imputed to the heat lost thro the Metallic Cylinder'. (6) This was also investigated by both Smeaton and Brindley who recognised the loss of heat by radiation from the surface of the cylinder. Even before that, Desaguliers had realised in 1744 the advantage of a brass cylinder for 'heating and cooling quick'. (7) Watt seems to have thought that, in a smaller cylinder, the volume of steam would be less in proportion to the external walls and so the cylinder walls would have a greater cooling effect. Larger engines would need proportionately less steam. (8) At this point, Watt appears to have abandoned trials on the University model. He may have felt that the next experiments he wished to carry out would have altered it so radically that it would no longer be the same engine which he had been asked to repair. He returned it and built his own.

Robison wrote that he returned to find Watt still settled in Glasgow, 'as fond of science as ever. Our acquaintance was renewed'. (9) Robison's interest in the University model was roused and he studied it as much as Watt did. He claimed how Watt,

> very early saw that an enormous quantity of steam was wasted. The great heat acquired in an instant by the cylinder, by the admission of a few grains of water in the form of steam, was an incontestible proof of the great quantity of fire contained in it; and as this could only come from the coals, Mr. Watt saw at once that the chief improvements that the engine was susceptible of must consist in contrivances for increasing the production [of steam], and for diminishing its waste. (10)

Here was the germ of a research programme in which Watt would try both to improve the performance of his boiler as well as to reduce the steam consumption of the engine. He would go far beyond just making the engine perform better and would seek to understand the basic principles on which it worked.

Watt was able to solve the questions which were raised during these trials by using three scientific instruments, the balance for determining weights, the barometer both for measuring atmospheric pressure as well as turned into a manometer for measuring steam pressures and, most important of all, the thermometer. Watt had constructed barometers as part of his scientific instrument business and likewise he was familiar with thermometers, several of which are preserved in his Garret Workshop but we have no way of determining their date. A book which he most certainly would have read was Martine's *Essays and Observations on the Construction and Graduation of Thermometers*, parts of which had been published at Edinburgh in the 1740s. Watt cleaned a balance for Anderson's class in November 1758. He may have received advice about balances from Black whose especial triumph in his doctoral thesis consisted in showing that chemical changes which occurred in his series of reactions could, without isolating the 'fixed air', be detected by weighing them at every stage on a balance. (11) As well as a thermometer, Roebuck wanted Watt to purchase for him a small pair of scales and weights in 1765. (12) One of Watt's balances is preserved in the Greenock Museum but, as this looks quite a good one in a wood and glass case and its style is later, it is unlikely to have been the one he had in 1763. At this period, the chemical balance was still fairly primitive and later ones would become much more accurate. (13)

Watt's thumb-nail sketch of his boiler with wooden sides. The fireplace was 7 inches diameter, the chimney three and the height of the part containing the steam and water fourteen inches.

The Boiler

According to Robison, Watt had seen the need for 'increasing the production', and, like Brindley before and Blakey and Joseph Hately after him, he turned his attention to the boiler. He began 'by Making the boyler perform with as little heat as possible'. (14) He tried a form of what we would call a flash steam boiler by injecting water only when the steam was wanted. From this arose an early experiment, possibly based on one of Black's in 1762, (15) with a small iron pan placed on a stove. Watt checked to see how long it took to evaporate a given measure of water in the pan and then timed the evaporation of the same volume when put in the pan drop by drop. The second method took marginally longer, 22 minutes as opposed to 20. After further experiments, such as blocking off part of the heating surface with a piece of wood, Watt came to the conclusion that 'the evapn of water is neither in proportion to the surface nor quantity of it but as the quantity of heat that enters it'. (16) He also realised that the heat source should be placed as close to the boiler as possible and that heat would be transferred more quickly through thin sheets of metal rather than thick. 'After many fruitless thoughts on the subject, I saw no boyler so perfect in that Respect as the Common tea kitchen (an Invention for which we are beholden to the Chinese)'. (17) This was more like a samovar rather than the western style kettle placed over a fire.

The boiler which Watt designed had the fire below the centre, surrounded by a thin iron casing resembling a modern milk bottle. The flue passed out through the top. The water was contained around this in an inverted truncated cone with a flat upper surface, the chimney protruding through the middle. The broader volume at the top would have given a large steam space. The outside of the cone was made of wood which was an excellent insulator. However, Watt found that those parts of the wooden casing which were above the water level soon became soft even when made of oak so that he could stick a piece of white iron into them. (18) While he could not have known that the lignin in wood will begin to soften at temperatures little above 100°C., it seems surprising that he did not recognise that wood can be softened by steaming which shipbuilders were doing by then. Watt said that his boiler 'answered my expectation evaporating prodigious Quantitys of Water & consuming little fuel'. (19) He would have used his balance to discover that it burnt four pounds of fuel an hour and evaporated twenty pounds of water (20) but we are given no other figures and therefore can make no comparisons. Yet this shows that he was aware of the fuel consumption. A drawing of his first separate condenser model, probably produced for the 1796 trials, shows a similar design of boiler on which the wooden casing has been replaced with sheet metal. Presumably Watt was well satisfied with his boiler design and continued to use it for some time. (21, see p367)

The Wooden Cylinder

Having succeeded in improving the boiler performance, Watt turned his attention to the engine itself. It seems it was Robison who 'hinted in some of our Conversations together that if the Cylinder was made of wood, it would not occasion the Loss of so much heat being less susceptible of heat or cold than Mettals'. (22) This marks the beginning of serious experiments which would change Watt from the mere mechanician into the natural philosopher. At that time he had not heard of Brindley's experiments. He looked for some substance with which to make the cylinder that would 'receive and give out heat slowly' so he constructed one from fir, 6 ins. diameter, 18 ins. long, hooped with iron, giving a 12 ins. stroke. (23) He tried to harden it with linseed oil. The size of this cylinder was quite large so it is possible that he followed Smeaton's warning about models. (See p326) However, after many experiments, the steam softened the wood and the cylinder bent inwards in the middle. But the performance must have been sufficiently encouraging for him to replace the fir with a harder type of wood which also became soft. He used his thermometer and observed that the water which came out of the eduction pipe was always nearly boiling and that, with steam at a low pressure, it took a long time before the snifting valve blew off and that the condensate in the cylinder could not be forced out until it was near boiling. What was happening was that the incoming steam had to reheat the cylinder, piston and remaining condensate to boiling point so that it could force out the contents against atmospheric pressure and prevent further condensation. Watt added 'I have since found that the Quantity of Steam Condensed would be many fulls of the Cylinder'. (24)

Here we see the concept of his perfect engine appearing. However, the consumption figures that Watt recorded did not correspond with those given by Desaguliers for large engines. This was another line of investigation to be followed up. He also found that when trying to improve the vacuum by throwing in more injection water, a disproportionate waste of steam resulted. Watt said that, on reflection, this seemed to be caused by the boiling of water in a vacuum and that it was necessary to cool the cylinder 'below the point at which water would boil in an exhausted vessel' if he were to achieve a perfect vacuum. (25) He must have been testing this with a manometer. Here were two further problems to be solved and therefore Watt's first model Newcomen engine with the wooden cylinder was very important in determining the line of his future enquiries.

After the failure of his wooden cylinders, Watt considered whether a series of Savery engines placed one above the other in a mineshaft working only with the suction might be a better solution. The steam would be supplied through lagged copper pipes from a boiler on the surface, but this idea was quickly abandoned (26) and he was forced to reconsider how to improve the Newcomen engine. Although he does not mention it, he must have made another model with a metal cylinder and continued his next experiments with that. He recognised that 'before anything could be known certainly about the best manner of Constructing Fire engines, there were several facts necessary to be determined'. (27) This would start him off on a voyage of scientific discovery reaching far beyond Newcomen engines. On these experiments he 'spent much time, and more money than was suitable to his circumstances, yet he made no advances towards a beneficial discovery'. (28)

Volumes of Steam and Water

Even with his wooden cylinder, Watt discovered that 'the steam condensed in filling it still exceeded the proportion of that required for large engines, according to the statements of Desaguliers'. (29) This may be the point where Watt began to realise the importance which heat played in the functioning of the steam engine as well as steam, even though he had been involved with heat in his boilers. Black said in his written evidence for the 1796 trial,

> After he [Watt] had been thus employed a considerable time, he perceived that by far the greatest waste of heat proceeded from the waste of steam in filling the cylinder with steam. In filling the cylinder with steam, for every stroke of the common engine a great part of the steam is chilled and condensed by the coldness of the cylinder, before this last is heated enough to qualify it for being filled with elastic vapour or perfect steam; he perceived, therefore, that by preventing this waste of steam, an incomparably greater saving of heat and fuel would be attained than by any other contrivance. (30)

Early experiments showed that three quarters of the steam were wasted during the ascent of the piston and later tests with better apparatus confirmed that the loss was much greater than this. (31)

Here we see the importance of Watt's concept of the perfect engine as a standard. Now he needed to know the actual steam consumption to discover just how many cylinders full of steam his engine was taking. He realised that he could determine the volume of steam from the volume of water, provided he knew the ratio of their volumes at boiling point. With this information, he could measure the volume of water evaporated from his boiler which was the easiest way of determining steam consumption. This question would lead him to doubt the figures printed by Desaguliers, who repeated many times that 'water in boiling is expanded 14000 times to generate steam as strong (i.e. as elastick) as common Air' and stated that this figure had been obtained through a great many observations made by himself and Henry Beighton. (32)

At Black's suggestion, (33) Watt took a Florence flask and inserted a glass tube which reached nearly to the bottom, carefully sealing the joint. He must have known how much water this flask contained when full. He poured about an ounce of water into it and placed it in an oven for over an hour to slowly boil the water and drive out both air and excess steam until all the water had been boiled away. It was allowed to cool and then weighed - the balance again. The flask was carefully dried and weighed again when it was found to be 4 grains lighter. Watt wrote, 'Consequently the water had expanded to above 1600 times its bulk'. (34) Watt later discovered that he had used the wrong figure for grains in a drop weight so that the expansion of steam should have been 1849, 'a Cubic Inch becoming a Cubic foot'. (35) This experiment was repeated with the Florence flask lying on its side without the glass tube. Watt later stated that he did 'not consider these experiments as extremely accurate, the scale beam of a proper size which I had then at my command not being very sensible'. (36)

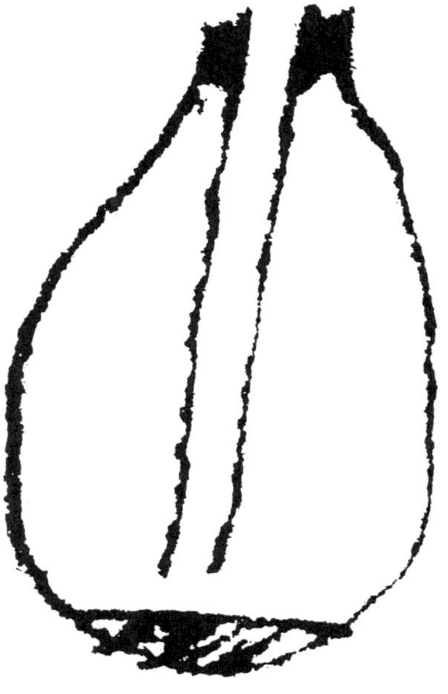

Watt's thumb-nail sketch of his Florence flask with a pipe inserted and the water at the bottom which would be turned into steam.

Watt had discovered the ratio of the volume of steam to that of water through determining the weight of the steam or water left in the flask. From then on he could calculate the steam consumption of an engine by measuring the amount of water evaporated from a boiler through either the weight with a balance or the volume of that water. He built a boiler

> which showed, by inspection, the quantity of water evaporated in any given time, and thereby ascertained the quantity of steam used in every stroke by the engine, which I found to be several times the full of the cylinder. (37)

Watt does not state how he measured this quantity. One way could have been by weighing the whole boiler before and after any trial. Another could have been by refilling the boiler to a level marked on it from a known volume of water. The volume of water would have been weighed first and then what was left weighed again. This shows the importance of the balance, which of course could have weighed the fuel as well.

Heat Capacity of Different Materials

Having confirmed that his engine was using many cylinders full of steam instead of one, Watt must have asked what was the cause of so much being wasted. We have seen that he attributed one reason to loss of heat through the radiation from the cylinder walls and hence his use of wood, but he also realised that the cylinder walls themselves absorbed a great deal of heat which was another reason for his trials with wooden cylinders. He quickly deduced that the cylinder walls became a source of heat when the steam was being cooled to create the vacuum. He therefore turned his attention to the heat capacities of different metals because he recognised that he had to revert to a cylinder made of some metal. He knew about Desaguliers's comments on cylinders of atmospheric engines.

> Some people make use of cast Iron Cylinders for their Fire-engines; but I would advise nobody to have them; because, tho' there are Workmen that can bore them very smooth, yet none of them can be cast less than an Inch thick, and therefore they can neither be heated nor cool'd so soon as others, which will make a Stroke or two in a Minute difference whereby an eighth or tenth less Water will be raise'd. A Brass Cylinder of the largest Size, has been cast under $^1/_3$ of an Inch in thickness; and, at the long run the advantage of heating and cooling quick will recompense the Difference in first Expense; especially when we consider the intrinsick Value of the Brass". (38)

Boerhaave had recorded Fahrenheit's experiments with water and mercury which had caught Black's attention and prompted his discovery that different substances have different capacities for heat. (39) Black could have discussed this with Watt. Robison wrote that Watt learnt the quantity by warming the cylinder and other parts to determine their heat capacities when he was investigating the amount of ineffective condensation. 'He made many experiments on various substances, in order to learn their properties in this respect'. (40) In 1780, Watt told J.J. Magellan that he 'was the first that tryed experiments to find the quantities of heat absorbed by wood and some other bodies in the year 1763 but invented no theories on the subject'. (41)

The only surviving record of these experiments, which would have used both the thermometer and the balance, is a short passage in Watt's notebook.

> a piece of Iron being heated to 120° & then plunged into water at 60° was found to have heated the water as much as an equall bulk of [water] heated to 120°
> Copper was found to do the same
> tin is said by Dr. Black to hold the half of the heat of an equall bulk of water. (42)

Robison said it was Watt who first 'considered the subject steadily and in a system' and drew Black's attention to the significance and practical importance of such research. Black carried on with this research and, through his students, (43) Robison claimed, 'had many experiments on the heats communicated to water by different solid bodies, and had completely established their regular and steady differences'. One result, according to Robison, was that Black recognised Watt's 'uncommon genius' and that this 'was the beginning of a friendship which lasted through life'. (44) While Black and Watt were close friends before this, it could well be that Watt's experiments on the steam engine showed him in a different light and may have subtly altered the form of their friendship.

From knowing the heat capacity of a metal and knowing the mass or weight of a cylinder, Watt could determine the amount of cold water at a given temperature which would be needed to cool that cylinder from boiling point to whatever temperature he wanted. For these experiments, he would use both the thermometer and the balance. At this period, he was assuming that the amount of heat in boiling water and in steam was the same. He already knew how much steam the cylinder contained. By the law of mixtures, he could calculate how much cold water should be needed to cool that steam down as well. But, when he came to add onto the figure for cooling the cylinder that for cooling the steam within it, his theoretical sums would not have given the same result as he obtained in practice. He found that the water coming out of his engine was hotter than he calculated. He was 'astonished at the great quantity of water required for the injection, and the great heat it had acquired from the small quantity of water in the form of steam which had been used in filling the cylinder'. (45) He wrote in his notebook, 'Injection in fire engines was much greater that I thought was necessary to Cool the Quantity of water contained in the Steam down to below boiling point'. (46) He would have to carry out different experiments to find out why.

The Boiling Points of Water

Before we examine these other experiments, the heat capacities of metals gave Watt the answer to another problem. He had quickly recognised that any hot water in the cylinder might be turned into steam as the pressure was reduced and that this would be exacerbated by the heat contained in the cylinder walls. This would partially vitiate the vacuum and give a residual pressure inside the cylinder which would resist the pressure of the atmosphere. (47) He knew of Cullen's work in this field and had probably read Cullen's paper of 1756 which dealt with the phenomenon of cooling by evaporation, particularly under reduced pressures. (48) Yet this essay does not mention any temperatures at which water will boil in a vacuum. Watt wrote in 1813 - 14,

> Experiments made long before by Dr. Cullen, John Robison, and others, in public classes, which proved that water, when placed in an exhausted receiver, boiled, and was converted into steam at the heat of 70° or 80° of Fahrenheit's thermometer, while it was well known that under the pressure of the atmosphere it required 212° of heat to make it boil, and emit steam capable of displacing the air. It was evident that, under intermediate pressures, intermediate degrees of heat would be required to make it boil. (49)

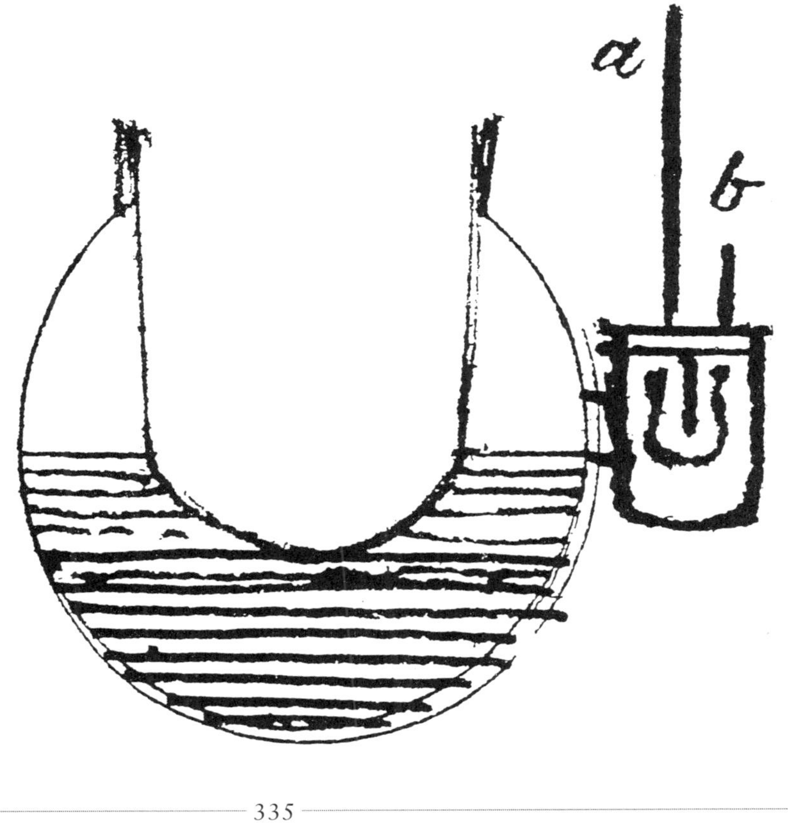

Watt's apparatus for determining the temperature of steam at various pressures. 'a' is the mercurial manometer and 'b' the thermometer. The circular vessel is the boiler.

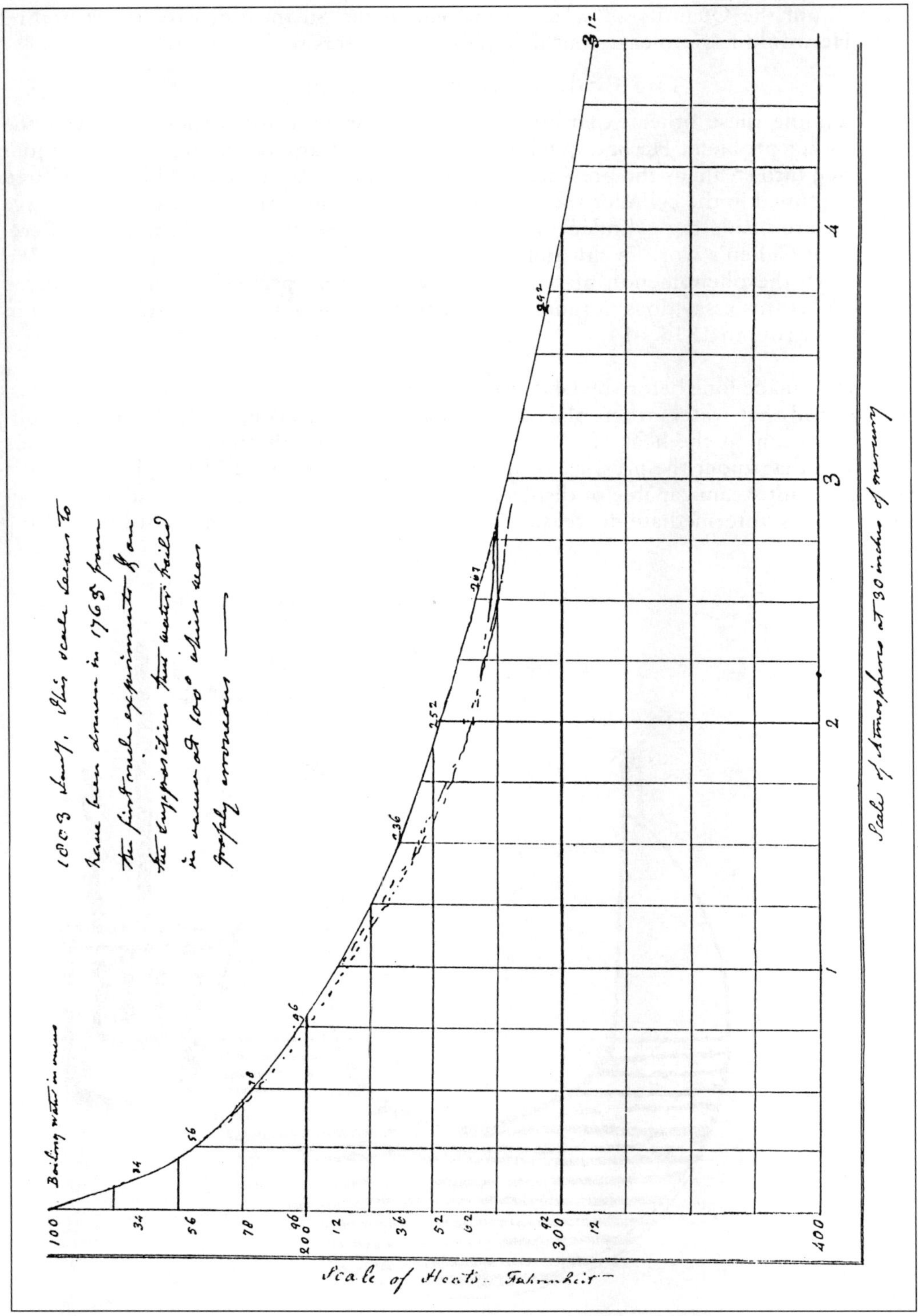

Watt's first graph showing the relationship of the heat of steam and its pressure drawn in 1765.

The extent of Watt's knowledge of this in 1764 is unknown, particularly the lower temperatures of water boiling under lower pressures, for he seems to have thought that this ceased at 100°F. (50) He proceeded to carry out his own experiments into the boiling points of water.

In Watt's notebook, there is an undated drawing of a spherical copper vessel which could be partially filled with water. (See p335) Connecting with this vessel was a smaller container into which were sealed the lower ends of a thermometer and a closed mercurial manometer 34 ins. long so that both could be read at the same time. The manometer gave readings above normal atmospheric pressure. In 1815, he told Sir Joseph Banks that he had recorded temperatures when the steam supported columns of mercury 15, 60 and 75 ins. high. (51) From his readings, Watt plotted a graph of pressure against temperature. This must be a very early example of such a method instead of the column or tabular arrangement more usual at the time and which we find, for example, in Smeaton's *Experimental Enquiry*. (52) A note added in January 1803 on a graph preserved at Birmingham reads, 'This scale seems to have been drawn in 1765 from the first rude experiments & on the supposition that water boiled in vacuo at 100° which was grossly erroneous'. (53) Watt wrote,

> By experiments which I then tried upon the heats at which water boils under several pressures greater than that of the atmosphere, it appeared that when the heats proceeded in an arithmetical, the elasticities proceeded in some geometrical ratio; and by laying down a curve from my data, I ascertained the particular one near enough for my purpose. (54)

If he were to achieve any accuracy when comparing these experiments on different days, Watt would have needed to know the prevailing atmospheric pressure from his barometer. He certainly did this on further experiments in 1774. (55) Watt interpolated the curve to below atmospheric pressure to see the relationship between pressure and temperature of the boiling point of water in a partial vacuum. His second curve showed him the necessity of cooling the steam down to a much lower temperature than the 100°F which he thought originally necessary if he were to achieve a perfect vacuum in his steam engine. The large volume of water he found he needed to do this was another problem that had to be solved.

Latent Heat

In the Newcomen engine, the vacuum was created by Newcomen's crucial discovery of direct injection of cold water into the cylinder. (56) We have seen that Watt was surprised by the amount of water needed, which he found 'was much greater than I thought necessary to Cool the Quantity of water contained in the Steam down to below boiling point'. (57) Even when trying his wooden cylinder model, he had 'perceived that the water issuing from the sinking pipe was always near boyling hot'. (58) If he had succeeded in reducing the steam consumption with the wooden cylinder, from where did all this heat come? This did not agree with the results he obtained by the law of mixtures of different amounts of water at different temperatures. He said, 'I mixed 1 part of boiling water with 30 parts of cold water I found it only heated to the arithmetical mean betwixt the two heats & that it was scarcely sensibly heated to the finger'. (59)

Watt then carried out his famous experiment with an ordinary type of kettle.

> I took a bent Glass tube & inverted it into the nose of a tea kettle the other end being Imersed in Cold Water I found on making the kettle boil that tho there was only a small increase of the water in the frigeratory that it was become boiling hot this I was surprized

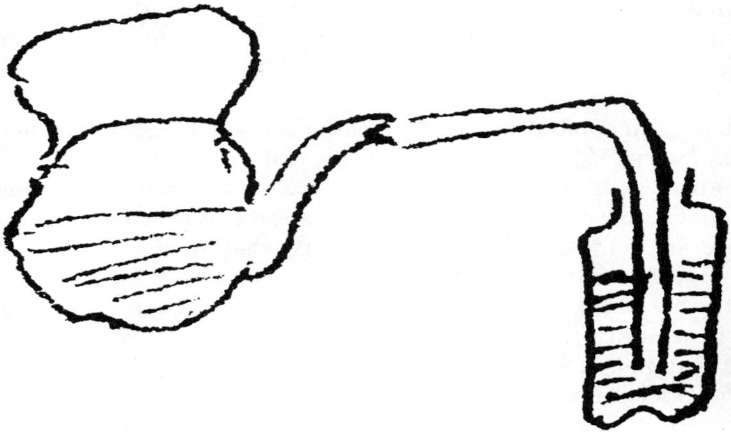

Watt's thumb-nail sketch of his kettle adapted to feed steam into a container of cold water to determine the value of the latent heat of steam.

> at & on telling it to Dr Black & asking him if it was possible that water under the form of steam could contain, more heat than it did when water it was heated to 212° or boiling hot it would receive no more heat tho the fire &c continued the same that the steams that go off do not appear sensibly hotter to the thermometer than 212 & that the Quantity of water that is evaporated in a minute is small in proportion to the heat that is certainly carried off or contained more heat than was sensible to the thermometer & that in Certain Circumstances they might part with that heat to other bodies. (60)

These experiments showed that the water in the frigeratory had gained about one-sixth in volume. In other words, water converted into steam can heat about six times its own weight of water from room temperature to boiling point. (61) Watt had tumbled upon the phenomenon of latent heat. The thermometer had played a crucial role. He mentioned measured volumes of water, suggesting that he had a container of a specific volume rather than weighing the water in this case. (62)

Watt himself was always very careful to ascribe the origin of the doctrine of latent heat to Black. (63) He later denied Robison's claim that he attended Black's lectures as well as being, in Robison's words, 'completely master of the subject'. (64) It seems to have been Watt's questioning Black about the amount of heat in steam that prompted Black to make further experiments to obtain more accurate figures for the vaporization of water. Watt also repeated his experiments in November 1764 with better apparatus, probably in the company of Lind, but did not record his results in his notebook. (65) In fact, Watt made two mistakes in his earlier calculations. He added the weight of the condensed steam to that of the cold water and he assumed that the steam to be condensed was heated up from the temperature of the cold water. But these errors did not seriously affect his results. (66)

Ascertaining the correct value for the latent heat of steam enabled Watt to calculate why, on his model engines, the injection water was always so hot. It would also have told him why so large a fire was necessary to raise steam and why steam contained so much heat. The doctrine of latent heat, together with knowledge of the heat capacities of materials, would give scientific explanations for the passage of steam through steam engines. Watt recognised that, while knowledge of latent heat was not necessary for the practical functioning of steam engines, it resolved the questions he had been unable to answer otherwise. In 1809, Watt junior told William Playfair that the doctrine of latent heat was not the cause of his father's improvements to the steam engine but that this doctrine gave a 'satisfactory solution of all the Phenomena'. (67) Watt confirmed this in a letter to Brewster in 1814.

> Although Dr. Black's theory of latent heat did not suggest my improvements on the steam-engine, yet the knowledge upon various subjects which he was pleased to communicate to me, and the correct mode of reasoning, and of making experiments of which he set me the example, certainly conduced very much to facilitate the progress of my inventions; and I still remember with respect and gratitude the notice he took of me when I very little merited it, and which continued throughout my life. (68)

Black's explanations gave Watt the final clues to the steam engine but equally Watt's experiments gave Black an understanding of the steam engine also, that it worked according to the scientific theories which he had promulgated. The Newcomen engine could be explained scientifically for the first time. This would also be true of Watt's separate condenser when he had invented it.

Watt's Dilemma and the Separate Condenser

Watt, having determined figures for the latent heat of steam and the heat capacity of his cylinder, could calculate the steam consumption of his engine accurately. He could 'judge the merits of Savarys engine in comparison with Newcomens' (69) but this showed him the impossibility of achieving the concept of his perfect engine. First, some steam would be destroyed, as he put it, to heat up the cylinder to boiling point before it could be filled with steam. Then the heat retained by the cylinder itself as well as the latent heat of the steam in the cylinder would have to be destroyed before a perfect vacuum could be obtained. He wrestled with the dilemma which presented itself and laid it down as an axiom:-

> That to make a perfect steam-engine, it was necessary that the cylinder should always be as hot as the steam which entered it, and that the steam should be cooled down below 100° in order to exert its full powers. The gain by such construction would be double:- first, no steam would be condensed on entering the cylinder; and secondly, the power exerted would be greater as the steam was more cooled. The *postulata*, however, seemed to him incompatible, and he continued to grope in the dark, misled by many an *ignis fatuus*. (70)

In 1817, Robert Hart and his brother questioned Watt in Glasgow about his invention of the separate condenser and his early steam engines. In 1859, Hart claimed that Watt gave him the following account of how the idea of the famous separate condenser originated.

> It was *in the Green of Glasgow*. I had gone to take a walk on a fine Sabbath afternoon... I was thinking upon the engine... and gone as far as the Herd's House, when *the idea came into my mind that as steam was an elastic body it would rush into a vacuum, and if a communication was made between the cylinder and an exhausted vessel, it would rush into it and might be there condensed without cooling the cylinder.* I then saw that I must get quit of the condensed steam and injection water, if I used a jet as in Newcomen's engine. Two ways of doing this occurred to me. First the water might be run off by a descending pipe, if an offlet could be got at the depth of 35 to 36 feet, and any air might be extracted by a small pump; the second was to make the pump large enough to extract both water and air... *I had not walked further than the Golf-house when the whole thing was arranged in my mind.* (71)

In another version of this story, Watt is reported to have said that it was several months between thinking of this and actually constructing a model which he made in the workshop in King Street. (72) None of this can be confirmed in surviving contemporary documents. The original flash of inspiration must have occurred before 29 April 1765 when Watt wrote a letter to Lind, giving details of experiments with his new engine. (73)

Watt himself described his invention.

> He considered that steam being an elastic fluid, it must follow the law of its kind; and that if there were two vessels, A and B, of equal or other dimensions, the one, A, filled with steam, and the other, B, exhausted, if a communication were opened between these vessels, the steam would rush from the full one into the empty one, and they would both remain half exhausted, (if the vessels were equal in size), or be filled with steam of half the density. If, then, into the second vessel, B, an injection of cold water were made, or cold water applied to the outside in sufficent quantity, the portion of steam which it contained would be condensed or reduced to water; and by the same law of nature that had operated before, more steam would issue from A into B until the whole was condensed, and a nearly perfect vacuum established in both vessels; yet as the cold water had not entered or touched A, that vessel would still retain its heat. (74)

Watt had at last discovered the basis for his perfect steam engine.

NOTES: WATT'S EXPERIMENTS ON NEWCOMEN ENGINE MODELS

1 Robinson & McKie, (1970), p. 434, J. Watt's Note Book, JWP, W/14, see p. 7 r, '1765'.
2 Muirhead, (1858), pp. 83 - 91.
3 Ibid, pp. 74 - 83 and Robison, (1822), Vol II, p. 113 and Law, (1969), pp. 9 & 45, note 6.
4 Robinson & McKie, (1970), p. 434, Watt's Note Book, JWP, W/14, p. 8 r.
5 B. & W. Col., MIII, 3/6, 'Ledger of Personal Accounts', f. 7.
6 Robinson & McKie, (1970), p. 434, J. Watt's Note Book, JWP, W/14, p. 7 r.
7 Desaguliers, (1763), Vol II, p. 536.
8 See Robinson & McKie, p. 418, Letter 259, 'Copy of Mr. Watt's Remarks on Mr. Robison's Edition of Dr. Black's Lectures', 4 Jan. 1809, but Law, (1969), p. 17, writes, 'Watt was mistaken in attributing the great consumption of steam to the scale effect in so far that in an atmospheric engine with a brass cylinder, the whole thickness of which assumes the same temperature as the steam, the volume of the cylinder contents and its own weight (and hence the thermal capacity) diminish in the same ratio'.
9 Muirhead, (1858), p. 62.
10 Ibid, p. 66.
11 Singer, (1941), p. 286.
12 JWP, 4/57.1, J. Roebuck to J. Watt, 13 Sept. 1765.
13 Daumas, (1989), p. 134 for notes on balances.
14 Robinson & McKie, (1970), p. 435, J. Watt's Note Book, JWP, W/14, p. 8 v.
15 Simpson, (1962), Guerlac, p. 17 rt.
16 Robinson & McKie, (1970), p. 435, J. Watt's Note Book, JWP, W/14, p.10 r.
17 Ibid.
18 Ibid, p. 436, 12 r.
19 Ibid, p. 435, 9 v.
20 Ibid, p. 436, 10 v.
21 B. & W. Col., Portfolio 1381. See Law, (1969), p. 27.
22 Robinson & McKie, (1970), p. 434, J. Watt's Note Book, JWP, W/14, p. 8 r.
23 Ibid, p. 436, 12 r. and Muirhead, (1858), p. 76.
24 Robinson & McKie, (1970), p. 436, J. Watt's Note Book, JWP, W/14, p. 13 r.
25 Ibid, p. 418, Letter 259, 'Copy of Mr. Watt's Remarks on Mr. Robison's Edition of Dr. Black's Lectures', 4 Jan. 1809 and Muirhead, (1858), p. 76.
26 Robinson & McKie, (1970) p. 436, J. Watt's Note Book, JWP, W/14, p. 13 r.
27 Ibid, p. 436, 14 r.
28 Muirhead, (1854), Vol I, p. lxxxvi.
29 Muirhead, (1858), p. 76.
30 Ibid, p. 59.
31 Ibid, p. 67.
32 Desaguliers, (1763), Vol II., p. 312. 'That Heat will add Elasticity to Fluids is evident from numberless Experiments, especially from Distilling and Chemistry. But what is needful to consider here is only, that it acts more powerfully on

Water than common Air; for the same Heat which rarefies Air only 2/3, will rarefy Water very near 14000 times, changing it into Steam or Vapour as it boils it'. See also pp. 338 - 339.

33 Muirhead, (1854), Vol I, p. lxxii and Simpson, (1962), Guerlac, p. 19 lft.
34 Robinson & McKie, (1970), p. 437, J. Watt's Note Book, JWP, W/14, p. 13 v.
35 Ibid, p. 437, 14 v.
36 Muirhead, (1858), p. 78.
37 Ibid.
38 Desaguliers, (1763), Vol II, p. 536.
39 Simpson, (1962), Guerlac, p. 18 lft.
40 Robison, (1803), Vol I, p. 504.
41 Robinson & McKie, (1970), p. 77, Letter 59, J. Watt to J. Magellan, 1 March 1780.
42 Ibid, p. 438, J. Watt's Note Book, JWP, W/14, p. 17 r.
43 Simpson, (1962), Guerlac, pp. 19 rt. & 20 lft.
44 Robison, (1803), Vol I, p. 504.
45 Muirhead, (1858), p. 78.
46 Robinson & McKie, (1970), p. 439, J. Watt's Note Book, JWP, W/14, p. 16 v.
47 Donovan, (1975), p. 261. Donovan calls this wrongly 'back pressure'.
48 *Essays and Observations*, (1796), W. Cullen, 'Essay on the Cold Produced by Evaporating Fluids...', p. 11 and Cardwell, (1971), p. 34.
49 Muirhead, (1858), p. 84.
50 Robinson & McKie, (1970), p. 474, J. Watt's Note Book, JWP, W/14, p. 47 r.
51 Ibid, p. 420, Letter 260, J. Watt to Sir J. Banks, 1 March 1815.
52 Ibid, p. 448, 29 r. and Smeaton, (1794), Table I, p. 15.
53 B. & W. Col., MI, 4/61, 'Graph of Boiling Points of Water'.
54 Muirhead, (1854), Vol I, p. lxx.
55 Robinson & McKie, (1970), p. 448, J. Watt's Note Book, JWP, W/14, p. 29 r.
56 Hills, (1989), p. 25.
57 Robinson & McKie, (1970), p. 439, J. Watt's Note Book, JWP, W/14, p. 16 v.
58 Ibid, p. 436, 12 r.
59 Ibid, p. 439, 16 v. Robinson and McKie have misunderstood this in their commentary on p. 427.
60 Ibid, p. 439, 17 v. & 18 v.
61 Dickinson, (1935), p. 35.
62 Robinson & McKie, (1970), p. 439, J. Watt's Note Book, JWP, W/14, p. 19 r.
63 Ibid, p. 76, Letter 59, J. Watt to J. Magellan, 1 March 1780 and Muirhead, (1854), Vol I, p. xlix. Note in J. Watt's handwriting; 'Dr. Robison is mistaken in this. I had not attended to Dr. Black's experiment or theory on latent heat, until I was led to it in the course of experiments upon the engines, when the fact proved a stumbling-block which the Doctor assisted me to get over'.
64 Robison, (1803), Vol I, p. viii.
65 Robinson & McKie, (1970), p. 439, J. Watt's Note Book, JWP, W/14, p. 19 r.
66 Cardwell, (1994), p. 176.

67 JWP, 3/37, J. Watt jun. to W. Playfair, 7 Jan. 1809.
68 Robison, (1822), Vol II, p. ix, J. Watt to D. Brewster, May 1814 and Muirhead, (1854), Vol II, p. 350, Letter 394, J. Watt to D. Brewster, May, 1814.
69 Robinson & McKie, (1970), p. 440, J. Watt's Note Book, JWP, W/14, p. 20 r.
70 Muirhead, (1858), p. 87.
71 Dickinson & Jenkins, (1927), p. 23, see note 3. The story was related by Mr. Hart to members of the Glasgow Archaeological Society, (*Transactions*, 1859, Ser. I, I.1). Dickinson and Jenkins found no contemporary record of this event among Watt's papers; neither has the present author.
72 Robinson & Musson, (1969), p. 42, No. 4, 'Report by Messrs. Hart of Glasgow of conversations with Mr. Watt in 1817. Communicated by John Smith, 19 March 1845'.
73 JWP, C1/15, J. Watt to J. Lind, 29 April 1765. See also Muirhead, (1854), Vol I, p. 3, Letter 1. Dickinson (1935), p. 36, gives May which must be wrong.
74 Muirhead, (1858), pp. 87 - 88.

Rival Engine Builders in Scotland

Watt's invention of his separate condenser did not mean that he abandoned the Newcomen engine. He must have realised quickly that, if he were to succeed with his new type of engine, he would need experience in erecting engines because, up to that period, the only type of engineering work with which he had been concerned was scientific instruments. The Newcomen engines he erected in Scotland are important in the development of his engineering practice because we shall see how once again he changed from being directly involved in both their design and construction to evolving into a consulting engineer, preparing only the plans and then giving general supervision. In this way he was not concerned with any manufacture of parts and had no need to supply capital in either fixed equipment to produce parts or in the actual erection. This was the pattern he and Boulton would principally follow at Birmingham. But first Watt had to break into a market in which there was already considerable competition. Here again he may have been favoured through his being a native Scotsman because, in this field, as in so many others in Scotland at this time, people brought their skills from south of the border to work in a country where there was a dearth of experience in the latest technologies.

The Carron Company

Scotland presented opportunities for exploiting its mineral wealth which included iron ore. The Carron Company turned this into castings, among which were some parts for steam engines. When Roebuck, Garbett and Cadell founded the Carron Company in 1759, it was the first time that Scotland had such a foundry and engineering works. (1) It is from the surviving Carron archives that we are able to glean additional information about engines erected by Watt as well as to see some of the competition he faced from other engineers. Even so, there must have been much more activity than we can discover through this source, particularly in respect to his rivals.

The Carron Company was engaged in the steam engine business at least by the begining of 1765 when, at the end of January, it was

> Resolved when the Loam Moulders are not employed with Goods wanted for ?North Brittain? or Newcastle to make 2 cylinders & 20 fathoms of pipes of the most suitable sizes to be sent to Cornwall for sale which appears to be the most probable method of introducing us into that trade. (2)

It is not clear whether 'that trade' refers to the steam engine business in general or to the Cornish section of it. During the later part of the summer in 1766, the Company cast a cylinder, possibly for Ostend, 46 ins. diameter by 10 ft. long, together with the bottom and piston. (3) However the Company never seems to have built any equipment for heavy forgings until well into the 1770s because, in 1770, it had to write a letter apologising to Charles MacDowall that it could not supply him with the cranks for his waterwheel pump at Drumlemble. (4. See p250) It was just after this that Smeaton improved the boring machines.

The part the Carron Company played in the supply of Newcomen engines is revealed in a letter to Peter Buchanan, a Glasgow brewer.

> We undertake the mounting of fire Engines so far as to make the Cylinders pipes Boyler Plates etc to draw[in]g & Dimentions given us, but we never attempt to put the pipes down or do any other kind of work about the Engine further than furnishing the apparatus - For that purpose you must apply to an engineer and in case you are not ac-

> quainted with a person sufficient for the undertaking we believe we can Recommend a Gentleman of Merit & Ingenuity to you If you favour us with the order you may depend we shall despatch it in as short time & upon as easy terms as any other founder. (5)

So we see that the Carron Company did not erect engines but supplied parts to the designs of other engineers. Watt would take advantage of this service, as did Smeaton and that 'Gentleman of Merit and Ingenuity', Hately, who became Watt's bitterest rival in the steam engine business.

Joseph Hately

The importance of Watt's scientific knowledge contributing to his success may be contrasted with Joseph Hately, whose patent in June 1768 (6, see p346) may have hastened Watt into taking out his famous one of 1769. At this period, Hately was in Scotland, working as an engine erector. As well as improvements to the boiler to give a bigger heating surface in order to evaporate 'the greatest quantity of water with the least quantity of fuel', Hately made his cylinder a 'double one, and the piston is also double, in order that cold water may constantly be circulating betwixt the two'. His object was 'to keep the interior as cool as possible'. (7) Pipes for the admission of steam and injection water as well as for the removal of the condensate were fitted in the bottom of the cylinder. While the vacuum must have been improved through greater condensation against the cylinder walls, steam consumption to reheat the cylinder must have been increased excessively. It was an instance of the right principle wrongly applied and enforces the importance of Watt being able to start in the right place through his contacts with scientific people and having the latest scientific knowledge. Hately replaced the conventional beam with a segment of a wheel and a small drum. The chain from the steam piston passed round the drum so that this piston had a shorter stroke than the pump. A quarter segment of a wheel, braced into this drum, supported the chain from the pump rod. Here again this seems to be the wrong way round, as it gives the greater advantage to the pump and not the steam piston. But it resulted a compact form of engine which was almost certainly copied from one of Watt's. (See p353)

Shortage of water for a waterwheel at Carron was delaying the supply of parts so that on, 23 June 1768, Hately contracted to alter the fire engine there. This may have been a Newcomen engine erected a little earlier by Mackell and Watt. (8, see p357) Presumably Hately changed it to work on his patent principles. It was finished by the autumn when the Carron Company paid Hately some money but Hately had failed to pay his workmen who were making excessive demands on the Company. There were also some problems with the engine because Carron wrote, 'We beg... to settle the future conduct on such a Plan as will be most likely to use the engine in the best way possible and prevent mistakes and misapprehensions which in a point of so much importance is absolutely necessary'. (9) A letter, probably to Hately, points out the desirability of lagging the boiler and pipes. 'The covering of the Boiler and clothing the pipes it is believed may save a great Waste of Steam and Power and be executed at a small Expense. We wish to know if you think it right'. (10) Such an early date as 1768 is interesting for recognition of the fuel economy that might be achieved by lagging. Hately's Carron engine may have been reasonably successful because he continued to be recommended by the Carron Company for other engines. Yet it may have been altered only a year later by another engine erector, Adam Smith.

Hately was engaged on some of Roebuck's pumping engines. The earliest date we have is probably October 1766 in connection with 'Doctor Roebuck's new Engine'. (11) Through his work for Roebuck, he must have learnt about Watt's patent application. Roebuck terminated his employment on 19 May 1769 when he wrote to Watt,

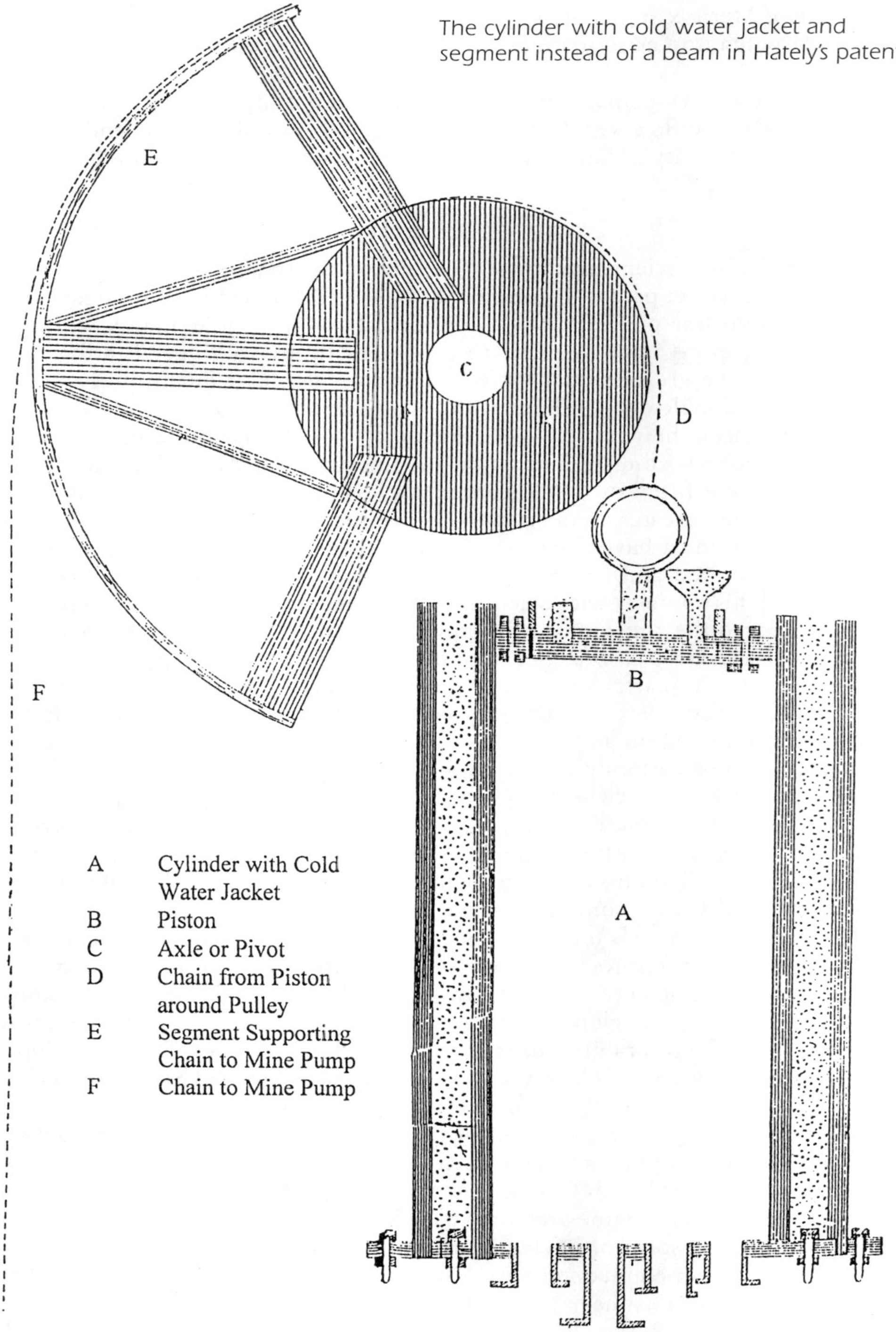

The cylinder with cold water jacket and segment instead of a beam in Hately's patent.

> Mr. Hately left this place today. I have closed all accounts with him. He is to call on your house and bring the contract with Ld Crawford which I ?signed? and you are desired to see my name scratched out. I am also to pay him eighty pounds at your house in three weeks which is in ?...? of the Two Hundred that I received when I met him at Glasgow - I hear that he has already sent for a coppy of your specification and that he charges me with revealing his secrets to you. I am very glad however that he cannot charge me with revealing your secrets to himself. (12)

Relations between these three seem to have been extensive and at first friendly. However that was to change with the publication of Watt's patent in 1769.

During August 1769, Hately visited Peggy in Glasgow and afterwards she sent the following letter to Watt while he was away from home, almost certainly working on his engine at Kinneil,

> Mr. Grea has imployed Heatly to make his machine but told him he had first offered it to you - but that you were ingaged with Dr Roebuck - he said he knew that was true and that you were both dammed mean and impertinent for you were making an ingine of which he was the contriver and had given a hint of it to the Doctor who he said was as great a scoundrel as yourself for doing such a thing he would make your pattant of no avail for the moment he knew it was going he would put a stop to it - he said a great deal more impertinent stuff about the Doctor's bad useage of him till Mr Grea said he knew your cariekter to be good and could not believe him - for if you would have maid his ingin he had never imployed another he had always harde the Doctor had an unexceptionable carikter and begged he would talk of the business he was sent about for - I wrote you this in case you and the Dr thought of any means of silencing him - for he may find a great many that will believe him. (13)

This letter has been quoted to give an instance of the hostility and rivalry Watt was facing during the development of his own steam engine in Scotland.

Hately's hostility to Watt and attempts to overturn Watt's patent did not end there. Roebuck junior later told Watt how Hately had ventured into the attic of Kinneil House to spy on Watt's engine and workshop. (14) It is possible that this happened in October 1769 when Hately visited Mr. Addison's brewery at Bo'ness and Roebuck junior had an argument with him about Watt's engine. (15) Hately boasted that his engine made between fifteen and eighteen strokes a minute and that it went better when surrounded with cold water. He claimed that Watt had copied some of his ideas while at the same time showing that he did not understand the principles of Watt's engine. He also slandered the characters of both Watt and Roebuck but his threats against them came to nothing. (16)

Hately continued to build Newcomen engines in Scotland for some years. The Carron Company papers show that he was erecting an engine for a Mr. Houston when his address was Balshagry. To Hately's enquiry for a cylinder and pipes which he wanted for the end of March, Carron replied on 28 January 1770 that he would have to place his order soon because they were discussing one for another engine. (17) While most of the parts for Hately had been cast by the middle of March, it was not until a month later that Carron invited him to see them; 'By Wednesday next we hope to have the bored parts finished on which day if convenient you may come over and inspect the whole'.

Hately's employment by Peter Buchanan must have started towards the end of 1772 because, during the latter part of December, he had written to Carron about parts of an engine which the Company undertook to supply within three months on receipt of order. 'You will please to be very minute when you confirm it'. (18) Buchanan had written as well as Hately and the order must have been placed early in January 1773. On 24 April, Carron wrote to Hately to see if he wished to inspect the engine parts which were nearly complete while they were still at Carron before they were sent to the Company's cellars or warehouse in Glasgow. There had been problems casting some pipes with a narrow bore which Carron was afraid might warp but this must have been resolved. (19) In February 1774, Hately was entitled 'Engineer' in a letter concerning some engine parts for Messrs. Montgomerie & Co. at Ayr. (20) Watt had supplied an engine to presumably the same Montgomerie in 1770. When Hately left Scotland is not known, but, in 1786, he was erecting a 'rotary reciprocal fire engine' in Cornwall. He had moved to Newcastle upon Tyne by 1790 and a couple of years later was in Bristol. (21) He was called as a witness against Boulton and Watt in the 1796 trial of their patent. (22)

The Smith Brothers

The Carron books show that two brothers, Adam and Peter Smith, were also erecting and maintaining steam engines in Scotland at this period. They were involved with steam engines for a considerable time because in May 1769, Carron 'Resolved that the Fire Engine be altered according to Adam Smith's Design and Estimate'. (23) Possibly Adam rectified Hately's patent engine. Their work shows another role which such engineers undertook, namely maintenance of engines after erection. One of the few contacts which Watt recorded with the Smiths was in June 1770 over repairs to Roebuck's Dunmore engine. (24) In June 1772, he inspected an engine erected by the Smiths before it was accepted by Orrs and Wilson which shows that he must have been recognised as a skilled engineer by that date. (25)

Peter Smith had quite an extensive practice looking after a number of engines. We learn about three from a letter Carron wrote to him in August 1773.

> The annexed work furnished you for Lord Cochranes Engine at Kincardine... stands at your Debit in our Books. Please to order the payment first time you come to this place, that your account may be closed before the Engine Work ordered for Bathgate is entered. The Boiler Plates are forwarded to Bathgate. We have your favour of yesterday's date, ordering 33 Boiler Plate for Mr Gray of Dalmarnock which are already put in hands. (26)

He was supplied with the pump barrels for another engine at Dysart that October. (27)

In 1774, during designing and making the engine which Smeaton sent to Russia for pumping out the Kronstadt dock at St. Petersburg, Carron consulted Peter Smith about making the links on the arch-heads out of cast iron.

> We will be much obliged to you if you will inform us whether any of the engines of your construction or under your inspection are worked with Cast Iron Chains [these had been proposed for Russia] if so we desire to know the *Dimensions of the Links* and whether there has been an instance of any of the Links breaking, likewise the *diameter* of the *Cylinder* and *Pumps* and *height* of the *column.* (28)

Adam Smith was to be sent to St. Petersburg to erect this engine. Some years later, Watt had arguments about cast iron chains on his pumping engines in Cornwall and it could be that his suspicion of them was based on Scottish experience.

Smeaton never competed with Watt in Scotland. Mackell was another person who might have become a rival but he formed a partnership with Watt to build engines which we must now examine. The number of engines installed at Scottish mines suggests that there were other engine erectors in addition to those mentioned but, as so little is known about either them or their engines, it is impossible to discover what share of the market Watt gained.

NOTES: RIVAL ENGINE BUILDERS IN SCOTLAND

1 Campbell, (1961), p. 7.
2 SRO, Carron Co., GD58/2/1/1, Minutes of General Court, 1760 - 1774, p. 162, 31 Jan. 1765.
3 Ibid, GD58/6/1/6, Letter Book, 1766 - 1767, p. 12, ? to F. Garbett, 1 Sept. 1766.
4 Ibid, GD58/6/1/11, Letter Book, 1770 - 1771, p. 69, to C. MacDowall, 23 Aug. 1770.
5 Ibid, GD58/6/1/12, Letter Book, 1772 - 1773, p. 298, to P. Buchanan, 30 Dec. 1772.
6 Patent 895, 30 June 1768.
7 Ibid.
8 SRO, Carron Co., GD58/6/1/7, Letter Book, 1767 - 1768, p. 512, to J. Hately, 24 Oct. 1768.
9 Ibid.
10 SRO, GD58/6/1/8, Letter Book, 1768 - 1769, p. 15, probably to J. Hately, (start of letter missing so no date).
11 B. & W. Col., MIII, 3/6, 'Ledger of Personal Accounts', f. 64, 20 Oct., no year, 'To new steam pipe for Mr. Hately'.
12 JWP, 4/57.12, J. Roebuck to J. Watt, 19 May 1769.
13 JWP, 4/4.15, Margaret Watt to J. Watt, 22 Aug. 1769.
14 JWP, 4/38, J. Roebuck jun. to J. Watt, 22 Nov. 1796.
15 JWP, 4/32.111, J. Roebuck jun. to J. Watt, 12 Oct. 1769.
16 JWP, 4/32.110, J. Roebuck jun. to J. Watt, 19 Oct. 1769.
17 SRO, Carron Co., GD58/6/1/9, Letter Book, 1769 - 1770, pt. 1, p. 373, 28 Jan. 1770; p. 453, 2 March 1770; p. 479, no date; GD58/6/1/10, Letter Book, 1769 - 1770, pt. 2, p. 200, 13 April 1770; p. 216, 20 April 1770, all to J. Hately.
18 Ibid, GD58/6/1/12, Letter Book, 1772 - 1773, p. 228, to J. Hately, 22 Dec. 1772.
19 Ibid, p. 389, 2 March and p. 483, 24 April 1773, to J. Hately.
20 Ibid, GD58/6/1/13, Letter Book, 1773 - 1774, p. 337, to J. Hately, 2 Feb. 1774.
21 Dickinson & Jenkins, (1927), pp. 315 - 316.
22 JWP, 3/10, Boulton v. Hornblower, 1796, 'Defence on behalf of Maberly and Hornblower'.
23 SRO, Carron Co., GD58/2/3/1, Committee Book No. 1, 1768 - 1785, p. 15, Resolution No. 58, 9 May 1769. See also GD58/6/1/13, Letter Book, 1773 - 1774, p. 408, to P. Smith, 17 March 1774.
24 B. & W. Col., MI, 2/42, 'Expenses on Surveys, 1769 - 1773', 28 - 29 June 1770.
25 JWP, 4/19.18, Orrs & Wilson to J. Watt, 21 April 1772.
26 SRO, GD58/6/1/13, Letter Book, 1773 - 1774, to P. Smith, 23 Aug. 1773.
27 Ibid, p. 161, to P. Smith, 11 Oct. 1773.
28 Ibid, p. 409, to P. Smith, 18 March 1774.

Newcomen Engines Built by Watt

An Engine Proposed for the Leadhills Mines (1)

It has been suggested that, to learn how full-size Newcomen engines worked, Watt took charge of one of Roebuck's engines at the School Yard pit. (2) The only mention of a visit down this pit in surviving records took place on 12 June 1770, long after he had commenced erecting engines, (3) But he did test another before October 1765 which must have given him practical experience. (4. See p375) Watt's first venture into the engine business of which we have any knowledge was unsuccessful. There is an laconic entry in his 'Cash Memorandum Book' for 12 April 1766 'Expences at Leadhills, 11/7' which conceals this story. (5) In addition, there are some undated drawings and estimates for a Newcomen type steam pumping engine for Leadhills among the James Watt Papers (6) which have led to the discovery that he tried to win an order for a steam engine here but failed. The story shows that the steam engine did not sweep away all rival technologies and was unsuitable in some cases.

The principal lead mines in Scotland were situated at Leadhills and Wanlockhead on the borders of Lanarkshire and Dumfriesshire. Those at Leadhills were the property of the Earl of Hopetoun, who had leased part of them to the Scots Mines Company. (7) Its shareholders were chiefly gentlemen in London who had appointed Stirling as manager at Leadhills in 1735. It is from his correspondence with the secretaries in London that the background to Watt's attempts can be discovered. Stirling had devised water courses, waterwheels, adits and drainage channels to help pump out and keep the mines clear of water. (8) However, by the middle 1760s, the mines had to be sunk deeper if new areas of ore were to be extracted. This would be costly because they were below existing adit levels. Although Stirling proposed building a third water-powered pumping engine, he pointed out to Hopetoun that they were already 'so much pinched for water', and resort had to be made to pumps operated by men. (9)

We can gain a picture of the layout of the mine and the problems of operating it from a report of a visit there in August 1779 by Smeaton. Springs and streams on the hillsides above the mine had been diverted through channels down a shaft to a twenty-four foot diameter overshot waterwheel inside the mine which operated lever pumps to drain lower levels. The water flowed out along an adit. However the vein was being mined at levels below the bottoms of these pumps and presumably the waterwheel was not powerful enough to drive pumps with a deeper lift. In 1779, the Company was planning either to install a second waterwheel to replace the hand pumps which drained the Susannah Level or to sink a new shaft and place a steam engine at the top of it. Smeaton considered building a steam engine at the mouth of the adit which could operate pumps inside the mine by 'sliding rods' along the adit but recommended installing a reciprocating water-pressure engine inside the mine instead. (10)

Watt became involved when the Scots Mines Company was trying to drive the Susannah vein deeper in 1765. The Company wanted to show that it was operating the mines properly because their lease, which may have been granted or taken out again in 1747, did not have long to run and they feared 'the improbability of its being renewed'. (11) One way of convincing Hopetoun to grant them a new lease was to raise more ore, which meant extending the mining galleries. In June 1765, the Secretary, James Farquharson, wrote to Stirling that, 'It would be a great pity if this Society should be deprived of the benefit of the Ore yet remaining for want of more Level to carry off the water'. (12)

The Company faced two problems with their water. One was water penetrating their mine workings which would flood them and prevent the miners extracting the ore. This could be solved by pumps, and, at that time, the most powerful pumps were driven by waterwheels.

So, in addition, the Company needed an adequate supply of running water outside the mine which could be diverted to turn waterwheels. Waterpower might also be used to operate stamps for crushing the ore and bellows for the smelt mill. But Smeaton noted that the water supply for the wheel in the mine was sometimes frozen up for four months in the winter which stopped both the wheel and the men working. (13) Work might come to a halt in the summer through a drought, as occurred in July 1765 when there was little activity except that lead was being carried. (14)

With waterwheels inoperative for long periods of the year, either through frost or drought, Stirling considered replacing or supplementing them with steam engines. At their meeting on 8 August 1765, the Court of Directors discussed an estimate sent by Stirling as well as the renewal of their lease.

> As to the former [the estimate] the Court observes your opinion, for which they will always have the highest regard, is quite against the Water Engine proposed, and that you rather incline to a Fire Engine, and are going to obtain the best Information you can whether a small one of the latter Sort may not be sufficiently effectual for draining the Water from the Susanna Vein. In the meantime, they wish you to avail yourself of all the advantage you can from Hand Pumps & will be glad to hear your further opinion on that important affair. (15)

Later correspondence shows that this part of the mine was being drained by twenty-four men daily operating lever pumps, so that it is hardly surprising that a more economical method was being sought.

Unfortunately Watt's estimates for his engines at Leadhills are not dated, but it is possible that Stirling sought his advice at this time from the following letter.

> They [the Committee] approve & commend your zeal & activity in the pains you are taking to obtain the best information possible in Scotland, touching the uses & improvements of Fire-Engines, that you may be the better enabled to get one constructed for the company at the Susanna Vein, where the use of it may be of the greatest advantage. But if you should not obtain all the Satisfaction you could wish it might perhaps be worthwhile to see what is done by the various machines used in the North of England for draining of Mines & Coal Pits, where they say great Improvements have of late years been made in that way. (16)

At this date, Watt was still experimenting with models of his separate condenser and was meeting only limited success. (17) There is nothing to show why Stirling should have contacted him when Watt had minimal experience with any steam engine. There may be one possible clue later because, in October 1767, Glasgow City Council ordered, 'The Dean of Guild and Brethren to admitt Mr. James Stirling Teacher of the Mathematicks in Glasgow Burgess and Guild Brother of the Burgh'. (18) Watt could have met Stirling through attending some lectures on mathematics.

During the autumn, Stirling must have proposed testing some other form of pump, for which approval was given 'provided it will not interfere with Erecting an Engine' (19) and he was asked to obtain the necessary information about the construction and expense of Fire Engines 'with all convenient dispatch'. William Hamilton, who had replaced Farquharson as Secretary, supposed that Stirling was away during November and the early part of December making his enquiries as there had been no communications from him. (20) Hand pumps had been used with success to drain the Susannah Level so presumably ore was being won which

Thomas Watts, the Treasurer, hoped would please Hopetoun and encourage him to renew their lease. (21) Watts was also 'glad to find that the Engine necessary to be erected should be done at so small an Expence, as the Gentleman whom you have consulted, imagines, but perhaps he may think otherwise, when he views the Works'. He may have seen the cheaper estimate submitted by Watt.

A letter from William Hamilton at the end of January confirms that the gentleman was Watt and William Hamilton urged 'as much dispatch in that affair as the nature of it will admit'. (22) But Watt had not visited Leadhills by early March when William Hamilton again enquired about the costs of erecting a steam engine as well as its fuel consumption and other annual running costs such as keeping it in repair. He wished to have these figures, as well as figures of the amount of water to be raised and the height, ready for the next meeting of the Committee in the middle of April because the Company had been approached by another person who had obtained a patent for a new machine for drawing off water worked by men. (23) The rival machine had been invented by a Mr. Erskine who demonstrated it to some members of the Committee. It is possible that this was Thomas, an active coal master who was related to Sir John Erskine, one of the promoters of the Scots Mines Company. (24) William Hamilton was hopeful that this machine would prove to be cheaper and better through its simplicity although it had not been tried. (25)

This letter was written on 10 April, at the same time as Watt spent four days surveying at Leadhills followed by a day's drawing. In addition to his 11s.7d. expenses, Watt charged 7s. for horse hire. (26) Watt could not send his estimates immediately because he had to work out the costs of erection and Stirling had to calculate the price of forming a chamber in the rock to house the engine underground. (27) In the meantime, Thomas Watts had 'seen some Modells of Machines belonging to the Man & particularly one which if it answers in the large, as well as in the Modell, will answer the Purposes better, & come much cheaper than that proposed by Mr. Watt'. (28) The model was 3 ft. in length, just over a foot wide, 4 ft. high and threw up water in a continuous stream to the height of 45 ft. If this type would not do, Erskine proposed one of another design so he was asked to go to Leadhills to inspect the site. He set out at the end of April.

Erskine arrived on 7 May and the next day was 'preparing himself for a long journey underground with the overseers whom I ordered to take time enough to shew him everything that can in anywise affect his Engine'. (29) Stirling's accounts of this visit tell us a great deal about Erskine's pump and the problems Watt had to face in the design of his engine. Stirling sent a sketch of the place where the engine was to be erected underground. It was not possible to transport there any parts of an engine longer than ten feet, or perhaps twelve if some sections of the rock were cut away. (30) Erskine's pump was wrought by a wheel, so must have been some form of rotary pump and not like the lever pumps used at Leadhills up to that time. It would not require much of the rock to be cut away and Erskine undertook to employ only eight men to pump up the water from twenty fathoms compared with the twenty-four then needed. To raise the water from thirty fathoms, he would employ only twelve men compared with the thirty-two in Stirling's way. Stirling foresaw one problem, that the mine might be flooded when it became necessary to lower Erskine's engine as the levels were sunk deeper because his engine had to be taken to pieces and re-erected to move it. Also, to repair it, parts would have to be ordered from Edinburgh. However, Erskine's engine would cost only £100.

This was much less that Watt's higher estimate for a steam engine.

> Mr Watt said that a Fire Engine sufficient for our business would come to £150, besides cutting out of some part of the rock. It would be a troublesome thing to supply it with

> coal underground, and bring the ashes up to the surface. It has one advantage that it would be fixt on the level never to be removed and could never be laid under water. Mr Watt is very remiss in sending some computations which he promised to send long ago and with which one must be furnished in order to estimate the whole expence of his machine. (31)

No consideration seems to have been given about water for the boiler or for condensation; presumably that for the waterwheel would suffice. Nor is there mention of how the smoke would escape, possibly up the shaft down which the water fell to the wheel. It seems hardly practical to have placed a Newcomen engine and boiler deep underground, although Smeaton mentioned that there was one in the coal mines of Whitehaven (32) and this was what Savery had intended to do. On the other hand, there would be no problem getting Erskine's machine 'through the subterraneous passages' and, on 24 May, Stirling said, 'I shal [sic] therefore write to Mr Watt to do nothing in that affair till he hear from me'. (33)

Stirling's comments show that he had received some estimates from Watt which may have been based on figures surviving in the James Watt Papers. (34) Watt must have been asked to design an engine to pump from the lower level of 30 fathoms or 180 ft. He observed that raising water from the existing 20 fathom level required five pumps, each 5 ins. diameter and a lift of 4 fathoms. The top pumps were worked by two men at a rate of 32 or 34 strokes a minute. The engine would work a pair of pumps, each of 15 fathoms, 9 $^{1}/_{2}$ ins. diameter, having a three foot stroke at about 20 strokes a minute. To raise this water, he proposed a cylinder of 44 ins. diameter, price £8, and a boiler between 6 to 7 ft. diameter, price £19. The total cost for the engine without the mine pumps was only £56. This must have been the first estimate which he submitted with a very low figure. However, there is a further set of calculations in which the boiler is priced £32.17s.6d. and the cylinder with bottom and piston at £26. The total for this second engine came to £104.2s.8d. and for the mine pumps £48.1s.6d, which presumably corresponds with Stirling's figure of £150. No allowance was made for transport of parts, but there was £5 for erection. These estimates seem very low, especially when compared with a slightly later engine at Carron for nearly £1,000. (35) No estimates have survived for either annual coal consumption or maintenance.

Watt's papers also include two drawings of the Leadhills engine which show a novel layout. The cylinder is upright, in the customary Newcomen engine position, but the 'hay-stack' boiler is to one side at a lower level and a hot well also at one side lower down. The cylinder is situated right on the edge of the mine shaft, down which the top of a pump and spear can be seen. So that his engine parts could be carried underground and to save making a large chamber in which to house his engine, Watt replaced the long lever beam with a wheel. Watt's piston inside the steam cylinder was connected to the pumps down the mine by a chain passing over this wheel. The pivots of the wheel were supported on wooden framing with one vertical leg and an inclined brace. The wood for this framing would have cost £2.6s.8d. and 10s. for the wheel. While he did not win this order, the Leadhills engine does show Watt's willingness to adapt designs and venture into new and original forms of construction. Yet we must query whether it would have been a success down a mine so perhaps it was fortunate that his delays in producing estimates may have saved him from a failure.

Here we may note that Smeaton also had the idea of using a wheel instead of a long beam at about the same time. He designed a form of portable Newcomen engine in 1765. To one side was the boiler. An 'A' frame of wood supported a wooden wheel over which passed a chain from the piston in the steam cylinder on one side to the pump on the other. (36) Farey commented that, 'It does not appear that this plan for a portable engine was put into execution by Mr. Smeaton till some years afterwards'. (37) Smeaton and Watt seem to have been

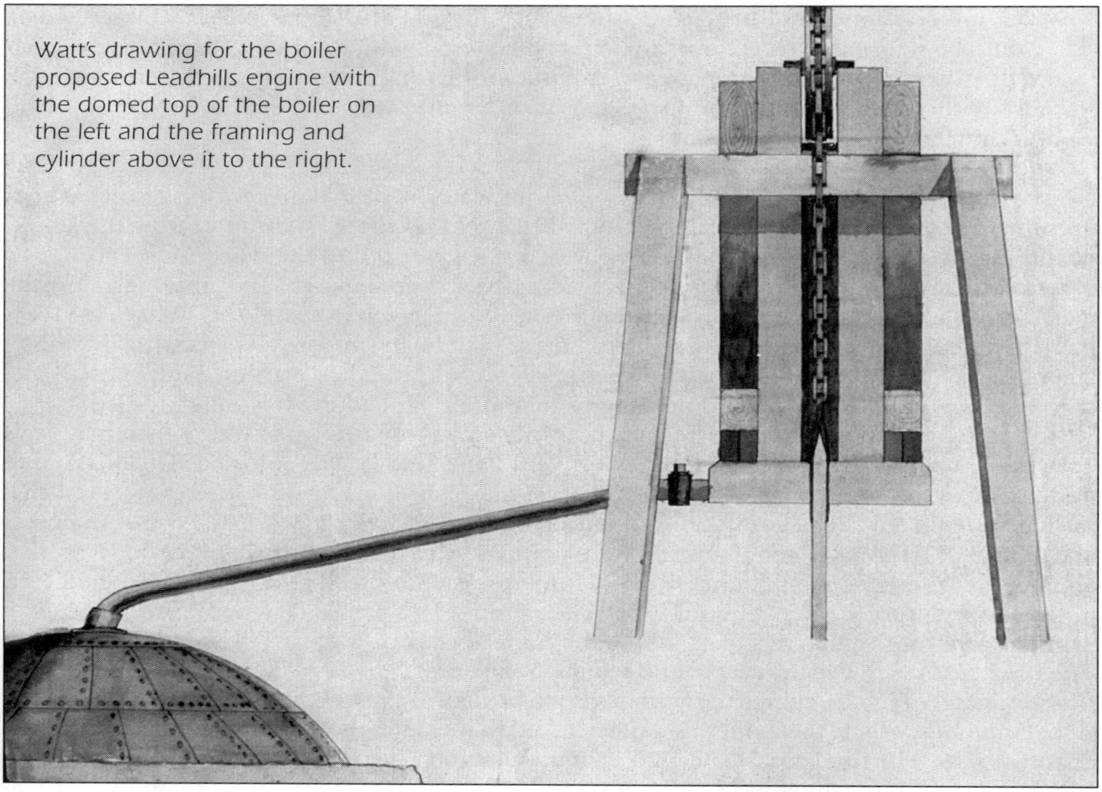

Watt's drawing for the boiler proposed Leadhills engine with the domed top of the boiler on the left and the framing and cylinder above it to the right.

thinking along parallel lines since we have no evidence that Watt could have known about Smeaton's design at this time.

Watt's Partnership with Robert Mackell to Build Steam Engines

After his failure at Leadhills, Watt may have realised that he needed the help of some more experienced person if he were to make any headway with his steam engines. Here Mackell again came to Watt's assistance and indeed this may be the original reason for the engineering partnership between the two because they formed some sort of agreement to build Newcomen engines. It could not have lasted long since Mackell was appointed 'Sub-Engineer' to Smeaton for building the Forth & Clyde Canal on 14 March 1768. (38) This position took up so much of his time that he virtually gave up his steam engine business but he was supplied with a cylinder from Carron for Lord Perth in the autumn of 1768. (39) Although the Carron Letter Books contain further correspondence with Mackell about canal and other matters, there is no other mention of steam engines for him.

Watt himself wrote a memorandum for Watt junior in 1808 concerning the Newcomen engines he had erected.

> The *common* Engines I erected, were a very small one to pump water to a turpentine still-tub at Carron Wharf 1765; a 24 Inch Cyl[inde]r for Mr. Bruce of Kennet in partnership with Mackell about 1766; a larger, I believe 45 Inch at Newton of Air, with the same in 1767 or 8, and these were all in which he was concerned with me. - One or two

> short stroke Engines 18 or 20 Inch Cylinders for Mr Colville of Torryburn, date uncertain, but before 1769. One or two of the same sort for Dr. Roebuck prior to 1770; and and [sic] these are all I recollect at present. (40)

Existing records confirm that this list is remarkably accurate since it was written nearly forty years after the events.

It is possible that the Carron Wharf engine was one which the Carron Company considered installing in 1765. They had employed Mackell that August to build a reservoir 'between the Paper Mill & Dunnipace' which they expected would be completed by the following May. (41) In December 1765, it was recorded, 'We will in short time sett abt erecting a small engine wch by means of our waste ?Parr? wood will throw up the Water we have occasion for... We are willing to trust the whole execution to R. Mackell'. (42) Watt may have been erecting this engine in early December when he told Roebuck that he hoped to see him 'as soon as I have given the necessary orders about setting up the engine'. (43) He recorded expenses at Carron in August, November and December 1765 as well as in the early part of 1766, (44) but there is no way of determining what these were for because he must have visited Carron around this period concerning parts for an engine with his separate condenser at Kinneil. In addition, there was a much larger water-returning engine at Carron to which we will return after looking at the one for Kennet.

Lord Kennet's Engine

Existing records show that the first engine which was built by the Watt and Mackell partnership, and so the first full-size steam engine built by Watt, was supplied to Lord Kennet who owned an estate between Clackmannan and Kilbagie beneath which lay extensive coal reserves. Sinclair commented in the 1790s that there were three collieries in the parish of Clackmannan, 'Clackmannan, Sauchy, and Kennet'. (45)

> The Kennet coal consists of two seams; one of 30 inches thick, another of 36. It was given up in 1726, and began again to be wrought in 1759. It has at present a small fire engine; but it is probable, it was once level free, as there is no trace remaining of any engine being upon it when it was first wrought. The distilleries in the parish consume a great part of this coal. It produces yearly about 6000 tons of great coal, besides a considerable quantity of chows. (46)

Stobie's map of Perthshire and Clackmannanshire published in 1783 has the word 'Engine' printed just north of the road between Clackmannan and Kilbagie where there still exist miners' houses at Kennet. Later maps show a colliery just to the north of this hamlet through which subsequently a railway line and today the modern by-pass road have been laid. The countryside is gently rolling and it is doubtful if the small streams in the vicinity could be diverted to power a pump driven by a waterwheel. This is confirmed by Sinclair's statement and so it would seem that, a few years after mining had recommenced in 1759, resort had to be made to introducing steam power.

There are a number of records about this engine which show the way Watt worked at first. There is a paper, 'Expence of Engine at Kennett', (47) which must have been for internal consumption only and contains two costings. The first is for a conventional Newcomen engine and house and the second for one of Watt's type based on the Leadhills engine with a wheel instead of a lever beam. The house for the conventional engine would be 12 ft. by 10 ft. by 26 ft. high, costing for the walls £27.10s. while that for Watt's was only 12 ft. by 10 ft. by 10 ft. high, costing just over £15. The 'walking beam', 36 ft. long, arch heads at 18 ft., stays, etc would cost probably just over £7 (one figure is indistinct) compared with £5.9s.3d. for Watt's wheel. The cylinder, 24 ins. diameter by 7 $^{1}/_{2}$ ft. long would be common to either at £28, the bottom £4.10s. and the piston £1.16s; the boiler, 9 $^{1}/_{2}$ ft. diameter £45 and its setting a further £12.10.0. The conventional type of Newcomen engine was estimated at £372.3s.7d. of which £83 was for mine pumps fifteen fathoms deep. Watt's design came to only £322.15s, so there was a potential saving of £50 but the document finishes, 'N.B. this [Watt's design] never having been executed must a larger allowance for incidents than the other which Mr McKell is acquainted with the execution off'. (48)

The draft agreement between Lord Kennet and Robert MacKell, Engineer at Falkirk, together with James Watt, Merchant in Glasgow, for the erection of this engine has no date. (49) They were to install a fire engine by 'the first of January next' at Pilmuire with the same dimensions as those given in the estimate quoted above. They were also to build 'a sufficient engine house' and all the other apparatus for this engine of 'sufficient strength and in proportion to the above Cylinder and Boyler and setting the same in going condition' at a cost of £400. Unfortunately this document gives no clue concerning which type of engine was to be erected since the price could be for either. Their work would be inspected before final acceptance and payment would be made in three stages, £100 on 1 August, £200 on 1 October and the final £100 on completion. Lord Kennet was to provide a smithy and 'furnish the same with all proper tools' and to 'furnish Ropes, Tackles Blocks and all other tackleing which are to be left with the Engine'. If indeed the metal parts were ordered from outside suppliers like Carron, for which we found no evidence in the Carron books, and the smithy was provided by Lord Kennet, then Mackell and Watt did not need to invest any capital in a workshop themselves. Also Lord Kennet would have paid directly for the parts supplied by other firms so the partnership had little financial outlay or involvement.

Beginning in May 1766, Watt's cash book contains notes for 'Expences at Kennet'. From then on, roughly every month until the end of December, his cash book records seven entries with total expenses of £5.13s.2d. and in addition £1.8s.6d. for horse hire but we cannot determine how long he may have spent there. (50) This ties up with supervising the erection of the engine as outlined in the draft of the agreement. We know that this engine was definitely erected because Lord Kennet wrote a letter of complaint to Watt on 15 May 1769. He had found that,

> Mr McKell has caused make the boiler vastly too slight, so that tho the engine goes very little commonly only five or six hours in the four and twenty yet the Boiler is already wore out, so that I am obliged to make a new one, this is owing to the plates being made too thin, so that they could not last the ordinary time... I do not blame you personally for this, because I know you trusted Mr McKell with that part of the work but as you was conjunct in the contract, it is necessary for me to complain to you of it as well as to him. (51)

It was Mackell who had insisted on the thin section of the boiler plates. Lord Kennet also complained about the engine house not being high enough and, in consequence, of the boiler not being set low enough. As it appears that Mackell was responsible for the boiler and Watt with other parts of the engine, and also, from the engine house being too low, it is likely that the engine had been built to Watt's design with a wheel instead of a beam. It must have performed satisfactorily if it was the one at work fifteen years later according to Sinclair. While silence descends about the outcome of this dispute, Mackell and Watt would use the design with a wheel for their next engine, a larger one at Carron.

The Carron Company Engine

With ever-expanding blast furnaces, boring mill, forges and other machinery, the Carron Company found itself short of water to power all its waterwheels. Therefore a steam engine was considered to pump the water back up from the tail race into the higher storage pond. The 'Plan & Estimate of fixing up a 72 inch Cylinder for drawing backwater at Carron...December 5, 1766' has survived. (52) This engine was a large one with a cylinder 72 ins. diameter and two 20 ft. diameter boilers. Carron was to supply the cast iron parts, but all the rest, such as building the engine house and the 'canal' for the water as well as erecting everything, was the responsibility of the two partners. Their total estimate came to £998.2s.6d, of which £63.10s. was for the canal and £100 for 'workmanship of the machinery and fitting the whole'. Then there was £100 for 'our own trouble' and £40 for 'our expences'. A drawing has survived showing one of Watt's wheel engines and the estimate includes £36.10s. for the wood and making the wheel. The drawing (53) also shows a form of steam valve in a circular casing with an oscillating flap valve but no valve gear. As this was in the vertical position away from the boiler, Watt had produced a design different from those usually found on Newcomen engines.

The remaining Watt papers are scarce for this period so no details have survived concerning the erection of this engine. Watt did not include it in his list and no references were found in the Carron Company archives. But one modern writer has said that Carron installed a steam engine in August 1767, which would link up with the date of the estimate after allowing for construction. (54) This engine is most likely to have been the inspiration for the segmental wheel instead of a beam in Hately's patent of 1768 (See p345) and thus the engine at Carron which Hately altered, so it must have been erected but perhaps did not perform as well as had been hoped.

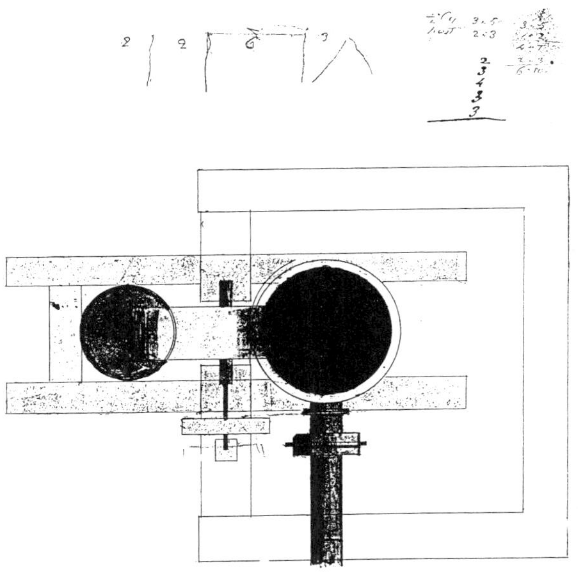

Plan of the Carron engine with the pump cylinder on the left and the steam cylinder on the right. The steam pipe enters from the bottom of the picture.

The Engine at Newton of Ayr

The Mackell and Watt partnership was involved with a third engine at 'Newton of Air' for a Mr. Dick of Glasgow at the end of December 1766 and early January 1767. (55) Dick had purchased the cylinder and other parts of an engine at Kilmarnock. Watt and Mackell were to use as much of the 'usefull' parts of the old engine in as 'frugal way' as possible. They were to build the engine house, supply such other parts as necessary, put in the pumps and 'direct the doing the whole according to the best of our judgement... for the sum of 160 pounds'. Payment was to be made in five instalments, commencing in February 1767 'by which time we expect to have the engine finished unless hindered by mason work not being finished'. (56) Dick was to supply the materials as well as a shop for the smiths with two fires and all the proper tackle. Here again, we see provision being made for what might be considered as the capital equipment necessary to construct the engine. From lack of other references and Watt's own statement, this appears to have been the final engine erected by the Mackell and Watt partnership.

Watt's Newcomen Engine Business

Watt developed his engine business in two ways, by repairing existing engines and by designing and erecting his own. He assisted with the maintenance of Roebuck's colliery engines, possibly because Roebuck was becoming more and more financially embarrassed. Indeed, this may be one reason why it took Watt so long to make his own engine run properly because his attention was diverted elsewhere. He told the Hart brothers in 1817 that he had erected a small engine on Taylors pit in order to pump out the coal workings there which were higher than those of the larger School Yard pit and so relieved that engine. (57) This could have been one of the small engines for Roebuck. Roebuck junior consulted Watt about problems they were having with the pumps at the West engine in November 1769. (58) This may be the same as the Wester engine where Watt sent his 'lads' to effect some repair the following April. (59) During the summer of 1770, Watt designed a new type of beam for this engine made out of four baulks of timber, braced together in an open framing, possibly based on something he had tried on his own engine at Kinneil. (60) Roebuck junior wanted Watt to look at the Dunmore engine in order to keep it running for a further couple of years while they extracted the coal reserves. During September 1770 Watt carried out tests on this engine. (61) All this gave Watt practical experience and familiarity with running Newcomen engines.

Engines for Peter Colville

An important customer for Watt was an influential colliery owner, Peter Colville, who had a mine at Torryburn in Fife by the shore of the Firth of Forth. Mining was long established in this region, dating back to possibly the sixteenth or early seventeenth century when dams were constructed to supply waterwheel pumps. (62) By our period, the mines must have been quite deep. Colville's standing in the community must have been high because he and Charles MacDowall were asked by the Carron Company at the beginning of 1767 to act as arbitrators in a dispute over how the Company was mining coal in a colliery it had leased. (63) Carron asked Colville to do the same again in 1770. (64) Watt went to Torryburn in October 1768 and paid additional visits later. (65) He told Cochrane that he had spent six weeks there one winter. (66) Starting in September 1768, Gardner spent twelve weeks at Torryburn and Osborne three weeks. (67) This must have been when they erected a couple of short stroke engines, possibly another original design of Watt's. (68) Watt carried out further work for Colville since, in August 1770, he was again at Torryburn and, during that autumn, Carron cast a new steam chest about which Colville was told, 'Your steam chest is in hand & you may be assured there shall not be a moments time lost in finishing & forwarding it to you'. (69) It was not until March 1771 that Colville wrote to Watt asking him to send Gardner because John Parker was afraid to install the steam chest by himself. (70) The first Boulton and Watt pumping engine in Scotland was erected for Colville at Torryburn in 1776 which shows how important a customer he was.

Engine for Robert Montgomerie of Ayr

During the autumn of 1770, Watt installed an engine for Robert Montgomerie at Ayr. Smeaton's map of Ayr harbour shows this situated a little to the north. (71. See p282) While a new engine house was built, some parts of an old engine were reused. Erection must have commenced in October for Montgomerie wrote on Christmas Day,

> I have the pleasure to inform you that on Friday 21s currt. our Engine was sett agoing to the utmost extent of our hopes, she goes admirably well & we are proceeding with our sinking...
> Our man John Scot I believe will go mad with Joy at his success, our Engine going so far superior to that of our Neighbours, I mean Mr Beaumonts, I must own he hath surpast my expectations both as to time & workmanship. (72)

From this letter, it sounds as if Watt had sorted out the teething problems with his Newcomen engines. He built this one after he had abandoned trials on his own engine at Kinneil.

Engine for Lord Cochrane

The last Newcomen engine erected by Watt in Scotland was one for Lord Cochrane at Culross. Discussions had started early in 1772 when Roebuck recommended Watt to undertake it but to see that he should 'be sure to gain handsomely'. (73) After Watt had judged the scheme for a tide mill impracticable, (See p286) the request in January 1773 that he should install a steam engine was countermanded very quickly because Cochrane though that he had found a second hand one at Dysart. However this proved to be unsuitable so that, at the beginning of February, Cochrane ordered Watt to build an engine with a cylinder 40 ins. diameter. The parts were supplied by the Carron Company to Watt's drawings. While the Carron Company queried some of the details on the drawings, it was remarked, 'The plans are very distinct'. (74) No doubt hastened by a letter from Cochrane himself, the parts were ready for Watt's inspection at Carron on 12 June. (75)

What is evident from Cochrane's letters is that Watt was acting in the capacity of a modern consulting engineer. He was to prepare the drawings and pass them on to appropriate firms to make the parts. He was also to prepare instructions for other workmen to carry out their respective functions. While Cochrane was going to be away,

> The charge of carrying on the works in my Absence will fall to Smeaton [presumably not John Smeaton]. Some Instructions on your part may help to make him both more expert in Carrying on the intended works and more trustworthy in my Absence. Please to send by bearer Directions for my Blacksmith as to the Boiler. (76)

Later when in Birmingham, Watt produced written instructions for his engine erectors. Cochrane expected that his engine would be running during the early part of August but the Carron Company had to threaten him with court action to obtain payment for what had been supplied. (77) Watt's Journals contain no record that he ever visited the site for the engine either while designing it, or while it was being erected and commissioned. Osborne was there for 19 days. Here was Watt's ideal role as a consulting engineer, for which he received his fees of £20. (78) In May 1774, Cochrane told Watt that 'his Culross engine burns 10 carts of 9 cwt each when kept perpetually going in 24 hours & the Kincardine engine 28 carts'. (79) This suggests that Watt's engine was much more economical. Was this due partly to the cylinder supplied by Carron having been bored on Smeaton's new boring machine?

These are probably all the Newcomen engines erected by Watt. He said in 1769 that he

> refused some common fire-engines, because they must have taken my attention so up as to hinder my going on with my own. However, if I cannot make it answer soon, I shall certainly undertake the next that offers; for I cannot afford to trifle away my whole life, which god knows may not be long. (80)

From this, it sounds as if Watt intended to pursue the mechanical engineering side of his career as actively as he would the civil. There is no doubt that, between 1766 and 1774, he had mastered 'thoroughly the details of the ordinary Newcomen engine, and ascertain[ed] the extent of its capabilities as well as of its imperfections'. (81) He changed his engineering methods between his first at Kennet where he was extensively involved with its erection to that at Culross which he never visited. He designed these engines and had their parts made by different contractors. Erection was carried out sometimes by his own men but, in the case of Cochrane's, he appears to have provided written instructions with some supervision from Osborne. He had no capital tied up either in workshops which were provided by the customers or in supplying parts which likewise was the responsibility of the customers. Here we can see the origins of the way in which he would organise the steam engine business when he moved to Birmingham in 1774 and formed his famous partnership with Boulton. But first he had to perfect his own engine.

Notes: Newcomen Engines Built by Watt

1 A longer account about J. Watt's failure to erect an engine at the Leadhills mines was published in *Mining History*, Vol 13, No. 6, Winter 1998, as R.L. Hills, 'James Watt's Steam Engine for the Leadhills Mines', pp. 25 - 28.

2 Smiles, (1878), p. 98 and Robinson & Musson, (1969), p. 41, 'Memoranda of James Watt Esquire from the Messrs. Harts. Communicated by John Smith, 19 March 1845'.

3 B. & W. Col., MI, 1/15, 'Journal', 12 June 1770.

4 Muirhead, (1854), Vol I, p. 7, Letter 6, J. Watt to J. Roebuck, 2 Oct. 1765; JWP, 4/32.36.

5 B. & W. Col., MI, 2/3, 'Cash Memorandum', 12 April 1766 and MIII, 3/6, 'Ledger of Personal Accounts', f. 59.

6 JWP, 4/52.4 - 6, Drawings and Estimates for the Leadhills engine.

7 Clow, (1952), pp. 364 - 365.

8 Riddell, (1979), p. 19.

9 Stirling of Garden Papers, Bundle 145, J. Stirling to The Earl of Hopetoun, 5 Jan. & 25 Feb. 1764.

10 Smeaton, (1812), Vol II, pp. 402 - 411, 'The Report of John Smeaton, engineer, upon the state of the powers employed, and capable of being employed at the Susannah Vein, at Lead Hills, for unwatering the low workings thereof'.

11 Clow, (1952), p. 365 and Stirling of Garden Papers, Bundle 107, T. Watts to J. Stirling, 30 May 1765.

12 Stirling of Garden Papers, Bundle 107, J. Farquharson to J. Stirling, 6 June 1765.

13 Smeaton, (1812), Vol II, p. 403.

14 Stirling of Garden Papers, Bundle 107, J. Farquharson to J. Stirling, 25 July 1765.
15 Ibid, Bundle 107, J. Farquharson to J. Stirling, 8 Aug. 1765.
16 Ibid, Bundle 107, J. Farquharson to J. Stirling, 29 Aug. 1765.
17 See JWP, 4/32.38, J. Watt to J. Roebuck, 23 Aug. 1765.
18 SRA, C1.1.33, Glasgow Council Minute Book, Jan. 1763 - Jan. 1768, p. 510, 5 Oct. 1767.
19 Stirling of Garden Papers, Bundle 107, W. Hamilton to J. Stirling, 31 Oct. 1765.
20 Ibid, Bundle 107, W. Hamilton to J. Stirling, 14 Nov. and 19 Dec. 1765.
21 Ibid, Bundle 108, T. Watts to J. Stirling, 2 Jan. 1766 and W. Hamilton to J. Stirling, 2 Jan. 1766.
22 Ibid, Bundle 108, W. Hamilton to J. Stirling, 30 Jan. 1766.
23 Ibid, Bundle 108, W. Hamilton to J. Stirling, 13 March 1766.
24 I wish to thank Mr. W.S. Harvey for this information.
25 Stirling of Garden Papers, Bundle 108, W. Hamilton to J. Stirling, 10 April 1766.
26 B. & W. Col., MI, 2/3, 'Cash Memorandum', 12 April 1766 and MIII, 3/6, 'Ledger of Personal Accounts', f. 59.
27 Stirling of Garden Papers, Bundle 108, W. Hamilton to J. Stirling, 24 April 1766.
28 Ibid, Bundle 108, T. Watts to J. Stirling, 17 April 1766.
29 Ibid, Bundle 607, J. Stirling to W. Hamilton, 8 May 1766.
30 Ibid, Bundle 607, J. Stirling to W. Hamilton, 3 May 1766.
31 Ibid, Bundle 607, J. Stirling to W. Hamilton, 10 May 1766.
32 Smeaton, (1812), Vol II, p. 408.
33 Stirling of Garden Papers, Bundle 607, J. Stirling to W. Hamilton, 24 May 1766.
34 JWP, 4/52.4 - 6, Drawings and estimates of the Leadhills engine.
35 JWP, 4/52.1, Estimate for an engine at Carron, 5 Dec. 1766.
36 Smeaton, (1950), p. 58.
37 Farey, (1827), p. 262.
38 Lindsay, (1968), p. 20.
39 SRO, Carron Co., GD58/6/1/7, Letter Book, 1767 - 1768, p. 435, to R. Mackell, 17 Oct. 1768.
40 Robinson & McKie, (1970), p. 419, 'Memorandum made at the request of J. Watt jun., in Jan. 1808'.
41 SRO, Carron Co., GD58/6/1/5, Letter Book, 1765 - 1766, p. 75, to R. Mackell, 24 Aug. 1765.
42 Ibid, GD58/6/1/5, Letter Book, 1765 - 1766, p. 187, to J. Munro, 2 Dec. 1765.
43 Muirhead, (1854), Vol I, p. 12, Letter 13, J. Watt to J. Roebuck, date is given as 4 Dec. 1765, but is probably later, see JWP, 4/32.32.
44 B. & W. Col., MI, 2/3, 'Cash Memorandum', 10 Aug., 16 Nov., 14 Dec., 1765 and 15 & 20 March, 12 April 1766.
45 Sinclair, (1978), p. 725.
46 Ibid, p. 726.
47 JWP, C7/7, 'Expence of Engine at Kennett'.
48 Ibid.
49 JWP, 4/52.7, 'Agreement for Kennet Engine'.
50 B. & W. Col., MI, 2/3, 'Cash Memorandum', 14 May, 3 June, 28 July, 11 Sept.,

10 Oct., 14 Nov. and 29 Dec. 1766.

51 JWP, 4/52.8, Ld. Kennet to J. Watt, 15 May 1769.

52 JWP, 4/52.1, 'Plan & Estimate of fixing up a 72 inch Cylinder for drawing back water at Carron by Messrs Watt & Mackell, December 5 1766'.

53 JWP, 4/52.3, Drawing of the Carron engine.

54 Shaw, (1984), p. 433 and Campbell, (1961), p. 42.

55 B. & W. Col., MI, 2/4, 'Cash Memorandum', 'Engine at Newton of Air, in company with Ro Mackele, 6 Jan. 1767'.

56 JWP, 4/52.9, J. Watt to Mr. Dick, 21 Dec. 1766.

57 Robinson & Musson, (1969), p. 41, 'Memoranda of James Watt Esquire from the Messrs. Harts. Communicated by John Smith, 19 March 1845'.

58 JWP, 4/32.106, J. Roebuck jun. to J. Watt, 4 Nov. 1769.

59 JWP, 4/32.14, J. Roebuck to J. Watt, 15 April 1770.

60 B. & W. Col., MI, 1/15, 'Journal', 4 June 1770 and JWP, 4/32.102, J. Roebuck to J. Watt, 31 May 1770.

61 JWP, 4/32.101, J. Roebuck jun. to J. Watt, 5 June 1770 et al. and B. & W. Col., MI, 1/15, 'Journal', 14 & 18 Sept. 1770.

62 Shaw, (1984), p. 69.

63 SRO, Carron Co., GD58/6/1/6, Letter Book, 1766 - 1767, p. 173, 23 Jan. and p. 198, 16 Feb. 1767 to P. Colville.

64 Ibid, GD58/6/1/10, Letter Book 1769 - 1770, p. 315, to P. Colville, 9 May 1770.

65 B. & W. Col., MI, 1/14, 'Journal', 16 Oct. 1768 et al.

66 Ibid, MI, 3/67, 'Lord Cochrane's Water Engine', no date.

67 Ibid, MIII, 3/2, 'Waste Book', Sept. 1768 et al. See also JWP 6/28.82, Various Accounts, 'Peter Colvile Esq. Glasgow Oct. 22, 1769'.

68 Smith, (1994), Bryden, p. 13 lft.

69 SRO, Carron Co., GD58/6/1/11, Letter Book, 1770 - 1771, p. 301, to P. Colville, Nov. 1770. See also GD58/6/1/11, p. 240, 22 Oct. and p. 241, 23 Oct. 1770 to J. Watt.

70 JWP, 4/19.17, P. Colville to J. Watt, 2 March 1771.

71 Smeaton, (1812), Vol III, p. 151, 'The Report of John Smeaton, upon the Harbour of Ayr'.

72 JWP, 4/19.8, J. Montgomerie to J. Watt, 25 Dec. 1770.

73 JWP, 4/32.80, J. Roebuck to J. Watt, 3 March 1772.

74 SRO, Carron Co., GD58/6/1/12, Letter Book, 1772 - 1773, p. 389, 2 March and p. 429, to J. Watt, 23 March 1773.

75 Ibid, p. 380, 23 Feb. and p. 584, 12 June 1773, to J. Watt.

76 JWP, 4/19.32, Ld. Cochrane to J. Watt, 14 Feb. 1773.

77 SRO, Carron Co., GD58/6/1/13, Letter Book 1773 - 1774, p. 394, to Ld. Cochrane, 18 Feb. 1774.

78 B. & W. Col., MI, 2/42, 'Expenses on Surveys', 12 July 1773.

79 Ibid, MI, 1/21, 'Journal', 19 May 1774.

80 Muirhead, (1854), Vol I, p. 71, Letter 61, J. Watt to W. Small, 20 Sept. 1769; Boulton 348.11.

81 Smiles, (1878), p. 98.

WATT'S OWN ENGINE: PARTS 1 & 2 WATT'S PERFECT ENGINE

While Watt's experiments with the University model engine and on his own Newcomen model engines appear to have followed some sort of sequence and been successful, those with his separate condenser do not seem for follow any similar pattern. Unlike Smeaton who altered only one parameter at a time, Watt made sweeping changes to his condensing apparatus, sometimes entirely replacing it. He admitted having difficulty in enlarging it from models to full-size machines but equally he may have caused problems for himself through not advancing slowly step by step. It may be noted that these experiments were carried out on his own without the help of either Robison or, perhaps more particularly, Black. Watt might have succeeded sooner if he had been able to receive advice from Black.

In his account of his invention of the separate condenser at the patent trial of 1796, Watt made no mention of the place but said,

> I did not invent this method piece meal but all at once in a few hours in 1765 I believe. The first step was the idea from the elastic nature of steam of condensing in a seperate [sic] vessel, 2d the getting out the water by a long pipe and the air by a pump, 3d that the pump would extract the water also 4th that grease might be used in place of water to keep the piston tight, 5th that Steam might be employed to press upon the piston in place of air 6 to keep the Cylinder warm.
> The next day I set about it. The boiler was ready, I took a large syringe of Tom Hamiltons 2 inches dia[mete]r, and two small pipes about 10 Inches long and 1/8 dia[mete]r immersed in a small round cistern, which I still have. I placed the Cyl[inde]r inverted tied a weight to the piston rod, which was hollow, and when I judged the cyl[inde]r filled with steam, I drew up the piston of the pumps and the weight immediately followed, to my great Joy, all this was done in a day or two after I had contrived it. (1)

While Watt was making his first model of his separate condenser, Robison was away, out of town. On his return, Robison wished to tell him about some observations he had made on the accounts concerning steam engines by Desaguliers and Belidor, so he

> ... came into Mr. Watt's parlour without ceremony, and found him sitting before the fire, having lying there on his knee a little tin cistern, which he was looking at. I entered into conversation on what we had been speaking of at last meeting, - something about steam. All the while, Mr. Watt kept looking at the fire, and laid down the cistern at the foot of his chair. At last he looked at me, and said briskly, 'You need not *fash* yourself any more about that, man; I have now made an engine that shall not waste a particle of steam. It shall all be boiling hot; - aye, and hot water injected if I please'. So saying, Mr. Watt looked with complacency at the little thing at his feet, and, seeing that I observed him, he shoved it away under a table with his foot. I put a question to him about the nature of his contrivance. He answered me rather drily...
> I was very anxious, however, to learn what Mr. Watt had contrived... I... found Mr. Alexander Brown, a very intimate acquaintance of Mr. Watt's, walking with another gentleman, (Mr. Craig, architect). Mr. Brown immediately accosted me with, 'Well, have you seen Jamie Watt?' - 'Yes.' - 'He'll be in high spirits now with his engine, isn't he?' 'Yes,' said I, 'very fine spirits.' 'Gad', says Mr. Brown, 'the condenser's the thing: keep it

but cold enough, and you may have a perfect vacuum, whatever be the heat of the cylinder.' The instant he said this, the whole flashed on my mind at once... I could only learn that there was a vessel called a condenser, which communicated with the cylinder, and that this condenser was immersed in cold water, and had a pump to clear it of the water which was formed in it. I also learned the great difficulty was to make the piston tight; and that leather and felt had been tried, and were found quite unable to stand the heat. I saw that the whole would be perfectly dry, and that Mr. Watt had used steam instead of air to press up his piston, which I thought by Mr. Brown's description was inverted. We parted, and I went home, a very silent companion. (2)

Robison had seen the condenser of Watt's perfect engine. Here again, we discover the importance of this concept. In our first surviving letter from Watt to Roebuck in August 1765, this concept is mentioned; 'On the faith of this, I have set about a larger and more perfect model, having now little doubt of its performing to satisfaction'. (3) Watt confirmed that his engine would 'not waste a particle of steam' a couple of weeks later when he wrote to Lind,

I have tried my small model of my perfect engine, which hitherto answers expectation, and gives great, I may say greatest, hopes of success... in greater model now far advanced; in short, I expect almost totally to prevent waste of steam, and consequently to bring the machine to its *ultimatum*. (4)

Was it this belief that he had actually discovered the perfect steam engine which kept Watt persevering with it against many set-backs and disappointments over the next ten years?

Having made his crucial invention based on scientific principles, Watt had to turn it into practical reality. After his short walk on the Green of Glasgow, he quickly realised that he would have to clear his separate condenser of the condensed steam, water and air. Two ways occurred to him. One was similar to that on some Newcomen engines; a long eduction pipe reaching to a depth of 35 or 36 ft. to allow the condensate to drain out against the vacuum, which was obviously impractical on models and many other engines. The other was to use a pump. (5) This was the key to his success. The other features of his new engine followed in quick succession.

In Newcomen's engine, the piston is kept tight by water, which could not be applicable in this new method; as, if any of it entered into a partially-exhausted and hot cylinder, it would boil, and prevent the production of a vacuum, and would also cool the cylinder by its evaporation during the descent of the piston. I proposed to remedy this defect by employing wax, tallow, or other grease, to lubricate and keep the piston tight. It next occurred to me, that the mouth of the cylinder being open, the air which entered to act on the piston would cool the cylinder, and condense some steam on again filling it. I therefore proposed to put an air-tight cover upon the cylinder, with a hole and stuffing-box for the piston-rod to slide through, and to admit steam above the piston to act upon it, instead of the atmosphere... There still remained another source of the destruction of steam, the cooling of the cylinder by the external air, which would produce an internal condensation whenever steam entered it, and which would be repeated every stroke; this I proposed to remedy by an external cylinder, containing steam, surrounded by another of wood, or of some other substance which would conduct heat slowly.

When once the idea of the separate condensation was started, all these improvements followed as corollaries in quick succession, so that in the course of one or two days the

> invention was thus far complete in my mind, and I immediately set about an experiment to verify it practically. (6)

This outlines the changes which Watt had to make to the Newcomen engine in order to turn it into his own type as well as the stages of development he would follow. However it was written over forty years after the event, possibly to support the claim that he had conceived the whole engine almost instantly.

Watt admitted to Small in 1769,

> I find that I am not the same person I was four years ago, when I invented the fire-engine, and foresaw, even before I made a model, almost every circumstance that has since occurred... The necessary experience in great was wanting; in acquiring it I have met with many disappointments. (7)

Some of his problems were caused when he was 'bewildering myself with mechanism foreign to the invention and above all perplexing myself with trying to condense without injection'. (8) He confirmed this in an account prepared for the 1796 patent trial entitled 'A Plain Story'.

> Had Watt been content with the mechanism of steam-engines as they then stood, his machine might soon have been brought before the public; but his mind ran upon making engines *cheap* as well as *good*, and he had a great hankering after inverted cylinders and other modifications of his invention, which his want of experience in the practice of mechanics in great, flattered him would prove more commodious than his matured experience has shown them to be. He tried, therefore, too many fruitless experiments on such variations. (9)

Here Watt fails to mention that it was also his first method of operating his separate condenser which caused him many fruitless experiments and must have cost a great deal of money as well as wasting much time.

While the basic principles may have been thought out quickly, it took far longer than a couple of days for Watt to invent some parts of the engine itself. This is confirmed by Lind writing to Watt in 1793 about acting as a witness in one of the patent trials,

> If I am prevented I shall assign my evidence to Mr De Luc - viz your letters to me on the your [sic] Fire Engine so long ago as 29th April 1765 etc about which time or soon after between [then] and September 1765 you informed me of the principal parts of your specification of the 6th of January 1769. (10)

In 1808, Watt admitted to Watt junior, 'The mechanism of the engine was not invented all at once, but has been in a course of improvement even to this day. I was ignorant of many things I know now and possibly of many things which were known to others'. (11)

Watt was both proud of his invention but secretive about showing it to even Robison, his closest friend. There is no mention of it either in his letters to his father until the following October when he said 'I am very busy with the fire engines which confining me to the workshop hindered me writing sooner'. (12) During the summer of 1765, Robison disappears from the story. Possibly Watt wished to prevent others discovering his great invention because, to his great annoyance, a form of valve gear, which he had invented, had been revealed by Robison to someone else. (13) There are examples later, such as the application of the crank, where Watt was extremely jealous of his discoveries. He said that he did not remember Robison being present at any of his experiments or improvements on the steam engine. (14) 'From the period of Mr. R.'s return to Glasgow, our interviews were rare: he had his duties to attend to and I, having become the father of a family, and loaded with cares of many kinds, had less time for mere philosophical conversations'. (15) It was Watt, and he alone, who would carry through this crucial development.

There were three periods when Watt was experimenting with his steam engine and separate condenser up to the time of his leaving Scotland. The first was from sometime in the spring of 1765 after the initial flash of inspiration for perhaps a year until the spring of 1766. This started with testing models in Glasgow and ended with the failure of his first attempt to build a large engine at Kinneil. Probably all these were inverted types. In the second period, Watt returned to more experiments on models in Glasgow from January 1768. Some of these must have had a conventional layout and were probably larger than the earlier series. This period culminated in the application for a patent later that year which was granted on 5 January 1769. (16) This led into the third phase when the decision was taken to erect another large engine at Kinneil, based on the performance of the earlier models and their continued testing. Erection of this second Kinneil engine started in the summer of 1769 and tests on it ended in June 1770 when Watt was called away to assist with a survey of the Clyde. There do not appear to have been any more tests after that, even on models, although Watt carried out more laboratory experiments on the properties of steam during the autumn of 1773 and spring of 1774.

Experiments on the First Model Engine

On his return from his Sabbath walk across the Green of Glasgow, Watt pondered over the apparatus he must construct to prove his theory. He needed privacy to prevent other people copying it and so his first engine was assembled in his own home. We have no contemporary accounts but those from 1796 are most likely correct in their outlines. A drawing, (17) probably prepared for the 1796 patent trials, shows that he used the boiler which he had developed during his experiments on Newcomen engines. He inverted the cylinder and hung a weight on the end of the piston rod, in a way similar to Robison's model engine. This arrangement would have saved space as well as the expense of a rocking beam and framing to support it. For a purely experimental model, there was no need for complex valve gear, so the steam inlet and exit valves were simply conical cocks turned by hand. Likewise the pump in the condenser unit was hand operated. It is not until 1768 that we find any hint of self-acting valve operating mechanisms.

The cylinder was made from Tom Hamilton's brass syringe with a diameter of 2 ins., the same as that on the University model Newcomen engine, but with a stroke around 10 to 12 ins., or more than double. One reason why Newcomen engines consumed so much steam was that air entered the top of the cylinder while the piston was descending and cooled the cylinder walls. Watt made tin-plate covers for both ends of his syringe and connected the lower end to the boiler so that he could use steam to drive his piston instead of the atmosphere, thus avoiding this source of waste. He made no attempt to place a steam jacket around the syringe which means that the rather crude model which Dickinson identified as being the first must be wrong. (18) That Watt's own recollection of the size of the syringe is correct is proved by the remarkable survival in the Garret Workshop of a piston of this diameter attached to a hollow piston rod with a plug at the bottom. (19) This also confirms that the engine was inverted because any condensate above the piston could be drained out down this rod by unscrewing the plug. (20)

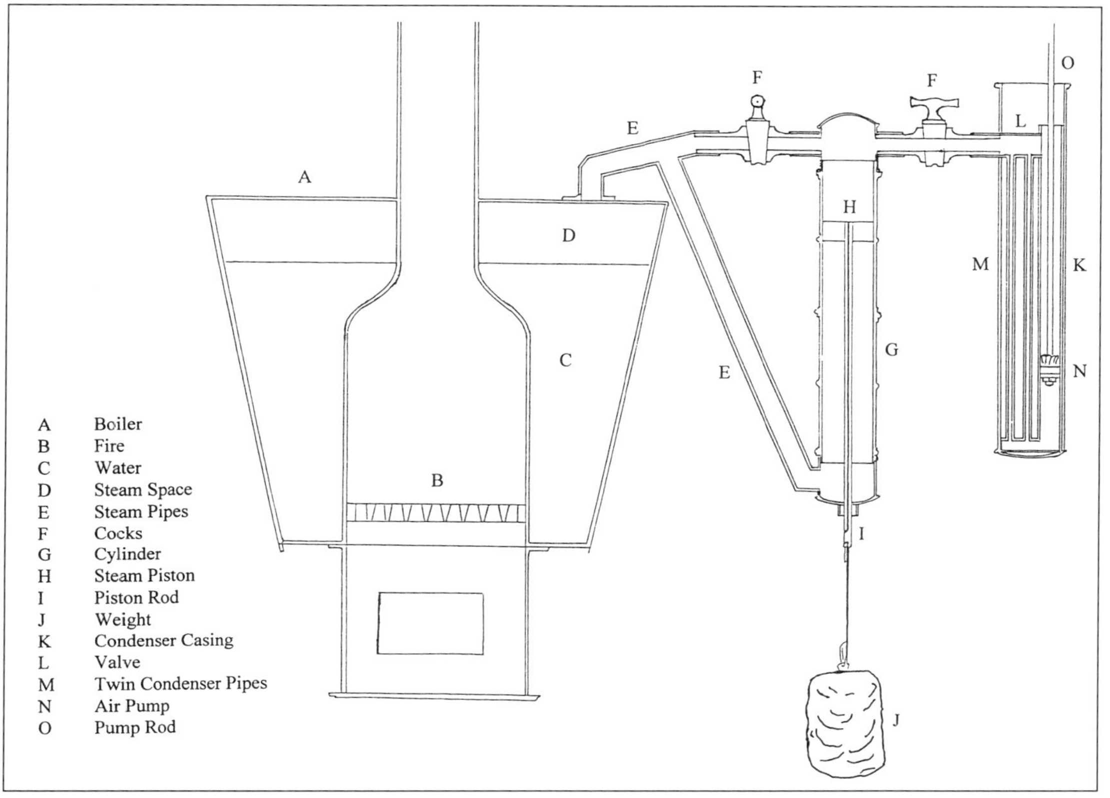

Watt's drawing of his first separate condenser which was probably prepared for one of the patent trials in the 1790s.

The Separate Condenser

Jardine was quite correct when he said,

> The last step of all was the most difficult, - the forming of the separate condensing vessel. The great knowledge Mr. Watt had acquired of the mechanical powers enabled him to construct it, but I have often heard him say this was a work of great difficulty, and that he met with many disappointments before he succeeded. (21)

Watt was never satisfied with his condensers all the time that he remained in Scotland. While he may have admitted his difficulties to his friends, he tried to hide them from other people and to create the myth that it was only badly bored cylinders which prevented his engine from working properly. He told Smeaton,

> Having built an Engine at Kinneal, they had for the present been brought to a stop, by a difficulty they had found out and had not satisfactorily removed; viz that of keeping the piston steam tight, without water, and Mr. Watt being wholly occupied in a different concern, & the engine laid aside for the present, to be resumed or not as it might happen. (22. See p323)

We shall see that it was only towards the end of his experiments in Scotland that Watt finally began to proceed along lines that were to lead to success in England and that he made frequent alterations to the forms of his separate condenser before radically altering its method of operation which was changed further after his arrival in Birmingham.

There were sound reasons why Watt decided to opt for surface condensing rather than continue with having cold water injected directly, which Newcomen had found to be so successful. This water would have carried air in with it which would have made the creation of a perfect vacuum impossible. Then a larger pump would be necessary to remove both injection water and condensate. On the other hand, Watt did not anticipate the difficulties he would face in constructing surface condensers. It was not until after 1850 that adequate ones were developed for use at sea. (23) While Watt ought to have been able to have calculated the amount of heat that would have to be exchanged, he does not seem to have realised how large a surface area would be needed to effect this. It was apparently Roebuck who recognised the significance.

Some of the problems may have been caused through Watt not understanding all the properties of steam. He probably followed Black in thinking that heat was some form of material substance which was added from the fire to the water to create steam. He knew that 'The great heat acquired in an instant by the cylinder, by the admission of a few grains of water in the form of steam, was an incontestible proof of the great quantity of fire in it'. (24) But it is difficult to establish what he thought happened to that heat in the condenser. In that first letter to Lind in April 1765, Watt talked about the steam being 'destroyed'. (25) About one of his models in April 1768, he reported, 'Now, I think, two of the most important facts are settled, - the perfection of the vacuum and the suddenty of the condensation'. (26) Even about the second Kinneil engine in March 1770, he said, 'The condenser seemed to despatch the steam instantly'. (27) What he thought happened to the heat is not clear. Seeing the apparent ease with which the vacuum was created may have caused him to underestimate the size needed for his surface condensers, particularly when the engines were working for long periods. Yet Robison would later point out how Watt knew that a great quantity of water was required to absorb the emerging heat. (28)

Much later Watt gave a description of that first condenser which Robison watched him shove away under the table.

> The condenser used upon this occasion consisted of two pipes of thin tin-plate, ten or twelve inches long, and about one-sixth inch diameter, standing perpendicular, and communicating at top with a short horizontal pipe of large diameter, having an aperture on its upper side, which was shut by a valve opening upwards. These pipes were joined at bottom to another perpendicular pipe of about an inch diameter, which served for the air and water-pump; and both the condensing pipes and the air-pump were placed in a small cistern filled with cold water. This construction of the condenser was employed from knowing that heat penetrated thin plates of metal very quickly, and considering that if no injection was thrown into an exhausted vessel, there would be only the water of which the steam had been composed, and the air which entered with the steam, or through the leaks to extract. (29)

This description fits both the drawing in the Boulton and Watt Collection (30) and an actual example in Watt's Garret Workshop. However the one in the Workshop is very well made for something knocked up quickly and may have been prepared to be shown at one of the patent trials in the 1790s because it has no valve to cover the hole in the horizontal pipe at the top of the condensing tubes. There is no valve either at the bottom of these pipes between them and the air pump cylinder. In the drawing, the piston of the air pump itself is shown as solid.

As Boulton pointed out, the air pump was the very soul of the engine, and, with it, Watt solved one of the enigmas of the condensing steam engine, how to remove the air and condensate. Smeaton thought only of snifting or blowing steam through any condensing vessel to remove the air, which would have meant heating it up and so negating all the advantages of a separate condenser. (31. See p323) Clues to Watt's source for his idea of the air pump with a solid piston can be discovered in one of the accounts of the trials of his 1769 patent when he commented about his specification,

> Every student of natural philosophy knows the effect of an air pump in extracting elastic fluids out of other vessels, and a pump for raising water is an appendage to almost every house, and as either of these kinds of pumps, or almost any other kind, will answer the present purpose, a description would have been superfluous. (32)

Watt, as a maker of and dealer in scientific instruments, would have been well aware of the normal pattern of air pumps used in laboratories for creating a vacuum and so known that these had solid pistons. The very name also suggests that he was more concerned about removing the large amount of air which he thought would be present rather than the condensate.

A replica of this condenser was made at the North Western Museum of Science and Industry in Manchester. There is no piston in the tin-plate condenser in Watt's Workshop so the one drawn was copied. We demonstrated this model by blowing steam through the main cylinder and having the vertical pipes empty. Then we closed the cock communicating with the boiler. A quick pump or two on the air pump piston was sufficient to draw the steam into the small pipes where it condensed, forming a vacuum. The piston in the steam cylinder rose quickly even though a 12 lbs. weight was suspended from it. We could understand how thrilled Watt must have been when his model worked even better as it lifted 18 lbs. (33)

Subsequent research has shown that we were operating this condenser on the wrong principle. At first, Watt followed the example of both Savery and Newcomen by creating a fresh vacuum for every stroke. This is another crucial point in understanding Watt's early problems. Condensing could be achieved inside a vessel either by pouring cold water over the outside, as Savery had done, or by standing the vessel in cold water. Both these were forms of surface condensing but needed to have steam inside the vessel to operate. Then there was Newcomen's method of direct injection of cold water into the steam. Watt first of all tried another method which accounts for the solid piston on his air pump. If the pipes of the condenser and the lower part of the air pump up to the piston were filled with water, when the solid piston was drawn up, the water would follow, pulling the water down the vertical condenser pipes, leaving a space at the top. The steam would follow the descending water and be condensed in the vacated cold space. When the solid air pump piston was pushed down, the water would be pushed back up the condenser pipes. If the cock between the steam cylinder and the condenser were shut, the hot condensate, together with any air brought over, would be driven upwards out of the small hole and valve in the connecting pipe from the steam cylinder, thus clearing the condenser. With this method of operation, the air pump piston must be solid because no water or condensate passes through it as with air pumps on later commercial engines. Once this method is understood, the layout and operation of all Watt's early condensers becomes clearer. However this is an idea which looks better in theory than it works in practice as it failed to give enough surface area. It led Watt up the wrong path and was one reason why he had to make so many unsuccessful experiments. Many problems were solved when he realised that the condenser could be left permanently exhausted.

More Early Experiments

The first surviving reference we have to Watt's invention is a letter to Lind on 29 April 1765. (34) It shows that he had already carried out tests, some time before the date of this crucial invention, given by Dickinson as the May of that year. (35) It also shows that Watt had discovered that steam consumption was much less than with engines of the Newcomen type, giving a better performance.

> I have now almost a certainty of the *facturum* of the fire-engine, having determined the following particulars: the quantity of steam produced; the ultimatum of the *lever engine*; the quantity of steam destroyed by the cold of its cylinder; the quantity destroyed in mine: and if there is not some devil in the hedge, mine ought to raise water to 44 feet with the same quantity of steam that theirs does to 32 (supposing my cylinder as thick as theirs), which I think I can demonstrate. (36)

Here Watt still considered that the thickness of the cylinder was significant because he had not yet added a steam jacket.

Watt continued, 'I can now make a cylinder of 2 feet diameter and 3 feet high only a 40th of an inch thick, and strong enough to resist the atmosphere; *sed tace*. In short I can think of nothing else but this machine'. (37) He had been experimenting with a cylinder for some engine and its diameter shows his ambitions to construct a large engine from the outset of his discovery. Such a thin-walled cylinder would have been formed out of sheet metal bent round and soldered up the seam. This form of construction points to the limits both of his workshop as well as his financial resources. While we do not know the size of Watt's great lathe, it would certainly not have had the capacity to bore cylinders of this size. This may be confirmed by Watt ordering a 'Bored Pipe' from Carron in April 1765 but no size is given. (38) Watt was probably anxious to manufacture as much as possible 'in house' to avoid others discovering his invention as well as to save costs and so he turned to ways of fabricating parts.

Watt continued to experiment with his first model during the summer of 1765 and, in his first letter to Roebuck that August, wrote,

> I have tried my new engine with good success; for though I have not been able to get it perfectly air-tight from its bad materials, yet, immediately on turning the exhausting cock, the piston, when not loaded, ascended as quick as the blow of a hammer, and as quick when loaded with 18 lbs. (being 7 lbs. on the inch) as it would have done if it had had an injection as usual. (39)

The performance was impressive, but it was being worked only by hand so probably made only a few strokes per minute. However this was sufficient to encourage Watt to 'set about a larger and more perfect model, having now little doubt of its performing to satisfaction'. (40) This helped to stimulate Roebuck's interest who became more and more involved with Watt's steam engines during the autumn of 1765. This large model was 'far advanced' when Watt wrote to Lind early in September and said, 'I expect almost totally to prevent waste of steam, and consequently bring the machine to its *ultimatum*'. (41)

Watt was faced with a new problem over his pistons. This was because he was using steam to force them up or down so he could no longer cover them with water as in the Newcomen engine, since water would have boiled away through the heat of the steam. Although much earlier the Marquis of Worcester and Leupold had driven pistons directly by steam, in the Newcomen engine, it was the pressure of the atmosphere which forced them down. To prevent air passing the piston and destroying the vacuum, the edge of the piston was packed with something like hemp rope kept in place by weights and kept supple through the hot water on top. This water also helped to create the seal because it was denser than air and did not pass through any gap between piston and cylinder as easily. The inability to procure a truly round and parallel cylinder bore caused Watt endless problems trying to obtain a gas-tight seal between it and the piston. From his earliest models in 1765 and even beyond the first engine sold in 1775, he never found a really satisfactory cure.

We see in another letter sent to Lind towards the middle of October that Watt may have tried a square piston but had to revert to a cylindrical one. Even so he had difficulty in keeping the piston steam-tight and tried rings of cloth. (42. See sketch below) He wrote in an early draft of his patent specification that he found common steam engines imperfect because the way of making the cylinder or piston tight was the cause of great friction. (43) He thought that rings of cloth would adapt themselves to a bad cylinder and make a flexible seal. Later in November he discovered what he claimed was the ideal substance for packing the piston - pasteboard - for if it were kept oiled, it had little friction. (44) He complained to Roebuck, 'There is no English pasteboard to be had here: the Scotch is made of old paper, the English of old ropes; it is it only that answers; so I wish you would order a cwt. from Edinburgh, and if not to be had there by first ship from London'. (45) Watt certainly anticipated using a great deal!

A sketch in a letter to Roebuck shows that this model had an inverted cylinder, with the piston rod connected to a beam below it. (See p372) The beam was pivoted at one end and the piston rod connected a little distance away from the pivot. It worked like a balance beam for weighing objects. Watt moved a weight along the outer portion of the beam to vary the load and calculate the pressure in the cylinder. He reported that he had better success with this model and that 'It readily works with 10 $^{1}/_{2}$ lbs. on the inch, and sometime I have made it lift 14 lbs.' (46) If he meant that the engine was developing a vacuum of 14 lbs., then it was performing very well. No wonder he was 'confident it will succeed to my utmost expectations... As to the steam consumed, it is very little'. (47) We learn that this model was quite large from a later annotation added by him.

> Inverted engine, with a new piston, the first not answering. The cylinder of this engine was five or six inches diameter, with a two feet stroke; the inner cylinder was made of copper, not bored but hammered, and not very true. It was enclosed in a wooden steam-case, and placed inverted; the piston working through a hole in the bottom of the steam-case. (48)

Watt's earliest sketch of the piston in his engine, showing it had an inverted cylinder.

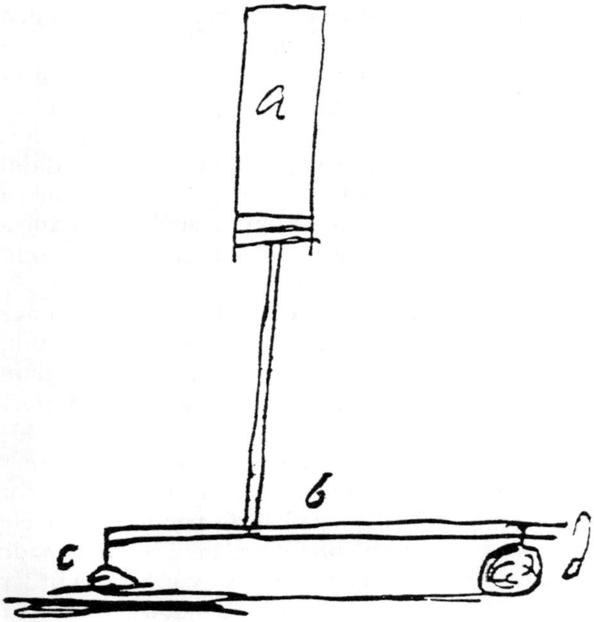

This is the most complete surviving sketch of any of Watt's early experimental models. It shows an inverted cylinder 'a', a fulcrum 'c' for a horizontal beam 'b' along which a weight 'd' could be moved to vary the load like a set of scales.

Although mentioned in this later annotation, reference to a steam jacket shows that one was definitely fitted to his second model. Robison claimed that one was added to the first model sometime in May 1765. 'I am certain that when I went out of town (May, I think,) 1765, he had not thought of the method of keeping the cylinder hot; and I am as certain that a fortnight after he had completed it, and confirmed it by experiment'. (49) The reason for fitting a wooden case or jacket around the cylinder and filling the space between them with steam was to keep the cylinder as hot as the steam itself. While wood was a good non-conducting material, it must have softened with the heat of the steam and been difficult to keep steam-tight because the jacket was made from staves like a barrel. But, once the jacket had been added, the specific heat of the material from which the cylinder was made became of no significance because the cylinder remained at a constant temperature while the engine was running. Watt continued to experiment until at least the middle of November 1765 on a model with an inverted cylinder. Through its size, it needed a large boiler. This was when he was delayed in writing to his father because he was confined to the workshop where he must have been experimenting. (50)

Experiments on the Properties of Steam

During the autumn of 1765, Roebuck helped Watt with his steam engine experiments in many ways. Not only did he procure materials for models but he carried out experiments with Watt into the properties of steam. He allowed Watt to test a Newcomen engine in one of his coal mines and asked him to design a boiler for another. Roebuck told Boulton that he had been convinced of the 'justness of the Principles' of Watt's separate condenser through the trial of a small one at Kinneil and so supported him. It was claimed much later that this had a 9 ins. cylinder. (51) Then he became involved with the construction of a large experimental engine in the grounds of Kinneil.

It appears that, at first, Roebuck took some convincing about the doctrine of latent heat and the other discoveries of Black and Watt. This may have delayed Roebuck's supporting Watt for

> Mr. Watt was obliged to yield for some time to the Doctor's confidence in his own great experience... But Dr. Roebuck had too much judgement not to see the conclusiveness of

> the experiments by which the doctrine of latent heat is established not to yield to their force'. (52)

Black said that Roebuck took a lot of persuading to understand fully 'the nature of steam, of the great quantity of heat, and consequently of fuel, necessary to produce it, and of the importance, therefore, of preventing the waste of it'. (53) It may not have been the concept of latent heat itself which Roebuck had difficulty in accepting but Watt's extension of this doctrine into what later became called Watt's Law.

In September 1765, Watt carried out an experiment with a further piece of apparatus, presumably devised by himself, which he described to Roebuck. (Plate 17) A vessel for water with two openings was placed in an open pan with more water in it on a stove so that the water boiled in both. When the air had been expelled from the vessel and the rest of the apparatus, a cork was placed in one opening. A pipe from the other led down to a second vessel made from a sealed pair of cones. All this would now be full of steam. The apparatus was removed from the stove and the boiling water in the pan replaced with cold to lower the heat in the closed vessel 'to that in which water boils in vacuo'. (54) The water in it would boil and this steam passed down to the conical one which had been placed in a 'Refrigeratory', another open pan containing cold water. Watt measured the change in temperature of the water in this pan and that of the condensate as well as its volume in the cones. From these figures, he was able to calculate the sensible and latent heats of the condensate.

> Care was taken to prevent the refrigeratory being heated by other things and Repetitions of the experiments agreed very nearly. any Objections that occur shall answer at meeting when we shall make the Experiment in Company. from this Experiment and some other facts I conclude that in proportion as the sensible heat of steam Increases its latent heat diminishes so that in the steam engine working with pressures above 15 lb must be more advantageous than below it. for not only the latent heat is diminished but the steam is considerably expanded by the sensible heat which is easily added. (55)

There are two important points here. First is the origin of Watt's Law. Watt was moving towards the fundamental idea of 'total heat', for he realised that, as the sensible heat varied, so the latent heat changed in the opposite direction. This led to the assertion that the total heat, or the sum of the latent and sensible heats was constant, Watt's Law, although much later this was shown to be inaccurate. (56) Watt was also beginning to stumble upon the advantages of higher pressure steam and its greater expansive powers through its greater temperature and pressure. Watt investigated the expansion of steam in various ways. We have seen that he used a Florence flask to determine how much water expanded when converted into steam at atmospheric pressure and so corrected the figure suggested by Desaguliers. In the case of his experiments for the basis of Watt's Law, he discovered that the addition of a small amount of heat to steam raised its expansive powers or pressure quicker than its temperature. He had already experimented with the small syringe to raise weights, creating a form of high pressure engine with the steam exhausting directly to atmosphere. The expansive power of steam was included in his 1769 patent specification. Then, starting from Watt's Law, he began to realise that, if the temperature of the steam were not reduced below 212°F, the latent heat boundary would not be crossed. He thought he might be able to make an improved engine with the 'forcible expansion of the steam' which would be drawn into a cold place and reduced in temperature from say 228°F down to 214°F so it would contract to five or six times its former bulk. The heat loss would be trifling. (57) In October 1770, he sent a drawing of such an engine to Small but it progressed no further. (58) This was different from the final expansive engine developed in Birmingham where Watt discovered that a more economical engine resulted from 'cutting off' the entry of steam to the cylinder at some point during the stroke and allowing it to expand for the remainder. The reasons behind the thermodynamic gains were unknown to him but the concept probably arose out of these earlier experiments.

In Newcomen engines, it is almost imposssible to use an early steam cut-off and allow the steam to expand. When steam expands and does work, condensation may be formed and Watt must have encountered this when, in some other experiments in 1765, he discovered that the cylinder walls were sometimes wet and sometimes dry. Not only would any water in the cylinder flash back into steam when a vacuum was created, but it would deprive fresh incoming steam of heat to convert it back into steam ready for the next stroke. Steam at higher pressures might have the advantage of remaining above this dew point when expanded and hence Watt's bewilderment for he could not have been aware of the thermodynamic background. This, together with the effects of Watt's Law, was the occasion when Roebuck found,

> The fact which you relate is so paradoxical, that, if less accurate observers of nature than yourself and Dr. Black had been witnesses of it, I should scarce have given credit to it... I shall be glad to try the experiment with you and make our observations on it afterwards. (59)

Watt and Roebuck carried out experiments together.

However before that, Roebuck had allowed Watt to run and test one of his Newcomen engines, said to be that at the Schoolyard pit. (60) The cylinder diameter of the engine there was 5 ft. 3 ins. diameter but that tested was smaller, with a cylinder diameter of 32 ins. and stroke of 5.2 ft. Watt calculated the evaporation rate of the boiler, the volume of the cylinder, and the surface area of the piston. He determined how much steam was condensed in rewarming the cylinder before the next stroke and found it was 97.2 'solid feet' while one cylinder full was only 28.6 'solid feet'. This confirmed how much steam was wasted reheating the cylinder. The pressure on the piston was 8.5 p.s.i. (61) This was the first of many trials on Newcomen engines which Watt would conduct and it may have helped to convince Roebuck of the correctness of Watt's theories because 'it served to show the great waste of fuel and steam'. (62)

About the same time, Watt became involved with Roebuck in the design of some boilers. One he described as a 'testudo' which had an arched top. It evaporated 25 lbs. of water with 4 lbs. of coal in about 44 minutes once boiling had commenced. The grate was formed from tubes filled with water while other tubes carried the heat through the water space. (63) Roebuck's boiler was to be 9 ft. diameter and deep at the crown to contain the steam. (64) At the same time, Watt was designing a much enlarged version of the boiler used for his Newcomen engine and separate condenser experiments with the fire in a central vertical tube. Now the outside was barrel-shaped and small pipes took the heat and flames from the top of the central fireplace up through the top of the barrel to the chimney. (65) He also experimented with methods of firing to produce as little smoke as possible by stoking with only a few coals at once and keeping the fire 'thoroughly inflame'. (66) When at Birmingham, he would patent a method of constructing the firegrate to reduce smoke pollution.

The First Kinneil Engine

Watt clearly received some remuneration for designing Roebuck's boiler and this may be how Roebuck first became involved financially. But, at the same time in November 1765,

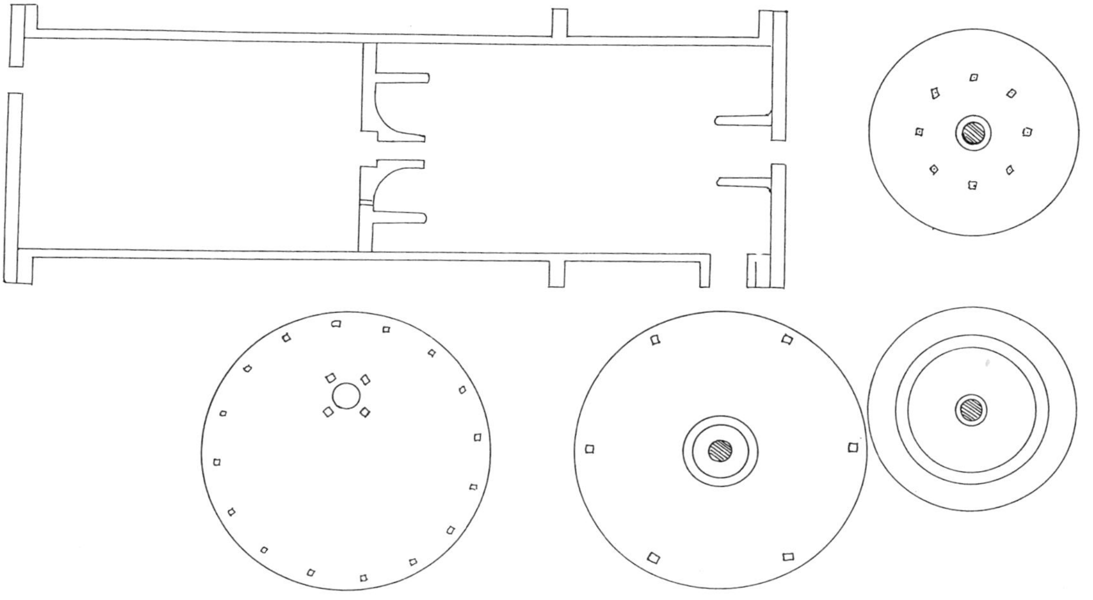

The scale drawing which Watt sent to Roebuck in November 1765 for a cylinder presumably to be cast at Carron for the first Kinneil engine. The inside length of the cylinder was to be '7 foot, diam. 24 in. to be made as truely round & equally wide as possible'. The hole for the piston rod was to be 2 $^{3}/_{4}$ ins. diameter.

Watt sent Roebuck a drawing of an inverted cylinder for an engine with a separate condenser. The cylinder was to be cast and bored at Carron for an engine to be erected at Kinneil. (67) As well as details of the cylinder design, there are sketches for covers and for the piston. The bore at 24 ins. was 'to be made as truely round & equally wide as possible'. The total length was 7 ft. but, allowing for packing flanges, etc., the stroke must have been nearer six. The diameter of the piston rod was 2 $^{3}/_{4}$ ins. Watt appears to have expected Roebuck to order this to be cast and machined at Carron but there are no details about payment. Roebuck obviously agreed to help finance Watt's first attempt to erect a large engine at Kinneil.

More details about this engine emerge in letters during December when Watt was at Kinneil. (68) The boiler was to be of the 'common kind' made with the thinnest plates. Watt enquired whether some pipes had arrived safely but was at a loss about constructing the condenser because his 'old white-iron man' had died. He proposed a plate type, sixteen double plates of a foot square set half an inch apart, which would be simpler to construct and have a greater surface area than 900 pipes of an eighth inch diameter. There had been some argument between Roebuck and Watt over the surface area for condensers, Roebuck advocating a large area for speed of condensation, while, as Robison pointed out, Watt knew it needed a great quantity of water. (69) Watt also had some new ideas for the air pump as well as pumps in the pit. (70)

The next we hear about the first engine for Kinneil is in February 1766 when Watt wrote to Roebuck,

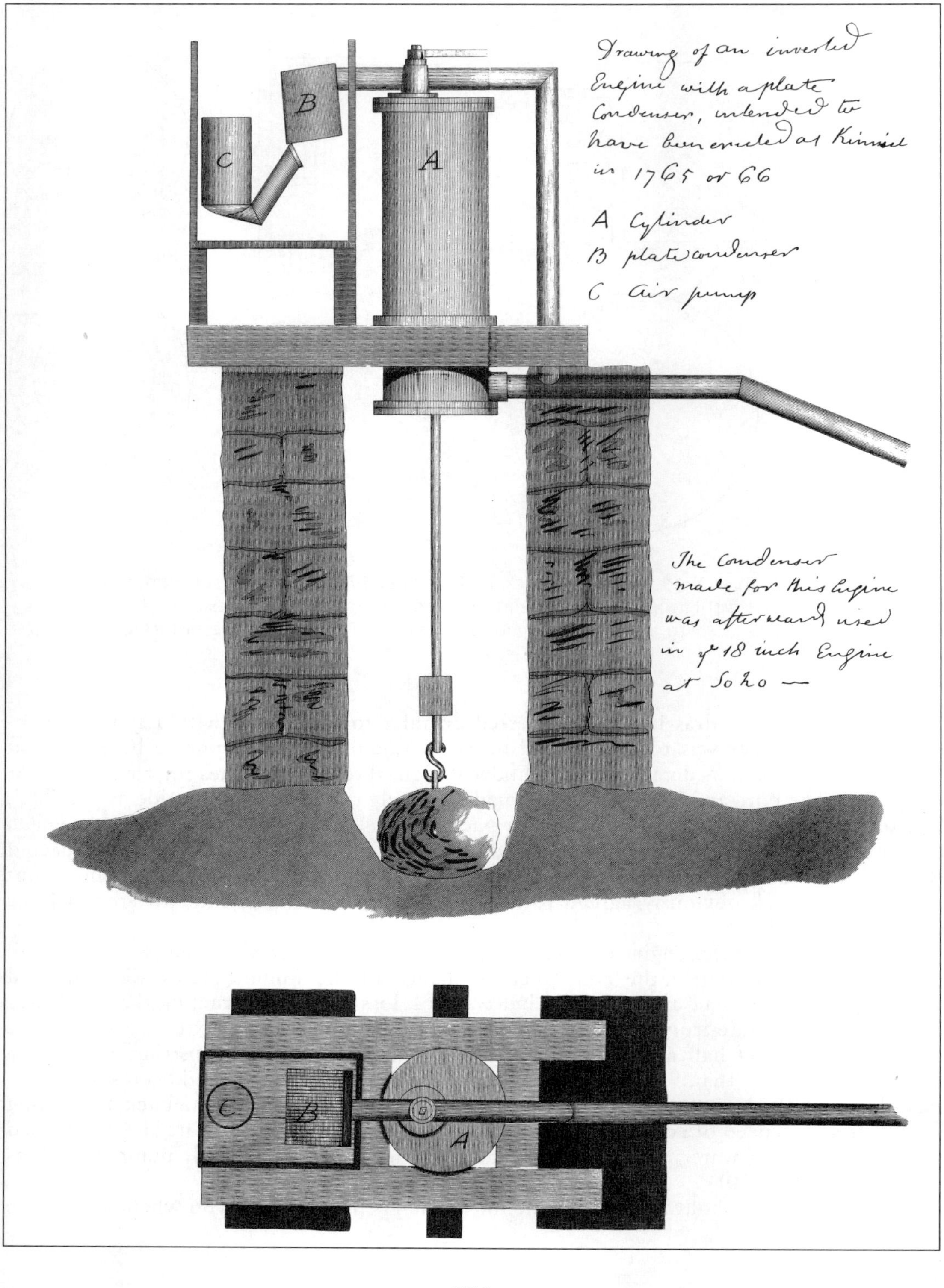
Drawing of an inverted Engine with a plate Condenser, intended to have been erected at Kinneil in 1765 or 66
A Cylinder
B plate condenser
C Air pump
B
C
A
The Condenser made for this Engine was afterward used in ye 18 inch Engine at Soho —
C
B
A

Opposite page: This engine must be the first one planned for Kinneil in 1766 but was never erected because the cylinder was so badly bored.

> The piston-rod for the large machine is near finished. I am somewhat at a loss about sending it, as if sent on a common cart Mr. Singleton and the rest of your people will unavoidably see it, and it being hollow will occasion speculation. I believe it will be best to send it in a box. (71)

The only reason for the problems and expense of making a hollow piston rod was for it to act as a drain for the condensate out of the cylinder in an inverted engine. It probably cost six guineas. (72) At the same time, Watt began designing another type of steam engine, a circular one to produce direct rotary action, the story of which will be told later. (See p424)

Watt must have been very confident of success to have ordered a cylinder for so large an engine so early in his experiments. The illustration in Muirhead shows a heavy weight hanging on the end of the piston rod and a condenser of very peculiar design which appears to be incomplete. There are no valve or pump operating mechanisms. The dimensions correspond

No mention can be found of any engine with a separate condenser and wheel instead of a beam having been erected in Scotland. Perhaps Watt drew this as an advertisement to show the layout of his new engine.

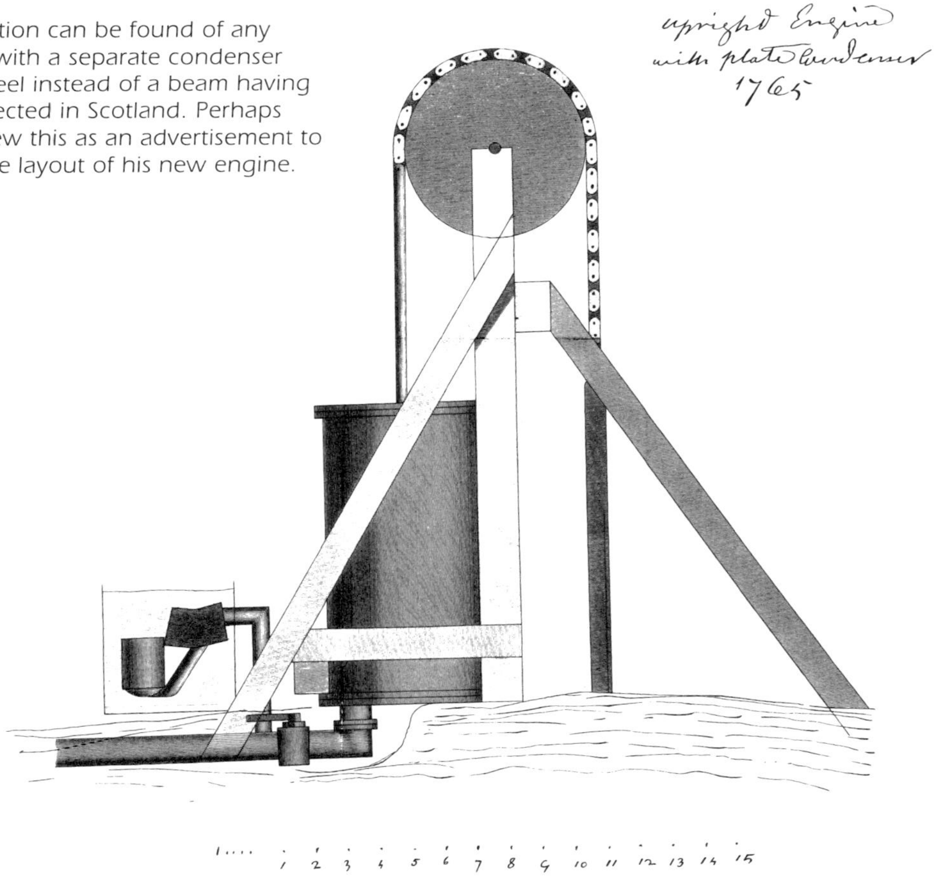

with the drawing of the cylinder which Watt sent to Roebuck. (73. See p374) The cylinder had not been cast at Carron by the middle of February. (74) Watt may have referred to it in April, 'Your cylinder is finished and sent to Carron wharf, with the ends on', but this suggests it was Roebuck's and not Watt's. (75) Whenever it did arrive, it was found to be too badly bored to be of any use, 'though the best Carron could make'. (76) The result was that this engine was abandoned. A undated list of expenses on 'Doctor Roebucks new Engine' (77) must refer to this first attempt at Kinneil because this would have been the only one needing a hollow piston rod. A partitioned condenser cost £4.4s. While there are costs for other small items and a steam chest at £1.6s, there is no price for the cylinder or framing. Gardner was paid £3.7s.6d. for 45 days work and Osborne 9s.4d. for 8 days. These were costs which Watt may have met himself. At this point, Watt ceased further experiments on any of his engines for well over a year.

Also shown in Muirhead is an engine with the cylinder in the usual position. (See p377) As the cylinder has a cover and there is a separate condenser, it may have been an early scheme for one of Watt's engines. The piston rod is connected to a chain passing over a wheel which is supported on a wooden framing similar to his engine for Leadhills. On this has been written, 'Upright engine with plate condenser'. (78) Perhaps Watt was dreaming of an engine he wished to construct but it must be conjectural because the dimensions on it show the bore around 42 ins. and the stroke 7 ft. The origin of the drawing has not been traced and no references have been found to an engine of such size with a separate condenser. It may show Watt's optimism for the sort of engine he hoped to build as early as this and, had his separate condenser worked, he might have built engines to his own designs like this in Scotland.

NOTES: WATT'S PERFECT ENGINE

1 Robinson & McKie, (1970), pp. 233 - 234, Letter 163, J. Watt to J. Robison, 24 Oct. 1796.
2 Muirhead, (1858), pp. 67 - 68.
3 Muirhead, (1854), Vol I, p. 4, Letter 2, J. Watt to J. Robison, 23 Aug. 1765; JWP, 4/32.38.
4 Ibid, Vol I, p. 5, Letter 3, J. Watt to J. Lind, 4 Sept. 1765; JWP, C1/15.
5 Dickinson & Jenkins, (1927), p. 23.
6 Muirhead, (1854), pp. lxxv - lxxvi.
7 Ibid, Vol I, p. 56, Letter 54, J. Watt to W. Small, 28 April 1769; Boulton 348.8.
8 Robinson & McKie, (1970), p. 234, Letter 163, J. Watt to J. Robison, 26 Oct. 1796.
9 Muirhead, (1854), Vol I, p. lxxxix.
10 JWP, 4/66.38, J. Lind to J. Watt, 21 June 1793.
11 Robinson & Musson, (1969), p. 109, No. 27, J. Watt to J. Watt jun., 12 Nov. 1808.
12 JWP, 4/11.131, J. Watt to J. Watt sen., 8 Oct. 1765.
13 Muirhead, (1858), p. 68.
14 Robinson & McKie, (1970), p. 411, Letter 257, 'J. Watt's Recollection of his Friend Dr. J. Robison, April 1805'.
15 Ibid, pp. 411 - 412.
16 Patent 913, 5 Jan. 1769.

17 B. & W. Col., Drawing PF 1381; see Law, (1969), p. 27.
18 Dickinson, (1935), p. 38, Plate IV and p. 70.
19 Law, (1969), p. 24.
20 This is shown wrongly on the drawing PF 1381 where it is depicted as a valve at the side; see Watt's comment Muirhead, (1854), Vol I, p. lxxvi.
21 Smiles, (1878), p. 86.
22 B. & W. Col., MII, 4/9.7b, J. Smeaton to J. Watt, 5 Feb. 1778.
23 Hills, (1989), p. 146.
24 Muirhead, (1854), Vol I, p. l.
25 Ibid, Vol I, p. 3, Letter 1, J. Watt to J. Lind, 29 April 1765; JWP, C1/15.
26 Ibid, Vol I, p. 22, Letter 27, J. Watt to J. Roebuck, 29 April 1768; JWP, 4/32.10.
27 Ibid, Vol I, p. 93, Letter 79, J. Watt to W. Small, 10 March 1770; Boulton 348.18.
28 Muirhead, (1858), p. 97.
29 Muirhead, (1854), Vol I, p. lxxvi.
30 B. & W. Col., Drawing PF 1381.
31 Ibid, MII, 4/9.7b, J. Smeaton to J. Watt, 5 Feb. 1778.
32 NLS, Rennie Papers, Ms. 19825.vi, 'Remarks by J. Watt upon the Specification of his Patent'.
33 Muirhead, (1854), Vol I, p. lxxvii.
34 Ibid, Vol I, p. 3, Letter 1, J. Watt to J. Lind, 29 April 1765; JWP, C1/15.
35 Dickinson, (1935), p. 36.
36 Muirhead, (1854), Vol I, p. 3, Letter 1, J. Watt to J. Lind, 29 April 1765; JWP, C1/15.
37 Ibid.
38 B. & W. Col., MII, 5/4, '19 April 1765, To the Carron Company, 1 Bored Pipe, £3.16.7'.
39 Ibid, Vol I, p. 4, Letter 2, J. Watt to J. Roebuck, 23 Aug. 1765; JWP, 4/32.38.
40 Ibid.
41 Ibid, Vol I, p. 5, Letter 3, J. Watt to J. Lind, 4 Sept. 1765; JWP, C1/15.
42 Ibid, Vol I, p. 8, Letter 8, J. Watt to J. Lind, 12 Oct. 1765; JWP, 4/32.13.
43 Robinson & Musson, (1969), p. 50, 'Draft for Patent Specification', no date; JWP, 4/24.5.
44 Muirhead, (1854), Vol I, p. 12, Letter 12, J. Watt to J. Roebuck, 18 Nov. 1765, JWP, 4/32.33.
45 Ibid.
46 Ibid, Vol I, p. 8, Letter 9, J. Watt to J. Roebuck, 16 Oct. 1765; JWP, 4/32.12.
47 Ibid.
48 Ibid.
49 Muirhead, (1858), p. 73.
50 JWP, 4/11.131, J. Watt to J. Watt sen., 8 Oct. 1765.
51 Boulton 334.5, J. Roebuck to M. Boulton, 12 Dec. 1768 and Robinson & Musson, (1969), pp. 41 - 2, 'Memoranda of James Watt Esquire, from the Messrs. Harts'.
52 Ibid, Vol I, p. lxi.
53 Ibid, Vol I, p. xxxviii.

54 Robinson & Musson, (1969), p. 46, No. 5 (ii); J. Watt to J. Roebuck, 9 Sept. 1765; JWP, 4/32.37.
55 Ibid.
56 Cardwell, (1971), p. 54.
57 JWP, 4/82, J. Watt 'Suppose to Roebuck', 8 July 1767.
58 B. & W. Col., MI, 1/16, 'Journal', 16 Oct. 1770.
59 Muirhead, (1854), Vol I, p. 6, Letter 5, J. Roebuck to J. Watt, 13 Sept. 1765; JWP, 4/57.1.
60 Robinson & Musson, (1969), p. 41, 'Memoranda of James Watt Esquire, from the Messrs. Harts. Communicated by John Smith. 19 March 1845' and Smiles, (1878), p. 98.
61 JWP, 4/32.36, J. Watt to J. Roebuck, 2 Oct. 1765. See Muirhead, (1854), Vol I, p. 7, Letter 6 for a synopsis.
62 Ibid.
63 Muirhead, (1854), Vol I, p. 10, Letter 10, J. Watt to J. Roebuck, 18 Oct. 1765; JWP, 4/32.11.
64 JWP, 4/32.31, J. Watt to J. Roebuck, 5 Dec. 1765.
65 JWP, 4/32.7, J. Watt to J. Roebuck, 13 Nov. 1765.
66 Ibid.
67 JWP, 4/32.34, J. Watt to J. Roebuck, 10 Nov. 1765.
68 JWP, 4/4.21, Margaret Watt to J. Watt, 17 Dec. 1765, 'To James Watt at Dr. Roebucks, Kinneil'.
69 Muirhead, (1854), Vol I, p. lxi.
70 Ibid, Vol I, pp. 12 - 13, Letter 14, J. Watt to J. Roebuck, 5 Dec. 1765; JWP, 4/32.31.
71 Ibid, Vol I, p. 14, Letter 16, J. Watt to J. Roebuck, 29 Feb. 1766; JWP, 4/32.29.
72 B. & W. Col., MIII, 3/6, 'Ledger of Personal Accounts', f. 64.
73 JWP, 4/32.34, J. Watt to J. Roebuck, 10 Nov. 1765.
74 Muirhead, (1854), Vol I, p. 14, Letter 15, J. Watt to J. Roebuck, 19 Feb. 1766; JWP, 4/32.30.
75 Ibid, Vol I, p. 14, Letter 17, J. Watt to J. Roebuck, 20 April 1766; JWP, 4/32.28.
76 Ibid, Vol I, p. 11, Letter 11, J. Watt to J. Roebuck, 10 Nov. 1766; 4/32.34, a note added by J. Watt.
77 B. & W. Col., MIII, 3/6, 'Ledger of Personal Accounts', f. 64.
78 Muirhead, (1854), Vol III, p. 1 for plate I.

Watt's Own Engine: Part 3
A Gap in Experiments

After the failure of Carron to bore accurately enough the cylinder for the first inverted engine at Kinneil, Watt stopped trials on his reciprocating engine but continued a little longer with his rotary engine during May 1766. (1) He had made considerable progress on the reciprocator since he had been able to keep the cylinder as hot as the steam entering it through surrounding it with a steam jacket. He had been able to prevent cold air entering the cylinder by placing a cover over it. This had additional advantages of using steam to move the piston, keeping the cylinder hot as well as introducing the possibility of a more powerful engine with steam at higher pressures. He had been able to fabricate a cylinder that was reasonably round and fit it with a piston that was reasonably steam-tight by having pasteboard packing. However it seems very doubtful whether pasteboard would have lasted with continuous running. It is most likely that so far Watt operated his models manually and that they did not work for long periods. He had proved that it was possible to condense in a separate vessel and that this showed promise of great economy in the consumption of steam. A pump removed the air and condensate. In one way, he was correct when he stated that he had thought of all the essentials almost immediately. Yet his best model was still probably an inverted type and obviously still needed a great deal of development to turn it into a large scale self-acting condensing steam engine.

Pressure of other work probably prevented Watt carrying out any more trials until the early part of 1768 because not only were there the scientific instrument business and the retail shop to supervise, but the experiments to obtain alkali from salt were continuing. As if all these were not enough, he was launching out into his engineering partnership with Mackell. We have definite evidence for the Newcomen engine at Kennet being erected during the second half of May 1766 until the end of the year, followed by the large one for Carron. Then, early in 1767, Watt was helping Mackell to survey the line of the 'Ditch' canal which resulted in a journey to London to help secure the passage of the bill through Parliament.

After the failure to secure their bill, Muirhead says Watt passed through Birmingham on his return journey to Scotland where he met Small for the first time. Boulton was away but Small and Boulton's partner, John Fothergill, showed Watt over the Soho Manufactory, completed in 1765. (2) Smiles says,

> Watt was much struck with the admirable arrangements... and recognised at a glance the admirable power of organisation which they displayed. Still plodding wearily with his model, and contending with the "villanous bad workmanship" of his Glasgow artisans, he could not but envy the precision of the Soho tools and the dexterity of the Soho workmen. (3)

While Watt might have been impressed with the organisation and dexterity of the workmen, Soho Manufactory specialised in small 'toys' and never had any machine tools capable boring steam engine cylinders. (4)

Although surviving archives leave a blank concerning this second visit to Birmingham, they do show that Watt passed through Lichfield (5) where he met another member of the Lunar Society, Dr. Erasmus Darwin. Watt certainly impressed Darwin who wished 'the Lord would send you to pass a week with me, and Mrs Watt along with you, - a Week! - a Month, a Year'. (6) The pair remained firm friends until Darwin's death in 1802. Their first discussions must have ranged over many topics for, in August 1767, Darwin sent Watt a long description of the method for pivoting the front wheels of his chaise designed by Richard Lovell Edgeworth. Darwin promised to keep secret Watt's steam engine improvements but began to foresee difficulties in its execution. Darwin must have shared his enthusiasm for his 'dear new Friend' with his fellow members of the Lunar Society, Boulton and Small. However, it must remain speculation how much this influenced later events.

From Lichfield, Watt went on to Ashbourne, Matlock and Ashford before reaching Sheffield. Then he passed quickly through to Newcastle upon Tyne before visiting Hartley, Morpeth, Walden Mill and travelling on to Scotland. He is most likely to have been visiting mining sites in these areas, particularly viewing Newcomen pumping engines. After his return home, Watt was asked by Cathcart to re-survey another line for the Forth and Clyde Canal via the northern Lomond Passage. This, with surveying associated river improvements and canals in the neighbourhood of Stirling and Alloa, kept Watt occupied until well into October. Therefore it was not until January 1768 that Watt wrote to Lind,

> I am going to be at home, God willing, for some time. I am going to try some things I am persuaded you will like to see (perpetual mobiles, the elixir magicum, and some other trifles of that kind)... What I knew about the steam-engine before you went away was but a triffle to what I know now. (7)

At last Watt had an opportunity to return to his own steam engine after all those months of hectic activity.

Experiments Start Again

While Watt's letter to Roebuck at the end of January 1768 was mainly concerned with the rotary engine, he concluded, 'the other model is going on'. (8) We will review progress over the next year until Watt was granted his patent. Smiles says that Watt hired an old cellar, situated in the first wide entry to the north of the beef-market in King Street, where he tried out his first model. Then there was a larger model, and for this Smiles says that Watt moved to the Delftfield pottery works which was disused at the time. (9) Watt may have been helped by Gardner who later said that he acted as Watt's assistant at all of his experiments both in Glasgow and Kinneil. (10) Gardner must have been the source for the account that Watt,

> Having procured an apartment in the Delf near the harbour of the Broomielaw, he shut himself up along with a single assistant, and commenced his experiments on the Steam Engine. During these experiments, the beam broke, and the machine, which had a nine inch cylinder, was for a time laid aside. Although the first experiments were not crowned with success, Mr. Watt, in the midst of varied avocations, never forgot the great object of his ambitions:- the improvement of the Steam Engine. (11)

But the move into the Delftfield may not have happened until February 1769, after Watt had become a partner and the pottery was certainly operating. Watt then reported to Roebuck, 'I have now set it up in a most convenient place, where the experiments can be made to your satisfaction'. (12)

In the middle of April 1768, Watt informed Roebuck that he had been testing his engine. From the comment that the steam would act above the piston, this one must have had a conventional layout with a beam. It was working much better though he had not got it perfectly tight for air seemed to be leaking through some 'imperceptible pores' in the metal. His piston was tight and lasted well. He must have been about to fit a steam jacket so it would be set 'wholly in steam' and he saw 'nothing else to hinder its performance, as the vacuum is formed very suddenly, and is best at the beginning, when it would be worst if no air got in'. (13) The comment that the vacuum was 'best at the beginning' may show that Watt was carrying out longer tests on this engine which were to reveal weakness in the design of his condenser. Certainly this was where the main thrust of his experiments over the next few months would lie. Roebuck replied that he would have rejoiced to hear more favourable news but hoped for better in a few days when he wished to meet Watt at Kilsyth, midway between Kinneil and Glasgow.

These models at Glasgow were a considerable size. In his Journal, which commences on 8 April 1768, there is an undated note about a little cast iron cylinder with a diameter of 8 $^1/_4$ ins. (14) This would seem to correspond with a letter to Roebuck that April when he wrote about the engine's performance. A later endorsement has added, 'Expt. on a large model of the engine 7 or 8 inches diameter'. (15) There may have been another cylinder since, in May, Watt mentioned one with a diameter 7 $^1/_2$ ins. and 12 ins. stroke. (16) Once again he was very pleased with the performance of this engine because he was achieving a vacuum of 14 p.s.i. with very little steam consumption.

The packing for the main piston was still pasteboard, baked with linseed oil and put on like the leather of a pump in the form of a coned washer. (17) This type of piston could be tightened up further. But, probably through working the engine for longer periods, it was found by October that pasteboard by itself was insufficient and that some lubricant was necessary.

A piston with pasteboard packing.

> The objections against the engine before the trials made last winter were, that it might be difficult to make a piston sufficiently steam-tight without water or some other fluid above it. Water cannot be used because if any kind of it got in, it would boil and occasion a loss of heat. Oil of some kind was the next thought of; but possibly it might be destroyed in time by heat... On trial, tallow being employed to keep the piston tight, answered perfectly well, and did not appear to be changed. (18)

So it would appear that, at least on these cylinders of an intermediate size, Watt had solved the problem of making his piston reasonably steam-tight.

But that spring he had an accident during a trial which 'succeeded much better, the vacuum being quickly formed and no air'. (19) He was achieving a vacuum of 24 ins., or about 11 $^1/_2$ p.s.i. He measured this with a manometer and, presumably through violent oscillations, some of the mercury escaped and 'played the devil with the solder'. (20) Mercury would form an amalgam with the metal in the solder. This happened in spite of taking precautions. He apologised to Roebuck, 'Dear Sir, I hope you will excuse this - I do not know whether to call it mistake or blundering - but am certain that I had done all I could, and hope that in the end good will come of it. Whether or not, I will do my duty to it'. (21)

This set him back about three days and, within a week, he carried out more trials when he found that his condenser, with a 2 ins. diameter pump, could support a weight of 42 lbs., 'which is coming as near as the whole pressure as we can reasonably expect'. (22) This was the diameter of the syringe used in his first experiments for the separate condenser and perhpas he used it for an air pump. As he decided he could not use a mercurial manometer for measuring pressures any longer, he began to calculate them by the size of the weight lifted.

From this, it sounds as if the engine was working reasonably well but Watt could not have been satisfied with it. In his next letter to Roebuck, Watt thought that some air was entering with the steam and proposed a different air pump.

> In order to get quit of that small quantity of air, I propose making two exhausting barrels instead of one, and giving them only half the stroke each that this has. I propose they should be worked by a double crank, the one ascending when the other descends, so that the pressure on the one that descends will nearly raise the one that ascends. (23)

Up to this point, Watt may have been operating the valves and the air pump manually because it is only at the end of this letter at the end of April 1768 that we find the first hint of any self-acting valve gear. 'I propose... making the engine go of itself, that the strokes may be counted, and the water measured that is boiled away. These things are in hand, and will soon be done'. (24) While the valve operating mechanism might have been driven off the main beam of the engine which moved up and down, we have just seen that his air pumps were to be worked by a double crank which must have necessitated a form of rotary motion. His aim of producing the perfect engine seems to have led him into further and further difficulties with his condenser unit so that he made more and more complex ones while he was in Scotland.

Watt described how this new condensing system should operate.

> Now, suppose the communication with the cylinder opened, and the piston of one of these [air pumps] raised, the steam will rush in and be condensed, and the greatest portion of the air will be in the condenser, because, being heaviest, it will go in with the first of the steam. Now, this piston, beginning to descend, shuts a valve that cuts off its communication with the cylinders; but the other piston ascending opens a fresh vacuum, in which there is no air, into which the remaining steam and air will go and be condensed, and the vacuum will be much more perfect than if there had been only one condenser having twice the stroke. If the vacuum is still not absolutely perfect, each of these cylinders may be made to take two strokes without any other loss than the friction of the exhausters as they will balance one another nearly. (25)

Watt thought he had established two facts, the perfection of the vacuum and the 'suddenty' of the condensation but he still needed to determine the steam consumption.

Accordingly he went on to construct his new condensing apparatus which, as well as twin air pumps must have had twin condensing units. On the top of each condenser, he fitted an outlet valve with a pipe which enabled him to measure the quantity of water condensed each stroke. (26) In the second half of May, he gave his new condensers a trial but did not have them quite tight. (27) However he had a further trial on 24 May when, although the new apparatus was still not quite tight, he achieved a vacuum of 14 p.s.i. The boiler was blowing off steam through the safety valve even when the engine was making twenty strokes a minute. He tried to work out how much steam the engine actually consumed and calculated that for

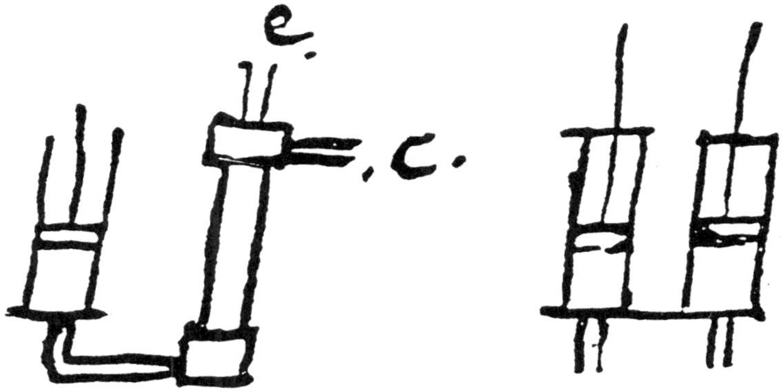

The first sketch of a condenser with twin air pumps having solid pistons. Steam entered the condenser through the pipe 'c' and the condensate was ejected through the vertical pipe 'e'.

every cubic foot of the cylinder, only one cubic inch of water needed to be evaporated. 'I am of opinion that, in a large engine, with a strong boiler and a hot steam, the half that quantity will be sufficient'. This gave him great optimism and he sincerely wished Roebuck 'joy of this successful result, and hope it will make you some return for the obligations I ever will remain under to you'. (28) Silence descends once more about any further trials that summer. These experiments during the first half of 1768 must have shown sufficient progress for Roebuck to agree that Watt should take out a patent. Threat of competition from Hately's patent granted on 30 June 1768 for his rival type of steam engine may have been a further spur. (29)

In addition to the scare of Hately's patent, there were urgent pleas from Robison who met Boulton in early July 1768. On Watt's recommendation, Robison called on Darwin where he happened to meet Boulton and Small. The following day, Boulton showed him round 'his vast works', the Soho Manufactory, and Robison reported to Watt,

> You are happy in the Esteem of such worthy people... I was much struck with many appearances of ingenuity in Boulton's works. It is partly at his desire that I now write to you. In the hopes that you had vigorously pursued the improvements of your fire Engine, and was getting your property secured, he has finished a building for one of your reciprocating Engines, and has waited delayd [sic] the erection of it on your account, because he says that he would be very sorry to put the world in possession of an improvement of which you are the Author. (30)

Watt could not have claimed a patent if he had sold a machine to someone else. Robison went on to warn that Edgeworth had conceived a circular engine somewhat similar to Watt's and that they had heard about someone, in fact Hately, claiming Watt's discoveries and patenting them. Therefore Robison urged Watt to bestir himself 'in getting your property secured' and hoped he would not neglect his interests.

In 1796, Robison claimed that, as he was being shown round the Soho Manufactory in 1768, Boulton, on learning that Watt was about to apply for a patent, said, '"I will stop that work", pointing to some brick pillars which lay on the other side of a large Water Course or Ditch. "I am about erecting a Steam Engine, very unlike what you described, but where I should have availed myself of what I learned from Mr. Watts Conversation. But this would not be right without his consent"'. (31) This implies that Boulton had actually designed his engine, using some of Watt's ideas.

NOTES: A GAP IN EXPERIMENTS

1 Muirhead, (1854), Vol I, p. 14, Letter 18, J. Watt to J. Roebuck, 14 May 1766; JWP, 4/32.27.
2 Ibid, Vol I, p. cxlvii.
3 Smiles, (1878), p. 133.
4 Dickinson & Jenkins, (1927), p. 29.
5 B. & W. Col., MI, 2/5, 'Cash Memorandum', pencil note on Expenses.
6 JWP, 4/87.4, E. Darwin to J. Watt, 18 Aug. 1767. For full text, see King-Hele, (1981), pp. 46 - 47.
7 Muirhead, (1854), Vol I, p. 15, Letter 20, J. Watt to Dr. J. Lind, 5 Jan. 1768.
8 Ibid, Vol I, p. 18, Letter 22, J. Watt to Dr. J. Roebuck, 28 Jan. 1768; JWP, 4/32.26.
9 Smiles, (1878), p. 88.
10 Cleland, (1825), p. 11.
11 Ibid, p. 10.
12 JWP, 4/32.20, J. Watt to J. Roebuck, 10 Feb. 1769.
13 Muirhead, (1854), Vol I, p. 18, Letter 23, J. Watt to J. Roebuck, 15 April 1768; JWP, 4/32.25.
14 B. & W. Col., MI, 1/14, 'Journal', no date.
15 JWP, 4/32.25, J. Watt to J. Roebuck, 15 April 1768.
16 JWP, 4/32.39, J. Watt to J. Roebuck, 24 May 1768.
17 Muirhead, (1854), Vol I, p. 21, Letter 27, J. Watt to J. Roebuck, 29 April 1768, JWP, 4/32.10, see illustration.
18 Ibid, Vol I, p. 31, Letter 34, J. Watt to M. Boulton, 20 Oct. 1768; Boulton 348.1.
19 Ibid, Vol I, p. 19, Letter 25, J. Watt to J. Roebuck, 19 April 1768; JWP, 4/32.4.
20 Ibid.
21 Ibid.
22 Ibid, Vol I, p. 20, Letter 26, J. Watt to J. Roebuck, 27 April 1768; JWP, 4/32.24.
23 Ibid, Vol I, p. 21, Letter 27, J. Watt to J. Roebuck, 29 April 1768; JWP, 4/32.10.
24 Ibid.
25 Ibid.
26 Ibid, Vol I, p. 23, Letter 28, J. Watt to J. Roebuck, 10 May 1768; JWP, 4/32.9.
27 Ibid, Vol I, p. 24, Letter 29, J. Watt to J. Roebuck, no date, JWP; 4/32.23.
28 Ibid, Vol I, p. 25, Letter 30, J. Watt to J. Roebuck, 24 May 1768; JWP, 4/32.39.
29 Patent 895, 30 June 1768.
30 Robinson & McKie, (1970), p. 13, Letter 6, J. Robison to J. Watt, 8 July 1768.
31 This letter written at the time in 1768 gives a very different account from Robison's later version of this meeting submitted during the Hornblower and Maberly patent trial in 1796. See Robinson & Musson, (1969), p. 31.

Watt's Own Engine: Part 4
Obtaining the Patent

Roebuck's coal mining ventures improved after the disastrous flooding in 1765. Early in 1766, Watt congratulated him on hearing that the 'Quality of your Coal, when found was equal to your most sanguine hopes & the richness little inferior to expected'. (1) The Bo'ness mines returned to profitability in 1768 (2. See p151) and Roebuck told Boulton that there was a 'good prospect of its repaying us all our expences'. (3) This must have encouraged him to support Watt's application for a patent. Watt had to make this in person so, at the beginning of August 1768, he once more set out from Glasgow bound for London, but this time in the luxury of a coach. The fare to Edinburgh was 12s, Edinburgh to Newcastle £2.13s.3d. and Newcastle to London £3.8s.6d. With incidental expenses, the total cost of the journey came to £9.13s.6d, a large sum for those days.

Watt may have arrived in London a day or two before he noted in his Journal for Tuesday 9 August the cost of the patent oath, 6s.6d. (4) The oath was only a minor part of the procedures to obtain a patent. After his return to Scotland, Watt employed Thomas Handley as his patent agent in London who attended to all the details of seeing that the application and specification were passed through the correct officers of the Crown. These included the Attorney General, Secretary of State, Privy Seal, Signet Office and Lord Chancellor. (5) Fees had to be paid to all of them and their clerks as well as gratuities. Handley's bill in May 1769 reached a total of £88.8s.9d. including ten guineas for 'his trouble'. (6) In addition, Watt's personal expenses in London amounted to £11.15s.9d, sundries there £36.4s.4d. and his return journey £18.8s.8d. (7) The would-be patentee had to prepare a petition to the Crown, briefly outlining the nature of his invention, its economic or social benefits and his particular claims to preferment. After the patent was granted, the full specification had to be submitted within four months. This necessitated Watt going to Berwick-on-Tweed towards the end of April 1769 to swear to his specification before a Master in Chancery which cost a further £6.11s.1d. (8) Hence it is possible to account for a total of over £170 as the cost of obtaining this patent, obviously something not to be undertaken lightly.

Watt bought a coach ticket for £1.2s. in London on Thursday 25 August 1768, and left for Birmingham that Saturday. (9) This was when he met Boulton in person for the first time while he was staying with Small. (10) Watt's journal is silent until Saturday 24 September when he hired John Hinton somewhere in the Newcastle-under-Lyme area for the Delftfield pottery. He had travelled through Lichfield and presumably must have been in the Potteries for a little while to discover someone who would be willing to emigrate to Scotland. He remained there until 2 October when he visited a copper works at Warrington before spending four days in Liverpool. After taking coaches to Kendal, Watt continued on horse back through Carlisle and Dumfries before reaching home on 11 October. (11) Almost immediately, he set off for Torryburn to erect the engine for Colville where he spent six weeks, (See p358) so he was away from home for the best part of four months.

Not only do these wanderings explain the opening remarks in a letter written 'Monday, I know not what day of September 1768' by Small when he said, 'I did imagine you had fallen into some bottomless pit in Derbyshire, or had been dipped into a cistern of air compounded with the inflammable principle, so I am very glad to hear you are alive'. (12) They must also explain Roebuck's impatience that autumn when he told Watt that he was allowing the most active part of his life 'glide away'. (See p154) Roebuck was anxious not to let 'a day, a moment' be lost in erecting a large engine according to the ideas Watt had at that time. (13) But Watt must have wanted to continue perfecting his model engine so Roebuck returned to the fray ten days later, 'I am as solicitous as yourself that every trial should be made of the small engine, which can be speedily executed, particularly the alteration of the condenser. But then I would not lose a single moment in completing what was resolved on'. (14)

It makes better sense if an undated letter from Jardine is placed at this point in time rather than where Muirhead has put it back in May. Jardine wished to give Watt a few friendly hints which reveal a great deal about the people involved.

> The doctor has been with you at Glasgow... I waited to find... if he had in any respect consented to the proposal from the South [a partnership with Boulton]; but understand, that the more he is convinced of the practicability of the scheme, the keener he is of carrying it to practice yourselves for your mutual advantage; and I am perfectly convinced that though he may mistake particulars, it is one of his strongest motives that you may not only have the honour but the advantage of your own ingenuity... My opinion is, James, that you will find it necessary, on account of your intimate connection, to fall in with his sentiments, and contribute as speedily, and as vigorously as you can, to bring it to some conclusion. He has proposed to erect one immediately at Bo'ness. I know very well the objections you have to this, and indeed to everything which will give you further trouble: but I apprehend when you consider deliberately the following particulars, you will endeavour to pluck up spirits for a few months longer. You are yourself perfectly satisfied as to the success you expect in practice, and I really apprehend that there is now so little to do for that purpose, that it is unnecessary to make any further division of your profit to a third person; for the very nature of your improvements is such that it is impossible it can fail to succeed much to your interest, even though it should not be carried to such perfection as might be expected from the gentleman in the south's assistance. I know you will think the erection of one at Bo'ness a further trouble, but I am really of opinion that even that trouble will be productive of considerable advantages. It would be hard to be disappointed in any respect in the first attempt that you make, and it would surely be better that you should have not only a model, but a finished engine, either to proceed upon as a pattern or to dispose of, as may be convenient. Nor do I apprehend that your trouble will be so great as you are ready to imagine. You judge of it too much from what is past. You are now in a very different situation, and can begin at principles which you had formerly to labour at by experiment and calculation. The person who is your operator has surely now learned to take some of the trouble off your hand; and I flatter myself that it would be now nothing more than mere amusement to attend to the proportions and accuracy of the execution. The Dr. proposes, not only in the present but also in future undertakings, to have everything that requires great nicety and exactness, to be made by the best workmen, and, as far as possible, to follow the method of watchmakers, who have little more to do than to adjust and combine materials. This may indeed be not such a perfect scheme as to have the whole conducted in the same manner, but under particular inspection; but it may nevertheless do well, and at least ease you of a part of your trouble; and all improvements in this respect will occur to you much better afterwards. I wish earnestly, therefore, that you would endeavour to keep away those plaguey headaches for a little longer. You have many motives, man, to have good spirits. You are surely very near to something that will be much to your advantage: the happiness, or at least the interest, of your family, - your own ease and amusement, - that life of ingenious indolence which you have often figured out to yourself, are all within prospect; - not to mention the honour of a discovery of so much importance, a circumstance which few would think so moderately of as yourself. I would, therefore, have you resolve immediately to spend a great part of the summer here... (15)

Jardine was a sort of bystander who could take a more disinterested view. Even though the engine was far from finished, Jardine was able to point out that its invention was significant and held out prospects for future prosperity. But, before this could be achieved, Watt had to prepare the patent specification and to construct an engine on a large scale that performed successfully, as well as discuss the expansion of his partnership with Roebuck possibly to include Boulton.

The 1769 Patent

Watt told Roebuck in November 1768 that he had 'written the specification two or three times over, but am not yet satisfied with it, though it is better than it was'. (16) Here again we can follow another case of Watt seeking perfection because some of these drafts have survived, showing many alterations. (17) Unfortunately they are undated so their order cannot be established. One problem must have been that the engine itself was far from being finished and a proved success which made any description in the specification difficult. This could explain Roebuck's reluctance in December to complete the patent because he wanted to keep it secret as long as possible and also his emphatic instruction to Boulton to be 'totally silent about this subject... I repeat it again that you will please not to speak of the Fire engine to any one *whatever* nor that I have any concern in it'. (18) Yet ten days later, he had changed his mind and instructed Watt 'by the first post write to your friend to take out the patent. I can spare the money without inconveniency'. (19)

The patent itself was sealed on 5 January 1769 which gave Watt till the beginning of May to submit his specification. (20) Handley warned him, 'You will be prepared when I see you to give the full proper Description & you must use great caution therein lest any other person who may see your specification sho'd endeavour at an improvement on your plan'. (21) It was not until nearly the end of January 1769 that Watt sent Small a copy of what he had prepared.

> The plan I pursued in drawing it out, was in the first part to lay down my principles in as few words as possible, so as to be intelligible and clear from double meanings; so that they alone should ascertain my invention in so far as related to myself. This I have declared in the next sentence; but as I have been informed that some patents have been defeated because the specification was not clear enough to enable other people to execute the schemes, I have added descriptions of the machines with drawings. However, if something after the nature of the first part could be contrived to answer the whole purpose, it would be better. (22)

Watt obtained a copy of Hately's patent which contained a drawing. (23. See p346) He faced a dilemma because he could not give a detailed description of an engine that so far existed only in model form and we should note his suggestion that he should concentrate on the principles. Small took the text to Boulton and they both agreed that Watt,

> ... should neither give drawings nor descriptions of any particular machinery (if such omissions would be allowed at the office), but specify in the clearest manner you can that you have discovered some principles, and thought of new applications of others, by means of both of which, joined together, you intend to construct steam-engines of much greater powers, and applicable to a much greater number of useful purposes, than any which hitherto have been constructed; that to effect each particular purpose, you design to employ particular machinery, every species of which may be ranged in two classes: one class for producing reciprocal motions, and another for producing motions round axes. (24)

Small went on to suggest a suitable form of specification which concentrated on the principles and told Watt, 'I am certain that, from such a specification as I have written, any skilful mechanic may make your engines, although it wants correction; and you are certainly not obliged to teach every blockhead in the nation to construct masterly engines'. (25)

Watt was very pleased to receive this advice which resolved his dilemma for he simplified Small's text and used it as the basis for his specification. (26) But Watt did not send his amended version until sometime in March. (27) It was almost the same as the one that was finally submitted. A copy was sent also to Watt's old friend, Cumming, for his comments. However in the middle of April, Small had still not heard from Cumming so he passed his copy on to Handley. (28) Some of Watt's later amendments were missing and Cumming altered the wording about the 'circulatory article' so that the final draft which Watt witnessed at Berwick on 25 April 1769 was the result of all these consultations. (29) Small was exceeding pleased to learn that they were in time. (30)

Watt declared that he 'hath, after much labour and expense, invented a Method of lessening the Consumption of Steam and Fuel in Fire Engines'. Here we see that, on Small's advice, he patented a method or principle rather than an actual machine which led to the arguments at the later trials of this patent. He stated that it,

> ... consists of the following principles:
>
> FIRST, that vessel in which the powers of steam are to be employed to work the engine, which is called *Cylinder* in common fire engines, and which I call *Steam Vessel*, must, during the whole time the engine is at work, be kept as hot as the steam that enters it; first by inclosing it in a case of wood, or any other materials that transmit heat slowly; secondly, by surrounding it with steam or other heated bodies; and, thirdly, by suffering neither water nor any other substance colder than the steam to enter or touch it during that time.
>
> SECONDLY, in Engines that are to be worked wholly or partially by condensation of steam, the steam is to be condensed in vessels distinct from the steam vessels or cylinders, whilst the engines are working, these condensers ought at least to be kept as cold as the air in the neighbourhood of the engines, by application of water, or other cold bodies.
>
> THIRDLY, whatever air, or other elastic vapour, is not condensed by the cold of the condenser, and may impede the working of the engine, is to be drawn out of the steam vessels or condensers by means of pumps, wrought by the engines themselves, or otherwise.
>
> FOURTHLY, I intend in many cases to employ the expansive force of steam to press on the pistons, or whatever may be used instead of them, in the same manner as the pressure of the atmosphere is now employed in common fire engines: in cases where cold water cannot be had in plenty, the engines may be wrought by this force of steam only, by discharging the steam into the open air after it has done its office.
>
> FIFTHLY, where motions round an axis are required, I make the steam vessels in form of hollow rings, or circular channels, with proper inlets and outlets for the steam, mounted on horizontal axles, like the wheels of a water-mill; within them are placed a number of valves, that suffer weights, so fitted to them as entirely to fill up a part or portion of their channels, yet rendered capable of moving freely in them by the means hereinafter mentioned or specified. When the steam is admitted in these engines, between these weights and the valves, it acts equally on both, so as to raise the weight to one side of the wheel, and by the re-action on the valves successively, to give a circular motion to the wheel, the valves opening in the direction in which the weights are pressed, but not in

> the contrary: as the steam vessel moves round, it is supplied with steam from the boiler, and that which has performed its office may either be discharged by means of condensers, or into the open air.
> SIXTHLY, I intend, in some cases, to apply a degree of cold not capable of reducing the steam to water, but of contracting it considerably, so that the engines shall be worked by the alternate expansion and contraction of the steam.
> LASTLY, instead of using water to render the piston, or other parts of the engines, air and steam-tight, I employ oils, wax, resinous bodies, fat of animals, quicksilver, and other metals, in their fluid state.
> BE IT REMEMBERED, that the said JAMES WATT doth not intend that anything in the Fourth Article shall be understood to extend to any engine where the water to be raised enters the steam vessel itself, or any vessel having an open communication with it. (31)

By concentrating on the principles and, except for the description of the steam wheel, not relating these principles to any specific type of steam engine, Boulton and Small had advised Watt to take out a patent which would cover not only Watt's first inventions but also subsequent developments of the steam engine over the next few years. It is doubtful if they could have foreseen how all-embracing this patent would be when upheld in the courts. Later Watt claimed that it covered his expansive engine with cut-off, his double-acting engine, jet condensing as well as surface condensing and the high pressure steam engine exhausting direct into the atmosphere. In other words his patent of 1769 held the key to many later improvements.

Notes: Obtaining the Patent

1 JWP, 4/32.30, J. Watt to J. Roebuck, 19 Feb. 1766.
2 Campbell, (1961), p. 13.
3 Boulton 334.5, J. Roebuck to M. Boulton, 12 Dec. 1768.
4 B. & W. Col., MI, 1/14, 'Journal', 9 Aug. 1768.
5 For full details of obtaining a patent, see MacLeod, (1988), pp. 40 - 41.
6 JWP, 4/41.2, T. Handley's account for 'Costs of Patent', 16 May 1769.
7 B. & W. Col., MI, 1/14, 'Journal', 'Expences of Journey to London, Aug. Sept. & Oct. 1768'.
8 JWP, 4/41.7, 'Acct. of Expence going to Berwick', April 1769.
9 B. & W. Col., MI, 1/14, 'Journal', 25 Aug. 1768.
10 Muirhead, (1854), Vol I, p. 32, Letter 35, J. Watt to J. Lind, 25 Oct. 1768; JWP, C1/15.
11 B. & W. Col., MI, 1/14, 'Journal', 11 Oct. 1768.
12 Muirhead, (1854), Vol I, p. 29, Letter 33, W. Small to J. Watt, Sept. 1768; Boulton 340.3.
13 Ibid, Vol I, p. 33, Letter 36, J. Roebuck to J. Watt, 30 Oct. 1768; JWP, 4/57.7.
14 Ibid, Vol I, p. 33, Letter 37, J. Roebuck to J. Watt, 9 Nov. 1768; JWP, 4/57.8.
15 Ibid, Vol I, pp. 25 - 28, Letter 31, G. Jardine to J. Watt, Monday 1768; JWP, 4/57.4.
16 Ibid, Vol I, p. 34, Letter 38, J. Watt to J. Roebuck, 9 Nov. 1768; JWP, 4/32.22.

17 See JWP 4/24 for a draft specification.
18 Muirhead, (1854), p. 34, Letter 37, J. Roebuck to J. Watt, 9 Nov. 1768; JWP 4/57.8 and Boulton 334.5, J. Roebuck to M. Boulton, 12 Dec. 1768.
19 See JWP, 4/32.122, J. Roebuck to J. Watt, 12 Dec. 1768 and Muirhead, (1854), p. 35, Letter 40, J. Roebuck to J. Watt, 22 Dec. 1768; JWP, 4/57.9.
20 Patent 913, 5 Jan. 1769 and JWP, 4/41.5, T. Handley to J. Watt, 6 Jan. 1769.
21 JWP, 4/41.3, T. Handley to J. Watt, 18 Feb. 1769.
22 Muirhead, (1854), Vol I, p. 35, Letter 41, J. Watt to W. Small, 28 Jan. 1769; Boulton 348.4.
23 JWP, 4/41.2, 'Costs of Patent', 16 May 1769.
24 Muirhead, (1854), Vol I, p. 37, Letter 42, W. Small to J. Watt, 5 Feb. 1769; JWP, 4/24.
25 Ibid,
26 Ibid, Vol I, p. 46, Letter 47, J. Watt to W. Small, 22 Feb. 1769; Boulton 348.5.
27 Ibid, Vol I, p. 49, Letter 50, J. Watt to W. Small, March 1769; Boulton 348.6.
28 Ibid, Vol I, p. 51, Letter 52, W. Small to J. Watt, 18 April 1769, Boulton 340.5.
29 Boulton 348.7, J. Watt to W. Small, April 1769.
30 Muirhead, (1854), p. 58, Letter 56, W. Small to J. Watt, 16 May 1769; Boulton 340.6.
31 Patent 913, enrolled 29 April 1769.

WATT'S OWN ENGINE: PART 5
THE FINAL PHASE OF EXPERIMENTS AT GLASGOW

While Watt was struggling with the text of his patent specification, he was also continuing to experiment with a model at Glasgow and preparing plans for a large engine at Kinneil. He told Boulton in the middle of December 1768 that he would advise him of the success of a complete model which he had almost finished. (1) Trials on this model were carried on until the following spring but there is no mention of what happened to this engine. The drawings and descriptions which Watt intended to submit with his patent have survived and show the lines along which he had developed his engine up to that time. While there are no identification letters on either the drawing or in the description of his reciprocating engine, some have been inserted in the following quotation to enable the reader to follow the text. (See p394) Watt wrote,

> What I have said is sufficient to enable the skilful mechanic to construct fire engines with the proposed properties - I have however annexed drawings with explanations to show how this may be reduced to practice and to describe some things which tho' less important are however necessary in the execution -
> The Cylinder [A] is covered with a lid [D] thro the center of which the Piston stalk [E] slides in a Collar of Pasteboard or Oakum steam tight. This end of the Cylinder communicates with the boiler by a Pipe joining the steam pipe at [F] and always allowing free passage for the steam to press on the Piston when wanted. At the lower end of the Cylinder is fixed a Regulating Valve [G] capable of making a communication either between the Boiler and the Cylinder or between the Cylinder and the Condenser [B]. The Condenser is not confined to any certain form tho the more surface that can be given to it the better. it should be very thin but must be air tight and strong enough to resist the pressure of the Atmosphere a part of it must be emptied both of air and water - This may be done by a Pump [C] in the following manner. The Piston [H] of this Pump being at the bottom of the Condenser is to be filled quite full of water which is prevented from going into the Cylinder by a Valve [I]. Then the regulating Valve [G] being open to steam and shut to the Condenser the Piston [H] is to be raised up the air contained in the Pipe [J] will then come into the Condenser thro the Valve [I] and the Piston [H] being allowed to descend will expel that air at the small Valve [K]. This operation repeated will leave the Condenser and pipe [J] perfectly free from air.
>
> The Piston [H] being raised up will leave the upper part of the Condenser free both of air and water. The Cylinder [A] is now to be filled with steam and the air and water let out at a Cock [L] which as soon as the Cylinder is thoroughly heated is to be shut by hand.
>
> The Regulating Valve [G] is now to be shut to the boiler and opened to the Condenser [B]; a part of the steam contained in the Cylinder [A] will immediately rush into the condenser which being immersed in cold water or having a stream of cold water running over it will instantly reduce it to Water more Steam will come from the Cylinder to be condensed in the same manner till the Cylinder is perfectly exhausted. The Piston [M] will then descend and raise the load required. The regulating valve will then be shut to the condenser and open to the boiler and the Cylinder filled with Steam as before and the Piston [H] of the Condenser raised ready for the next stroke - NB As no water can in

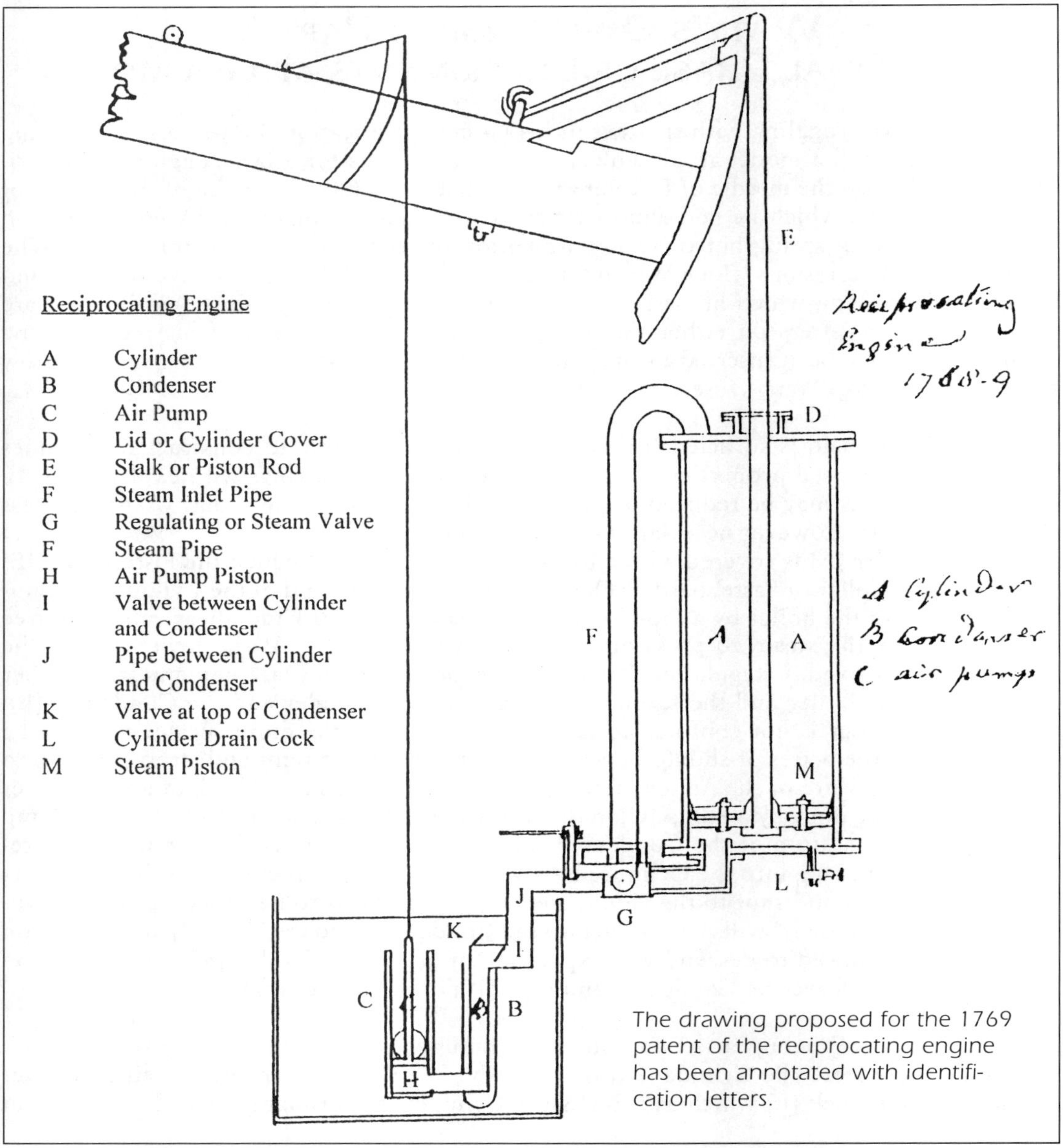

The drawing proposed for the 1769 patent of the reciprocating engine has been annotated with identification letters.

this engine be made use of to keep the Piston of the Cylinder tight its place must be supplied by oil or melted tallow - By the Cylinder I always mean the principal working Vessel of the machine whatever its form may be. (2)

The cylinder is shown without a jacket. This may date the drawing to January 1769 when Watt told Small about his model engine, 'I have also done wrong in removing the outer cylinder, which kept the inner one always surrounded by steam, ready to supply any loss of heat that might happen by evaporation from the inside surface'. (3) He may have had difficulty keeping a wooden steam jacket tight. (4) He had made a cast-iron cylinder cover which condensed steam, causing water to lie on the top of the piston. He found that, without a jacket, he was losing a great deal of heat from the sides of the cylinder, causing condensate to form inside. This prevented formation of a proper vacuum. (5) Therefore, during February, he was busy making an 'outer case' of wood to act as a jacket once more. (6) He tried a copper bottom to the cylinder, again jacketed, and even encased his regulator valve and oil pump in steam jackets. (7) The description shows that the gland for the piston rod was packed with either pasteboard or oakum which appear to have been successful for we never hear of steam leaking here.

According to Watt's later annotations in March 1769, he had a cast iron cylinder 8 ins. diameter but it was tapered, being one eighth of an inch wider at one end than the other. (8) Watt also mentioned a block-tin one he had been using earlier. He said, 'The best and cheapest cylinders will be of block-tin; they will be defended from external injuries by the wooden case, and have nothing to fear from the inside.' (9) This is the first mention of block-tin, or pure tin, the material which Watt would use for the cylinder of the second Kinneil engine. Some oils had a corrosive effect on cast-iron which was the reason for choosing block-tin. It would have been harder than copper but much more expensive. Watt explained how these cylinders were to be made.

> The inner cylinder will be best made of block-tin, because it is not liable to be dissolved by oil. It can be made very equally wide by flattening a sheet of it and making it equally broad, and then turning it up and soldering. It does not wear the pistons, nor seem to be worn out by them itself; secondly, when worn out it will not lose much of the value of the metal, of which it need not con[tain much], as it need not exceed a quarter of an inch thickness'. (10)

In his patent, Watt specified 'employing oils, wax, resinous bodies, fat of animals, quicksilver, and other metals, in their fluid state' to render the piston and other parts of his engines air- and steam-tight. (11) Oil was in fact tried in a quite extraordinary manner. It seems that the piston was covered with a layer of oil in place of the water in Newcomen engines. Some of this leaked past the piston and collected in the bottom of the cylinder from where it was recirculated to the top of the piston again by a small Savery type steam pump. He told Roebuck in March 1769, 'I therefore make use of the oil pump and train-oil, which

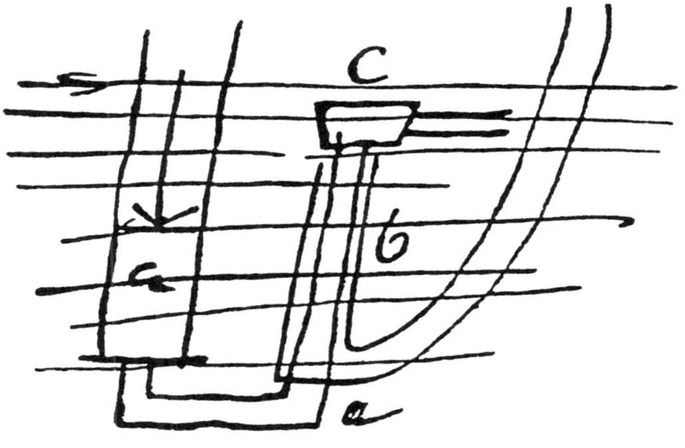

This is the earliest drawing with flap valves in the bucket of the air pump 'a'. The condenser pipe was surrounded by an outer pipe 'b' so cold water ran between them. At 'c' a valve opened when the air pump bucket was at its lowest position so cold water ran in and helped to flush out the hot.

answers perfectly well, keeping a constant circulation'. (12) Watt must have been running his experimental model for quite long periods. The drawing not submitted for the 1769 patent has no details of any self-acting valve gear. He prepared drawings for a new regulator or steam valve in April which must have been for the model engine. (13) He had found that steam was being trapped in part of the mechanism and cooled which then caused problems when next creating a vacuum.

In February, Watt was pleased with alterations which he had made to his condenser, perhaps those described in the text not submitted for the 1769 patent. This shows that his condenser, which had a single pipe and air pump, was still full of water. He had fitted a valve at the top of the body of the condenser which could be briefly opened so that, when the air pump piston was in its lowest position, cold water could flow through the condenser and out of the top to flush out any heated water. (See p395) The air pump piston would then be raised to create a vacuum as well as lifting the condensate out of the top of its barrel. This is the first time we find valves fitted to the air pump piston 'as in a common pump' to allow water to pass through it, something abandoned in his next condensers. He proposed to work this condenser by a waterwheel, presumably anticipating what he would do at Kinneil. (14) While at Glasgow, he worked it by a handle. (15)

The proposed patent drawing shows the air pump piston connected to the main beam of the engine which is almost certainly incorrect at this period of development. There are two reasons for this being wrong, one is the timing of its operation, which will be discussed later, and the other is that he described how this piston must be used to clear water and air out of the condenser before making a stroke with the steam piston. This must mean that the air pump could be worked separately from the engine as indeed we have seen already. Also we have seen that he was using two air pumps by this time while his drawing shows only one.

However this condenser did not satisfy him for, in April, he improved it with frictionless pumps which were 5 ins. diameter but had only one inch stroke! He tried to simplify the construction and increase its size. (16) But he found a place where air could lodge so he was not obtaining a proper condensation and again set about designing a new condenser. (17) Clearly it was this new one which was tried on 28 May 1769. (See below)

Twin air pumps worked by a rocking beam. The valves in the bottom of the condenser were opened by triggers so that cold water rushed in when the appropriate piston was at the bottom of its stroke, giving rapid condensation.

> I have this day made a conclusive trial of my new condenser... On trial, the condensation is more rapid than ever, the force necessary to move the condenser much less than you could well imagine, and the ease of execution quadruple what it ever was of any former condenser, for instance. This one is made of wood and block-tin. (18)

He sent a sketch to Small which shows that this condenser had two air pumps, which he said were 3 $^1/_2$ ins. diameter and 1 $^1/_4$ ins. stroke. They were moved by a beam,

> made to vibrate by means of two spiral [cams] wheels that press down first one end then the other. This spiral puts them down very quick in the first part of its motion, but during the last fourth of its revolution it only moves them 1 $^1/_4$ inch...[Presumably he means one quarter of an inch.] During this part of the revolution, the valve at bottom is opened by a trigger... then the water surrounding the condenser being higher by six inches than the top of the pumps, rushes up through the condenser and pump and cools it. (19)

Watt's condensers were becoming more and more complicated. This description shows that these air pumps must have been operated by some type of rotary drive and not by the engine itself. Also it shows that he was trying to flush his condenser with cold water to remove any remaining hot condensate. He would continue to develop both these features at Kinneil.

Watt was still using a model engine and, whatever success he had with it, he had to carry out many more experiments to make a satisfactory large scale one. We can imagine some of the problems he would face, such as air gathering under the solid piston of the air pump, obtaining a sufficient area in the condenser to cool the steam from an engine running continuously, as well as constructing a surface condenser that would not leak. While he was carrying out trials in Glasgow, whether in some cellar or at Delftfield, there are hints that he had trouble in obtaining sufficient cooling water. (20) At least at Kinneil there was a valley behind the house with a small stream running down it which could be used to supply cold water as well as drive a waterwheel which, in February 1769, he proposed should work his condenser. (21) But, in the meantime, he continued with experiments in Glasgow until the end of May 1769.

NOTES: THE FINAL PHASE OF EXPERIMENTS AT GLASGOW

1 Muirhead, (1854), p. 34, Letter 39, J. Watt to M. Boulton, 12 Dec. 1768; Boulton 348.2.
2 Robinson & Musson, (1969), pp. 52 - 53, draft Steam Engine specification, JWP, 4/24.
3 Muirhead, (1854), Vol I, p. 36, Letter 41, J. Watt to W. Small, 28 Jan. 1769; Boulton 348.4.
4 Ibid, Vol I, p. 48, Letter 48, J. Watt to J. Roebuck, 14 March 1769; JWP, 4/32.19.
5 Ibid, Vol I, p. 45, Letter 45, J. Watt to J. Roebuck, 11 Feb. 1769; JWP, 4/32.21.
6 Ibid, Vol I, p. 44, Letter 44, J. Watt to J. Roebuck, 10 Feb, 1769; JWP, 4/32.20 and Vol I, p. 45, Letter 45, J. Watt to J. Roebuck, 11 Feb. 1769; JWP, 4/32.21.
7 Ibid, Vol I, p. 54, Letter 54, J. Watt to W. Small, 28 April 1769; Boulton 348.8.
8 Ibid, Vol I, p. 48, Letter 48, J. Watt to J. Roebuck, 14 March 1769; JWP, 4/32.19 and Vol I, p. 49, Letter 49, J. Watt to J. Roebuck, 16 March 1769; JWP 4/32.18.
9 Ibid, Vol I, p. 49, Letter 49, J. Watt to J. Roebuck, 16 March 1769; JWP, 4/32.18.
10 Ibid, Vol I, p. 50, Letter 50, J. Watt to W. Small, March 1769; Boulton 348.6.
11 Patent 913, 5 Jan. 1769.
12 Muirhead, (1854), Vol I, p. 49, Letter 49, J. Watt to J. Roebuck, 16 March 1769; JWP, 4/32.18 and p. 50, Letter 50, J. Watt to W. Small, March 1769; Boulton 348.6 which has a sketch.
13 Ibid, Vol I, p. 52, Letter 53, J. Watt to J. Roebuck, 28 April 1769; JWP, 4/32.17.
14 Ibid, Vol I, p. 47, Letter 47, J. Watt to W. Small, 22 Feb. 1769; Boulton 348.5.
15 Ibid, Vol I, p. 48, Letter 49, J. Watt to J. Roebuck, 16 March 1769; JWP, 4/32.18.
16 Ibid, Vol I, p. 51, Letter 51, J. Watt to W. Small, April 1769; Boulton 348.7.
17 Ibid, Vol I, p. 52, Letter 53, J. Watt to J. Roebuck, 28 April 1769; JWP, 4/32.17.
18 Ibid, Vol I, p. 61, Letter 58, J. Watt to W. Small, 28 May 1769; Boulton 348.9.
19 Ibid.
20 Ibid, Vol I, p. 50, Lettter 50, J. Watt to W. Small, March 1769; Boulton 348.6.
21 Ibid, Vol I, p. 47, Letter 47, J. Watt to W. Small, 22 Feb. 1769; Boulton 348.5.

Watt's Own Engine: Part 6 The Final Kinneil Engine

During the autumn of 1768, Watt began designing his much larger engine to be erected at Kinneil and wrote to Roebuck on 9 November, 'On considering the engine to be erected with you, I think the best place will be to erect a small house in the glen behind Kinneil'. (1) The walls of that house are still standing. He was preparing drawings at the end of April and into May 1769. (2) The new Kinneil engine was being erected during the summer although progress was slow. The first serious experiments there were made a couple of weeks before Watt wrote to Small on 20 September, when he had to admit that, 'The trial has not been decisive; but I am still allowed to flatter myself with hopes'. (3) Modifications were made continually until the beginning of April 1770 when Watt left to survey the line of the Strathmore Canal. He sent Roebuck instructions about further improvements (4) which were carried out by Gardner and Osborne. Watt called at Kinneil on his way back from Perth in the middle of May and returned a month later. He was in the course of more experiments when he was summoned to Glasgow to meet Golborne over surveys of the River Clyde and the Monkland Canal.

13 June 1770 was indeed a black day for the steam engine because Watt wrote in his Journal, 'Made some experiments, inconclusive, on engine in the forenoon at 2 o'clock set out for Glasgow where I arrived about 10 o'clock'. (5) The engine was never to run again at Kinneil. The Carron Company papers recorded

Watt's workshop behind Kinneil House.
(See also plate 18)

in the June Committee Minutes that, 'Doctor John Roebuck has Assigned over his Effects and Management of his affairs...' (6) Roebuck's bankruptcy stopped any further work so it was not until Boulton made arrangements to take over his share in the patent and Watt's debts that any further progress could be made. In the middle of May 1773, Watt began to dismantle the engine at Kinneil for shipment to London and so to Birmingham. (7) He followed only in May 1774, finally arriving there on the 31st. (8) The engine was re-erected at Boulton's Soho Manufactory where tests started again.

On 28 April 1769, Watt announced to Small, 'I have begun to erect an engine of an 18 inch cylinder and 5 foot stroke, at Kinneil, the doctor's place. However, it is to be done as privately as we can, as we do not propose to puff; therefore we expect your silence on this head'. (9) This announcement was a little premature because he did not send the plan and elevation to Roebuck until the first of May with the request that 'I wish the wood of the ['Pit head' crossed out] frame to be got ready, but not to be put up till I come'. (10) Unfortunately the drawing has not survived but the engine was to have the normal layout of a reciprocating beam engine. As the smith work might cause delays, Watt suggested that a start might be made on the boiler as well as the chains which were to have three rows of links made from the best tough iron.

He told Small at the end of May that he hoped to have the engine going within a month (11) but then thought in early July that, while the great beam was hung, the boiler 5 $^1/_2$ ft. (presumably diameter at the bottom) set and the condenser finished, it could take possibly three weeks to complete. (12) This condenser, which was made partly of lead with a wooden frame, was much to Watt's liking and had block tin air pumps, 9 ins. diameter. The beams supporting the main beam were 14 ins. square. He originally proposed mounting the main beam which was about 10 ft. long with unequal centres at 6 ft. and 4 ft., with the longer length on the steam side. (13) The first pump would have been 18 ins. square and 15 ft. depth but a later one had a diameter of 18 $^1/_2$ ins. with a depth of 25 ft.

It was a very depressed Watt who finally wrote to Small on 20 September, a couple of weeks after what must have been a disastrous initial trial of his Kinneil engine, and told him,

> This [Clyde survey] I would not have meddled with had I been certain of being able to bring the engine to bear; but I cannot, on an uncertainty, refuse every piece of business that offers... However, if I cannot make it answer soon, shall certainly undertake the next [Newcomen engine] that offers; for I cannot afford to trifle away my whole life, which God knows may not be long. (14)

He admitted that fitting all the parts together and adjusting them had taken much longer than he had anticipated. At the first trial, having pumped the air out of the condenser, the main piston descended only a couple of feet and then, on the second attempt, only a few inches. The leather in the water pump had turned inside out and the pasteboard packing on the steam piston had torn. After repairs, on the second trial, the engine made a few brisk strokes but soon went more and more slowy. One fault may have been excessive friction in the air pump. Then the steam piston failed and oil came into the condenser. Cork packing was tried this time, but the cylinder itself was found to be oval in one place from some injury received in erecting it. Next, the cork also failed. There were many more tribulations but the condenser performed well enough although it was worked by a man because the waterwheel was not ready. In addition, the boiler made sufficient steam with a small fire in spite of there being many leaks. The conclusion he drew from this trial was that his engine 'would work easily with 8 lbs. on the inch, and would not consume above half the steam used by a common engine. Even this I will not positively affirm, although I think there is reason to believe it'. (15) Watt had to redesign his condenser unit which being prepared that November.

But it would be a little while before Watt could return to his engine because, during the winter, he was asked to prepare a survey for the Monkland Canal as well as improvements to Port Glasgow harbour. Therefore it was not until February 1770 that he could concentrate on the Kinneil engine again. After this long hiatus, he promised Roebuck that he would go to Kinneil and 'not to leave this place till the whole of the alternate engine has been effectually tried'. (16) Roebuck hoped that he could report better success to Boulton before the end of the month. (17)

The second Kinneil engine was fitted with a wooden jacket around the cylinder which Watt said would protect the inner one made of block-tin. (18) There is no doubt that the jacket fulfilled Watt's intention and kept the main cylinder hot. But Watt did not understand the true thermodynamic principles of the steam engine in which, to do work, the steam must both expand and lose heat. Just as the cylinder should be as hot as the incoming steam, so the outgoing steam ought to have expanded and dropped in temperature, finally being as cold as the condenser in the most efficient engine. Sadi Carnot pointed out in 1824,

> 1) The temperature of the fluid should be made as high as possible, in order to obtain a great fall of caloric, and consequently a large production of motive power.
> 2) For the same reason the cooling should be carried as far as possible. (19)

Carnot pointed out that there should be no sudden falls in the temperature and that the caloric had to pass to the second reservoir, i.e. the condenser, at a lower temperature. The steam in Watt's cylinder would have received heat from the steam in the jacket and so passed to the condenser at a comparatively high temperature. This defect would have increased the fuel consumption and caused the condenser to consume more cold water than the optimum, reducing the efficiency of the engine. It was a point never understood by Watt (through no fault of his). His engines might have performed better with just wooden lagging round the cylinders.

A few invoices have survived from this period which show some of the people who supplied Watt with parts. In May, Daniel Biggs provided a 'stem' valve, a double pump, a 'stim' pipe, a frame for the condenser and a packing 'cass' for the 'steem' cover. (20) In July, there came from John Wilson a variety of iron work including some for a 'Modell of Fire Engine', a piece of rolled plate iron, iron cut circular for valves, as well as '269 $^1/_2$ pounds of iron work for Fire Engine'. (21) Stephen Maxwell submitted accounts from 1766 to 1771 for copper pipes and other copper work amongst which was a 'white iron cilender' in January 1768 and the copper bottom for a cylinder in February 1769. But the most interesting item he supplied was a 'sheet of Block Tin, £13.4s.7d.' on 10 May 1769. (22) This sheet sounds like the one which may have been formed into the cylinder for the Kinneil engine.

This cylinder was damaged in course of erection. It was found to be slightly oval which Watt proposed to rectify where it stood. (23) It would be made true and hardened by 'hammering it with a mallet on a piece of hard wood fitted to it', (24) a method Watt had used in 1765 which does not sound very satisfactory. (25) It must have been the same cylinder which, in October, Watt told Small, 'The cylinder was a full quarter inch thick and 18 inch diameter. At the worst place the long diameter exceeded the short about $^3/_8$; it was made of the best tin, not hammered. If the engine succeeds, I propose making them of copper'. (26) With such inaccuracy, it is not surprising that Watt had difficulty obtaining a good seal for his piston. He was stretching the technology then available in Scotland to its limits.

Even after the block-tin cylinder had been repaired and appeared to be quite true in 1769, Roebuck still thought that the material might be too soft and that the piston might distort it so it would have to be replaced with one of cast iron. (27) This must have been the cylinder which Watt packed up at Kinneil for Birmingham in 1773 and which was re-erected at Soho with a wooden casing. (28) Here Roebuck's fears were proved true because, just before Christmas Day 1774, an inspection of the engine revealed that 'The cylinder had been forced in by the steam'. (29) It must have been repaired and continued in use until it could be replaced with a cast iron one supplied by John Wilkinson. The new cylinder was ordered by 26 January 1775 and arrived on 24 April (30) but even this one, bored on Wilkinson's new machine, was 'not perfect, yet there doth not appear any very gross error'. (31) At last, Watt had a cylinder that was reasonably round and he could make this part of his engine run satisfactorily but this confirms that here was one problem which he never properly solved while he was in Scotland.

We have seen how the packing on the steam piston failed during the first tests so that steam was heard constantly rushing past the piston into the condenser. One cause may have lain with the oil pump where some of the valves and openings were too small and had to be modified. Some oil was carried through into the condenser and became mixed with the water there so it became useless. Various ideas were proposed to prevent this, including mixing flour in the oil, and Watt asked Small whether a substitute for oil could be found. (32) Two-ply pasteboard was tried for the packing but that failed as well as cork. In the following January, Small endeavoured to have a 'hat to be made of a large piece of cotton' (33) but the hat maker could not fashion it without glueing or weaving it. In addition, Watt had thought of leather, tow and horse-dung mixed, or possibly mercury or Newton's metal for sealing the piston. (34) In March 1770, he had considerable success by reverting to cork and found no oil passed through the condenser. He commented, 'There are therefore two essential points ascertained, a true cylinder, and a piston tight without oil, and seemingly durable. So far well'. (35) There the problem of sealing the piston seems to have rested until after the engine was re-erected in Birmingham.

We are given no hint about the form of any self-acting valve gear until the very end of the trials when Watt sent Roebuck instructions about modifications to be undertaken while he was away surveying the Strathmore Canal. 'The plug-frame is to be got fitted up as soon as the patterns for the new condenser are finished and sent off. When the plug is finished complete, the cast-iron snout is to be opened, and a strict examination passed upon the regulator and its snout'. (36) This work must have been carried out ready for the last attempt to run this engine in Scotland on 13 June. We are not told either whether the waterwheel was ever commissioned to drive the air pumps. Co-ordination between a man or a waterwheel driving the air pumps and the movement of the piston in the steam cylinder would seem to be impossible. Such an arrangement may well account for another of Watt's difficulties in achieving success.

When he was describing the first trial of the final Kinneil engine in September 1769, he seems to have been referring to the condensing vessel when he said, 'The condenser performed well enough, and seemed to condense the steam readily... I am inclined to try my former condenser again, having thought of some improvements on it'. (37) But he could not have been satisfied and was considering using the condenser unit he had prepared for the aborted Kinneil engine in 1766 which would have been similar in size. This one may not have been used either for it was later sent to Soho where many experiments were tried with it. (38) In 1774, Roebuck expressed surprise that it should work at all at Birmingham, 'considering the slightness of the materials and its long exposure to the injuries of the weather'. (39)

A drawing of an improved condenser layout appeared in a letter to Small in November 1769. This had two condensers, each consisting of eight vertical pipes, joined at the top

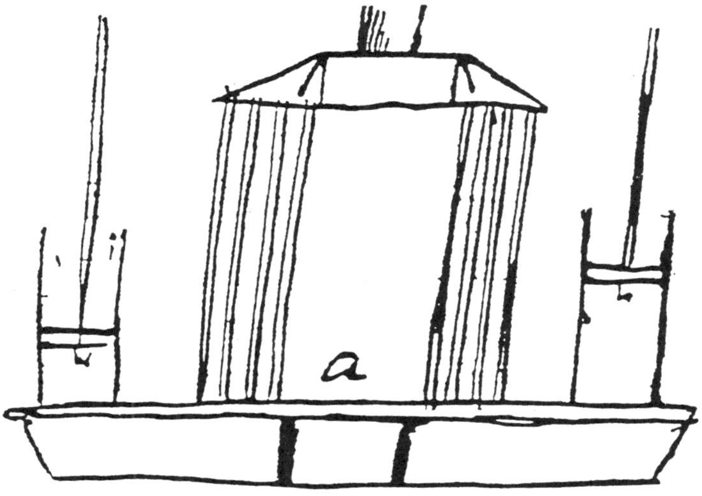

Each condenser had eight vertical pipes, 18 inches long, 3/4 in. diameter, 1/2 in. apart. The appropriate pump drew the water and condensate down them 16 inches.

through non-return valves to a common steam pipe, but emptying at the bottom into separate chambers for the air pumps. He described this layout,

> They [the pipes] are each 3/4 inch diameter and 18 inches long, 16 inches of which will be evacuated each stroke of the pump; they are to be 1/2 an inch distant from one another in all directions; each set is to be surrounded at 1/2 an inch distant in a box of wood, through which cold water can be made to run at pleasure. They are joined at top by a thin cast-iron box, through which they communicate with the steam. It is made sloping at the ends, that as little useless water as possible may be in circulation. They are joined at bottom by another cast-iron box, through which they communicate with the pumps, which are of copper, 5 inches diameter; two diaphragms of wood prevent their communicating with one another, the box being continued only for strength. (40)

One interesting point is that he had calculated how far down the pipes the pumps would draw the water at each stroke, suggesting that he was investigating heat transfer problems. Joints in a composite structure of wood and metal would have been difficult to keep sealed.

In the collections at the Science Museum, there is a well-made model of a surface condenser with 140 small vertical tubes which bears some similarity to Watt's pipe condenser. That this model has actually worked is suggested by some of these tubes being blocked off with solder as if to repair leaks. This condenser empties through a valve into a rectangular pipe at the side. This pipe is connected at its lower end to a vertical air pump with a solid piston while, at its upper end, it terminates in another flap valve. The air pump piston would draw condensate into the rectangular pipe on its up-stroke and expell both air and condensate through the upper flap valve on its down-stroke. Unfortunately this model cannot be dated but could have been designed in late 1769.

By November 1769, Watt was experiencing difficulty with the condenser itself heating up, probably through running the Kinneil engine for longer periods. It is doubtful whether his idea of drawing the water down the body of the condenser would have exposed enough surface area. In January 1770, he was designing a new condenser to remedy this problem (41) but even this did not answer, for, during trials in March, while 'the condenser seemed to despatch the steam instantly', the engine did not start to move for two or three seconds after the exhaust valve had opened. Watt ascribed this to the middle part of the condenser retaining heat which was given out to the water lying on it.

Roebuck realised the real cause when he told Boulton in March 1770 that they had remedied the imperfections of the cylinder and piston.

> But the Condenser was not so effectual as in our small Engine. Mr. Watt had contracted it too much. The next week we shall have a more complete Condenser and a further trial. I now consider the affair of the Engine as absolutely certain, though it may still take some time to complete it to the perfection it is capable of. (42)

Watt was trying to make his condenser walls as thin as possible to transfer heat quickly but found problems of enlarging their scale.

> The pipe condensers I had in the models were made of white-iron, but it is not easy to make this one of this size sufficiently thin without risking it being crushed. Copper is, to be sure, the proper substance, but the difficulty of getting its joints tight persuaded me to make it of iron, which has rendered it useless. I am going to try more experiments on it to-morrow, and if I cannot persuade it to perform better, must return to the old condenser, which, I think, I can greatly improve. (43)

We can confirm that Watt was still keeping his condenser full of water and creating a fresh vacuum for each stroke because he told Small in January 1770 that 'at first setting out, all is full of water'. (44) But in the middle of March he discovered that the condenser, once the vacuum had been created, could be kept exhausted all the time, as happened in later engines. He told Small his reasons for his earlier designs.

> The reason why this condenser was not tried sooner was that I once imagined that there was much air mixed with steam; I therefore intended my condenser to be perfectly exhausted each time by filling it with water; now this condenser required two large pumps, and consequently was too great a burthen, but now I find little air to be mixed with steam, and I see by calculation that it will make little difference whether the condenser be filled with water or no... Therefore a pipe condenser, if used, may be kept perpetually exhausted, like the receiver of the air-pump. (45)

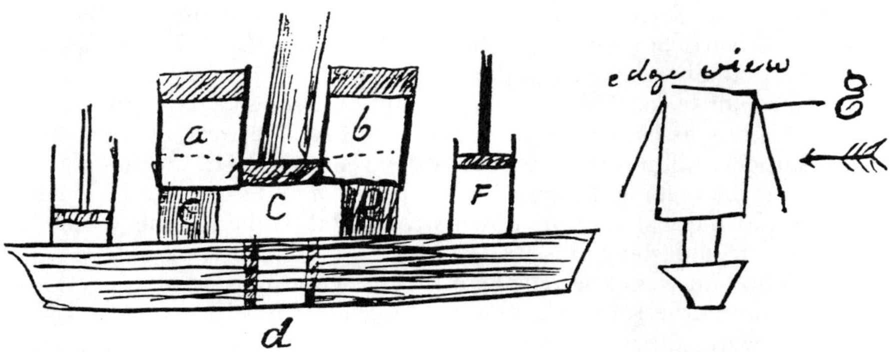

An improved condenser with air pumps 5 inches diameter. 'a' and 'b' are cooling boxes with flaps at the side, joined by a copper box 'c'. One pump was worked ahead of the other so at 'd' were wooden partitions to act as separators.

A drum condenser with horizontal pipes. Air and condensate were driven out through valves 'c' by the respective pumps 'd'. Valves 'e' prevented the condensate flowing back into the drum.

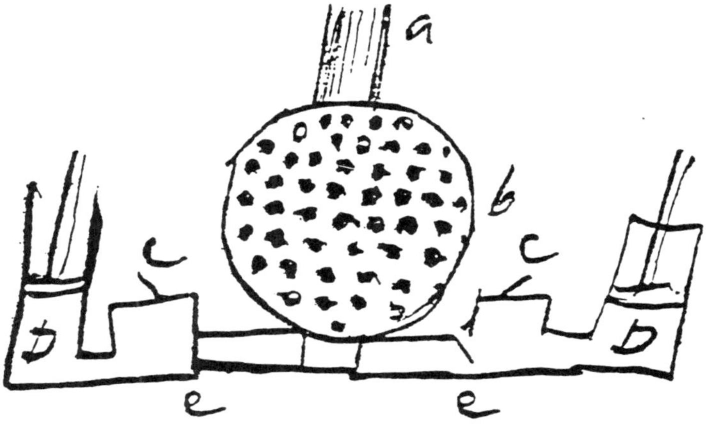

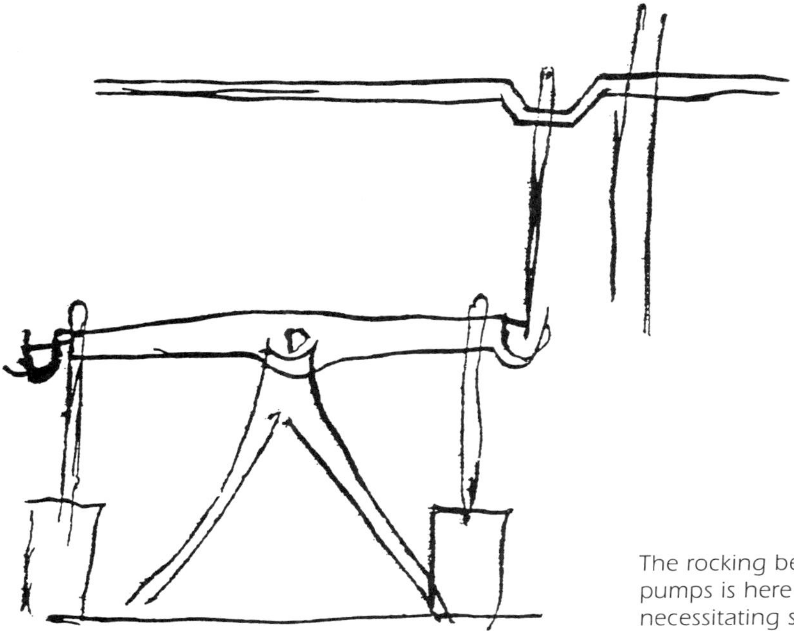

The rocking beam operating the air pumps is here driven through a crank, necessitating some form of rotary motion.

Just when the change was made is difficult to establish because, in one of Watt's note books, some jottings on 7 April 1770 suggest that the condenser and air pump may still have been full of water. (46) The middle of March could well mark one of the critical turning points in the development of the separate condenser when Watt realised that he did not need to create a fresh vacuum for each stroke. If this is correct, then one of the crucial keys to a successful engine was not discovered until just before trials were abandoned in Scotland.

Instead of returning to his old condenser, Watt designed two new ones. The first was quite extraordinary because the vertical pipes were replaced with boxes, of which two sides and the top consisted of flap valves. (See p404) The steam was drawn into the bottom of a box by the up-stroke of its respective air pump where it would be condensed by the water in it. Any air would pass to the top. On the down-stroke of the pump, the condensate would be forced out through the side valves and the air through the top. Watt was still trying to flush out hot water from his condenser at each stroke. If this condenser failed to work, Watt was going to return to an earlier one made of 'thin parallelepipeds of white iron'. He also proposed yet another form of condenser, the third in a single letter. In this instance, it was a horizontal cylinder or drum with many small pipes running through it. (See p 405) Here, for the first time, we find a calculation of the size of surface condenser needed. 'Remember that it requires 23 cubic inches of water, at 50°, to cool 90° one cubic foot of steam; and that it will not be proper to allow the water to be heated above $^1/_{16}$ inch over the whole surface of the condenser'. (47) The second Kinneil engine had twin condensers and air pumps which were finished in July 1769 with 9 ins. diameter pistons. (48) To work them, Watt substituted a crank because the original 'spiral wheels' made a thumping noise. A later drawing shows a condenser with twin pumps driven by an oscillating beam from a crank. (49. See p405) These pumps were 5 ins. diameter and 4 ins. stroke. He continued with twin pumps at Birmingham. When he left Kinneil to start his Strathmore canal survey, he asked Roebuck to take exact dimensions for a new condenser, so the form of the condenser was by no means settled when trials ceased in Scotland. (50)

Operating the Engine

Creating a fresh vacuum for each stroke of the engine would mean that Watt had to develop a method of doing this quickly and at the right time. Newcomen had made his crucial discovery of direct injection probably by accident but it enabled a vacuum to be created rapidly. Even so, there are accounts of the piston pausing at the top of the stroke while the cold water injection created a low enough internal pressure for the atmosphere to force the piston down and overcome the opposing weight of the pump rods and water at the other end of the beam. (51) In the Newcomen engine, any air that was carried in by the steam or injection water or through leaks, would be expelled out of the snifting valve by the fresh incoming steam on the next stroke. This was impossible with the surface condenser so Watt had to clear his condenser both of hot condensate and air. To begin with, he thought that a lot of air passed through the engine which he probably reduced unintentionally in two ways. The first was by surrounding the cylinder with a steam jacket where the pressure of the steam kept out external air, particularly during the time there was a vacuum in the cylinder. Second, the condensing apparatus was submerged in cold water so that, if there were any leaks, cold water would be drawn in and not air.

At the start of his experiments, he was working with his condenser and air pump filled with water. When the solid piston was raised, the water would be drawn down the condenser, leaving a void at the top. This was connected to the steam cylinder through a valve so that, when this valve was opened, steam would rush in and fill the void. Since the condenser and its walls were cold, the steam would be condensed and, as Watt realised, this would continue until all the steam was condensed. The hot condensate would lie on top of the cold water in the condenser with any air above it. When the valve connecting the condenser with the main cylinder had been closed, the piston of the air pump could be pushed down, raising the water in the condenser. A suitable valve in the top of the condenser allowed the air and hot water to pass through and run away. There could be no foot valve between the condenser and air pump because the original water had to oscillate to and fro quite freely. If Watt could gauge the volume of the condensate, then he could arrange the stroke of the air pump to be such a length that it pushed out all the hot condensate and air each time.

Such a method of working could have given a good vacuum quickly but suffered from several disadvantages. One would have been transfer of heat into the top part of the condenser and the water there. He attempted to remedy this by flushing the system through with cold water. Then it is not clear how the level of water inside the condenser was maintained nor how any air that collected below the piston of the air pump could be expelled. But the main difficulty would have been working the air pump piston at such time as it would create the vacuum when the steam piston had reached the top of its stroke. None of these problems may have been crucial when models were being operated with the valves controlled by hand, but, when later engines were being run for longer periods with automatic valve gear, Watt had to modify his designs considerably.

Timing of the Air Pump

The problem of timing becomes apparent when a study is made of the drawing (See p394) which was prepared but never submitted for the 1769 patent. It shows one side of a beam engine with the pistons of both the steam cylinder and air pump in their lowest positions. At the top of the condenser, there is a flap valve to admit steam from the cylinder into the condenser and, close by it, another flap valve to let out the air and condensate from the top of the condenser. The air pump piston is solid. The top of the pump barrel is open and appears to lie below the level of water in the cistern which surrounds the condensing apparatus. In the side of the air pump barrel, just above the piston, is a circle, possibly to show the diameter. If it is a hole, it cannot be explained unless it is to allow some circulation of water within the pump itself. The whole of the air pump and condenser should be full of water.

The position of the steam piston at the bottom of the cylinder in this drawing shows that the power stroke with the vacuum underneath has just been completed. Next steam should be admitted to the cylinder to allow its piston to rise. The air pump will also rise, evacuating the condenser and forming a vacuum. But, at that point of the engine cycle, the air pump should be descending to push water into the condenser and eject the condensate out of the top. This could be achieved if the air pump were situated on the other side of the wall supporting the beam, as was done on some later pumping engines. But this still would not have solved the problem of how to create the vacuum to enable the steam piston to descend which must be done before that piston can move. In other words, the air pump piston must be moving ahead of the steam piston so that the vacuum can be created just as the steam piston finishes its upward stroke. While this is possible with hand-operated valve gear and air pump, it is impossible on non-rotative pumping engines because these have no way in which the air pump piston can move at a different cycle to the main piston. As long as Watt was trying to create a fresh vacuum with his air pump for each stroke, he needed some means of moving the air pump piston other than the engine itself. This is presumably why he introduced men or a waterwheel to operate the air pump. But, once he was operating with a permanent vacuum, the difficulty of timing was overcome, providing the condenser had adequate capacity.

Capacity of the Condenser

Judging by the numerous different types of condensers with which Watt experimented, this unit could not have functioned as well as he desired. Robison said that,

> The Doctor [Roebuck] thought to produce the condensation, with sufficient rapidity and accuracy, by a very extensive surface; and Mr. Watt knew that it also required a great quantity of water, or other matter to receive the emerging heat. I know that these differences of opinion retarded the completion of the engine. (52)

Both were right! One disadvantage of filling the condenser with water and uncovering a portion of it by withdrawing the water with the air pump would have been that this limited the surface area exposed. This would have delayed the formation of the vacuum so that the steam piston did not move quickly. Watt wrote to Boulton about this, 'The next objection was that the vacuum might not be formed by exhaustion sufficiently sudden or sufficiently perfect'. (53) However it is difficult to determine the causes of some of these problems because in this instance the packings on both the steam piston and the water pump piston were not working properly.

Complaints of heat in the condenser occur when trials were being made on the larger model in 1769 (54) and then at Kinneil, suggesting that bigger engines and longer periods of operation may have shown up deficiencies. On one condenser at Kinneil, Watt tried to pass the steam up the water-filled pipes to ensure it was all condensed but ran into difficulties with the heat being transferred into the air pumps where the water might boil.

> On condensation the steam, whenever it entered (which would be as soon as any part of the pipes was empty), would rise, or attempt to rise, through the water in the pipes, and would heat it; but the piston of the pump still ascending, that water would descend into the box below, and part might get into the pump, and there be converted into steam; not being exposed to much condensation at best, if it got no further than the box it would remain warm until the cold of the pipes made a vacuum, when it would boil and go to them to be condensed, and it is much to be feared that this heating and cooling might prolong the time of condensation. I have sometimes thought that something of this kind happened in the other condenser, and concurred with other devils to plague me. (55)

Even by admitting steam at the top of the condenser, Watt did not entirely overcome avoiding hot water remaining in parts of the pipes which connected the condensers nor could he avoid the water in the condensers becoming hot.

> At first setting out, all is full of water:- now, that water will be heated by the steam and forced by it into the pump, where it will be cooled, but some water will still remain in the middle part, which cannot come out nor be cooled but by boiling. This may take some time, as the heat the metal had acquired will be transmitted to it gradually: all this will be greatly remedied in the new condenser. (56)

Problems with the condenser are recounted in one of the few dated records in Watt's note books. It is difficult to determine precisely what was happening because the account cannot be tied up with the apparatus, but, on 7 April 1770, we can see that the engine still was not working properly. 'When the water flowed over the little cistern & the condenser was wrought slow she made about 10 strokes a minute with a strong steam. When the engine stood a little & the condenser was freed of air & cool'd by standing she descended without hesitation'. (57) Therefore, shortly before trials ceased at Kinneil, Watt had not overcome his problems with the separate condenser and so it was not only the cylinder and piston which would have to be improved at Soho but the condenser as well.

The Performance

With his engine still not performing to his satisfaction, Watt had to make that difficult decision during the summer of 1770 whether to continue with experiments or to accept the position of overseeing the construction of the Monkland Canal. We have few indications of the costs of all the different condensers and other equipment which he had been designing and producing for his engine but, what with family ties and his own lack of resources coupled with Roebuck's insolvency, he realised that he had to abandon his cherished dream of his steam engine.

It is difficult to determine how much Watt's seeking perfection caused delays. When he was contemplating his first tin-plate condenser and Robison burst into the room unexpectedly, Watt said to him, 'I have now made an engine that shall not waste a particle of steam'. (58) To Lind, he wrote a little later that his small model of his perfect engine showed great hopes of success and that he expected 'almost totally to prevent waste of steam, and consequently to bring the machine to its *ultimatum*'. (59) In the first few months of experimenting in 1765, he said to Roebuck, 'I must remedy everything before I send it away'. (60) For example, the tightness of the piston was essential for the success of the engine because slack packing would enable steam to pass through wastefully. Hence the importance of both an accurately bored cylinder and a well-fitting piston. (61) The reason for his twin condensers and air pumps was to achieve a perfect vacuum. (62) There was a fear that forming the vacuum by 'exhaustion', in other words emptying water out of the condenser by the air pump might not create a sufficiently sudden or perfect vacuum. (63) While seeking perfection is admirable, Watt probably carried this to greater lengths than others and so delayed the success of his engine.

Watt's interest in the efficiency of the engine and hence its perfection is shown by his measuring the steam consumption. He was uneasy over the performance of his new twin condensers 'till I know the exact quantity of steam consumed'. (64) He intended to do this by measuring the condensate but then found he could not in the way he had proposed so he measured the amount evaporated in the boiler. (65) These figures would have shown how many cylinders full of steam the engine was consuming to make each stroke. He carried out more laboratory trials on steam during August 1773 and the spring of 1774 to determine this more accurately. He repeated his former experiments on water boiling in a vacuum and under different pressures. He found that the 'ascent was regular' up to 268°F which corresponded to 52 ins. of mercury or just over 25 p.s.i. (66) After trying to determine the expansion of mercury in thermometers in the autumn of 1773, he turned back to steam. Once again, he 'began a course of experiments upon the heat which water boils from vacuo to air' (67) which cost him a great deal of time and labour. (68) A second note book contains accounts of these experiments. (69) He discovered why his condensers were not performing as he had anticipated and found the remedies which would form the next stage of improvements at Birmingham. (70)

Watt had to overcome many difficulties with the materials he employed but the choice he made may have caused further problems, particularly with his cylinders. The cast iron cylinders he obtained were very inaccurately bored, with one for even a model engine being $^1/_8$ ins. larger at one end. (71) While the difficulties of machining cast iron with its hard outer skin must have been immense at that time, Watt's other choices of fabricating cylinders from sheets of either block tin or copper, which would not have worn well, are very strange to us. (72) Likewise making multi-tubular and plate condensers was ambitious, which is shown by their not being regularly employed until the 1850s. Then the piston packing even with hemp was not really satisfactory until piston rings were developed in the 1830s. Possibly Watt shied away from a direct metal to metal rubbing contact through the difficulty of keeping the surfaces oiled. Adequate cylinder lubrication remained a problem until oil could be injected into the incoming steam.

His method of operating the condenser by keeping it full of water can be understood when we realise that he thought he had to clear a great deal of air from the system as well as recreating a vacuum at each stroke. However this, together with surface condensing, certainly brought him many problems. How he could have co-ordinated the strokes of the engine with air pumps worked by men or a waterwheel is difficult to understand. Certainly, if his engine had needed an external power source to operate these pumps, it would never have been successful. But Watt persevered and showed great ingenuity in all his various attempts to overcome the difficulties. Yet it is clear that it was not only the problem of keeping the piston steam tight which he could not 'satisfactorily remove' (73) while he was in Scotland. A new accurately bored cylinder by Wilkinson solved merely one of the problems. Development at Birmingham included a different type of condenser, jet and by surface, where the vacuum was maintained all the time, and reversion to a single air pump with a bucket piston through which the condensate flowed. These changes in the condensing apparatus were some of the important improvements leading to a much simpler engine which turned out to be commercially successful.

Yet from the very first, Watt's engine had shown much greater economy than others then existing. In the earliest letter we have about the engine, that to Lind on 29 April 1765, Watt had claimed, 'And if there is not some devil in the hedge, mine ought to raise water to 44 feet with the same quantity of steam as theirs does to 32'. (74) This economy was shown in later trials as well. For example, on 16 October 1765, Watt reported to Roebuck, 'As to the steam consumed it is very little: my little boiler fills the cylinder in less than half a second after it has been exhausted'. (75) At the first trial of the final Kinneil engine in September 1769, Watt told Small that, 'The boiler with a small fire easily supplied more steam than we could destroy, although there were many outlets for it, which we took no care to stop... It... would not consume above half the steam used by a common engine'. (76) The following March, just before trials ceased, he reported, 'The consumption of steam is as moderate as expected'. (77) This low consumption must have been one of the factors which convinced Roebuck of the soundness of Watt's concepts and that it was worthwhile backing him to apply for a patent. Presumably it also encouraged Watt to go on experimenting.

Watt's engine developed more power, size for size, than the Newcomen type because it achieved a better vacuum. Newcomen engines were run with about 8 p.s.i. pressure on the piston. On one of Watt's models in April 1768, a vacuum of 24 ins., or just under 12 p.s.i., was reported (78) and he hoped this would rise higher after some alterations. At a trial of the double condensers a month later, there was a vacuum of 14 p.s.i. (79) With his larger model and a steam pressure barely above atmospheric, he succeeded in lifting 620 lbs. weight and then increased this to 704 lbs. which it raised easily. (80) If the cylinder were 8 ins. diameter, this would give a pressure of 12 lbs. increasing to over 13 $^{1}/_{2}$ p.s.i. Although Watt gave the theoretical maximum for his final Kinneil engine, no actual performance figures have been found for the time it was being tested in Scotland. After it had been running for a while at Soho during June 1775, Boulton reported a vacuum of 28 and sometimes 28 $^{1}/_{2}$ ins., 14 p.s.i., which fell about 3 ins. each stroke. The temperature of the cooling water was 68°F going into the condenser and only 80°F leaving. (81) These are very good figures which show that the engine was beginning to function properly and this roused many people to take an interest in it.

> Wilkinson hath been here, and says all his neighbours are impatient to see the event of his engine. Some, he says, have suspended their new erections until his is finished; and all of them, he is sure, will have their engines altered, which he says will be a better trade than new erections, and that work alone will be sufficient for our lives. (82)

Notes: The Final Kinneil Engine

1 Muirhead, Vol I, p. 34, Letter 38, J. Watt to J. Roebuck, 9 Nov. 1768; JWP, 4/32.22.

2 Ibid, Vol I, p. 52, Letter 53, J. Watt to J. Roebuck, 28 April 1769; JWP 4/32.17 and Vol I, p. 57, Letter 55, J. Watt to J. Roebuck, 1 May 1769; JWP, 4/32.16.

3 Ibid, Vol I, p. 66, Letter 61, J. Watt to W. Small, 20 Sept. 1769; Boulton 348.11.

4 Ibid, Vol I, p. 101, Letter 81, J. Watt to J. Roebuck, 8 April 1770; JWP, 4/32.8.

5 B. & W. Col., MI, 1/15, 'Journal', 13 June 1770.

6 SRO, Carron Co., GD58/2/3/1, Committee Minute Book, No. 1, 1768 - 1785, p. 89, June 1770.

7 B. & W. Col., MI, 1/20, 'Journal', 17 - 21 May 1773 and Muirhead, (1854), Vol II, p. 49, Letter 114, J. Watt to W. Small, 20 May 1773; Boulton 348.31.

8 B. & W. Col., MI, 1/21, 'Journal', 31 May 1774.

9 Muirhead, (1854), Vol I, pp. 54 - 55, Letter 54, J. Watt to W. Small, 28 April 1769; Boulton 348.8.

10 Ibid, Vol I, p. 57, Letter 55, J. Watt to J. Roebuck, 1 May 1769; JWP, 4/32.16.

11 Ibid, Vol I, p. 62, Letter 58, J. Watt to W. Small, 28 May 1769; Boulton 348.9.

12 Ibid, Vol I, p. 64, Letter 59, J. Watt to W. Small, 5 July 1769; Boulton 348.10.

13 Ibid, Vol I, p. 34, Letter 38, J. Watt to J. Roebuck, 9 Nov. 1768; JWP, 4/32.22.

14 Ibid, Vol I, p. 71, Letter 61, J. Watt to W. Small, 20 Sept. 1769; Boulton 348.11.

15 Ibid.

16 Ibid, Vol I, pp. 88 - 89, Letter 76, J. Roebuck to M. Boulton, 7 Feb. 1770; JWP, 6/44.10.

17 JWP, 6/44.8, J. Roebuck to M. Boulton, 10 Feb. 1770.

18 Muirhead, (1854), Vol I, p. 49, Letter 49, J. Watt to J. Roebuck, 16 March 1769; JWP, 4/32.18.

19 Hills, (1989), p. 166.

20 JWP, 6/28.35, Invoice from D. Biggs, May 1769.

21 JWP, 6/28.28, Invoice from John Wilson, May - July 1769.

22 JWP, 6/28.27, 'James Watt Bt. of Stephen Maxwell', 10 May 1769.

23 Muirhead, (1854), Vol I, p. 71, Letter 61, J. Watt to W. Small, 20 Sept. 1769; Boulton 348.11.

24 Ibid, Vol I, p. 77, Letter 64, J. Watt to W. Small, 21 Oct. 1769; Boulton 348.12.

25 Ibid, Vol I, p. 10, Letter 9, J. Watt to J. Roebuck, 16 Oct. 1765; JWP, 4/32.12.

26 Ibid, Vol I, p. 77, Letter 64, J. Watt to W. Small, 21 Oct. 1769; Boulton 348.12.

27 Boulton 344.9, J. Roebuck to M. Boulton, Feb. 1770.

28 Muirhead, (1854), Vol II, p. 85, Letter 146, M. Boulton to J. Watt, 24 April 1775; JWP, 6/19.22.

29 Dickinson & Jenkins, (1927), p. 108.

30 Muirhead, (1854), Vol II, p. 80, Letter 139, M. Boulton to J. Watt, 26 Jan. 1775; JWP, 6/19.21 and p. 84, Letter 146, M. Boulton to J. Watt, 24 April 1775; JWP, 6/19.22.

31 Ibid, Vol II, p. 87, Letter 147, M. Boulton to J. Watt, May 1775; JWP, 6/19.6.
32 Ibid, Vol I, p. 76, Letter 64, J. Watt to W. Small, 21 Oct. 1769; Boulton 348.12.
33 Ibid, Vol I, p. 86, Letter 74, W. Small to J. Watt, 23 Jan. 1770; Boulton 340.12.
34 Ibid, Vol I, p. 87, Letter 75, J. Watt to W. Small, 31 Jan. 1770; Boulton 348.16.
35 Ibid, Vol I, p. 93, Letter 79, J. Watt to W. Small, 10 March 1770; Boulton 348.18.
36 Ibid, Vol I, p. 101, Letter 81, J. Watt to J. Roebuck, 8 April 1770; JWP, 4/32.8.
37 Ibid, Vol I, p. 70, Letter 61, J. Watt to W. Small, 20 Sept. 1769; Boulton 348.11.
38 Ibid, Vol I, p. 13, Letter 14, J. Watt to J. Roebuck, 5 Dec. 1765; JWP, 4/32.31, see annotation.
39 JWP, 4/57.23, Dr. J. Roebuck to J. Watt, 12 Nov. 1774.
40 Ibid, Vol I, p. 78, Letter 66, J. Watt to W. Small, 3 Nov. 1769; Boulton 348.13.
41 Ibid, Vol I, p. 85, Letter 72, J. Watt to W. Small, 3 Jan. 1770; Boulton 348.15.
42 Boulton 334.12, J. Roebuck to M. Boulton, 15 March 1770.
43 Muirhead, (1854), Vol I, p. 94, Letter 79, J. Watt to W. Small, 10 March 1769; Boulton 348.18.
44 Ibid, Vol I, p. 85, Letter 72, J. Watt to W. Small, 3 Jan. 1770; Boulton 348.15.
45 Ibid, Vol I, p. 99, Letter 80, J. Watt to W. Small, 16 March 1770; Boulton 348.19.
46 B. & W. Col., MI, 1/14, 'Journal', 7 April 1770.
47 Muirhead, (1854), Vol I, p. 100, Letter 80, J. Watt to W. Small, 16 March 1770; Boulton 348.19.
48 Ibid, Vol I, p. 64, Letter 59, J. Watt to W. Small, 5 July 1769; Boulton 348.10.
49 Ibid, Vol I, p. 50, Letter 67, J. Watt to W. Small, 3 Nov. 1769; Boulton 348.13.
50 Ibid, Vol I, p. 101, Letter 81, J. Watt to J. Roebuck, 8 April 1770; JWP, 4/32.8
51 Hills, (1989), p. 9.
52 Muirhead, (1858), p. 97.
53 Muirhead, (1854), Vol I, p. 31, Letter 34, J. Watt to M. Boulton, 20 Oct. 1768; Boulton 348.1.
54 Ibid, Vol I, p. 47, Letter 47, J. Watt to W. Small, 22 Feb. 1769; Boulton 348.5.
55 Ibid, Vol I, p. 79, Letter 66, J. Watt to W. Small, 3 Nov. 1769; Boulton 348.13.
56 Ibid, Vol I, p. 85, Letter 72, J. Watt to W. Small, 3 Jan. 1770; Boulton 348.15.
57 B. & W. Col., MI, 1/14, 'Journal', 7 April 1770.
58 Muirhead, (1858), pp. 67 - 68.
59 Muirhead, (1854), Vol I, p. 5, Letter 3, J. Watt to J. Lind, 4 Sept. 1765; JWP, C1/15.
60 Ibid, Vol I, p. 8, Letter 7, J. Watt to J. Roebuck, 11 Oct. 1765; JWP, 4/32.35.
61 Ibid, Vol I, p. 12, Letter 12, J. Watt to J. Roebuck, 18 Nov. 1765; JWP 4/32.33 and p. 18, Letter 23, J. Watt to J. Roebuck, 15 April 1768; JWP, 4/32.25.
62 Ibid, Vol I, p. 22, Letter 27, J. Watt to J. Roebuck, 29 April 1768; JWP, 4/32.10.
63 Ibid, Vol I, p. 31, Letter 34, J. Watt to M. Boulton, 20 Oct. 1768; Boulton 348.1.
64 Ibid, Vol I, p. 23, Letter 28, J. Watt to J. Roebuck, 10 May 1768; JWP, 4/32.9.
65 Ibid, Vol I, p. 24, Letter 30, J. Watt to J. Roebuck, 24 May 1768; JWP, 4/32.39.
66 B. & W. Col., MI, 1/20, 'Journal', 23 Aug. 1773.

67 Ibid, 27 Feb. - 9 March 1774.
68 Muirhead, (1854), Vol II, p. 75, Letter 132, J. Watt to W. Small, March 1774; Boulton 348.41.
69 B. & W. Col., MI, 2/8, 'Note Book of James Watt, 1770 - 1781'.
70 Muirhead, (1854), Vol II, p. 60, Letter 121, J. Watt to W. Small, 24 Aug. 1773; Boulton 348.37.
71 JWP, 4/32.18, J. Watt to J. Roebuck, 10 March 1769.
72 Muirhead, (1854), Vol I, p. 77, Letter 64, J. Watt to W. Small, 21 Oct. 1769; Boulton 348.12.
73 B. & W. Col., MII, 4/9.7b, J. Smeaton to J. Watt, 5 Feb. 1778.
74 Muirhead, (1854), Vol I, p. 3, Letter 1, J. Watt to J. Lind, 29 April 1765; JWP, C1/15.
75 Ibid, Vol I, p. 9, Letter 9, J. Watt to J. Roebuck, 16 Oct. 1765; JWP, 4/32.12.
76 Ibid, Vol I, p. 70, Letter 61, J. Watt to W. Small, 20 Sept. 1769; Boulton 348.11.
77 Ibid, Vol I, p. 94, Letter 79, J. Watt to W. Small, 10 March 1770; Boulton 348.18.
78 Ibid, Vol I, p. 19, Letter 25, J. Watt to J. Roebuck, 19 April 1768; JWP, 4/32.4.
79 Ibid, Vol I, p. 24, Letter 30, J. Watt to J. Roebuck, 24 May 1768; JWP, 4/32.39.
80 Ibid, Vol I, p. 48, Letter 49, J. Watt to J. Roebuck, 16 March 1769; JWP, 4/32.18.
81 Ibid, Vol II, p. 90, Letter 150, M. Boulton to J. Watt, June 1775; JWP, 6/19.4.
82 Ibid, Vol II, p. 94, Letter 153, M. Boulton to J. Watt, [end of Aug. or Sept.] 1775; Boulton 348.52.

6
ENTER MATTHEW BOULTON

Matthew Boulton, F.R.S., 1728 - 1809.

Proposed Partnership with Boulton

Matthew Boulton, 1728 - 1809

Matthew Boulton was born in Birmingham on 3 September 1728. (1) He left school at fourteen and went straight to work for his father, also called Matthew. Their premises were on Snow Hill just to the north of the centre of Birmingham where they made what were then called 'toys'. These might include trinkets, seals, snuff boxes, filigree work, inkstands, cork screws, snuffers, watch chains, hooks and other small metal items. The Boultons specialised in steel articles, and most probably concentrated on buckles. Buckles were fashionable at that period on shoes or boots, knee breeches and even on men's socks. When only seventeen, Boulton is said to have produced an enamelled or inlaid buckle which helped greatly to enlarge the output of his father's business. He became a partner when he came of age and his father entrusted him with the entire management from then on. The young man had good looks, an engaging manner, was a good mixer and so able to get on well with his fellow manufacturers.

When he was possibly twenty-eight, he married Mary, a distant relative and the elder daughter of Luke Robinson, an opulent merchant of Lichfield. They had no children and had not been together many years when she died, possibly in 1759 or 1760. Sometime after this, he married Anne, the younger daughter, which was forbidden by ecclesiastical law at that time. The fortune they brought with them would have enabled Boulton to retire from business but instead he chose to invest it in expansion of his works. His interest in natural philosophy was stimulated by a visit of Benjamin Franklin to Birmingham in 1758 as well as by meeting Darwin who had moved his medical practice to Lichfield in 1756. (2) Boulton, Darwin, together with Small, founded the famous Lunar Society where they and their friends met on the night of a full moon to discuss the latest scientific discoveries. He shared these interests with Roebuck, with whom he attempted to make accurate thermometers, possibly in 1764. Although Roebuck may have left Birmingham before the meetings of the Lunar Society were established, he and Boulton continued to correspond but distance and then Roebuck's financial problems after 1773 reduced their contacts. (3)

Boulton's father died in 1759 when he inherited the business. His restless energy drove him to expand it and continually improve his products. The cramped premises on Snow Hill, driven by horse power, soon became too small and, in 1761, Boulton began to negotiate for a site to the north of Birmingham in Handsworth where the Hockley Brook could be dammed, forming a reservoir to provide waterpower for a new factory. In the following year, Boulton formed a partnership with John Fothergill and so the famous Soho Manufactory was launched. Each partner put £5,000 into the venture which Fothergill had to borrow but Boulton was able to find from his wife's fortune. Boulton attended to the organization and production while Fothergill directed his attention to establishing trade outlets both at home and overseas.

The Soho Manufactory was opened in 1765, consisting of buildings around four squares, with shops, warehouses and houses 'for a thousand workmen', to quote a Directory of 1774, although 600 was more likely the figure to begin with. (4) Certainly it was laid out on a lavish scale for Boulton started to manufacture a wide variety of small metal products such as silver plate, buttons, snuff boxes and artistic objects. It was these latter which brought him into contact with members of the aristocracy. Through people like the third Duke of Richmond and Lennox, the first Duke of Northumberland, the second Earl of Shelburne and Horace Walpole, Boulton was able to borrow or make drawings of vases, candelabra, statuary and other works which he reproduced at Soho. He was later able to obtain audiences

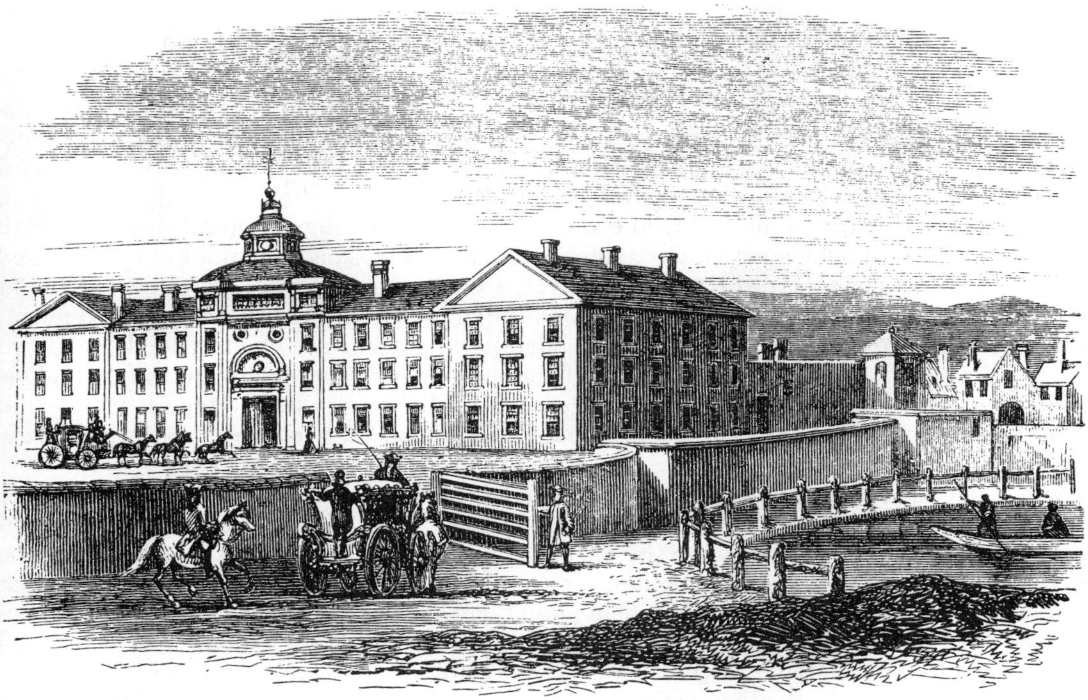

Soho Manufactory.

with George III and Queen Charlotte who purchased his objects. Boulton aimed to establish a reputation for excellence in the design and production of his products and quickly rose to the forefront of Birmingham manufacturers.

Therefore we should not be surprised that Watt was entranced when Small showed him round the Soho Manufactory in 1767. He recorded his first impressions many years later in a memoir after Boulton's death in 1809.

> The goods then manufactured there were steel gilt & fancy buttons, steel watch chains & sword hilts, plated wares, ornamental works in Or moulu, Tortoise shell snuff boxes, Bath metal buttons inlaid with steel & various other articles which I have now forgot. A mill with a water wheel was employed in Laminating metal for the buttons, plated goods etc and to turn Laps for grinding & polishing steel work & I was informed that Mr Boulton was the first inventor of the inlaid buttons, & the first who had applied a mill to turn the Laps. Mr. B. at that [time] also carried on a very considerable trade in the manufacture of buckle chapes, in the making of which he had made several very ingenious improvements. Besides the Laps in the mill, I saw an ingenious lap turned by a handwheel for cutting & polishing the steel studs for ornamenting buttons, chains, and sword hilts etc, and a shaking box put in motion by the mill for scowering button blanks & other small pieces of metal which was also a thought of Mr B. - there was also a steelhouse for converting iron into steel, which was frequently employed to convert the

> cuttings & scraps of the chapes & other small iron wares into steel which was afterwards melted & made into cast steel for various uses. (5)

Boulton must have been a good inventor and manager to have designed all these new machines as well as the layout of the factory. In his Memoir, Watt described his friend as,

> ... not only an ingenious mechanick, well skilled in all the practices of the Birmingham manufacturers but possessed in a high degree the faculty of rendering any new invention of his own or others useful to the publick by organizing & arranging the processes by which it could be carried on, as well as of promoting the sale by his own exertions & by his numerous friends & correspondents.
> His conception of the nature of any invention was quick & he was not less quick in perceiving the uses to which it might be applied & the profits which might accrue from it.
> When he took any scheme in hand, he was rapid in executing it & on these occasions neither spared his own trouble [n]or expence. He was a liberal encourager of merit in others & to him the country is indebted for various improvements brought forward under his auspices. (6)

Boulton was the perfect partner for the timid, irresolute Watt and to Boulton the world must be indebted for bringing forward Watt's invention of the separate condenser.

The scale of Boulton's enterprise at Soho soon caused difficulties with the water supply in the Hockley Brook which proved to be inadequate to drive all these machines, so he considered employing a steam engine to lift the water from the tail race of his waterwheel back to the upper reservoir in order that the water might be used over and over again. (7) He had a model made of a Savery engine towards the end of 1765 or early 1766 which he sent to Franklin for his comments. In a covering letter Boulton wrote,

> My engagements since Christmas have not permitted me to make any further progress with my fire engine but as the thirsty season is approaching apace necessity will oblige me to set about it in good earnest... I therefore beg, if any thought occurs to your fertile genius which you think may be useful or preserve me from error in the execution of this engine you'll be so kind as to communicate it to me. (8)

Franklin made little constructive comment, but Darwin had heard about it and wrote in his usual enthusiastic vein, 'Your Model of a Steam Engine I am told gain'd so much approbation in London, that I cannot but congratulate you on the mechanical fame you have acquired by it'. (9) Earlier Darwin had discussed with Boulton a scheme for driving a 'fiery Chariot' with a reciprocating Newcomen engine but nothing came of it. (10) Now both these men were looking for steam engines which would produce rotary motion even though Boulton could have used the intermediary of a waterwheel. This will be a point of great significance in the story of Boulton and Watt. Boulton's model was received back in Soho where he continued experiments until he heard about the engine on which Watt was working. After Robison and Watt had both visited Soho in 1768, Boulton abandoned his trials and decided to wait until Watt's engine had been perfected.

Potential Partnership

When Watt first met Boulton, probably both saw advantages in some form of partnership and may not have expected that this would turn into a life-long personal friendship. For

Boulton, Watt appeared on his doorstep with a patent for two types of perfect steam engine. How much Boulton may have been swayed by the concept of this engine being supported by the scientific discoveries of Black is now impossible to know. But we can imagine that he would have been keen to acquire a more economical type, especially one that might be able to drive his polishing and other machinery directly. For Watt, here was a possible chance to demonstrate his rotary engine, but, more certainly, a chance to gain the backing of an important, influential English industrialist and so gain openings for his engines south of the Scottish border. After all, his patent was valid for 'England, our Dominion of Wales, and Town of Berwick upon Tweed, and also in our Colonies and Plantations', (11) but not Scotland. Perhaps he and Roebuck thought that the major market would be in England.

At this point we have a puzzle over a letter written by Small to Watt with the date 7 January 1768. That is definitely the date but its contents make better sense if Small made the mistake which we all often make and forgot that he was actually in 1769. Therefore we will not follow Muirhead's placing where it is the first of Small's letters to Watt but put it in this better sequence in 1769. (12) Watt possibly mentioned to Small something about his steam engine in 1767 for Small was pleased that Watt was in London during August the following year 'taking the steps necessary to secure proper advantage from your excellent invention'. (13) While Boulton was delaying the erection of his engine, Small promised that he, Boulton and Darwin would tell their neighbours about how Watt could help them and invited Watt to Birmingham.

Watt was introduced to Boulton in the summer of 1768 by 'Mr Garbett or Dr Small' and took the opportunity to explain to him 'my invention of the Steam engine & several other schemes of which my head was then full, in the success of which he expressed a friendly interest'. (14) Watt's description of his steam engine must have included his rotary engine because Small, in his next letter, said that he 'hoped soon to travel in a fiery chariot of your invention'. (15) The rotary engine clearly captured Boulton's interest and it may have been this which prompted him to discuss a partnership with Watt and 'express a desire to be concerned in my fire-engine'. (16)

Watt returned to Glasgow full of enthusiasm about forming a partnership and spoke about it immediately to Roebuck who was willing to discuss terms so Watt explained the situation to Boulton in October 1768.

> The case is this:- By several unsuccessful projects and expensive experiments I had involved myself in a considerable debt before I had brought the theory of the fire-engine to its present state. About three years ago, a gentleman who was concerned with me died. As I had at that time conceived a very clear idea of my present improvements, and had even made some trial of them, though not so satisfactory as has been done since, Doctor Roebuck agreed to take my debts upon him, and to lay out whatever more money was necessary either for experiments or securing the invention; for which cause I made over to him two-thirds of the property of the invention. The debts and expenses are now about 1200£. I have been since that time employed in constructing several working fire-engines on the common principles, as well as in trying experiments to verify the theory. As the Doctor, from his engagements at Bo'ness, and other business, cannot pay much attention to the executive part of this, the greatest part of it must devolve on me, who am, from my natural inactivity and want of health and resolution, incapable of it. It gave me great joy when you seemed to think so favourably of our scheme, as to wish to engage in it; I therefore made it my business, as soon as I got home, to wait on the Doctor and propose you as one I wished he would make an offer to, which he agreed to with a great deal of pleasure, and will write you in a few days, that if agreeable you may

> be a third part concerned, on paying the half of the cost and whatever you may think the risk he has run deserves, which last he leaves to yourself. If you should not choose to engage on these terms, we will make you an offer when the whole is more perfect, which I hope it will soon be. (17)

Watt also expressed the hope that Small might join them as well. Sometimes Watt could be rather economical with the truth for we see here that he made no mention of how he had incurred most of these debts, not with his steam engine but with his rather unsuccessful shop. His debts were around £1,000 and expenses on the steam engine £200. (18)

For some reason, Roebuck did not send his offer until the middle of December and, when it came, it must have surprised Boulton because it was quite different from the terms suggested by Watt.

> Mr. Watt wrote to you sometime agoe about his New Plan of an Engine. It is some Years since a tryal of a very small one was made at Kinneil which convinced me of the justness of the Principals on which it was constructed and encouraged me to take a Burthen of Mr. Watt's hands which his loss of time and numerous expensive experiments had incurred - In a week or two we shall make a more compleat tryal with a larger Engine and if the result is answerable to our expectations I shall with pleasure communicate a share of the property to yourself in the three Counties in your neighbourhood Warwickshire, Stafford^{s}. & Derby^{r}. and the terms you and I shall not differ about. (19)

The first response from Birmingham was written by Small, if our dating of his January 1768/9 letter is correct. He had not realised Watt's connection with Roebuck and had hoped that Watt would move to Birmingham where he and Boulton could assist him. While he was willing to employ any sum of money in a partnership which he liked, he felt he was not well enough acquainted with Roebuck to know if they would suit each other. More significantly, he finished his letter with, 'Whether it would be possible to manage the wheel and reciprocating engines by separate partnerships without their interfering, I am not certain; but if it is, Boulton and I would engage *with you* in either, *provided you will live here*'. (20) This shows the lines on which Boulton and Small were thinking about the development of Watt's rotary engines.

Boulton excused his slowness in replying to Watt's letter of 20 October and Roebuck's of 12 December by explaining that he had been travelling all over the country so he could not write his famous answer until 7 February 1769.

> I note what you say in respect to your connection with Dr. Roebuck, from whom I received a letter, dated the 12th December, offering me a share of his property in your engine, as far as respects the counties of Warwick, Stafford, and Derby. I am obliged to you and him for thinking of me as a partner in any degree; but the plan proposed to me is so very different from that which I had conceived at the time I talked with you upon that subject, that I cannot think it a proper one for me to meddle with, as I do not intend turning engineer. I was excited by two motives to offer you my assistance - which were, love of you, and love of a money-getting, ingenious project. I presumed that your engine would require money, very accurate workmanship, and extensive correspondence, to make it turn out to the best advantage; and that the best means of keeping up the reputation, and doing the invention justice, would be to keep the executive part out of the hands of the multitude of empirical engineers, who, from ignorance, want of experience, and want of necessary convenience, would be very liable to produce bad and inac-

> curate workmanship; all which deficiencies would affect the reputation of the invention. To remedy which, and to produce the most profit, my idea was to settle a manufactory near to my own, by the side of our canal, where I would erect all the conveniences necessary for the completion of engines, and from which manufactory we would serve all the world with engines of all sizes. By these means and your assistance we could engage and instruct some excellent workman, who (with more excellent tools than would be worth any man's while to procure for one single engine) could execute the invention 20 per cent cheaper than it would be otherwise executed, and with as great a difference of accuracy as there is between the blacksmith and the mathematical-instrument-maker. It would not be worth my while to make for three counties only; but I find it very well worth my while to make for all the world.
>
> What led me to drop the hint I did to you was the possessing an idea that you wanted a midwife to ease you of your burthen, and to introduce your brat into the world, which I should not have thought of if I had known of your pre-engagement; but as I am determined never to embark in any trade that I have not the inspection of myself, and as my engagements here will not permit me to attend any business in Scotland, and as the Doctor's engagements in Scotland will not, I presume, permit his attendance here, and as I am almost saturated with undertakings, I think I must conclude to -. No, you shall draw the conclusion... Yet I live in hopes that you or I may hit upon some scheme or other that may associate us in this part of the world...
>
> Query - is the wheel-engine invention also included in your transfer to the Doctor, or can it be separated from it with propriety ? I hope it may not be long before you make this place in your way to London; and pray remember, when you come, always to put up at the Hotel d'Amitié sur Handsworth Heath, where you will always find a friendly reception. (21)

Once again, we must note the interest in Birmingham in Watt's rotary engines. In the event, it was the sons of Boulton and Watt who established Soho Foundry for producing complete engines in 1795. The fathers remained essentially consulting engineers, as indeed Watt had become while in Scotland. Boulton's visionary concept of an engineering works where complete engines could be prepared, situated by the side of a canal for ease of transport to destinations throughout the world, makes more sense if both types of Watt's engines were being constructed there. Special machine tools for quantity production of parts would be economical only with large scale output.

Failure to Form a Partnership

If Boulton found the partnership terms very different from those Watt had outlined, this may have been through Roebuck becoming less enamoured of such a scheme. Jardine gave Watt a hint of this when he wrote, 'Understand, that the more he [Roebuck] is convinced of the practicability of the scheme, the keener he is of carrying it to practice yourselves for your mutual advantage'. (22) Progress with trials on models during the spring of 1769 may have made Roebuck even more reluctant to reduce his interest in what at last had prospects of proving to be a profitable investment. Watt told Small that April that he wished,

> Mr. Boulton and you had entered into some negotiation with the Doctor about coming in [as] partners. I am afraid it is now too late; for the nearer it approaches to certainty, he grows the more tenacious of it. For my part, I still think as I did, that it would be for our mutual advantage. His expectations are solely from the reciprocator; possibly he may be tempted to part with the half of the circulator to you. This I say of myself. Mr.

> Boulton asked if the circulator was contrived since our agreement; it was, but it is a part of the scheme, and virtually included in it. (23)

But, in this same letter, Watt wrote seriously to Small because he realised that, if the reciprocator were successful, then he would not have time to develop the rotary one as well. He told Small,

> You want a steam-chaise; pray make one. I give you leave, and will also give you advice... Seriously, you will oblige me if you will negotiate the following affair: I find that if the engine succeeds, my whole time will be taken up in planning and erecting the reciprocating engines, and the circulator must stand still, unless I do what I have done too often, neglect certainty for hope. Now Mr. Boulton wants one or more for his own use. If he will make a model of it, of 20 inches diameter, at least, I will give him my advice and as much assistance as I can. He shall have liberty to erect one of any size for his own use, and if he should choose more, the terms will be easy, and I shall consider myself as much obliged to him.. If it should answer, and he should not think himself repaid for his trouble by the use of it, let me know, and I will propose it to the Doctor, and doubt not of his consent. (24)

Small consulted with Boulton about this proposal and replied in the middle of May that they had 'sufficient confidence in the wheel-engine to undertake it' (25) and that they were both willing to assist in making models of it. However he was afraid that the two types might compete with each other and yet still felt that they ought to be pursued together in a single manufactory. This would mean that the partners would need to live near each other and proper agreements would have to be drawn up.

By this time, construction had started on the final engine at Kinneil which could have made Roebuck even more reluctant to enter into any agreement that might lower his profits. Watt hoped that negotiations would not be broken off at least until he had the pleasure of seeing Small which he hoped would be soon after the completion of this engine. (26) Roebuck met Boulton that July but we have no record of their discussions. (27) However, Roebuck's attitude must have changed after the disastrous trial in September when it became obvious that the engine would need a great deal more expensive development. Watt's letter to Small of 20 September 1769 revealed what had happened so they could

> ... judge how far it may be your interest to engage in it on its own account. As to the Doctor, he has been to me a most sincere generous friend, and is a truly worthy man. As for myself, I shall say nothing, but that if you three can agree among yourselves, you may appoint me what share you please, and shall find me willing to do my best to advance the good of the whole; or, if this should not succeed, to do any other thing I can to make you all amends, only reserving to myself the liberty of grumbling when I am in an ill-humour. (28)

In his depression and desperation, Watt suggested to Roebuck that they make an agreement with Boulton even if it were hard on themselves because,

> 1st. From Mr. Boulton's own character as an ingenious, honest, and rich man. You know him much better than I do, but the worst I have ever heard of him amounted to his being a projector. 2ndly, From the difficulty and expense there would be of procuring accurate and honest workmen, and providing them in proper utensils, and getting a proper

> overseer or overseers. If, to avoid this inconvenience, you were to contract for the work to be done by a master-workman, you must give up a great share of the profit. 3rdly. The success of the engine is yet far from being verified. If Mr. Boulton takes his chance of success from the account I shall write Dr. Small, and pays you any adequate share of the money laid out, it lessens your risk, and in a greater proportion than I think it will lessen your profits. 4thly. The assistance of Mr. Boulton's and Dr. Small's ingenuity (if the latter engage in it) in improving and perfecting the machine may be very considerable, and may enable us to get the better of difficulties that might otherwise damn it. Lastly. Consider my uncertain health, my irresolute and inactive disposition, my inability to bargain and struggle for my own with mankind; all which disqualify me for any great undertaking. On our side, consider the first outlay and interest; the patent; the present engine, about 200£. (although there would not be much loss in making it into a common engine); two years of my time, and the expense of models. (29)

Roebuck met Boulton but once again we have no record of their discussions. Small declined to participate in any partnership because he had just tied up all his money elsewhere. (30) Boulton on the other hand took a while to come to his decision so that it was not until 2 November that Roebuck junior told Watt the joyful news that 'Boulton has accepted of a share in the engine'. (31) Accordingly Roebuck himself sent to Small and Boulton terms of an agreement on 28 November 1769.

> Whereas Mr. James Watt has assigned to me two-thirds of the patent of the steam-engine, which he took out some time in the course of the last winter, I hereby offer you one-half of the above two-thirds, or one third of the whole patent, on condition that you pay to me such a sum, not less than one thousand pounds, as you, after the experiments of the engine shall be completed, shall think just and reasonable; and twelve months from this date you are to take your final resolution. I also oblige myself to procure Mr. Watt's assent to this agreement'. (32)

Small wrote a couple of days later in haste to say that he and Boulton had agreed with Roebuck. (33) Watt responded a few days afterwards, 'I received yours, and shake hands with you and Mr. Boulton on our connection, which, I hope, will prove agreeable to us all. I have just got the new alterations of the engine ready to carry to Kinneil'. (34) Perhaps trials recommenced on the Kinneil engine in anticipation of financial support from Boulton and Small.

Small wished Watt that 'with all your attention you may be able to succeed perfectly in this engine for Dr. R. without a new cylinder' but at the same time he said that he was 'not very anxious about reciprocating engines since I have considered more carefully the circulating'. (35) Roebuck told Boulton in February 1770 that he had great confidence that 'the present Engine will work completely' (36) and a little later in March, 'I now consider the affair of the Engine as absolutely certain, though it may still take some time to complete it to the perfection it is capable of'. (37) We do not know just how close Watt may have been to success during that spring of 1770 but it seems that he was at last beginning to operate his condenser in a better way. If that is true, he must have felt very frustrated when he had to abandon trials on 13 June 1770 due to Roebuck's first bankruptcy because his engine at Kinneil was never to run again in Scotland, fulfilling Roebuck's prophecy that it would take some time to bring it to its ultimate perfection. The collapse of Roebuck's affairs meant that neither Boulton nor Small exercised their right to take up any share in Roebuck's offer. In the meantime we must return to Small's interest in the 'circulating' one.

NOTES: PROPOSED PARTNERSHIP WITH BOULTON

1 Dickinson, (1936), for the most recent biography. See also Smiles, (1878).
2 King-Hele, (1963), p. 15.
3 Schofield, (1963), pp. 112 - 113.
4 Dickinson & Jenkins, (1927), p. 26 and Dickinson, (1936), p. 48.
5 Dickinson, (1936), p. 203.
6 Ibid, p. 208.
7 *Transactions of the Newcomen Society*, Vol 58, 1986 - 1987, R.L. Hills, 'A Steam Chimera, A Review of the History of the Savery Engine', pp. 27 - 44.
8 Dickinson, (1936), p. 76, M. Boulton to B. Franklin, 22 Feb. 1766.
9 King-Hele, (1981), p. 38, Letter 66B, E. Darwin to M. Boulton, 11 March, 1766.
10 Ibid, p. 27, Letter 64E, E. Darwin to M. Boulton, [1764?].
11 Patent 913, 5 Jan. 1769.
12 Muirhead, (1854), Vol I, p. 16, Letter 21, W. Small to J. Watt, 7 Jan. 1768; Boulton 340.1.
13 Ibid, Vol I, p. 28, Letter 32, W. Small to J. Watt, 12 Aug. 1768; Boulton 340.2.
14 Dickinson, (1936), p. 203.
15 Muirhead, (1854), Vol I, p. 29, Letter 33, W. Small to J. Watt, Monday, no date, Sept. 1768; Boulton 340.3.
16 Ibid, Vol I, p. 30, Letter 34, J. Watt to M. Boulton, 20 Oct. 1768; Boulton 348.1.
17 Ibid.
18 Ibid, Vol I, p. 74, Letter 63, J. Watt to J. Roebuck, 24 Sept 1769; JWP, 4/32.15.
19 Boulton 334.5, J. Roebuck to M. Boulton, 12 Dec. 1768.
20 Muirhead, (1854), Vol I, p. 17, Letter 21, W. Small to J. Watt, 7 Jan. 1768; Boulton 340.1.
21 Ibid, Vol I, pp. 41 - 43, Letter 43, M. Boulton to J. Watt, 7 Feb. 1769; JWP, 6/19.23.
22 Ibid, Vol I, p. 26, Letter 31, G. Jardine to J. Watt, Monday, no date, 1768; JWP, 4/57.4.
23 Ibid, Vol I, pp. 55 - 56, Letter 52, J. Watt to W. Small, 18 April 1769; Boulton 340.5.
24 Ibid.
25 Ibid, Vol I, p. 59, Letter 56, W. Small to J. Watt, 16 May 1769; Boulton 340.6.
26 Ibid, Vol I, p. 64, Letter 59, J. Watt to W. Small, 5 July 1769; Boulton 348.10.
27 Ibid, Vol I, p. 65, Letter 60, W. Small to J. Watt, 26 July 1769; Boulton 340.7.
28 Ibid, Vol I, p. 72, Letter 61, J. Watt to W. Small, 20 Sept. 1769; Boulton 348.11.
29 Ibid, Vol I, p. 74, Letter 62, J. Watt to J. Roebuck, 24 Sept. 1769; JWP, 4/32.15.
30 Ibid, Vol I, p. 76, Letter 63, W. Small to J. Watt, 10 Oct. 1769; Boulton 340.8.
31 Ibid, Vol I, p. 77, Letter 65, J. Roebuck jun. to J. Watt, 2 Nov. 1769; Boulton 4/57.13.
32 Ibid, Vol I, p. 82, Letter 68, J. Roebuck to W. Small and M. Boulton, 28 Nov. 1769; JWP, 6/44.9.
33 Ibid, Vol I, p. 82, Letter 69, W. Small to J. Watt, 30 Nov. 1769; Boulton 340.10.
34 Ibid, Vol I, p. 83, Letter 70, J. Watt to W. Small, 12 Dec. 1769; Boulton 348.14.
35 Ibid, Vol I, p. 84, Letter 71, W. Small to J. Watt, 26 Dec. 1769; Boulton 340.11.
36 Boulton 334.9, J. Roebuck to M. Boulton, Feb. 1770.
37 Boulton 334.12, J. Roebuck to M. Boulton, 15 March 1770.

Rotary Steam Engines and the Steam-Wheel

Watt's First Attempts (1)

We must return to an earlier period in Watt's life to trace the development of his rotary engines and the steam-wheel. The latter proved to be one of the crucial links which helped to deepen the friendship between Watt, Boulton and Small and indeed was one of the main reasons for Watt's moving to Birmingham in 1774 in order to carry out trials on the one Boulton and Small had constructed there. Elijah Galloway, in his *History and Progress of the Steam Engine*, concluded his account of Watt's rotary engine or steam-wheel with the words, 'Such was the plan designed as the first rotatory engine. It is because it *was* the first, and because it was the invention of Watt, that we place it here'. (2) Galloway's claim for Watt's priority is probably correct if we exclude toys such as Hero of Alexandria's 'Aeolipile' invented around AD 100 or Giovanni de Branca's crude impulse turbine of 1629. Marten Triewald was unsuccessful a little before 1728 when trying to drive hoisting machinery with the Newcomen engine he had installed at the Dannemora Mines in Sweden. (3) Otherwise Savery or Newcomen engines only raised water for a waterwheel and produced rotary motion in this indirect way.

The first indication that Watt was thinking about obtaining a direct rotary motion from a steam engine can be found in a letter to Roebuck dated 19 February 1766. 'I have thought on a simpler circular steam-engine than what I mentioned to you, and which I expect will be practicable'. (4) Ten days later, Watt had 'set about the circular engine, and I expect to have it done in about a week'. (5) As usual, Watt was over-optimistic for he found that his circular engine took a great deal more work than he expected since it was not finished by the middle of May. (6) We are given no indication of the designs for these first rotary engines and no more references to them have been found during the next eighteen months. However, later evidence and drawings show that Watt invented two direct rotary engines, a semi-rotary engine and what became called his steam-wheel. The designs of the two rotaries and the semi-rotary will be examined last because our knowledge of them is principally obtained from Watt's patent of 1782 and correspondence at that time. (7)

The Steam-Wheel

Watt must have thought that his steam-wheel showed the most promise and he finished some trials before the end of January 1768 when he wrote to Roebuck that he had,

> not yet got the little wheel completed. It was, as I suspected, a quantity of the putty that had got loose, and, in the form of little balls, hindered the valves from shutting. On consideration I found that it would be very difficult to guard against that inconvenience if putty was used in the model; I therefore endeavoured to use some other substance, such as paper, cork, etc., that would remain where they were put; but from the complication of the grooves I could not make it tight in that way. I have therefore made some considerable alterations on it which will allow me to use sulphur for making it tight, and shall have a trial of it this night or to-morrow forenoon (8)

Watt had to seal not only flap valves but also rotating joints where he may have used the putty. He had made a model of wood, 9 ins. diameter, which he turned round by blowing air into the steam pipe. (9) This model was non-condensing.

Its design must have been similar to that shown in the drawing prepared for submission with his 1769 patent, judging by a description he wrote.

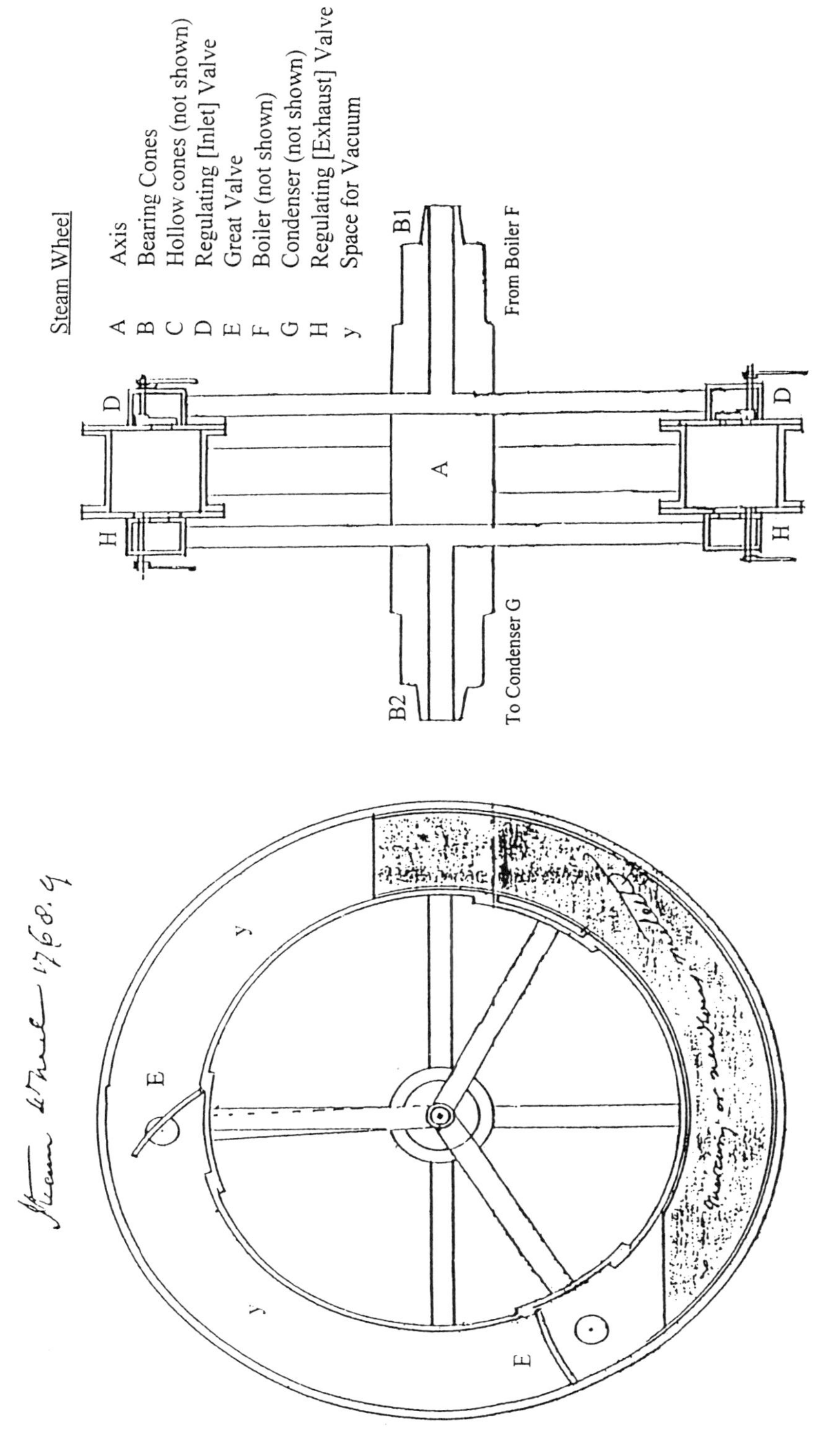

The drawing proposed for the 1769 patent of Watt's steam-wheel has been annotated with identification letters.

> A Wheel is to be made with a hollow channel (*close on all sides*) round its circumference - The size of this wheel and the channel round it are regulated by the power required - In this channel at proper Intervals are placed a number of valves not fewer than three which will allow any body to pass round the wheel in one direction but will allow nothing to go the contrary way. The axis [A] of this wheel is hollow for some space at each end and terminates in cones [B] which are fitted into hollow Cones [C]. The one [B1] communicating with the boiler [F] and the other [B2] with the Condenser [G] - From the hollow part of the axis on each side of the wheel proceed as many Tubes as there are Valves in the Channel round the wheel the other end of each of these tubes is inserted into this Channel near the Valves and is there shut by a regulating Valve [D]. A quantity of Quicksilver or oil or any other fluid which does not boil so easily as water must be poured into the Channel and the external air being now cut of[f] and the Channel exhausted of air a Regulating Valve [D] is to be opened to allow the Steam to enter betwixt a great Valve [E] and the surface of the fluid this Valve being shut the steam will press equaly on it and the surface of the fluid in the Channel the fluid will retreat from the Valve and rise on the other side of the wheel till the excess of its weight on that side is equal to the power required as the wheel will turn round in the Opposite direction more Steam will enter and support the fluid in the Channel to the height to which it was raised, another great Valve [E] will emerge from the fluid and be shut, another regulating Valve [D] is at the same time opened the first regulating Valve must now be shut the corresponding great Valve [E] which will now be in the position [x] and a regulating Valve [H] communicating with the Condenser opened the Steam contained in the space [y] will rush into the condenser and leave that part of the Channel exhausted - The wheel will continue to go round and the same operation will be repeated on each of the Valves in Succession. (10)

Watt claimed that three valves and three sections gave a steady turning motion. At this period, he seems to have preferred a solid weight sealed in the rim by mercury or some form of packing.

Links with Birmingham

Watt's ideas about rotary engines must have been known to his immediate circle of friends because Robison wrote to him from Watford in July 1768 after meeting Darwin, Small and Watt's 'very warm friend Mr. Boulton'. Boulton wished to urge Watt on with his experiments because he needed a steam engine for his Soho Manufactory. Robison warned Watt to waste no time,

> This he [Boulton] further says is the more necessary as one Edgeworth has hit upon a thought pretty similar to yours, and has also conceived a circular Engine on your principles, and has even thought of employing [mercury] as a preventative of friction in the same manner as yours. (11)

Edgeworth never patented his engine but could have been a dangerous rival. Watt met Boulton soon afterwards and presumably took the opportunity to give glowing accounts of both his reciprocating and rotary engines. As far as Boulton and Small were concerned at this period, it was most likely to have been Watt's rotary engines or steam-wheel which roused their enthusiasm. (12) In addition to driving vehicles, other large markets were envisaged for steam-wheels to propel canal barges on the Birmingham canals and drive sugar mills in the West Indies. No wonder Boulton and Small were excited by the prospects. Indeed,

Watt's steam-wheel may have played a role of crucial importance in cementing the relationship between himself and Boulton which would eventually lead to the successful development of the reciprocating engine while ironically the rotaries and steam-wheel failed.

In the ensuing correspondence between Watt, Roebuck, Boulton and Small over their joining the partnership for the 1769 patent, the steam-wheel was frequently mentioned. Boulton and Small wished to see if they could enter into a separate agreement for this engine alone. (13) They also enquired whether the two might be built in the same factory and this was when Boulton outlined his idea for an engine works situated by the side of his canal. (14) His vision for supplying steam engines to all the world makes greater sense if the rotaries and steam-wheel are included with the reciprocators. This connection was made by Small in a letter three months later. (15)

Watt despatched drawings as well as a description of his steam-wheel in the draft (16) for his patent to Small at the end of January, admitting that 'The description of the wheel is very defective'. (17) Small's suggestion for the part concerning the steam-wheel was much more theoretical than Watt's and concentrated on the principles. (18) The final specification described the steam-wheel in much more detail than the reciprocating engine but the rotaries were not included. In the patent text, we see Watt struggling to find terms explaining his new forms of steam engine which had never been written down before. (19) On the drawing that was never submitted by Watt, (20) the weight is described as 'mercury or Newton's metal' which Watt specified as 'equal parts tin, lead, bismuth, 1/9 mercury'. (21) The modern composition is 50% bismuth, 31.2% lead, 18.8% tin giving a melting point of 98^0 Celsius.

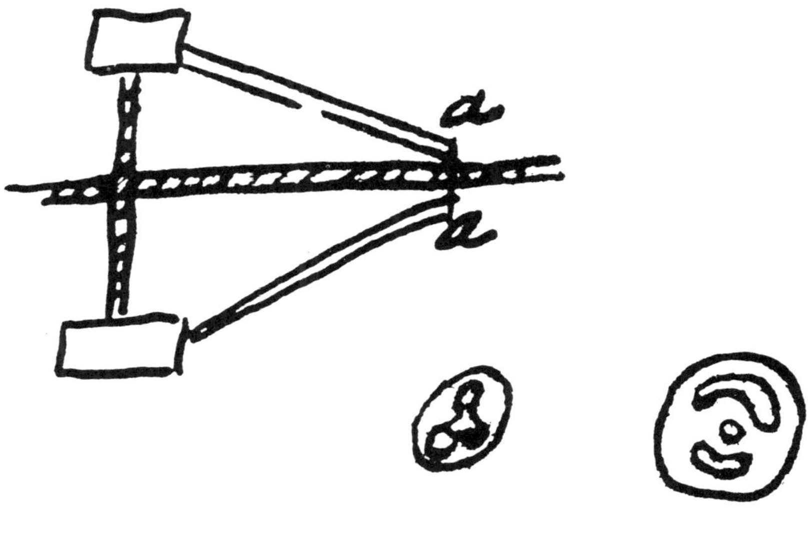

Small's sketch of his method of steam admission and exhaust through pipes and a flat rotary valve uncovering ports.

Small's interest in the steam-wheel is shown by his next letter to Watt which was mostly about copal varnish but included a passage which suggests that he had tried a model. (See Illus above)

> Nothing in the wheel has puzzled me more than has exclusion of the fluid-piston from the steam pipes. And at last the method I have been obliged to take up with is not sufficiently simple. The rest is simple & easy to be executed, requiring no cocks, nor any apparatus whatever, excepting three internal valves'. (22)

Small's suggestion for a simpler wheel consisted of one steam pipe acting as both inlet and exhaust, terminating in a hole in a central face-plate. This plate would be ground true to a stationary plate with a small steam inlet hole and an elongated exhaust hole. This would avoid employing Watt's valves at the rim although it would mean that the steam in the pipes would be wasted. (23) Small had experimented with both mercury and Newton's metal and thought he could avoid their 'calcination'. He also doubted whether the wheel would rotate very fast. Here he seems to have shown a greater understanding of the problems than Watt.

Watt had found trouble with the liquid metal escaping into the steam pipes. Presumably the pressure of the steam ought never to be great enough to raise the level of this liquid weight above the centre of the wheel, or the exit point of the steam. But the sudden admission of steam might have driven the metal up a pipe and its inertia carried it into the outlet. Watt had developed a form of mercury trap in the shape of a hollow ball in which the inlet pipe would protrude from the bottom partly into the ball, leaving a space at the bottom where the mercury could be collected, while the outlet was at the top. He also tried spiral arms. (24)

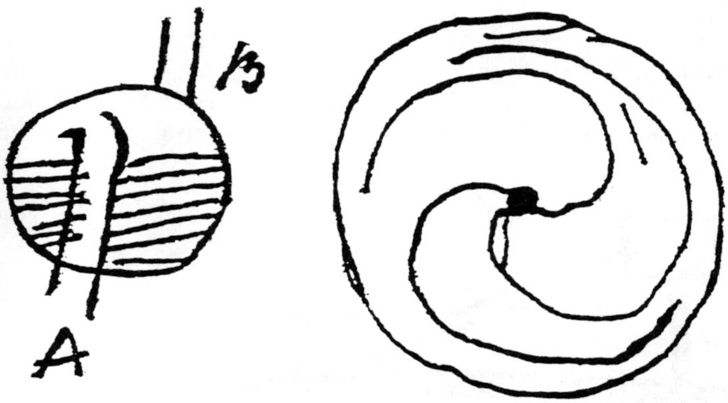

(A) Watt's suggestion for a mercury trap with steam entering through 'B' and out to the condenser at 'A'.

(B) Watt thought spiral pipes might prevent mercury escaping.

In April 1769, Watt received a letter from Small mainly about the text of the patent but also warning Watt of another source of competition.

> A linen-draper at London, one Moore, has taken out a patent for moving wheel-carriages by steam. This comes of thy delays. I dare say he has heard of your inventions. Do come to England with all possible speed. At this moment how I could scold you for negligence! However if you will come hither soon, I will promise to be very civil and buy a steam-chaise of you and not of Moore. (25)

Watt, who, as we have seen, intended that his steam-wheel could have driven chaises, was suspicious that Hately may have told Moore something about his designs. (26) Small thought that Watt, with his patent already lodged, ought to ask Moore for a large sum, £12,000, not the £1,200 which he must have said his engine had cost, because, if wheeled carriages and boats could be moved by steam, then Watt's invention was worth this sort of money. (27)

It may have been this which made Watt realise that he could not develop both his reciprocating engine and steam-wheel together. Accordingly he wrote to Small at the end of April, 'You want a steam-chaise; pray make one', (28) also suggesting that he should help Boulton to make a model 20 ins. in diameter. Small and Boulton had sufficient confidence in Watt's designs to proceed but needed to settle what form any partnership would take with Roebuck. (29) Watt realised that his reciprocating and rotary engines would be applied to different purposes so that the two would not be in competition, although he would have liked their development to be undertaken together. (30)

Construction at Birmingham

Small wrote to Watt in November 1769 that he had thought of an easy and obvious way of moving carriages by a reciprocating engine but gave no indication how. (31) It was not until after Christmas that the steam-wheel was mentioned again, during which period Small must have been reconsidering it and wrote,

> I am not very anxious about reciprocating engines since I have considered more carefully the circulating; one of which, equal to four horses, I can get made for fifteen guineas or less. Dr. R. is still in London, Pray let us lose as little time as possible after his return to you, for fourteen years will soon be gone. (32)

Some provisional agreement must have been reached quickly because on 23 January 1770, Small told Watt that a carpenter was making a pattern for a wheel to be cast at a price of £15. He also thought that a liquid weight would be preferable to a solid one because the valves could be made less accurately. (33) This pattern was to be sent off to Coalbrookdale on 1 March. (34) Watt agreed, 'Now for the exclusion of solid pistons', so that mercury was used after that. (35) Watt also outlined his method of making the flap valves steam tight. This was to face them with 'broad-cloth' soaked in linseed oil or shoe varnish, dried in a stove, and then painted with white lead. This cloth would be mounted on the surface of a metal plate thin enough to flex and shape itself to the valve seat. (36) Other joints would be made with pasteboard. Watt suggested shutting the valves by springs but he never seems to have considered the effects of centrifugal force on them if his wheel rotated at any speed. This might have acted against the spring pressure and, depending upon how the valve was hung, prevented it from either opening or closing. The hammering of the valve on its seat would have quickly destroyed a cloth facing which Boulton and Small later replaced with metal surfaces. It is difficult to see how such valves would have lasted long with either a solid weight or the mercury constantly forcing them open.

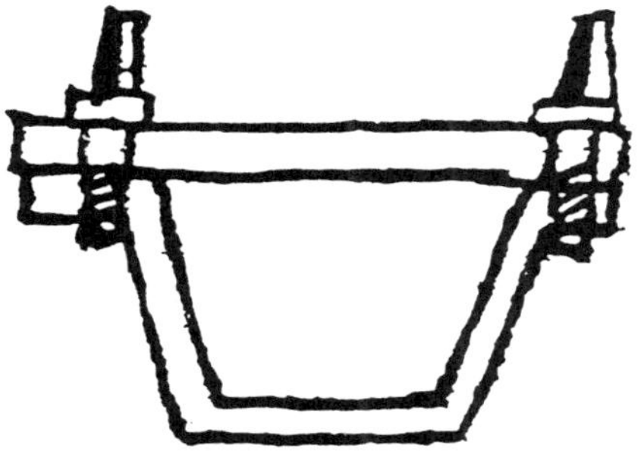

Section of Groove.

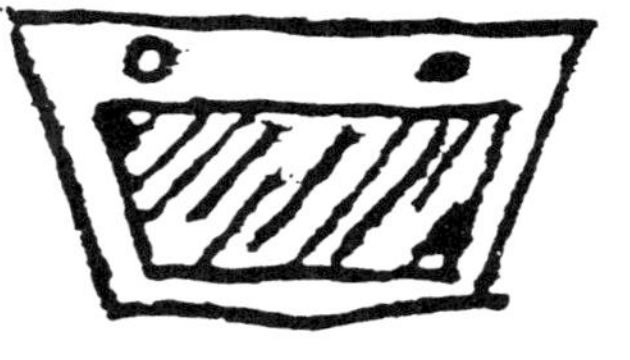

Valve frame.

Flap valve and frame for the main valves in the steam-wheel.

Through starting his Scottish career as civil engineer in earnest, Watt made little progress on the steam-wheel and the scene shifted to Birmingham. Small told him in the middle of September 1770 that their interest 'has continued, but has been ineffectual, from most ridiculous causes'. (37) He had written many letters urging Coalbrookdale to send the castings which, when they finally arrived, were unsound and totally useless. Moreover the blemishes had been covered over with something to conceal them. However they had found another caster at Bilston who promised the parts in a few days. Boulton and Small had decided to make quite a large model since they had hopes of moving canal boats by a non-condensing form of this engine. This was owing to Watt's greater experience in operating a condenser and in any case coals were cheap at Birmingham. They saw a broad field of operation because 150 boats were employed 'in these new waveless canals' (38) where the coal for the boats would be free. (39) Therefore they wished to settle the partnership arrangements.

Small's letter prompted a renewal of their correspondence with Watt replying on the last day of September 1770. He recognised that, now Small was beginning to make experiments with models, he would understand better some of the time and pains that he, Watt, had spent on his. He wished to share with Small some of his observations to spare Small trouble, but thought that Small would become master of them sooner than himself. A large part of this letter went into greater detail about his method of making the valves with cloth facing which he was uncertain whether he had explained to Small earlier. He also mentioned the problem of condensation and suggested a cock to drain water off from the top of the mercury. Once the engine was running, he imagined there would not be much condensate. Without a condenser, they would need a boiler to generate steam with a pressure of 30 ft. of water (13 p.s.i.) which would give about half the effect of a condenser. At this period, Watt was willing to contemplate the use of steam at much higher pressures than he was later. Also this was when Watt suggested his 'spiral oar' which has been seen as his invention of the screw propeller but in fact both he and Small soon found this idea was useless and quickly abandoned it. (40) Watt tried to find scientific explanations about the speed which his engine might attain and suggested that Small might apply calculus to solve the problem, 'put your pen to it...' (41)

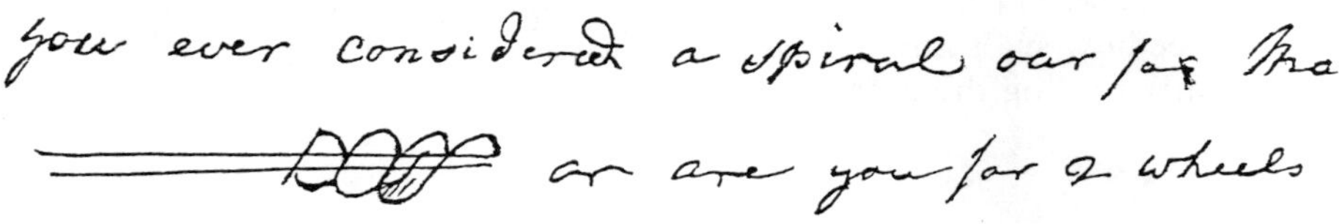

Watt's sketch of his 'spiral oar'.

Small responded with another long letter in which he outlined some of the experiments he had conducted to determine the performance of the steam-wheel. He had tried the fluid metal for the weight, presumably Newton's metal, in a phial to see the effects of steam on it. He had tried models of spiral oars and found them ineffective. He suggested making the engine of large size to prevent the mercury entering the steam pipes. He recommended heating the wheel and then pouring tallow into it, presumably before starting, in which way he hoped to prevent collection of water, calcination of the metal, as well as getting the valves to fit better. For his boiler, he suggested using a dozen retorts, or, if they were insufficient, a couple of dozen, in a reverberatory furnace, with one safety valve for all. He still came to the same conclusion about the speed of the wheel, that it would not rotate very fast. (42) Watt tried to reassure Small about the use of cloth and asked him how he was going to refill the retorts. Instead he recommended a boiler with a hammered iron bottom and a copper top with plenty of space for steam. He asked Small to let him know of any further problems. (43)

When Small wrote on 3 February 1771, he reported, 'I am ashamed to tell you that the wheel-engine is not yet finished, from the idleness and folly of workmen. I have daily a better and better opinion of it, though I still think it cannot go fast, chiefly from effects of centrifugal force'. (44) The hollow rim was cast in sections. Presumably to make it stronger, one end of each section was cast with a spigot to fit into the next on which the flap valve was fitted. The workmen found it difficult to fit the valves on the ends of these spigots and Small proposed to cut them off. Silence descended for seven months until the middle of October 1771, when Small reported that everything had been ready for over two months except the millwright's work who had caused continual disappointments. He and Boulton had been lucky to employ a clockmaker who had ground the valves flat with great care to fit properly. Quicksilver or mercury had been purchased for the weight because this would retain its resale value should the machine not succeed. The present delay was caused by Boulton being so ill with a fever for five weeks that Small had not even shown him Watt's last two letters. (45)

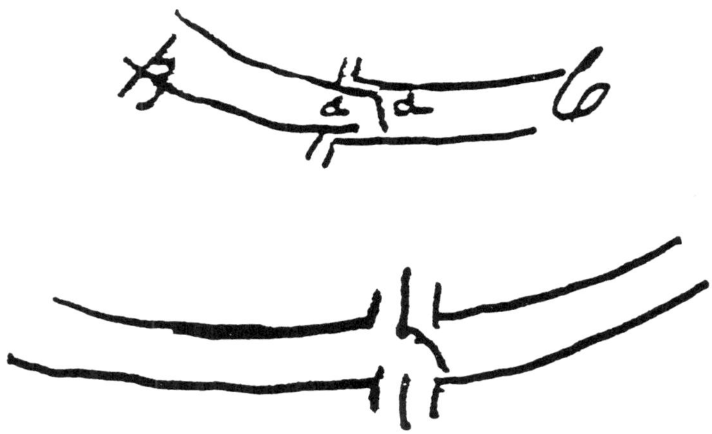

Small's explanation of his proposal for cutting off the spigots on the ends of the rim sections of the steam-wheel to give easier access to the valves.

Two months later, the engine was still not ready although Small hoped that it would be in a couple of weeks for a trial in 'the forcing way', without the condenser. Where coals were cheap, such as around their area, simplicity of construction would be appreciated while a condenser could always be added later. He continued about practical problems such as rust or the calx of mercury preventing the valves from shutting. (46) This letter must have crossed with one from Watt enquiring after Boulton's health and progress on the engine. (47) Possibly it was never sent for Watt was to repeat most of its contents in another letter on Christmas Eve 1771.

> I advise you to begin your trial of the engine with water in place of mercury, and by blowing into it with a bellows. You may thus find out and remedy some faults more easily than with mercury and steam. Place a tight wooden or iron vessel like a grind-stone-trough under your wheel whenever you begin with mercury, for some will un-avoidably spill at first. (48)

People at that time were completely unaware of the health hazards of mercury and used large quantities without taking any precautions.

At that point, work ceased again. In August 1772, Roebuck reported from London, that Boulton and Small were waiting for the mercury to arrive from London so they could try the engine. (49) Roebuck told Boulton to order three barrels of mercury from Messrs Grote Son & Co. which would weigh about 561 lbs. at a cost of 3s.10½d. per pound or about a total of £87. It could be resold at 3s.10d. per pound after use. (50) This shows the sort of expense which Small and Roebuck incurred on these experiments but no further figures have been found. All Watt could do was to beg Small 'to go on with the experiments as fast as convenient'. (51) In September, Roebuck thought Boulton and Small would be trying the circular engine and, if they operated it non-condensing, would need a pressure of around 30 p.s.i., much higher than any used with reciprocating engines. If necessary, a condenser could be added as in Watt's original plan. (52)

Probably the trial never took place because, in November 1772, Small explained that Boulton had lost money by the failure of the bankers, Neale, Fordyce and Downe. (53) While they had done everything in their power to hasten the trial, they were still waiting for a boiler from Wilkinson which he had promised solemnly would be ready the following week. Small thought Watt should come to Birmingham immediately when they were ready. (54) A month later, Small said, 'If the wheel engine should go, we intend to apply it instantly to a boat, and send it on a trading voyage along as many of the canals as now communicate'. (55) It was not until the end of January 1773, that he reported,

> I have lived to see the discovery of the longitude; the perfecting of the steam engine; and indeed everything but the perpetual motion, the stone of the philosophers, and Spring eternal!
>
> At last your boiler is ready, and we are adjusting everything for trial, which will be retarded by Mr. B.'s being obliged to pass several weeks in London, where he now is. (56)

It would be a long time before any trial actually took place.

Boulton and Small must have spent a large sum of money constructing their steam-wheel and were anxious for Watt to come and help them undertake trials with it. Some of the delays were caused by the discussions settling Roebuck's affairs and some caused by Watt's commitments to his surveying projects. Watt had sent parts of his reciprocating engine from Kinneil to Birmingham in May 1773 but Roebuck's settlement with his creditors had still not been concluded by August so that Small urged Watt, once agreement had been reached with Roebuck, 'If ... you are tolerably at leisure, pray come to us and complete the engines, for they are now both at Soho... The wheel engine might be tried in a very few days after your arrival, at least in the forcing way'. (57) Nothing further happened and no trials were made.

Smeaton visited Birmingham and saw the circular engine when he told Boulton that 'It would not do'. (58) Perhaps that was why Small told Watt in February 1774,

> We have tried no experiment with the circular engine. The engine is mounted, the boiler is ready, the piece that connects the boiler to the steam-pipes is also ready, but not ground to exclude air; lastly, the quicksilver is ready. The valves were ground very carefully, but whether they are steam-tight we know not.
>
> Probably nothing more will be done before you come, for neither Mr. B. nor I have leisure to attend the trials or the workmen; and to employ workmen on articles of this kind, without superintendence, we have experienced to be too expensive. Besides, you ought to see all the facts that may occur. (59)

At the same time in Scotland, Watt was carrying out further experiments to determine more accurately the temperature of steam both in a partial vacuum and at different temperatures above atmospheric pressure. (60) To this Small's response was, 'I am most sensible of the great importance of the course of experiments you are engaged in, both to the engine and to the improvement of Natural Philosophy'. (61)

Watt Moves to Birmingham

Watt reached Birmingham at the end of May 1774. (62) He must at first have concentrated on his reciprocating engine because he had been so near to success with it at Kinneil. However there were definitely some trials with the steam-wheel because his friend Hutton wrote to George Clerk, Commissioner of Customs, probably that July,

> I have a letter from Watt - he has brought his curious wheel to go, it works by steam - he says it appears to be right in all essentials and goes with an equable motion & great power. In short I believe it will answer (his own words) - This will raise his fame yonder it being so new a thing for that is what catches the multitude; tho I think the improvement of the reciprocating fire engine is the thing of greater utility but everyone will not be sensible of the merits of that invention. (63)

Roebuck asked Watt in November, 'When you have tried the fusible metal in your circular machine, I shall wish to be informed of the experiment'. (64)

In spite of Hutton's perceptive preference for the reciprocating engine, the potential of the steam-wheel may have seemed greater to the others involved for Watt told Boulton, 'I understand there will be an almost unlimited sale for wheel-engines to the West Indies, at the rate of 100£. for each horse's power'. (65) In one of Watt's notebooks, there is an entry given only the date 1775 about an enquiry from the West Indies from a Dr. Alex Wilson concerning sugar mills. Four mules were needed at a time to drive a mill. The mules had to be changed every hour and a half so that 24 animals were required. The initial cost was probably £600 with £200 annual running expenses. (66) This was probably the cause of Watt's calculating figures for a large circular engine.

> **Dimensions of Circular Engine.**
> 24 feet diameter Stroke 6 feet area of bore 12 inches power equal 52 in[ch] cylinder about 6 tun iron in the two sides 216 feet in the inner & outer Circle 144 feet will weigh about 4 tun

at £16	£	64. 0. 0
Screws		10. 0. 0
Valves		15. 0. 0
Regulators		6. 0. 0
Pipes		6. 0. 0
Axes		10. 0. 0
Plunger		8. 0. 0
Lead		84. 0. 0
Mercury		20. 0. 0
		213. 0. 0
Wood of wheel & supports £50.		(67)

The large size, 24 ft. in diameter, and the weight of over 10 tons should be noted because they show Watt's ambitions for his steam-wheel.

In the spring of 1775, application was made to Parliament for an Act to extend Watt's 1769 patent and it would seem that every effort was being made to get one or other engine running properly. Boulton noted Watt's comments and regretted, 'that we have not one of each species ready to work, as two working models would clear their way into the House of Commons better than blind words can do'. (68) Boulton carried on with the reciprocating engine while Watt was in London that spring and, with the new cast iron cylinder arriving from Wilkinson towards the end of April, that engine at last began to perform reasonably. (69) However, the steam-wheel was not abandoned completely for Boulton told Watt in May 1775, 'The pattern for the wheel-engine will be finished in a few hours'. (70)

Trials on the Steam-Wheel

The steam-wheel was of course included in the Act extending Watt's 1769 patent because it was mentioned in the full text of that patent which was ratified again. There is an eight-page printed leaflet outlining Watt's engines which is thought to have been prepared at this time and mentions the steam-wheel.

> The second species of the improved engines is a machine which produces a circulating or rotative motion by the immediate action of steam; it consists of a wheel placed vertically, which has a hollow channel round its circumference, as if a tube were bent into a circle and its two ends were joined together.

Watt then described how the steam-wheel worked and concluded,

> The advantages of this particular construction are evident. It gives the means of producing rotative motions of all kinds by steam, and by the equality of its velocity is better adopted to many purposes than those engines which have a reciprocating or alternative motion. In point of power gained from a given quantity of steam, it has all the advantages of the reciprocating species, as the steam vessel is kept uniformly hot, and the steam is condensed in a distinct vessel in the same manner as in the other. (71)

In another printed pamphlet which he circulated to gain support probably after the first hearing of his plea when there was great opposition, Watt claimed that he had completed in 1774 'working engines of each of the two species he has invented'. (72)

One of Watt's employees at Soho, Joseph Harrison, gave evidence,

> A. I have seen one of Mr. Watt's Wheel Engines at work - The wheel was 6 feet Diameter - worked by steam - it moved with great velocity.
> Q. What was the Engine employed in?
> A. It was erected for experiments to prove what could be done - myself and another man while it was at rest attempted to hold it - but as soon as the steam was let into it - it was out of our powers to hold it - the wheel weighed upwards of a Ton ([in pencil at side] This applies to the wheel engine, see evidence of Mr. S. Garbett below - remarkable as the first method of producing rotative motion by steam). (73)

Garbett was called to give evidence on 12 April and said that he had seen 'Mr. Watt's wheel engine' when he went out of curiosity. He saw that two men could not prevent its starting and that it could easily be connected to any other machine. However he had no knowledge of how it operated other than by taking steam from a small boiler. He made no mention of a condenser so it is probable none was fitted. (74)

Robert Mylne was also questioned about the steam-wheel but he most probably had never seen one. He was questioned whether one machine joined to another, as in the case of the steam-wheel moving a pump or other mechanism, ought to be considered a single machine, the fear being that a monopoly granted for the steam-wheel would cover the other as well. Mylne thought that the merit of Watt's invention lay in the application of steam to create a circular motion and that a monopoly, when granted, he did not think would be a sufficient reward for such a meritorious invention. He was asked whether quicksilver was the fluid body used and whether the steam had to be of the density of two atmospheres to raise the quicksilver 14 ins. but this he could not answer. (75) What these witnesses show is that some experiments were carried out on Watt's steam-wheel and that it did work to some extent.

Although priority must have been given to erecting the reciprocating engines as orders were received, development work continued on the steam-wheel. Boulton wrote to Watt at the end of February 1776 that, through lack of work in the smith's shop, he intended to put some men onto the steam-wheel. Through not knowing the specific gravity of the weight of metal for the weight or piston, he proposed running with 'weak steam' and suggested inserting bent pipes to prevent the fluid metal escaping along the steam pipes. This is the only indication we have of the problems encountered when running a steam-wheel. Boulton doubted whether the machine could be set to work by the end of the week as Watt hoped. (76) From the drawing, this modification seems unlikely to have succeeded and must point to one reason for the failure of the steam-wheel.

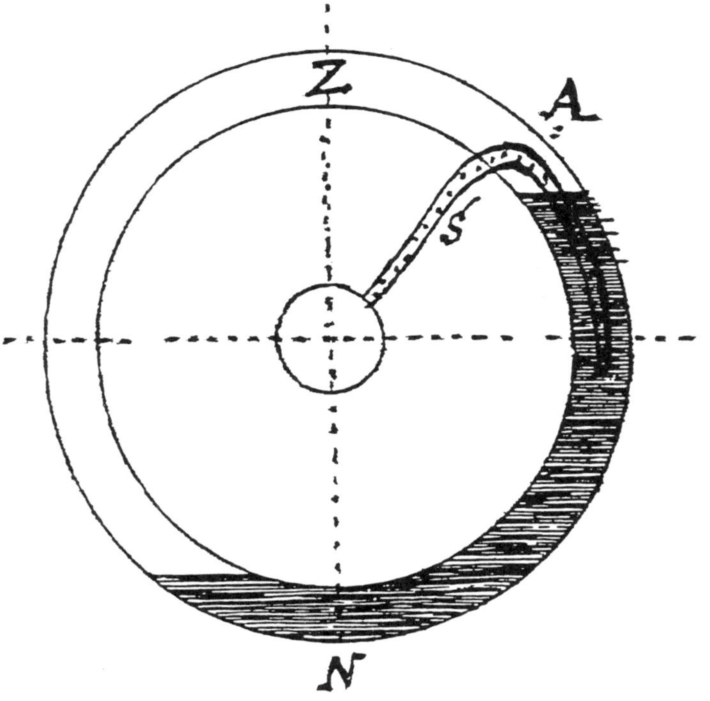

Boulton's suggestion for 'spiral arms'.

But presumably this alteration was tried for Boulton told Watt at the end of July 1776, 'The wheel-engine is ready for trial, except the steam-pipe; the boiler is set, and many wheels will be wanted so soon as they are ready for sale. I did not sleep last night, my mind being absorbed in STEAM'. (77) Boulton said in an undated letter to Watt but which has been placed in 1776,

> If we had a hundred wheels ready made, and a hundred small engines like Bow engine, and twenty large ones executed, we could readily dispose of them. Therefore let us make hay whilst the sun shines, and gather our barns full before the dark cloud of age lowers upon us, and before any more Tubal-Cains, or Watts, or Dr. Fausts, or Gainsboroughs, arise, with serpents like Moses', that devour all others. (78)

Boulton obviously saw great potential for Watt's steam-wheels and this may explain his anxiety a little later for Watt to develop a reciprocating rotative type when it had become evident that the steam-wheel was a failure. Boulton and Small must have spent a great deal of time and money on trying to develop the steam-wheel which shows how seriously they thought it a practical proposition.

Boulton and Watt received enquiries from people who wished to purchase a steam-wheel. The first was in June 1775 from Samuel Roe of Macclesfield who wanted one to turn his silk mill. This was followed in October of the same year by John Collet in London who inquired about one for turning laps, grinding stones and polishing steel work. However Watt suggested that Proctor and Beilby at Sheffield should use a reciprocating engine to raise water for a waterwheel until the steam-wheel was perfected. There were a further four enquiries in 1777. (79)

Rotary Engines

It seems to have been Boulton who kept pressing Watt to develop rotary or rotative engines. While Watt was in Cornwall during 1779, Boulton sent him a letter about circular engines. (80) In 1781 and 1782, Watt decided to patent all the inventions relating to the steam engine which he had made previously, including his early rotary ones which he had conceived in 1765 or 1766. The 1782 patent (81) covered his ideas for rotary engines but not the steam-wheel, presumably because this was contained in that of 1769, but also possibly because it did not work. He clearly intended that these patents should block anybody else trying to develop steam engines. Three designs were described, one for an oscillating rotary engine, another for an engine with a single fixed-bladed rotor and the last for a rotor with collapsible blades.

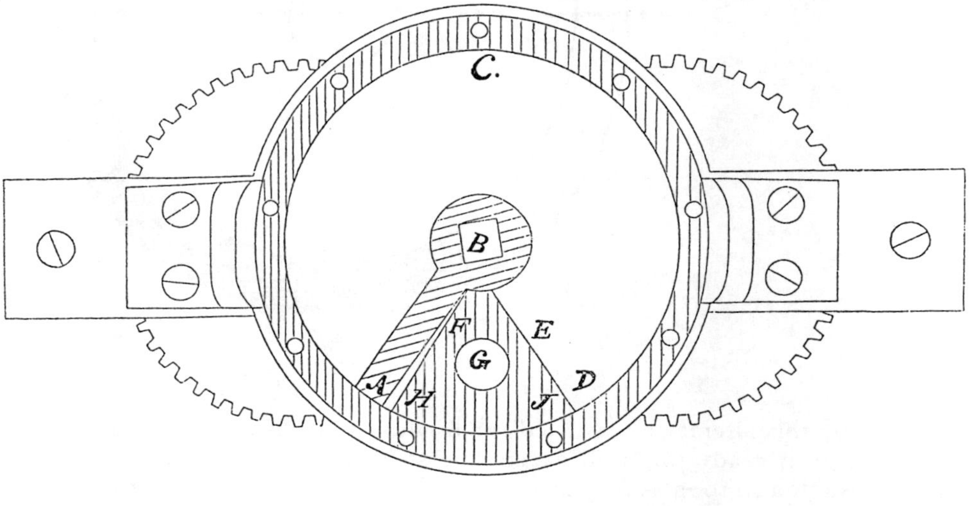

Watt's semi-rotative steam engine with a vane 'A' that was pushed by the steam round to 'E,D' and then back again. (Fig 1 in text, p439)

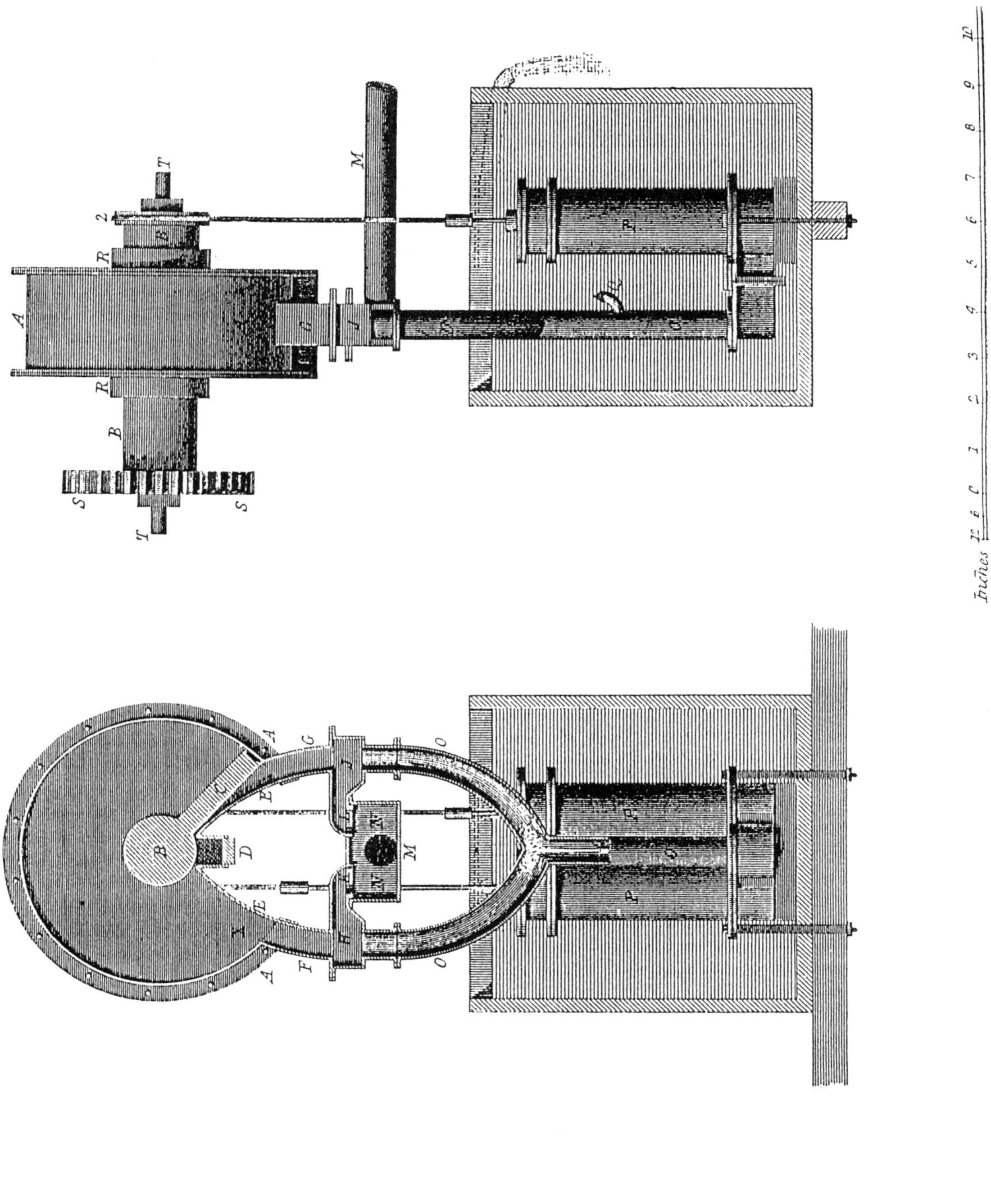

A presumably conjectural drawing of the semi-rotative engine connected to a condenser. (Fig 1 in text, p439)

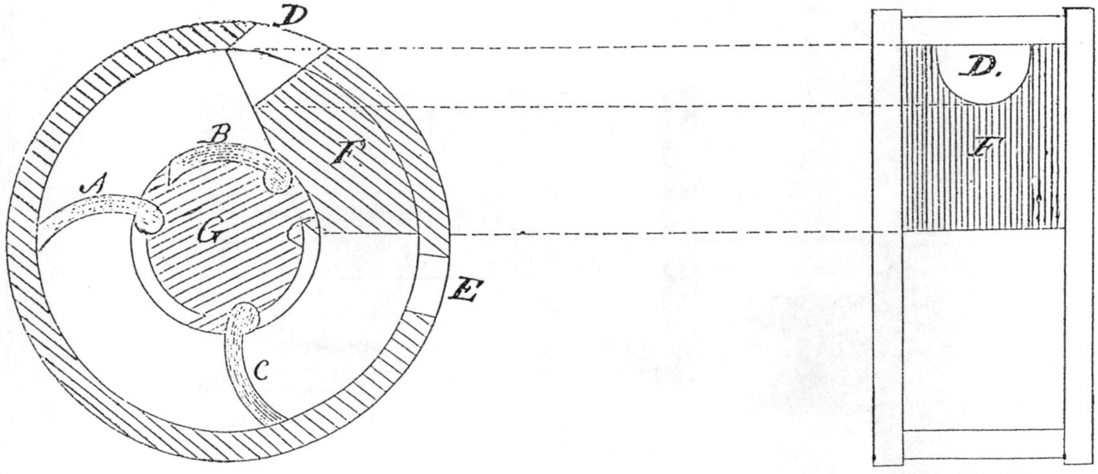

A rotary engine with three movable vanes 'A,B,C' on the rotor 'G'. The steam enters at 'E'. (Fig 2 in text, p439)

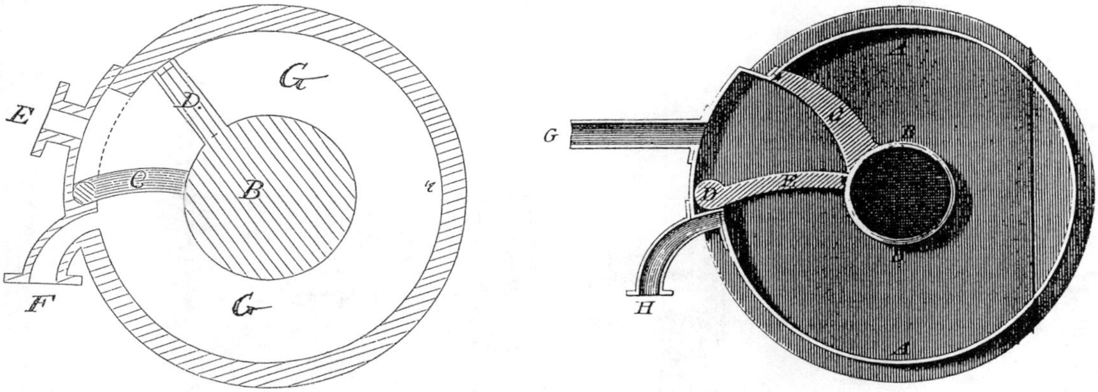

A rotary engine with one fixed vane 'D' on the rotor and a hinged valve 'C'. The steam enters at 'E'. (Fig 4 in text, p439)

Watt's description in his patent gave a very lengthy account of his rotary engines. He was not prepared to be accused again of giving an inadequate specification. Watt sent Hamilton in September 1782 a much briefer description of the models which he had built while still at Glasgow soon after he had invented his separate condenser. Parts survive in the Garret Workshop.

> I remember seeing the patterns of these internal parts not long before I left Glasgow and perhaps Osborne who got my patterns may have them yet or may remember them. The model was made by Daniel McMillan who is not now in Britain... The fig 2nd. is the thing most necessary to be proved as being in itself a continued rotative motion which the other is not Fig 4th was only an idea together with many others of its brothers and these ideas were suffered to sleep...

> A model of a semi-rotative steam engine, executed in 1765 or 1766 by J.W. at Glasgow. The drawing is half of real size of ye model which was made in brass & is about 2 inches deep.
> By means of the cock G the steam is admitted by the opening F between the piston A and the first obstacle HFJ and there being vacuum in ye cylinder ACD the steam forces the piston to revolve on ye axis B until A arrives at D then the steam is admitted by the opening E and discharged by the opening F and the piston revolves in ye contrary direction, but by means of wheels on ye other side gives a continued rotative motion to machinery or works reciprocating pumps as may be required.
> The Fig 2 is another way in which the same model was intended to be used to produce a continued rotative motion A B & C are thence crooked valves which when folded down form a part of the axle G and when opened out perfectly fill the space between it and the cylr. The steam is admitted by the passage E and there being vacuum behind A the steam acts against the valves A & C and causes them and the axle to which they are fastened to turn round; when the valves come to the place B they are folded down by the first obstacle F and the steam issues at D and the motion is continued by the valves succeeding one another.
> Fig 4 is a drawing of another method of a continued rotative motion. The steam being admitted by E between the valve C & the piston or radius D, and there being vacuum in GG - the piston revolves towards G until it arrives at the valve C which is then opened by proper machinery. The pipe E which admits the steam being first shut; when the piston passes the pipe F the steam rushes out through that pipe into the condenser and the piston having past the valve E is opened and the steam acts as before. The motion during the absence of the steam is maintained by means of a fly or a weight. (82)

On his first day home from Cornwall in November 1782, Watt seems to have spent most of the time 'with Mr Martin from Truro to whom showed the revolving engine and explained the up and down one'. (83) The last we hear of them is in February 1783 when Watt noted that he 'Drew 10 inch circular engine with round or cock valves'. (84) By this time, Watt was busy with his reciprocating rotative engines and by 30 June 1784 had conceived his famous parallel motion. (85) This solved one of the problems of converting reciprocating motion into rotary. On the other hand, sealing the vanes and valves in Watt's rotary engines would have been almost impossible and they would never have worked in practice.

The steam-wheel became a subject of dispute during the various patent trials in the 1790s. At Edward Bull's trial in 1793, Joseph Bramah, speaking on behalf of Bull, was scathing about it and thought it utterly impracticable because, owing to the friction of the weight, it would never produce 'any salutary effect'. (86) In the 1796 trial of Jabez Hornblower and Stephen Maberley, the defendants claimed that 'Watt has not described his rotary engine properly There is no plan or drawing'. (87) In his 'Answers to the Objections made to his Specification', Watt said about the fifth article of his 1769 patent that he hoped to

> ... prove that the Variety of his Invention there described, is in the first place intelligible, and secondly that it has been made and used in effect. - It was found however that the present Mode of producing rotative motions from reciprocating Engines was less expensive, and in several respects preferable, therefore the other was abandoned. (88)

This was an easy way of avoiding admission that his steam-wheel and rotary engines were failures.

NOTES: ROTARY STEAM ENGINES AND THE STEAM-WHEEL

1 A longer version of this chapter was printed in *Transactions of the Newcomen Society*, Vol 70, No. 1, 1998 - 1999, R.L. Hills, 'James Watt and his Rotary Engines', pp. 89 - 108.
2 Galloway, (1830), p. 76.
3 Lindqvist, (1984), p. 234 f.
4 Muirhead, (1854), Vol I, p. 13, Letter 15, J. Watt to J. Roebuck, 19 Feb. 1766; JWP, 4/32.30.
5 Ibid, Vol I, p. 14, Letter 16, J. Watt to J. Roebuck, 29 Feb. 1766; JWP, 4/32.29.
6 Ibid, Vol I, p. 14, Letter 18, J. Watt to J. Roebuck, 14 May 1766; JWP, 4/32.27.
7 Patent 1,321, enrolled 4 July 1782.
8 Muirhead, (1854), Vol I, p. 17, Letter 22, J. Watt to J. Roebuck, 28 Jan. 1768; JWP, 4/32.26.
9 Ibid, Vol I, p. 89, Letter 77, J. Watt to W. Small, 25 Feb. 1770; Boulton 348.17.
10 JWP, 4/24, Draft Steam-engine Specification prior to 1769 Patent, see Robinson & Musson, (1969), p. 53.
11 Robinson & McKie, (1970), p. 13, Letter 6, J. Robsion to J. Watt, 8 July 1768.
12 Muirhead, (1854), p. 29, Letter 33, W. Small to J. Watt, Monday, no date, Sept. 1768; Boulton 340.3. Small hoped to 'travel in a fiery chariot of your invention'.
13 Ibid, Vol I, p. 16, Letter 21, W. Small to J. Watt, 7 Jan. 1768 [should be 1769]; Boulton 340.1.
14 Ibid, Vol I, p. 43, Letter 43, M. Boulton to J. Watt, 7 Feb. 1769; JWP, 6/19.23.
15 Ibid, Vol I, p. 59, Letter 56, W. Small to J. Watt 1769; Boulton 340.6.
16 JWP, C1/35, 'Steam-Wheel, suppose prior to 1769'.
17 Muirhead, (1854), Vol I, p. 35, Letter 41, J. Watt to W. Small; Boulton 348.4.
18 Ibid, Vol I, p. 37, Letter 42, W. Small to J. Watt, 5 Feb. 1769; JWP, 4/24.
19 Patent 913, 5 Jan. 1769.
20 JWP, 6/42, 'Drawing intended for Specification, 1769'.
21 Muirhead, (1854), Vol I, p. 87, Letter 75, J. Watt to W. Small, 31 Jan. 1770; Boulton 348.16.
22 Boulton 340.4, W. Small to J. Watt, 12 or 19 Feb. 1769.
23 Ibid.
24 Muirhead, (1854), p. 89, Letter 77, J. Watt to W. Small, 25 Feb. 1770; Boulton 348.17.
25 Ibid, Vol I, p. 52, Letter 52, W. Small to J. Watt, 18 April 1769; Boulton 340.5.
26 Ibid, Vol I, p. 53, Letter 53, J. Watt to J. Roebuck, 28 April 1769; JWP, 4/32.17.
27 JWP, C4/C1, 19 Feb. 1769, 'Extracts from Dr. Small's Correspondence with Mr. Watt'. See also Muirhead, (1854),Vol I, p. 58, Letter 56, W. Small to J. Watt, 16 May 1769; Boulton 340.6.
28 Muirhead, (1854), Vol I, p. 55, Letter 54, J. Watt to W. Small, 28 April 1769; Boulton 348.8.
29 Ibid, Vol I, p. 59, Letter 56, W. Small to J. Watt, 16 May 1769; Boulton 340.6.
30 Ibid, Vol I, p. 60, Letter 58, J. Watt to W. Small, 28 May 1769; Boulton 348.9
31 Ibid, Vol I, p. 82, Letter 67, W. Small to J. Watt, 5 Nov. 1769; Boulton 340.9.

32 Ibid, Vol I, p. 84, Letter 71, W. Small to J. Watt, 26 Dec. 1769; Boulton 340.11.
33 Ibid, Vol I, p. 86, Letter 74, W. Small to J. Watt, 23 Jan. 1770; Boulton 340.12.
34 Ibid, Vol I. p. 92, Letter 78, W. Small to J. Watt, 28 Feb. 1770; Boulton 340.13.
35 Ibid, Vol I, p. 89, Letter 77, J. Watt to W. Small, 25 Feb. 1770; Boulton 348.17.
36 Ibid.
37 Ibid, Vol II, p. 4, Letter 84, W. Small to J. Watt, 17 Sept. 1770; Boulton 340.14.
38 Ibid.
39 Ibid, Vol II, p. 11, Letter 86, W. Small to J. Watt, 5 Oct. 1770; Boulton 340.15.
40 Ibid, Vol II, p. 8, Letter 85, J. Watt to W. Small, 30 Sept. 1770 and Muirhead, (1858), p. 208.
41 Muirhead, (1854), Vol II, p. 6, Letter 85, J. Watt to W. Small, 30 Sept. 1770. See also B. & W. Col, MI, 1/15, 'Journal', 29 Sept. 1770 for notes on this letter.
42 Muirhead, (1854), Vol II, p. 11, Letter 86, W. Small to J. Watt, 5 Oct. 1770; Boulton 340.15.
43 Ibid, Vol II, p. 13, Letter 87, J. Watt to W. Small, 20 Oct. 1770; Boulton 348.21.
44 Ibid, Vol II, p. 15, Letter 85, W. Small to J. Watt, 3 Feb. 1771; Boulton 340.16. Muirhead gives the date wrongly as 14 Feb.
45 Ibid, Vol II, p. 17, Letter 91, W. Small to J. Watt, 19 Oct. 1771; Boulton 340.17.
46 Ibid, Vol II, p. 19, Letter 92, W. Small to J. Watt, 16 Dec. 1771; Boulton 340.18.
47 JWP, 4/59/5. J. Watt to W. Small, suppose 15 Dec. 1771.
48 Muirhead, (1854), Vol II, p. 20, Letter 93, J. Watt to W. Small, 24 Dec. 1771; Boulton 348.23.
49 Ibid, Vol II, p. 22, Letter 95, J. Roebuck to J. Watt, 20 Aug. 1772; JWP, 4/57.18.
50 Boulton 334.17, J. Roebuck to M. Boulton, 22 Aug. 1772.
51 Muirhead, (1854), Vol II, p. 23, Letter 96, J. Watt to W. Small, Aug. 1772; JWP, 4/59.6. See also B. & W. Col., MI, 1/19, 'Journal', 25 Aug. 1772.
52 JWP, 4/57.19, J. Roebuck to J. Watt, 5 Sept. 1772. See also B. & W. Col., MI, 1/19, "Journal", 8 Sept. 1772.
53 B. & W. Col., MI, 1/19, 'Journal', 7 Nov. 1772 and Dickinson, (1936), p. 83.
54 Muirhead, (1854), Vol II, p. 31, Letter 101, W. Small to J. Watt, 16 Nov. 1772; Boulton 340.24.
55 Ibid, Vol II, p. 35, Letter 103, W. Small to J. Watt, 3 Dec. 1772; Boulton 340.25.
56 Ibid, Vol II, p. 37, Letter 105, W. Small to J. Watt, 27 Jan. 1773; Boulton 340.26.
57 Ibid, Vol II, p. 57, Letter 119, W. Small to J. Watt, 8 Aug. 1773; Boulton 340.33.
58 Ibid, Vol II, p. 64, Letter 124, W. Small to J. Watt, 8 Oct. 1773; Boulton 340.35.
59 Ibid, Vol II, p. 74, Letter 131, W. Small to J. Watt, 23 Feb. 1774; Boulton 340.39.
60 Ibid, Vol II, p. 75, Letter 132, J. Watt to W. Small, 3 March 1774; Boulton 348.41. See also B. & W. Col., MI, 1/20, 'Journal', 20 Feb. - 9 March 1774.
61 Muirhead, (1854), Vol II, p. 76, Letter 133, W. Small to J. Watt, 29 March 1774; Boulton 340.40.
62 See B. & W. Col., MI, 1/21, 'Journal', 31 May 1774.
63 SRO, GD18/859373/3, Dr. Hutton to G. Clerk, ? July 1774.
64 Muirhead, (1854), Vol II, p. 79, Letter 136, J. Roebuck to J. Watt, 12 Nov. 1774; JWP, 4/57.23.

65 Ibid, Vol II, p. 81, Letter 140, J. Watt to M. Boulton, 31 Jan. 1775; Boulton 348.45.
66 B. & W. Col., MI, 2/8, 'Note Book of James Watt, 1770 - 1781', London 1775.
67 Ibid, MI, 2/5, 'Cash Memorandum', no date.
68 Muirhead, (1854), Vol II, p. 81, Letter 141, M. Boulton to J. Watt, 2 Feb. 1775; JWP, 6/19.8.
69 Ibid, Vol II, p. 84, Letter 144, M. Boulton to J. Watt, 24 April 1775; JWP, 6/19.22.
70 Ibid, Vol II, p. 87, Letter 147, M. Boulton to J. Watt, May 1775; JWP, 6/19.6.
71 B. & W. Col., MI, 8/25, 'An Account of James Watt's Improvements upon the Steam or Fire Engine', no date but ascribed to 1775.
72 Boulton 348.54, 'Watt's Case Before Parliament'.
73 JWP, 4/53, 'Copy of Minutes on Recommitment of Mr. Watt's Engine Bill', 11 April 1775.
74 Ibid, S. Garbett's Evidence, 12 April 1775.
75 Ibid, R. Mylne's Evidence, 12 April 1775.
76 Muirhead, (1854), Vol II, p. 95, Letter 154, M. Boulton to J. Watt, 24 Feb. 1776; Boulton 348.57.
77 Ibid, Vol II, p. 101, Letter 159, M. Boulton to J. Watt, 25 July 1776; Boulton 348.63.
78 Ibid, Vol II, p. 102, Letter 160, M. Boulton to J. Watt, [1776]; Boulton 348.69.
79 Lord, (1932), pp. 152 - 153 and 236 - 237.
80 JWP, C3/5, 'Journal', 4 Oct. 1779.
81 Patent 1,321, enrolled 4 July 1782.
82 JWP, Letter Book 1, J. Watt to G. Hamilton, 24 Sept. 1782.
83 JWP, C3/6, 'Diary', 20 Nov. 1782.
84 JWP, C3/7, 'Journal', 19 Feb. 1783.
85 Dickinson, (1935), p. 136.
86 JWP, 3/14, Patent trial 'Boulton v. Bull', June 1793.
87 JWP, 3/29, 'Boulton and Watt v. Hornblower and Maberly', Michaelmas Term, 1796, Bundle 2, p. 23.
88 JWP, 4/26, 'Answers by Mr Watt to the Objections made to his Specification', 27 Sept. 1796, p. 24.

Watt Moves to Birmingham at Last

Years of Frustration

On that fatal 13 June 1770, when Watt was summoned to Glasgow to attend Golborne on a survey of the Clyde, he could not have realised that he would never test his engine again in Scotland. At that time, he must have felt he was very close to success because, in one of his letters to Roebuck with instructions for altering the engine, he said, 'I grudge confoundedly the time Mackell made me lose at Glasgow:- my experiments on this condenser might otherwise have been complete before I left it. But the Lord reward him for it in a proper manner'. (1) On his return from the Strathmore survey, when he was offered the position as resident engineer on the Monkland Canal, he was faced with the agonizing decision. Either he could continue with experiments on his engine, the outcome of which were uncertain, or he could accept 'an honourable and perhaps profitable employment' (2) as civil engineer which would entail three or four days a week supervision. The apparent certainty of £200 salary per annum was not lightly to be refused, so he accepted, 'though at the same time I resolved not to drop the engine, but to prosecute it the first time I could spare'. (3)

He still hoped 'soon to have another touch at the engine' (4) and, whatever frustrations Watt may have felt, it is obvious that he intended to try and make his engine work successfully. He must have looked at it whenever he visited Roebuck at Kinneil during the next few years even though Roebuck's situation continued to deteriorate so that the chances of finishing the engine became remote. Just when Watt decided that he must move the reciprocating engine to Birmingham is not clear. As early as January 1769, Small had been pressing Watt 'that you should settle here, and that Boulton and I should assist you as much as we could, which in any case we will most certainly do'. (5) They were already investigating the possible separation of the 'wheel and reciprocating engines', the former to be made in Birmingham. (6) The warmth of Small's interest and friendship must have given encouragement to Watt, especially when he recommended Watt to

> let nothing divert you from the business of engineering. You are sensible that both Boulton and I engaged in the patent scheme much more from inclination to be in some degree useful to you than from any other principle, so that, if you are prosperous and happy, we do not care whether you find the scheme worth prosecuting or not. (7)

Watt may have felt very lonely in Glasgow at this time because his closest friends, Black and Robison, had both left. Small was the person to whom he poured out his heart and shared his troubles. In 1771, Small pointed out possible openings for sales of engines in the south where people were inquiring about Watt's engine.

> A friend of Boulton and me in Cornwall sent us word four days ago that four or five copper-mines are just going to be abandoned, because of the high price of coals, and begs me to apply to them instantly. The York Building Company delay rebuilding their engine, with great inconvenience to themselves, waiting for yours. Yesterday application was made to me, by a mining company in Derbyshire, to know when you are to be in England about fire-engines; because they must quit their mine if you cannot relieve them. (8)

The English grass across the border to the south must have been beginning to appear much greener than that in Scotland.

But first Watt tried to see if his friends could help him raise money to carry on his engine experiments. Some letters written in April 1771 have not survived for Watt explained how he could not immediately attend to experiments on the engine but that he would be willing to devote himself to it for a year if a suitable subscription could be raised. (9) This suggestion was also put to Roebuck. However Watt's hopes were dashed that autumn when Small replied,

> Nothing but aversion to tell you disagreeable news, and daily hopes of being able to communicate such as should be very much otherwise, could have rendered me silent to you. I hated to tell you that Boulton and I found we could not serve you effectually in the subscription way, which was your last proposal; because people, as no engine had been made upon your principles, doubted whether any could be made. This we found so frequently, that we feared to hurt you by pressing even our most intimate acquaintances. On the other hand, everything has been ready for above two months in the circulating engine, excepting what depends on millwrights. (10)

Here was the usual dilemma. Nobody would support experiments on an engine that was not working so no experiments could be made to complete it. The outlook must have seemed bleak except for that ray of hope with the rotary engine. Boulton and Small's progress with the steam-wheel may have prompted Watt to suggest that the reciprocating engine should be taken to Birmingham also but, once again, his hopes were dashed because Boulton thought this should be done later. (11) Then Roebuck gave Watt full power to treat with them 'to settle all matters... about the engine'. (12) Had Watt but known it, the fact that his engine was incomplete and so thought to be valueless worked to his advantage in the next turn of events.

The Financial Situation Worsens

The failure of the Ayr Bank in June 1772 made a general financial crisis much worse in Scotland so that Roebuck's affairs took a turn for the worse. By July, even Boulton and Small were adversely affected and Small was cautioning, 'In the present times ready money cannot be preserved too carefully... The news you have sent me is terrible. What I most dread is the ruin of some of the infant manufacturers in your part of the world'. (13) Boulton himself nearly went bankrupt at this time through failure of his bankers in London (14) so could not have been in a position to help Watt for a while. (15) Although that August Roebuck hoped to raise some £200 out of £500 which Watt must have thought necessary probably to complete the experiments, (16) it must have become clear that his position was dire. Watt had suggested to Roebuck that he should 'dispose of the whole property of the engine for £1000 if he can do no better'. (17)

Watt pressed Boulton and Small to come to an agreement with Roebuck, probably not realising their desperate situation. He said that he never thought of receiving money for any part of his share which he was willing to reduce, even though he was out of pocket to a much greater sum than his proportion of the property of the engine. He regarded this as money 'spent on his education' (18) for he had been financing most if not all of the later expenditure on the trials. (19) In his view,

> The Doctor's affairs being yet far from being reinstated give me little hope of help from that quarter; in the meantime, the time of the patent is running on. It is a matter of great vexation to me that the Doctor should be out so great a sum upon this affair, while he has otherwise such pressing occasion for the money. I find myself unable to give him

> such help as his situation requires; and what little I can do for him is purchased by denying myself the conveniences of life my station requires, or by remaining in debt where it galls me to the bone to owe.
> Notwithstanding my natural despondence, I am convinced that the machine may be made to answer in a very considerable degree, and in more forms than one, but that I am by no means a proper person to carry it into execution. (20)

We see at the same time Watt's optimism for his engine yet his doubts about his ability. He felt genuine remorse that he was partly responsible for Roebuck's situation but obviously saw that his only hope lay in Birmingham and asked Small for advice about how to proceed. He was hoping that they would be more closely concerned with the engine having as much as half the share. (21)

While Watt's comments made Small more deeply interested in the success of the reciprocating engine, their own financial position meant they could not advance with either type. He felt that, whatever the merits of either, they could not do justice to Watt or to his inventions. Everybody was retrenching. (22) But, in November 1772, we find Watt suggesting a way out of the situation. Whatever pain Watt may have felt about involving Roebuck so deeply in the steam engine venture, now the businessman started to come to the fore. Until his collieries began to run into difficulties once more, Roebuck must have been thinking about experimenting with a small engine himself but Watt had had to point out that his own experiments on small engines had been sufficiently successful and that it was principally with the larger final one at Kinneil on which difficulties had appeared. How could Roebuck sort out in a few days all the problems which had cost Watt himself so much labour? Now Roebuck was willing to part with part or the whole of the engine property 'upon such terms as, I dare say, in better times, you and Mr. B. would have had no hesitation of accepting'. (23)

Watt was very anxious that Boulton and Small should consider Roebuck's proposition. Watt was suggesting that, in return for a definite share of at least half, they should take charge of future experiments and 'of finding money to carry on the business in case of success'. Watt was anxious that 'the experiments may go on, and the matter be concluded'. (24) Possibly Small was having to conceal the true state of Boulton's and his financial situation for the answer he sent must have thrown Watt into another state of despair.

> Besides the difficulties in money, you must be sensible there is another, which is insuperable at the present. It is impossible for Mr. Boulton, or me, or any other honest man, to purchase, especially from two particular friends, what has no market price, and at a time when they might be inclined to part with the commodity at an under value. We have, and have had from the beginning, no other wish in this affair, as I have often said to you, except that of assisting you and the Doctor, and of promoting the introduction of a capital invention...
> I hope Dr. R. will soon recover his spirits. A few months will probably restore mercantile matters to their former ease, and he has infinite resources. (25)

Watt admired their delicacy and apologised for urging them to take over Roebuck's share in the patent, and promised not to give any further trouble on this account until he saw them, when 'you must expect another onset, unless you shall positively say that you do not think it a practicable or profitable scheme'. (26) But he would not so easily give up what he must have considered as his last chance to get his engine experiments going again.

Financial Discussions and Watt Settles with Roebuck

Watt's chance came through a meeting of Roebuck's creditors called for 2 April 1773. These people were 'alarmed that under the management of the present trustees a very large additional debt has been incurred, instead of a fund raised from the profits of Borrowstoneness to discharge his debts'. (27) Among these creditors were both Boulton and Watt who had both lent Roebuck money. We will follow Watt's story first before turning to Boulton. Watt's Journals show that he lent Roebuck £100 through a six months bill on 19 August 1770. (28) There are other similar entries for money changing hands in both directions so that, by April 1773, Roebuck had borrowed £180. (29) While Roebuck had paid Watt's debts to the tune of £1000 as agreed in their original partnership, (just what debts did this sum cover?) he had not paid for all the costs of the engine experiments and so had failed to implement his agreement. Therefore Watt was in a strong position to enforce his claims for these costs as well as the money lent subsequently. (30) Watt's proposal was that he would pay off Dr. Black and 'all debts contracted by me in name of the compy and to give up all claims to the £180 pd the doctor'. Presumably this covered anything Watt had paid out on the steam engine trials and possibly the alkali from salt experiments as well. Watt later said that he 'was out £500 on that concern', probably meaning the steam engine. (31) In return, Roebuck was to 'give up all claims to the shop, goods - & to give each other discharge, he retaining his $^{2}/_{3}$ and I $^{1}/_{3}$ of the engine'. (32) Some of the shop goods still remained in Watt's house so presumably he now took formal possession of them.

Watt met Roebuck at Kilsyth where they settled their accounts on 15 May 1773 and discharged each other. Knowing a little bit about the financial background, we can see that Watt was once again being rather economical with the truth in the discharge he gave Roebuck on 17 May.

> In the year 1767, Doctor John Roebuck at Kinneil entered into partnership with James Watt at Glasgow, to verify and carry into practice an improved fire-engine invented by the said James Watt. Doctor Roebuck was to pay a debt of 1000£. incurred by the said James Watt in making the experiments tending to the invention of the engine, and also to pay the expense of the patent and further experiments.
> James Watt was to attend and conduct the experiments, and assigned to the Doctor two-thirds of the property of the said invention, retaining one-third for his own use.
> Dr. Roebuck has paid the thousand pounds, but the expense of the other things has been principally paid by James Watt.
> In consideration of the mutual friendship subsisting between Dr. Roebuck and myself, and because I think the thousand pounds he has paid more than the value of the property of the two-thirds of the inventions, I hereby take upon myself all other sums I have laid out or paid upon it, also all other debts I have contracted upon that head, relieving the Doctor from the same, and meaning this as an absolute discharge for all sums he may have been owing me before this date. (33)

Watt had at last escaped from the millstone of the debts which were threatening to be hung round his neck!

Roebuck allowed the £178 which Watt had advanced him to be in payment of the Kinneil engine. So, armed with his discharge, Watt set off immediately to Kinneil to rescue what was now his own property, his engine, which he found perishing. Obviously he wanted to secure it before there could be any change of mind by Roebuck or complaints by Roebuck's creditors. He spent four days dismantling it and packing it into crates for shipment first to

London and then on to Birmingham. The wooden framing he left because he felt it was not worth the price of the carriage but the cylinder, iron work, pumps and other parts totalling four tons were packed in a dozen crates and sent by the ship the *Duchess of Hamilton*. (34) All this was done in such haste that he had not been able to warn Small who later said that the parts ought to have been sent to Gainsborough on the Trent. (35) Watt presented the other creditors with a fait accompli at their meeting on 29 May. (36)

Boulton Settles with Roebuck

Watt had secured his engine and also retained his one third share in the patent so he schemed how to transfer the remaining two thirds into Boulton's hands. But here once again he had to tread carefully. On the one hand he had to convince Boulton as to the certainty of success with his engine, while he could not show any such enthusiasm to Roebuck or his creditors who would want a share in any eventual profits. Roebuck must have known just how close Watt thought he was to success and, in his dire financial plight, would have been searching for any possible source of income. In November 1772, Watt had suggested a scheme whereby Boulton could make a deal with Roebuck. Watt was urging Small that he and Boulton should acquire Roebuck's share in the patent.

> Consider also that Dr. R. owes Mr. Boulton money, which will go in part of the price, which can never be so low as at present. I am sorry that there is occasion to ask a price, but it cannot be helped; the Doctor's circumstances oblige him to demand it if he parts with any great part of the property. (37)

Obviously Watt was seeking to drive a bargain whereby Boulton would receive Roebuck's share in the patent in exchange for the loans Boulton had advanced.

Roebuck's correspondence with Boulton is full of financial dealings. Normally these involved Boulton's partnership with Fothergill and so, for these, Boulton would have to obtain his partner's consent which might mean that Fothergill would become a partner in the steam engine business. But Boulton must also have lent probably £500 in his private capacity on the strength of Roebuck's proposed assignment of half of his two thirds interest in the engine patent in November 1769. (38) While at the time, this was not in fact considered as part of this agreement, Roebuck was willing that it should be in October 1771 and Watt raised this point later. (39) Roebuck's total indebtedness to Boulton was about £1,200. (40)

Boulton had heard from Roebuck that there would be a meeting of his creditors on 2 April 1773 so he wrote to Watt empowering him 'to vote and act for me in what relates to the debt owing from Doctor Roebuck to me and my partner Mr. John Fothergill'. (41) Boulton was willing to reimburse Watt any expenses he incurred and sent a list of the debts and interest which came to a total of £630.11s.3d. In a covering letter to Watt, he also said that this was

> ... all I shall claim but at the same time I must tell you *in confidence* that I advanced the money to the Dr some years ago a sum nearly equal to the aforesaid merely upon the good opinion I had of the Drs Honor & honesty (which opinion I have not yet changed) *without entering into any written agreement*, what Interest or advantage I should receive for it, having more in view the Interest of the Dr than myself, but as this sum hath never appeared in the D[ebto]r Books nor to the C[redito]rs I will not now claim it if I never receive a farthing of it. (42)

Unfortunately the continuation of this letter is torn just where Boulton is discussing the proposals about the engine partnership which he realised would not now be carried out but it looks as if he was suggesting that Roebuck should 'relinquish or make over to me [his shares]'. (43)

Armed with this letter, Watt discussed the situation with Roebuck's sons who must have agreed to the transfer providing that their father received one ninth of the clear profit in the event of the engine proving successful. (44) Roebuck junior told Watt that his father agreed to this, 'that is cancelling the whole debt, and reserving a small share to himself of the profits as they may turn out'. (45) Roebuck gave his acceptance of the terms to Watt for transmission to Boulton on 15 May 1773. (46) Watt must have thought himself home and dry. While he was trying to help Roebuck and his sons who were in a precarious financial position at this time, he realised that he must consider his own interests as well as his family and could help Roebuck no longer. (47) However during August, Roebuck received some personal financial assistance so his position improved. (48)

On the other hand Watt's hopes were dashed again because Boulton would not accept the terms of Roebuck's offer which included Roebuck being paid part of any eventual profits. Boulton told Watt,

> Although there is nothing in Dr. R.'s letter but what is agreeable to the spirit of my proposal, yet the explicit declaration of the advantages I proposed to allow him in case of success in the engine, must in all probability become known to his creditors; in consequence of which they will possess themselves of those advantages as part of his property, and thereby my intentions of benefitting the Doctor must be defeated and the creditors will receive no benefit, as it cannot be supposed that they will co-operate in forwarding the engine, or that I should forward it singly, whilst part is the property of strangers. (49)

Small supported Boulton by saying that he wished they could meet to discuss the situation for they had no idea when the other creditors might meet to make any settlement. He was afraid that the other creditors might think that he and Boulton were trying to arrange a neat bargain for themselves. He also realised that they might be able to buy the patent rights very cheaply because 'you know how cheap the generality of mankind hold unexecuted projects, however valuable'. (50)

There can be little doubt that Boulton was trying to avoid accepting the clause which could mean sharing the profits, particularly with Roebuck's creditors. He was also having difficulty with Fothergill who had an interest in the money owed them by Roebuck but did not want to have anything to do with either the patent or the engine. (51) Eventually Fothergill was paid his share of Roebuck's debts but, afterwards when the engine became a success, this led to much acrimony on his part. Watt thought Boulton's reasons for refusing 'honourable and perhaps prudent'. (52) This left Boulton's claims on the £630 unsettled so Watt suggested that he should propose at the next meeting of the creditors,

> ... that you will discharge the debt of 630£. if they will empower the Doctor to make over to you his property in the engine invention; for all the models, etc. are mine... My heart bleeds for his situation, and I can do nothing to help him. I stuck by him till I have much hurt myself. (53)

The discussion between Watt, Small and Boulton continued during that summer. While none of Roebuck's creditors valued the engine at a farthing, (54) Watt was considering whether he could take out a law suit against Roebuck to force him to fulfil the terms of their original 1767 agreement or alter those on the most recent one suggested to Boulton. Small urged

Watt to go to Soho to complete both engines and to 'endeavour to find some plan that may be beneficial to yourself and the Doctor, and just to his creditors'. (55) Watt intended to meet Roebuck to effect some settlement (56) but this probably did not happen.

A meeting of the creditors was held during August 1773 when the 'composition agreed to by his creditors is very low'. (57) Roebuck junior hoped for some peace at last. (58) This obviously affected the value of Boulton's claims but it might suggest that Roebuck should be paid more out of any eventual profits. Watt pressed Boulton to come to a settlement and write Roebuck a letter so that, by the time he had finished the survey of the canal from Inverness to Fort William, Boulton might have full possession. Watt was unwilling to make any further experiments 'until the whole property is settled' and mentioned that he had 'tried some more experiments upon heats of water under various pressures, and I think I see plainly the whole faults of the condenser, and their remedies'. (59)

Watt Leaves Scotland for Birmingham

Watt was deeply affected when Peggy died while he was away surveying the canal in September 1773 and his gloom must have been further deepened by Small's letter at the end of the month saying that Boulton had not written to Roebuck. Small dared not enquire why, because it was 'an affair of too much delicacy'. (60) The situation had not altered just over a week later (61) and even in December they 'were still where we were about the engine'. (62) Watt must have been in the depths of despair for he replied to Small,

> I am heart-sick of this country; I am indolent to excess, and, what alarms me most, I grow the longer the stupider. My memory fails me so as often totally to forget occurrences of no very ancient dates. I see myself condemned to a life of business; nothing can be more disagreeable to me; I tremble when I hear the name of a man I have any transactions to settle with.
>
> What I am fittest for is a surveying engineer. Is there any business in that way? (63)

Small tried to encourage him for he thought that Watt might have a good business in England once he could show an example of his work. (64) But what had Watt achieved, with only a few contracts completed, no working engine and the Monkland Canal partly finished which nobody in the south could see?

Our records do not reveal that Boulton ever approached Roebuck but there must have been some progress because, in another letter which has not survived, Watt wrote to Small on 6 February 1774, telling him 'to get things ready for trial of engine'. (65) It was not until the end of March that Watt received a reply from Small saying, 'If you judge the title to the engine to be clear, and your experiments to be sufficiently advanced for a reasonably good construction of it, come hither as soon as possible'. (66) From this it sounds as if Boulton had not made any agreement and that the decision was left to Watt. Small was well aware of the importance of Watt's experiments on the boiling points of water in a vacuum and under pressure at that time for both the engine and 'to the improvement of Natural Philosophy' but was also 'solicitous you should convert your discoveries into fame and money as fast as possible'. (67) Watt still had to complete his reports on his various surveys before he could consider leaving Scotland so it was not until the end of April that he could tell Small, 'I begin now to see daylight through the affairs that have detained me so long, and think of setting out for you in a fortnight at furthest'. (68) He was sorting out papers during the first week of May when he finished his report on the Caledonian Canal and on the fifteenth packed up his belongings. (69) The two children, James and Margaret, were sent to stay with their grandfather so Watt set out alone on 16 May for Kinneil. Perhaps he made some settlement with

Roebuck then. He went on to Edinburgh where he had to settle various business transactions about surveying and other matters. So he did not leave Edinburgh until 20 May when he went via Selkirk and Longtown, reaching Cockermouth on Sunday 22 May. He took the attractive route through Keswick, past 'Winander mere' to Kendal before going on to Preston, Bolton and Manchester. But Manchester was full owing to the races so he looked at the locks at Runcorn before travelling to Newcastle-under-Lyme and Lichfield. The last entry in this Journal was for 31 May 1774 when he noted 'as approach Birmingham stone grows softer and gravel harder'. (70) So Watt at last arrived to join his engines in Birmingham. Watt never lived in Scotland again.

NOTES: BOULTON AND WATT

1 Muirhead, (1854), Vol I, p. 104, Letter 82, J. Watt to J. Roebuck, 15 April 1770; JWP, 4/32.14.
2 Ibid, Vol II, pp. 2 - 3, Letter 83, J. Watt to W. Small, 9 Sept. 1770; Boulton 348.20.
3 Ibid.
4 Ibid, p. 4.
5 Ibid, Vol I, p. 17, Letter 21, W. Small to J. Watt, 7 Jan. 1768[9]; Boulton 340.1.
6 Ibid.
7 Ibid, Vol II, p. 5, Letter 84, W. Small to J. Watt, 17 Sept. 1770; Boulton 340.14.
8 Ibid, Vol II, pp. 15 - 16, Letter 89, W. Small to J. Watt, 14 Feb. 1771; Boulton 340.16.
9 B. & W. Col., MI, 1/14, 'Journal', 14 April 1771.
10 Muirhead, (1854), Vol II, p. 17, Letter 91, W. Small to J. Watt, 19 Oct. 1771; Boulton 340.17.
11 Ibid, Vol II, p. 19, Letter 92, W. Small to J. Watt, 16 Dec. 1771; Boulton 340.18.
12 Ibid, Vol II, p. 21, Letter 93, J. Watt to W. Small, 24 Dec. 1771; Boulton 348.23.
13 Ibid, Vol II, pp. 21 - 22, Letter 94, W. Small to J. Watt, 11 July 1772; Boulton 340.22.
14 Dickinson, (1936), p. 83 and B. & W. Col., MI, 1/19, 'Journal', 7 Nov. 1772.
15 Schofield, (1963), p. 86.
16 Muirhead, (1854), Vol II, p. 22, Letter 95, J. Roebuck to J. Watt, 20 Aug. 1772; JWP, 4/57.18.
17 B. & W. Col., MI, 1/19, 'Journal', 5 Aug. 1772.
18 Muirhead, (1854), Vol II, p. 23, Letter 96, J. Watt to W. Small, Aug. 1772; JWP, 4/59.6. See also B. & W. Col., MI, 1/19, 'Journal', 27 Aug. 1772.
19 Muirhead, (1854), Vol II, p. 24, Letter 97, J. Watt to W. Small, 30 Aug. 1772; JWP, 4/59.4.
20 Ibid.
21 B. & W. Col., MI, 1/19, 'Journal', 7 Nov. 1772.
22 Muirhead, (1854), Vol II, p. 26, Letter 99, W. Small to J. Watt, 29 Oct. 1772; Boulton 340.23.

23 Ibid, Vol II, p. 28, Letter 100, J. Watt to W. Small, 7 Nov. 1772; Boulton 348.24.
24 Ibid, p. 29.
25 Ibid, Vol II, p. 32, Letter 101, W. Small to J. Watt, 16 Nov. 1772; Boulton 340.24.
26 Ibid, Vol II, p. 33, Letter 102, J. Watt to W. Small, 24 Nov. 1772; Boulton 348.25.
27 Ibid, Vol II, pp. 43 - 44, Letter 109, M. Boulton to J. Watt, 29 March 1773; Boulton 348.29.
28 B. & W. Col., MI, 1/15, 'Journal', 19 Aug. 1770.
29 Ibid, MI, 1/19, 'Journal', 13 April 1773.
30 Ibid, MI, 1/20, 'Journal', 31 May 1773.
31 Ibid.
32 Ibid, MI, 1/19, 'Journal', 13 April 1773.
33 Muirhead, (1854), Vol II, pp. 47 - 48, Letter 113, 'Discharge by Mr. Watt to Dr. Roebuck', 17 May 1773; JWP, 4/57.21.
34 JWP, 4/57.22, J. Roebuck to J. Watt, 6 July 1773, see note in another hand at end.
35 Muirhead, (1854), Vol II, p. 49, Letter 114, J. Watt to W. Small, 20 May 1773; Boulton 348.32 and Vol II, p. 56, Letter 118, J. Watt to W. Small, 25 July 1773; Boulton 348.35.
36 B. & W. Col., MI, 1/20, 'Journal', 29 May 1773.
37 Muirhead, (1854), Vol II, p. 29, Letter 100, J. Watt to W. Small, 7 Nov. 1772; Boulton 348.24.
38 Ibid, Vol I, p. 82, Letter 68, J. Watt to W. Small and M. Boulton, 28 Nov. 1769; JWP, 6/44.9.
39 JWP, 4/57.17, J. Roebuck to J. Watt, 18 Oct. 1771.
40 B. & W. Col., MI, 1/19, 'Journal', 8 April 1773.
41 Muirhead, (1854), Vol II, p. 44, Letter 109, M. Boulton to J. Watt, 29 March 1773; Boulton 348.29.
42 Boulton 348.30, M. Boulton to J. Watt, 29 March 1773.
43 Ibid.
44 B. & W. Col., MI, 1/19, 'Journal', 13 April 1773.
45 Muirhead, (1854), Vol II, p. 46, Letter 111, J. Roebuck jun. to J. Watt, 18 April 1773; JWP, 4/57.20.
46 B. & W. Col., MI, 1/20, 'Journal', 15 May 1773.
47 Muirhead, (1854), Vol II, p. 55, Letter 118, J. Watt to W. Small, 25 July 1773; Boulton 348.35.
48 Ibid, Vol II, p. 58, Letter 120, J. Watt to W. Small, 17 Aug. 1773; Boulton 348.36.
49 Ibid, Vol II, pp. 51 - 52, Letter 116, M. Boulton to J. Watt, 12 June 1773; Boulton 348.33.
50 Ibid, Vol II, p. 53.
51 Ibid, Vol II, p. 52 and Dickinson, (1936), p. 84.
52 Muirhead, (1854), Vol II, p. 54, Letter 117, J. Watt to M. Boulton, 4 July 1773; Boulton 348.34.
53 Ibid, Vol II, p. 55.
54 Ibid, Vol II, p. 56, Letter 118, J. Watt to W. Small, 25 July 1773; Boulton 348.35.
55 Ibid, Vol II, p. 57, Letter 119, W. Small to J. Watt, 8 Aug. 1773; Boulton 340.33.
56 Ibid, Vol II, p. 58, Letter 120, J. Watt to W. Small, 17 Aug. 1773; Boulton 348.36.

57 Ibid, Vol II, p. 59, Letter 121, J. Watt to W. Small, 24 Aug. 1773; Boulton 348.37.
58 JWP, 4/32.43, J. Roebuck jun. to J. Watt, 24 Aug. 1773.
59 Muirhead, (1854), Vol II, p. 60, Letter 121, J. Watt to W. Small, 24 Aug. 1773; Boulton 348.37.
60 Ibid, Vol II, p. 60, Letter 122, W. Small to J. Watt, 29 Sept. 1773; Boulton 340.34.
61 Ibid, Vol II, p. 63, Letter 124, W. Small to J. Watt, 8 Oct. 1773; Boulton 340.35.
62 Ibid, Vol II, p. 72, Letter 129, W. Small to J. Watt, Dec. 1773; Boulton 340.38.
63 Ibid, Vol II, p. 69, Letter 127, J. Watt to W. Small, 11 Dec. 1773; Boulton 348.38.
64 Ibid, Vol II, p. 72, Letter 129, W. Small to J. Watt, Dec. 1773; Boulton 340.38.
65 B. & W. Col., MI, 1/20, 'Journal', 6 Feb. 1774.
66 Muirhead, (1854), Vol II, p. 76, Letter 133, W. Small to J. Watt, 29 March 1774; Boulton 340.40.
67 Ibid, p. 77.
68 Ibid, Vol II, p. 77, Letter 134, J. Watt to W. Small, [2]9April 1774; Boulton 348.42.
69 B. & W. Col., MI, 1/21, 'Journal', 6 - 8 and 15 May 1774.
70 Ibid, MI, 1/21, 'Journal', 31 May 1774.

BIBLIOGRAPHY

Adamson, John, *Sauchie and Alloa, A People's History*, Clackmannan District Libraries, 1988.

Anderson, R.G.W., 'Joseph Black: An Outline Biography', in Simpson, A.D.C. Ed., *Joseph Black: 1728 - 1799, A Commemorative Symposium*, Royal Scottish Museum, Edinburgh, 1982.

Arago, D.F.J., *Historical Eloge of James Watt*, Translated from the French with additional notes and an appendix, by James Patrick Muirhead, Esq., M.A. J. Murray, London, 1839.

Aspin, C.S. & Chapman, S.D., *James Hargreaves and the Spinning Jenny*, Helmshore Local History Society, 1964.

Bald, Robert, *A General View of the Coal Trade of Scotland, Chiefly that of the River Forth and Midlothian*, Oliphant & Brown, Edinburgh, 1808.

Balston, Thomas, *James Whatman, Father & Son*, Methuen, London, 1957.

Bauman, Ian, 'Coalmining at Culross in the 16th and 17th Centuries', *Forth Naturalist and Historian*, Vol 7, 1984.

de Belidor, Forest, *Architecture Hydraulique, ou L'Art de Conduire d'Élever et de Ménager les Eaux pour les Différents Besoins de la Vie*, C.A. Jombert, Paris, 1737 - 53.

Black, Joseph, see Robison, John, Ed., *Joseph Black, Lectures on the Elements of Chemistry, delivered in the University of Edinburgh*, Edinburgh, 1803.

Boerhaave, H., *A New Method of Chemistry; including the Theory and Practice of that Art: Laid down on Mechanical Principles, and Accommodated to the Uses of Life*, J. Osborn & T. Longman, London, 1727.

Bootsgezel, J.J., 'William Blakey - A Rival to Newcomen', *Trans. Newcomen Society*, Vol XVI, 1935 - 36.

Boucher, Cyril T. G., *James Brindley, Engineer, 1716 - 1772*, Goose, Norwich, 1968.

Bremner, David, *The Industries of Scotland, Their Rise, Progress and Present Condition*, A. & C. Black, 1869, reprint David & Charles, Newton Abbot, 1969

Brewster, D. Ed., Robison, J., *A System of Mechanical Philosophy*, Edinburgh, 1822.

Brock, William H., *The Fontana History of Chemistry*, Fontana Press, London, 1992.

Bryden, D.J., 'The Jamaican Observatories of Colin Campbell, F.R.S. and Alexander MacFarlane, F.R.S.', *Notes and Records of the Royal Society*, Vol 24, 1970.

Bryden, D.J., *James Short and his Telescopes*, Royal Scottish Museum, Edinburgh, 1968.

Bryden, D.J., 'James Watt Merchant: The Glasgow Years, 1754 - 1774', in Smith, Dennis, Ed., *Perceptions of Great Engineers: Fact and Fantasy*, Science Museum, London, 1994.

Bryden, D.J., *Scottish Scientific Instrument Makers, 1600 - 1900*, Royal Scottish Museum, Edinburgh, 1972.

Burt's Letters from the North of Scotland, with an Introduction by R. Jamieson, W. Paterson, Edinburgh, 1876.

Butt, John, *Industrial Archaeology of Scotland*, David & Charles, Newton Abbot, 1967.

Campbell, R.H., *Carron Company, Founded 1759*, Oliver & Boyd, Edinburgh, 1961.
Cardwell, D.S.L., *The Fontana History of Technology*, Fontana Press, London, 1994.

Cardwell, Donald S.L., *From Watt to Clausius, The Rise of Thermodynamics in the Early Industrial Age*, Heinemann, London, 1971.

Chambers, R., *Lives of Illustrious and Distinguished Scotsmen forming a Complete Biographical Dictionary*, Blackie & Son, Glasgow, 1834.

Clarke, T.N., Morrison Low, A.D., & Simpson, A.D.C., *Brass and Glass, Scientific Instrument Making Workshops in Scotland as Illustrated by Instruments from the Arthur Frank Collection at the Royal Museum of Scotland*, National Museums of Scotland, 1989.

Cleland, James, *Annals of Glasgow, Comprising An Account of the Public Buildings, Charities, and the Rise and Progress of the City*, Glasgow, 1816, another edn. 1829.

Cleland, James, *Historical Account of the Steam Engine and its Application in Propelling Vessels: with an Account of the Proceedings at London, Glasgow, and Greenock for Erecting Monuments to Mr. Watt; also a Detailed Account of the Number and Power of Steam Engines in Glasgow*, Blackie, Glasgow, 1825.

Clow, Archibald & Clow, N.L., *The Chemical Revolution, A Contribution to Social Technology*, Batchworth Press, London, 1952.

Cochrane, Thomas, Ed. D. McKie, *Notes from Dr. Black's Lectures on Chemistry*, Wilmslow, 1966.

Copeland, Robert, *A Short History of Pottery Raw Materials and the Cheddleton Flint Mill*, Cheddleton Flint Mill Industrial Heritage Trust, 1972.

Cosh, Mary, 'Clockmaker Extraordinary: The Career of Alexander Cumming', *Country Life*, 12 June 1969.

Coutts, J., *History of the University of Glasgow*, Glasgow, 1909.

Crowther, J.G., *Scientists of the Industrial Revolution, Joseph Black, James Watt, Joseph Priestley, Henry Cavendish*, Cresset Press, London 1962.

Cullen, W., 'Essay on the Cold Produced by Evaporating Fluids, and of Some Other Means of Producing Cold', *Essays and Observations, Physical and Literary*, Read before a Society in Edinburgh, Anno 1755, W. Creech, Edinburgh, 1796, (see also under Black).

Daumas, Maurice, *Scientific Instruments of the 17th & 18th Centuries and their Makers*, Batsford, 1972, new edn. Portman Books, London, 1989.

Day, Lance & McNeil, Ian, Ed., *Biographical Dictionary of the History of Technology*, Routledge, London, 1996.

Desaguliers, John T., *A Course of Experimental Philosophy*, A. Millar, 3 edn., London, 1763.

Devine, T.M., *The Tobacco Lords, A Study of the Tobacco Merchants of Glasgow and their Trading Activities, c. 1740 - 1790*, J. Donald, Edinburgh, 1975.

Devine, T.M. & Jackson, G., Ed., *Glasgow, Vol I, Beginnings to 1830*, Manchester University Press, 1995.

Dickie, D.M. & Forster, C.W., Ed., *Mines and Minerals of the Ochils*, Clackmannanshire Field

Studies Society, 1974, reprint 1994.

Dickinson, H.W., *James Watt, Craftsman and Engineer*, Cambridge University Press, 1935.

Dickinson, H.W., *Matthew Boulton*, Cambridge University Press, 1936.

Dickinson, H.W. & Jenkins, R., *James Watt and the Steam Engine*, Oxford, 1927, reprint Moorland, Ashbourne, 1981.

Dictionary of National Biography, see Lee, S., Ed., Smith Elder & Co., London 1897.

Doig, Andrew, 'Dr. Black, a Remarkable Physician', in Simpson, A.D.C., Ed., *Joseph Black, 1728 - 1799, A Commemorative Symposium*, Royal Scottish Museum, Edinburgh, 1982.

Donovan, A.L., *Philosophical Chemistry in the Scottish Enlightenment, The Doctrines and Discoveries of William Cullen and Joseph Black*, Edinburgh University Press, 1975.

Drummond, A.I.R., *Old Clackmannanshire*, Alva, 1953.

Duckham, Baron F., *A History of the Scottish Coal Industry, Vol I, 1700 - 1815*, David & Charles, Newton Abbot, 1970.

Durie, Alastair J., *The Scottish Linen Industry in the Eighteenth Century*, J. Donald, Edinburgh, 1979.

Farey, John, 'Steam Engine' in Rees, A., Ed., *The Cyclopaedia; or Universal Dictionary of Arts, Sciences, and Literature*, Vol XXXIV, London, 1816.

Farey, John, *A Treatise on the Steam Engine, Historical, Practical and Descriptive*, London, 1827.

Farr, A.D., *The Campbeltown & Machrihanish Light Railway*, Oakwood Press, Abingdon, 1969, reprint 1987.

Fitton, R.S., *The Arkwrights, Spinners of Fortune*, Manchester University Press, 1989.

Fleming, James Arnold, *Scottish Pottery*, Glasgow, 1923.

Galloway, Elijah, *History and Progress of the Steam Engine*, T. Kelly, London, 1830.

Gibbs, F.W., *Joseph Priestley, Adventurer in Science and Champion of Truth*, Nelson, London, 1965.

Goodison, Norman, *English Barometers, 1680 - 1860*, Cassell, London, 1969, reprint Antique Collectors Club, Woodbridge, 1992.

Goring, Rosemary, Ed., *Chambers Scottish Biographical Dictionary*, Chambers, Edinburgh, 1992.

Graham, Patrick, *A General View of the Agriculture of the Counties of Kinross and Clackmannan*, Edinburgh, 1814.

Guerlac, Henry, 'Joseph Black's Work on Heat', in Simpson, A.D.C., Ed., *Joseph Black, 1728 - 1799, A Commemorative Symposium*, Royal Scottish Museum, Edinburgh, 1982.

Hamilton, Henry, *The Industrial Revolution in Scotland*, 1932, reprint F. Cass, London, 1966.

Henson, G., *History of the Framework Knitters*, 1831, reprint David & Charles, Newton

Abbot, 1970.

Hills, R.L., 'From Cocoon to Cloth', *Istituto Internazionale di Storia Economica "F. Datini"*, Serie II, Vol 24, 'La Seta in Europa, Secc. XIII - XX'.

Hills, R.L., 'How James Watt Invented the Separate Condenser', *Scientific Instrument Society*, Bulletin 57 & 58, 1998.

Hills, R.L., 'James Watt at Campbeltown', *The Kintyre Magazine*, The Kintyre Antiquarian and Natural History Society, Nos. 41 - 43, 1997 - 1998.

Hills, R.L., 'James Watt, Mechanical Engineer', *History of Technology*, Vol 18, 1996.

Hills, R.L., 'James Watt, 1736 - 1819, Pioneer of Scientific Engineering', *Manchester Memoirs*, Manchester Literary and Philosophical Society, Vol 134, 1995 - 1996.

Hills, R.L., 'James Watt's Barometers', *Scientific Instrument Society*, Bulletin 60, March 1999.

Hills, R.L., 'James Watt and his Rotary Engines', *Transactions of the Newcomen Society*, Vol 70, No. 1, 1998 - 1999.

Hills, R.L., 'James Watt's Steam Engine for the Leadhills Mines', *Mining History*, Vol 13, No. 6, Winter 1998.

Hills, R.L., 'James Watt's Surveys around Stirling', *Forth Naturalist and Historian*, Vol 21, 1998.

Hills, R.L., 'The Origins of James Watt's Perfect Engine', *Transactions of the Newcomen Society*, Vol 68, 1996 - 1997.

Hills, R.L., 'John Watt's Map of the Clyde', *Notes and Records of the Royal Society*, Vol 52, No. 1, 1998.

Hills, R.L., *Power from Wind, A History of Windmill Technology*, Cambridge University Press, 1994.

Hills, R.L., *Power from Steam, A History of the Stationary Steam Engine*, Cambridge University Press, 1989.

Hills, R.L., 'A Steam Chimera, A Review of the History of the Savery Engine', *Transactions of the Newcomen Society*, Vol 58, 1986 - 1987.

Hills, R.L., 'William Lee and his Knitting Machine', *Journal of the Textile Institute*, Vol 80, No. 2, 1989.

Hills, R.L. & Reece, J., 'British Papermills: Ruthven and Methven', *The Quarterly*, British Association of Paper Historians, No. 14, 1995.

Irvine, William, *Essays Chiefly on Chemical Subjects*, J. Mawman, London, 1805.

Jardine, G., 'Account of John Roebuck, M.D., F.R.S. Edin.', *Transactions of the Royal Society of Edinburgh*, Vol. 4, pt. 1, 1796.

Kelly, H.E., 'Documents relating to Delftfield', *Scottish Pottery*, 15th Historical Review, 1993.

King-Hele, Desmond, *Erasmus Darwin, 1731 - 1802*, Macmillan, London, 1963.

King-Hele, Desmond, Ed., *The Letters of Erasmus Darwin*, Cambridge University Press, 1981.

Kinghorn, Jonathan & Quail, Gerard, *Delftfield, A Glasgow Pottery, 1748 - 1823*, Glasgow Museum & Art Galleries, 1986.

Labarre, E.J., *Dictionary and Encyclopaedia of Paper and Paper-making*, 2 edn., Swets & Zeitlinger, Amsterdam, 1952.

Law, R.J., *James Watt and the Separate Condenser*, H.M.S.O., London, 1969.

Lee, Sidney, Ed., *Dictionary of National Biography*, Smith Elder & Co., London, 1897.

Leupold, J., *Theatrum Machinarum Hydrotechnicarium*, Leipzig, 1724.

Lewis, Michael J.T., *Early Wooden Railways*, Routledge & Kegan Paul, London, 1970.

Lindqvist, Svante, *Technology on Trial, the Introduction of Steam Power Technology into Sweden, 1715 - 1736*, Almqvist & Wiksell International, Stockholm, 1984.

Lindsay, Ian G. & Cosh, Mary, *Inveraray and the Dukes of Argyll*, Edinburgh University Press, 1973.

Lindsay, Jean, *The Canals of Scotland*, David & Charles, Newton Abbot, 1968.

Lindsay, Jean, 'Robert Mackell and the Forth and Clyde Canal', *Transport History*, Vol I, 1968.

Lythe, S.G.E, 'James Watt and the Strathmore Canal Project', *Transport History*, Vol 1, 1968.

Lythe, S.G.E. & Butt, J., *An Economic History of Scotland, 1100 - 1939*, Blackie, Glasgow, 1975.

Mackell, Robert & Watt, James, *An Account of the navigable Canal, proposed to be cut from the River Clyde to the River Carron, as surveyed by Robert Mackell and James Watt*, London, 1767.

MacLeod, Christine, *Inventing the Industrial Revolution, The English Patent System, 1660 - 1800*, Cambridge University Press, 1988.

MacMillan, Nigel S.C., *The Campbeltown & Machrihanish Light Railway*, David & Charles, Newton Abbot, 1970.

Mactaggart, C., 'Life in Campbeltown', Lecture to Kintyre Antiquarian Society, Campbeltown, 1923.

Martine, George, *Essays and Observations on the Construction and Graduation of Thermometers and on the Heating and Cooling of Bodies*, 3 edn., A. Donaldson, Edinburgh, 1780.

Middleton, W.E. Knowles, *The History of the Barometer*, Johns Hopkins University, Baltimore, 1964.

Middleton, W.E. Knowles, *A History of the Thermometer and Its Uses in Meteorology*, Johns Hopkins University, Baltimore, 1966.

Moore, John N., 'Many Years Servant to the Town: James Barrie and the eighteenth century mapping of Glasgow', *Scottish Geographical Magazine*, Vol 113, No. 2, 1997.

Moore, John N., *The Maps of Glasgow, A History and Cartobibliography to 1865*, Glasgow

University Library, 1996.

Morrison-Low, Alison, 'Scottish Barometers', A talk prepared for the Royal Meteorological Society Historical Group, 8 March 2000.

Muirhead, James Patrick, *The Life of James Watt, with Selections from his Correspondence*, J. Murray, London, 1858.

Muirhead, James Patrick, *The Origin and Progress of the Mechanical Inventions of James Watt*, J. Murray, London, 1854.

Murdock, A., *The People Above: Politics and Administration in Mid-eighteenth Century Scotland*, Edinburgh, 1980.

Musson, A.E. & Robinson, E, *Science and Technology in the Industrial Revolution*, Manchester University Press, 1969.

Newton, Norman S., *Campbeltown's Distilleries, a Guide for Visitors*, 1991.

Nugent, Charles & Walker, Nichola, *Thomas Hearne Rediscovered, A Fresh Look at Landscape Watercolours*, Whitworth Art Gallery, Manchester, 1996.

Pacey, A.J., 'Vis Viva and Power', typescript in the Department of History of Science and Technology, UMIST.

Page, F.G., 'Chemical and Analytical Aspects of the Early Alkali and Bleaching Industries in Britain', Ph.D. Thesis, University of Leicester, 1999.

Partington, J.R., *A Short History of Chemistry*, Harper, New York, 1937, 3 edn., 1957.

Paul, Sir James Balfour, *The Scots Peerage, Founded on Wood's Edition of Sir Robert Douglas's Peerage of Scotland*, D. Douglas, Edinburgh, 1906.

Phillips, R., *Public Characters for 1802 - 1803*, London, 1803.

Pollitt, A. Gerald, *Historic Inverness*, Melven Press, Perth, 1981.

Priestley, J., *Historical Account of the Navigable Rivers, Canals, and Railways, throughout Great Britain*, 1831, reprint David & Charles, Newton Abbot, 1969.

Quail, Gerard, 'James Watt at Delftfield: the work of James Watt as a potter at Delftfield, Glasgow, and the development of cream-coloured earthenware', *Scottish Pottery Historical Review*, No. 6, 1981.

Ramsay, W., *The Life and Letters of Joseph Black*, Constable, London, 1918.

Rees, Abraham, *The Cyclopaedia; or Universal Dictionary of Arts, Sciences, and Literature*, London, 1819, and extracts reprint David & Charles, 1972.

Reilly, Robin, *Josiah Wedgwood, 1730 - 1795*, Macmillan, London, 1992.

Renwick, Robert, Ed., *Extracts from the Records of the Burgh of Glasgow, 1739 - 1759*, Glasgow, 1911.

Riddell, John F., *Clyde Navigation, A History of the Development and Deepening of the River Clyde*, J. Donald, Edinburgh, 1979.

Riley, P.W.J., *The English Ministers and Scotland, 1707 - 1727*, London, 1964.

Robinson, Eric & McKie, Douglas, *Partners in Science, James Watt and Joseph Black*, Constable, London, 1970.

Robinson, E. & Musson, A.E., *James Watt and the Steam Revolution*, Adams & Dart, London, 1969.

Robison, John, Ed., *Joseph Black, Lectures on the Elements of Chemistry, delivered in the University of Edinburgh*, Edinburgh, 1803.

Robison, John, Letter in *The Universal Magazine of Knowledge & Pleasure*, Vol 21, November, 1757.

Robison, John, 'Steam' and 'Steam Engine', *Encyclopaedia Britannica*, 3 edn., Edinburgh, 1797.

Robison, John, *A System of Mechanical Philosophy*, with Notes by Sir David Brewster, Edinburgh, 1822.

Rolt, L.T.C. & Allen, J.S., *The Steam Engine of Thomas Newcomen*, Moorland, Hartington, 1977.

Ross, Alex, 'Notes on the Formation of the Caledonian Canal, and Its Effects on the Highlands', *Transactions of the Gaelic Society of Inverness*, Vol XIII, 1886 - 1887.

Rousseau, G.S. & Porter, R., Eds., *The Ferment of Knowledge: Studies in the Historiography of Eighteenth-century Science*, Cambridge University Press, 1980.

Royal Commission on the Ancient and Historical Monuments of Scotland, *Argyll, Vol I, Kintyre*, 1971.

Savery, T. *The Miners Friend or an Engine to Raise Water by Fire*, London, 1702.

Schofield, Robert E., *The Lunar Society, A Social History of Provincial Science and Industry in Eighteenth Century England*, Clarendon Press, Oxford, 1963.

Shaw, John, *Water Power in Scotland, 1550 - 1870*, J. Donald, Edinburgh, 1984.

Simpson, A.D.C., Ed., *Joseph Black, 1728 - 1799, A Commemorative Symposium*, Royal Scottish Museum, Edinburgh, 1982.

Sinclair, Sir John Ed., *The Statistical Account of Scotland, 1791 - 1799, Vol 1, General Appendix*, reprint E.P. Publishing, Wakefield, 1977.

Sinclair, Sir John, Ed., *The Statistical Account of Scotland, 1791 - 1799, Vol IX, Dunbartonshire, Stirlingshire & Clackmannanshire*, reprint E.P. Publishing, Wakefield, 1978.

Sinclair, Sir John, Ed., *The Statistical Account of Scotland, 1791 - 1799, Vol XII, North & West Perthshire*, reprint E.P. Publishing, Wakefield, 1977.

Singer, Charles, *A Short History of Science from the Earliest Times to A.D. 1900*, Clarendon Press, Oxford, 1941.

Singer, Charles & Underwood, E. Ashworth, *A Short History of Medicine*, Clarendon Press, Oxford, 1941.

Sked, P. Ed., *Culross, A Short History of the Royal Burgh*, National Trust for Scotland, 1995.

Skempton, A.W., Ed., *John Smeaton, F.R.S.*, T. Telford Ltd., London, 1981.

Smeaton, J., *A Catalogue of the Civil and Mechanical Engineering Designs, 1741 - 1792, of John Smeaton, F.R.S.*, Newcomen Society, Extra Publication No. 5, London, 1950.

Smeaton, J., *Experimental Enquiry concerning the Natural Powers of Wind and Water to turn Mills and Other Machines depending on a Circular Motion...*, I. & J. Taylor, London, 1794.

Smeaton, J., *John Smeaton's Diary of his Journey to the Low Countries, 1755*, Newcomen Society, Extra Publication No. 4, London, 1938.

Smeaton, J., *Reports of the Late John Smeaton, F.R.S., made on Various Occasions in the Course of his Employment as a Civil Engineer*, Longman, Hurst, Rees, Orme & Brown, London, 1812.

Smiles, Samuel, *Lives of the Engineers. Early Engineering. Vermuyden - Myddleton - Perry - James Brindley*, J. Murray, London, 1874.

Smiles, Samuel, *Lives of the Engineers. Harbours, Lighthouses, Bridges. Smeaton and Rennie*, J. Murray, London, 1874.

Smiles, Samuel, *Lives of the Engineers. The Steam-Engine. Boulton and Watt*, J. Murray, London, 1878.

Smith, Annette M., *Jacobite Estates of the Forty-five*, J. Donald, Edinburgh, 1982.

Smith, Denis, Ed., *Perceptions of Great Engineers, Fact and Fancy*, Science Museum, London, 1994.

Snoddy, T.G., *Round About Greenock*, 1937, reprint Allen Lithographic Co., Kirkcaldy, 1950.

Sobel, Dava, *Longitude, The True Story of a Lone Genius who Solved the Greatest Scientific Problem of his Time*, 1995, in Britain, Fourth estate, London, 1996.

Soutter, David C., 'Additional Information Relating to John Gardner, senior (1734 - 1822), Maltman, Scientific Instrument Maker, Town Surveyor to the City of Glasgow', typescript, Glasgow, 1997.

Stuart, Robert, *Historical and Descriptive Anecdotes of Steam Engines, and of their Inventors and Improvers*, Wightman & Cramp, London, 1829.

Swinbank, P., 'Experimental Science in the University of Glasgow at the Time of Joseph Black', in Simpson, A.D.C., Ed., *Joseph Black: 1728 - 1799, A Commemorative Symposium*, Royal Scottish Museum Edinburgh, 1982.

Swinbank, P., 'James Watt and his Shop', *Glasgow University Gazette*, c. 1968.

Swinbank, P., 'A University Partnership', typescript, Glasgow, 1969.

Switzer, Stephen, *An Introduction to a General System of Hydrostatics and Hydraulics, Philosophical and Practical*, T. Astley, London, 1729.

Thomson, George, 'James Watt and the Monkland Canal', *Scottish Historical Review*, Vol XXIX, No. 108, October 1950.

Thomson, George, 'The Monkland Canal, A Sketch of the Early History', typescript,

Monkland Library Services Dept., 1945.

Turnbull, Jill, 'Scottish Cobalt and Nicholas Crisp', *Transactions English Ceramic Circle*, Vol 16, pt. 2, 1997.

Turner, Gerard L'E., *Antique Scientific Instruments*, Blandford Press, Poole, 1980.

Watt, James, *An account of the Scheme for rendering navigable the Rivers Forth and Devon with estimates of the Expence by James Watt*, Edinburgh, 1774. (see Sinclair, Vol I)

Watt, James, *An Account of James Watt's Improvements upon the Steam or Fire Engine*, London, c. 1775.

Watt, James, *Report concerning the Harbour of Port-Glasgow, made to the Magistrates of Glasgow, by James Watt, Engineer, and submitted to the consideration of the Merchants*, Glasgow, 1771.

Watt, James, *A Report to the Honourable His Majesty's Commissioners for managing the Annexed Estates in Scotland concerning the Isthmuses of Tarbert and Crinan*, Glasgow, 1772.

Watt, James, *A Report concerning the Possibility and Expence of Carrying an Artificial Navigation into Strathmore; made to the Honourable His Majesty's Commissioners for Managing the Annexed Estates in Scotland*, Glasgow, 1771.

Watt, James, *A Scheme for making a navigable Canal from the City of Glasgow to the Monkland Coalierys*, Glasgow, 1769.

Weatherill, L., *The Pottery Trade and North Staffordshire, 1660 - 1760*, Manchester University Press, 1971.

Williamson, George, *Memorials of the Lineage, Early Life, Education, and Development of the Genius of James Watt*, T. Constable, 1856.

Wilson, Patrick, 'Biographical Account of Alexander Wilson, M.D., late Professor of Practical Astronomy in Glasgow', *Transactions Royal Society of Edinburgh*, Vol 10, 1806.

Books Consulted but Not Used in References.

Anderson, R.G.W., *The Playfair Collection and the Teaching of Chemistry at the University of Edinburgh, 1713 - 1858*, Royal Scottish Museum, Edinburgh, 1978.

Bald, Robert, *A General View of the Coal Trade of Scotland, chiefly that of the River Forth & Midlothian*, Oliphant & Broan, Edinburgh, 1808.

Black, Joseph, 'Experiments upon Magnesia Alba, Quick-Lime, and Other Alcaline Substances; to which is annexed An Essay on the Cold produced by Evaporating Fluids and of Some Other Means of producing Cold'. Extracted from William Cullen, *Essays and Observations, Physical and Literary*, Read before a Society in Edinburgh, Anno 1755, W. Creech, Edinburgh, 1796.

Cameron, A.D., *The Caledonian Canal*, Melvin Press, Perth, 1972, reprint 1983.

Campbell, R.H., *Scotland since 1707, The Rise of An Industrial Society*, J. Donald, Edinburgh, 1965, 2 edn. 1985.

Carvel, John L., *One Hundred Years in Coal, The History of The Alloa Coal Company*, T. & A. Constable, Edinburgh University Press, 1944.

Crouther Gordon, T., *The History of Clackmannan*, Civic Press, Glasgow, 1936.

Crouther Gordon, T., 'James Watt's Bold Plan for the Devon', *The Alloa Journal*, Sept. 1960.

Dickinson, H.W., *A Short History of the Steam Engine*, Cambridge University Press, 1938.

Fellowes, Gordon Ian, *Famous Scottish Lives*, Odhams Books, London, 1967.

Hamilton, Henry, *An Economic History of Scotland in the Eighteenth Century*, Clarendon Press, Oxford, 1963.

Hooykaas, R., *Religion and the Rise of Modern Science*, Scottish Academic Press, Edinburgh, 1972.

M'Lean, Hugh A., *Robert Urie, Printer in Glasgow, c. 1711 - 1771*, J. Maclehose, Glasgow, 1914.

MacLeod, Christine, 'James Watt, Heroic Invention and the Idea of the Industrial Revolution', in Berg, M. & Bruland, Kristine, Ed., *Technological Revolutions in Europe, Historical Perspectives*, E. Elgar, Cheltenham, 1998.

McKie, D., & Heathcote, Niels H. de V., *The Discoveries of Specific and Latent Heats*, E. Arnold, London, 1935.

Marshall, T.H., *James Watt (1736 - 1819)*, L. Parsons, London, 1925.

Moir, D.G., Ed., *The Early Maps of Scotland to 1856*, Royal Scottish Geographical Society, Edinburgh, 1983.

Morzer Bruyns, W.F.J., *The Cross-Staff, History and Development of a Navigational Instrument*, Walburg Institute, Amsterdam, 1994.

Newton, Norman S., *Machrihanish, A Guide for Visitors*, May 1989.

Paget-Tomlinson, Edward W., *The Complete Book of Canal and River Navigations*, Waine Research Publications, Albrighton, 1978.

Park, Brian A., *The Woollen Mill Buildings in the Hillfoots Area*, Forth Naturalist and Historian, Stirling, 1984.

Pollitt, A. Gerald, *Historical Inverness*, Melvin Press, Perth, 1981.

Robertson, James, *General View of the Agriculture in the Southern Districts of the County of Perth*, J. Nichols, London, 1794.

Rolt, L.T.C., *James Watt*, Batsford, London, 1962.

Soutter, David C., 'The Gardner Family of Glasgow, Maltmen and Scientific Instrument Makers', typescript, Glasgow, c. 1977.

Taylor, James, *The Great Historic Families of Scotland*, J.S. Virtue, London 1887.

Thomas, J., *The Rise of the Staffordshire Potteries*, c. 1971.

Thomson, Thomas, Ed., *A Biographical Dictionary of Eminent Scotsmen originally edited by Robert Chambers*, new edn. Blackie & Son, London, 1870.

Whetstone, Ann E., *Scottish County Government in the Eighteenth and Nineteenth Centuries*, J. Donald, Edinburgh, 1981.

Archival Sources.

Argyll and Bute Record Office;
McTavish of Dunardy Papers,
Maps.

Birmingham Central Library;
Boulton and Watt Collection,
James Watt Papers,
Matthew Boulton Papers,
Muirhead Papers.

British Map Library.

Edinburgh University;
Joseph Black Papers.

National Library of Scotland;
Erskine Murray Papers,
John Rennie Papers.

Register House, Edinburgh;
Greenock Parish Registers.

Royal Society, London;
John Smeaton's Designs, 1741 - 1792,
James Watt Papers.

Scottish Record Office;
Carron Company Records,
Duke of Argyll's List of Papers,
Forfeited Estates Papers,
Maps,
Mar & Kellie Papers.

Stirling Record Office;
Stirling of Garden Papers.

Strathclyde Regional Archives;
Clyde Port Authority Maps,
Glasgow City Archives,
Maps.

Index – People

Y

Index – Places & Subjects

D

Q

R

S